Basic Genetics

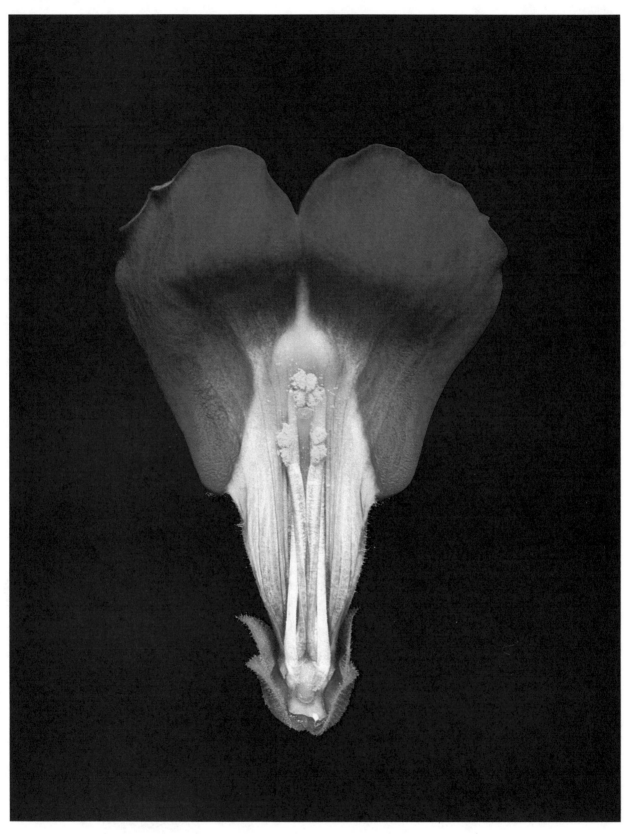

A *wildtype flower of* Antirrhinum majus (*snapdragon*), *a plant used for genetic and molecular studies of the mechanisms of flower development. The lower parts of the corolla leaves have been removed, revealing the reproductive organs (the pistil and four stamens).*

THE JONES AND BARTLETT SERIES IN BIOLOGY

Basic Genetics

SECOND EDITION

Daniel L. Hartl

Washington University School of Medicine

Jones and Bartlett Publishers

BOSTON

TO THE MEMORY OF
David Freifelder
Leon A. Snyder

Editorial, Sales, and Customer Service Offices:
Jones and Bartlett Publishers
20 Park Plaza
Boston, MA 02116

Library of Congress Cataloging-in-Publication Data

Hartl, Daniel L.
    Basic genetics / Daniel L. Hartl.—2nd ed.
        p.   cm.
    Includes bibliographical references and index.
    ISBN 0-86720-173-8
    1. Genetics.   I. Title.
    [DNLM;   1. Genetics, Medical.   QZ 50 H331b]
QM430.H373   1991
575.1—dc20
DNLM/DLC
for Library of Congress   90-15661
                                                                CIP

Cover illustration   Pseudocolor image of a ^{32}P-labeled human RFLP pedigree analysis for cystic fibrosis generated on a Betagen Nucleic Acid Blot Analyzer. (Courtesy of D. E. Sullivan.)

Frontispiece   Cover of *Science*, vol. 250, 16 November 1990. From "Genetic Control of Flower Development by Homeotic Genes in *Antirrhinum majus*," by Z. Schwarz-Sommer, P. Huijser, W. Nacken, H. Saedler, and H. Sommer, pp. 931–936. Copyright 1990 by the American Association for the Advancement of Science.

Production   Bookman Productions
Designer   Hal Lockwood
Editor   Patricia Zimmerman
Illustrator   Illustrious, Inc.
Composition   The Clarinda Company

Printed in the United States of America

10 9 8 7 6 5 4 3 2 1

C O N T E N T S

C H A P T E R   1

The Elements of Heredity and Variation   1

C H A P T E R   2

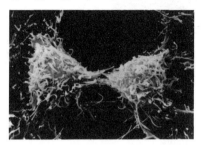

Genes and Chromosomes   27

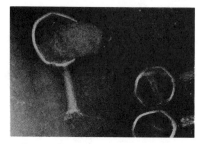
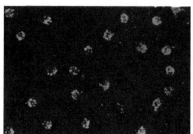

C H A P T E R   6

Variation in Chromosome Number and Structure 143

C H A P T E R   7

Extranuclear Inheritance 167

C H A P T E R   8

Population Genetics and Evolution 183

C H A P T E R 9

Quantitative Genetics 207

C H A P T E R 10

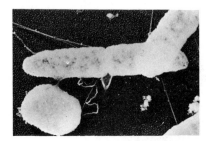

Genetics of Bacteria and Viruses 235

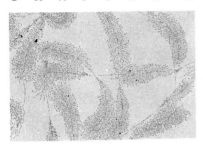

Gene Expression 265

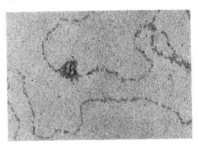

Regulation of Gene Activity 299

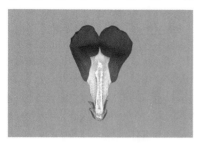

CHAPTER 15

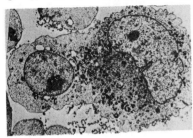

Somatic-Cell Genetics and Immunogenetics 397

CHAPTER 16

Genetic Engineering 419

The philosophy embodied in the first edition of *Basic Genetics*—that is, to focus on the genetic implications of biological phenomena and to minimize, as much as possible, the details of cell biology and biochemistry—has proved to be very popular with a large number of instructors. A new edition is necessary because of new discoveries and changing perspectives in the field. The book is intended to meet the needs of the shorter, less encyclopedic, course in introductory genetics. It addresses what can be reasonably covered in a one-quarter or one-semester course. The text therefore concentrates on the essentials of genetics, omitting the encumbering detail of many advanced and specialized topics. The focus is on the basic principles of genetics and on explaining these fundamentals to students in as clear and concise a way as possible.

Throughout this book, I have integrated classical and molecular principles. The first four chapters serve as a general introduction to genetics. Chapter 1 includes classical Mendelian genetics, and Chapter 2 emphasizes the implications of the presence of genes in chromosomes. Chapter 3 focuses on the principles of gene mapping by means of testcrosses or tetrad analysis. Chapter 4 is concerned with DNA as the genetic material and its structure and mode of replication.

Some important specialized topics are introduced in Chapters 5 through 7. In Chapter 5, the emphasis is on the structure and molecular organization of chromosomes and on the types of DNA sequences found in eukaryotic chromosomes. The genetic consequences of variation in chromosome number and structure are discussed in Chapter 6, which also includes the human chromosome complement and common abnormalities. Chapter 7 concerns extranuclear inheritance associated with genetic determinants transmitted through the cytoplasm.

Chapters 8 and 9 deal with genetics at the population level. The basic principles of population genetics and their implications for the process of evolution are introduced in Chapter 8. Chapter 9 covers the important subject of quantitative genetics and the ways in which the principles of quantitative genetics are applied to the improvement of agricultural animals and plants. Chapter 9 also includes examples and pitfalls of the application of quantitative genetics to human behavioral traits. Although population and evolutionary genetics are generally regarded as highly mathematical subjects, most students will find the mathematics to be easily manageable.

Chapters 10 through 16 pick up the thread of molecular genetics introduced earlier and develop it in greater detail. Chapter 10 focuses on the special genetic features of bacteria and viruses and emphasizes their importance in revealing many key principles of genetics. Chapter 11 provides the details of gene expression through gene transcription, RNA processing, and translation. Chapter 12 considers the mechanisms that govern gene regulation in prokaryotes and eukaryotes. Learning more about these mechanisms provides the impetus of much current research in genetics.

Chapter 13 is an entirely new addition to *Basic Genetics*. It focuses on developmental genetics—a field in which knowledge has blossomed in the past few years. Developmental genetics has advanced beyond the mere collection of interesting mutants and anecdotes to a stage at which fundamental principles of development can be enunciated.

The processes of mutation, mutagenesis, and recombination form the subject of Chapter 14. Chapter 15 deals with genetic studies of somatic cells in culture and with the immunological subjects of blood groups, antibody and T-cell receptor variability, and histocompatibility genes that affect transplant success.

In Chapter 16, emphasis is on the technical side of molecular genetics through the use of recombinant DNA in genetic engineering. These important procedures have revolutionized genetic analysis and have resulted in many commercial applications in medicine and agriculture. The chapter also includes an introduction to some of the new, large-molecule cloning systems, such as yeast artificial chromosomes (YAC's), and a discussion of how these approaches have made possible the large-scale analysis of complex genomes, including the human genome.

Basic Genetics contains a number of special features to aid learning. Each chapter includes a summary of its major points. In addition to the key terms listed at the end of each chapter, there is a glossary at the end of the book. Each chapter contains several demonstration problems and their solutions with the reasoning explained in detail. These worked problems exemplify the application of key concepts and principles. Each chapter also contains a set of problems, graded in difficulty, for the students to test their understanding, and the methods of solution and answers are given in full at the end of the book.

I am indebted to the many reviewers, too numerous to mention individually, who advised me in the preparation of this book and who read and criticized all or parts of it. Among them are specialists in various aspects of genetics who checked the text for accuracy, instructors in genetics who evaluated the material for suitability in teaching, and, most importantly, the undergraduate and graduate students who identified cases in which the explanations of difficult concepts were successful and those in which they were opaque and in need of further clarification. Special thanks go to several colleagues who reviewed the manuscript of the second edition: Dr. Celeste Berg (University of Washington), Dr. Rowland H. Davis (University of California, Irvine), Dr. David Evans (University of Guelph), Dr. James L. Farmer (Brigham Young University), Dr. Jonathan Gallant (University of Washington), Dr. Michael A. Goldman (San Francisco State University), Dr. Philip

Hanawalt (Stanford University), Dr. Philip Hastings (University of Alberta), Dr. Mark F. Sanders (University of California, Davis), and Dr. D. A. Whited (North Dakota State University). Dr. Rayla Temin (University of Wisconsin) was very generous in allowing me to inspect the detailed marginal comments and suggestions that she made for improving the first edition.

I also wish to acknowledge the support and encouragement of Patricia Zimmerman, manuscript editor, and Hal Lockwood, who designed and produced the book. Much of the credit for the attractiveness and readability of the book should go to them. Thanks also to Jones and Bartlett, the publishers, for the high quality of the book's production. I am also grateful to the many people who contributed photographs, drawings, and micrographs from their own research and publications.

The first edition of this book was written in collaboration with David Freifelder and Leon A. Snyder. In the interim, both of these dear friends succumbed to cancer. Although thoroughly revised, *Basic Genetics* still embodies their very substantial creative impulses, and it is to David and Leon that this edition is very respectfully dedicated.

January, 1991   Daniel L. Hartl

So, here you are, taking a course in genetics, and probably wondering what you are going to learn. Will the material be interesting? Is there any reason to be taking this course other than to satisfy an academic requirement? At the end of the course, will you feel glad that you took it? Will there be any practical value to what you will learn? This brief introduction is designed to reassure you that the answer to each question is Yes.

Genetics began as the study of heredity. It is an ancient discipline. At least 4000 years ago in Sumeria, Egypt, and other parts of the world, farmers recognized that they could improve their crops and their animals by selective breeding. Their knowledge was based on experience and very incomplete, but they did recognize that many features of plants and animals were passed from generation to generation. Furthermore, they discovered that desirable traits—such as size, speed, and weight of animals—could sometimes be combined by controlled mating and that, in plants, crop yield and resistance to arid conditions could be combined by cross-pollination. The ancient breeding programs were not based on much solid information, because nothing was known of genes or any of the principles of heredity. Genetics has come a long way, mainly in the twentieth century, and in this book you will learn the rules followed by genes and chromosomes as they pass from generation to generation and you will be able to calculate in many instances the probabilities by which organisms with particular traits will be produced. The calculations require only simple arithmetic, and so there is no reason to be intimidated, even if higher mathematics is not a comfortable part of your repertory.

Modern genetics began in the mid-nineteenth century with Gregor Mendel's careful analyses of inheritance in peas, which will be examined in detail in Chapter 1. Mendel's experiments were simple and direct and brought forth the most-significant principles that determine how traits are passed from one generation to the next. His kind of experiments, which occupied most of genetic research until the middle of the twentieth century, is called **transmission genetics**. Some people have called it formal genetics, because the subject can be understood and the rules clearly seen without any reference to the biochemical

nature of genes or gene products. Geneticists of that period concerned themselves with a variety of organisms that might seem odd to the student. Certainly, the most-important one was the fruit fly, *Drosophila melanogaster*. Others were corn (maize), chickens, mice, and, later, various fungi. Why on earth do some geneticists choose to spend their days examining bottles of flies or working in hot cornfields? The main reasons are the following:

1. Genetic analysis needs traits that are easily detected and a large collection of variants (called mutants). *Drosophila* is extraordinary in this respect because natural populations include variants with different eye color, wing shape, bristle type, and so forth, and new mutations are readily detected in the laboratory. Corn is a good choice, too, because variants of kernel type and color are easily seen.

2. Genetic analysis requires sizable numbers of individuals. In genetic experiments, a sufficient number of organisms must be examined to determine the frequencies with which traits appear in each generation and to compare theoretical probabilities with observations. Again, *Drosophila* is excellent, because hundreds of flies will live happily in bottles and hundreds of bottles can be maintained in a laboratory. Corn is good also, because a large number of plants can be grown on a fairly small plot of land. Colonies of mice can also be maintained, though in smaller number, but mice compensate for this limitation by possessing traits present in higher organisms and of direct relevance to human disease; for example, different mouse strains with hereditary diabetes, cancer, and various immunological disorders have been developed.

3. Ideally, generation times should be short, so that the transmission of traits from one generation to the next can be followed in a reasonable period of time. From this point of view, Mendel's original choice of peas was good, but not the best, because he could grow only one or two crops per year. In *Drosophila*,

the interval between fertilization of an egg and production of a fertile adult is only a few weeks. Corn has its disadvantages in this respect (only one crop per year), but one compensation is its importance as an agricultural crop.

4. The organisms should be easy to maintain in the laboratory or field. Clearly, a population of fruit flies is more manageable than a pride of lions, and mice are more convenient than elephants.

Ultimately, the *Drosophila* geneticist is usually interested in organisms of greater intrinsic importance than flies (for example, wheat, rice, and human beings), and so one might wonder about the value of analyzing *Drosophila* (or almost any other experimental organism) in the first place. However, genetics is one of the unifying features of the living world, and in fact the Mendelian principles of heredity apply quite generally to most higher organisms. Thus, we study *Drosophila* because it is efficient to use as a model organism, and we then apply the principles learned to wheat, rice, and people.

In the mid-1890s, as biochemistry was becoming a highly developed science, geneticists began to wonder more intensely than ever before about the biochemical nature of the gene. How could information be maintained in molecules, and how could this information be transmitted from one generation to the next? In what way is the information different in a mutant? Biochemists floundered for some time, because there was no logical starting point for such an investigation, no experimental "handle." However, beginning in the 1920s and later in the 1940s, a few critical observations were made (they are documented in Chapter 4), which implicated a molecule discovered in 1869, deoxyribonucleic acid (DNA). The timing of these observations was not right though, and the observations were initially ignored. This was, ironically, the second time in the development of genetics that important findings were overlooked. Mendel's experiments were not appreciated because the thinking of the time did not allow the possibility that genes were discrete objects; the concept of DNA as the genetic material was rejected because the structure of the molecule was unknown and the consensus was that genes were made of proteins. However, critical experiments in 1952 and the discovery of the structure of DNA in 1953 by Watson and Crick changed all this. Genetics entered the DNA age and, in the next decade, we came to understand the chemical nature of genes and how genetic information is stored, released to a cell, and transmitted from

one generation to the next. It has been estimated that the overall body of scientific knowledge doubles roughly every ten years; however, during the first three decades after the discovery of DNA structure, the body of genetic knowledge grew with a two-year doubling time. These were exciting times, indeed, and you will be presented with a distillation of these findings in the few molecularly-oriented chapters of this book. Just to give a sample, we know with clarity how a gene is copied, how a mutation arises, how genes are turned on and off when their activity is needed or not needed; we also know, by direct isolation of macromolecules, the chemical products of thousands of genes and, by elegant chemical procedures, the precise sequence of the DNA bases for many genes. These topics constitute the branch of genetics called **molecular genetics.**

In 1970, genetics underwent its most-recent revolution: the development of recombinant DNA technology. This technology is a collection of techniques that enable genes to be transferred, at the will of the molecular geneticist, from one organism to another. It has had an enormous effect on genetic research, particularly in our ability to understand gene expression and its regulation in plants and animals. Topics theretofore unapproachable suddenly became amenable to experimental investigation. Currently, it is giving us new tools of great economic importance and of value in medical practice. This branch of genetics is known as **genetic engineering.** Current projects of great interest include the genetic modification of plants and domesticated animals and the production of clinically active substances. New recombinant DNA strategies have made it possible to study whole genomes (the totality of genetic information in an organism) rather than single genes. The complete set of DNA instructions has been determined by direct DNA sequencing in a number of viruses (and cellular organelles, such as mitochondria), and programs are underway to determine the complete DNA sequence of bacteria, yeast, nematodes, and *Drosophila.* Just on the horizon is the capability of determining the complete DNA sequence in the human genome.

For the past fifty years, by far the greatest practical effect of genetics has been in the field of agriculture and animal husbandry. Studies of the genetic composition of economically important plants has enabled plant breeders to institute rational programs for developing new varieties. Among the more-important plants that have been developed are high-yield strains of corn and dwarf wheat, disease-resistant rice, corn with an altered and more nutritious amino acid composition (high-lysine corn), and wheat that grows faster, allowing crops to be grown in short-season regions such as Canada and Sweden. The techniques for developing some of these strains will be explained in this book. Often new plant varieties have new deficiencies, such as a requirement for increased amounts of fertilizer or decreased resistance to certain pests. However, to eliminate these deficiencies is a problem for the modern geneticist, who, armed with the classical techniques, has the job of manipulating the inherited traits. Genetic engineering is also providing new procedures for such manipulations, and quite recently there have been dramatic successes.

Genetics has for some time made important contributions to medicine and modern clinical practice, but once again progress is accelerating because of the increased emphasis on genomic analysis. Genetic studies of the immune system have yielded information about immunological diseases and are enabling physicians to develop programs for organ transplantation. Recent genetic experiments have revealed thousands of new genetic markers in the human genome and have given us new methods for the detection of mutant genes, not only in affected persons, but also in their relatives and in members of the population at large. These methods have given genetic counseling new meaning. Human beings are heir to several thousand different inherited diseases. Married couples can be informed of the possibility of their producing an affected offspring and can now make choices between childbearing and adoption. Consider the relief of a man and woman who learn that neither of them carries a particular defective gene and that they can produce a child without worry. Furthermore, even with the knowledge that an offspring might be affected with a genetic disorder, techniques are available to determine *in utero* if a fetus is, in fact, affected.

I hope that the small number of examples presented in this introduction have convinced you that genetics is well worth learning about, and even enjoyable. The study of genetics is relevant not only to geneticists and biologists, but to all members of our modern, complex, technological society. Understanding the principles of genetics will help you to make informed decisions about certain matters of political, scientific, and personal concern. This should be the goal of taking a genetics course and studying a genetics text.

As a pedagogical aid, important terms are printed in **boldface** in the text. These terms are collected at the end of each chapter in a section titled "Key Terms." You should know their meanings because they form the basic vocabulary of genetics. Each chapter also includes a summary at the end of the text. Sample problems are solved in a section titled "Examples of Worked Problems." Each chapter ends with a fairly large collection of problems. The first few problems test your knowledge of the vocabulary in the text and the more-elementary facts. The problems then increase in difficulty. It is essential that you work as many of the problems as you can, because experience has shown that practice with problems is a good way to learn genetics and to identify particular points or concepts that have been misunderstood. Sometimes, it is not even necessary to solve a problem completely, but only to read the problem and decide whether you could solve it if asked to do so. The answers to all of the problems, and full explanations, are given at the back of the book. A problem will be more useful to you if you take a fair shot at it before turning to the answer.

The Elements of Heredity and Variation

In 1866, Gregor Mendel published the results of his experiments on inheritance in garden peas. From these studies, he discovered the existence of discrete hereditary elements and the rules determining their transmission from parent to offspring. The principles of inheritance that Mendel recognized ultimately

---

Above: The ancient practice of artificial pollination. An Assyrian relief, dating from the ninth century B.C., showing masked priests pollinating date palms. (Courtesy of Hans Stubbe.)

1

became the foundation for genetics and a major factor in the development of modern biology.

Our consideration of genetics begins with the development of the fundamental concept of the science—namely, that the hereditary element deduced by Mendel and now called a **gene** is a unit of inheritance transmitted from parent to offspring in the course of reproduction. The ideas and methods of analysis that were initiated by Mendel are the subject of this chapter.

1-1   Mendel and His Experiments

During the period in which Mendel developed his theory of the basis of heredity, he was a monk and a teacher of mathematics in a local school. However, the study of natural phenomena, especially inheritance, was his main interest. The principal difference between Mendel's approach and that of other scientists who were interested in inheritance is that he thought in quantitative terms about traits that could be classified into a small number of categories. He proceeded by posing simple questions to be answered by experiments and then looking for statistical regularities that might identify general rules.

Mendel selected peas for his experiments for two reasons: (1) he had access to varieties that differed in observable alternative characteristics, and (2) his earlier studies of flower structure (Figure 1-1) indicated that peas usually reproduce by self-pollination, in which pollen produced in a flower is transferred to the stigma of the same flower. To produce hybrids by cross-pollination, he needed only to open the keel petal (enclosing the reproductive structures), remove the immature anthers before they had shed pollen, and dust the stigma with pollen taken from a flower on another plant. A **hybrid** is the offspring of a cross betweeen inherently unlike individuals.

Mendel recognized the need for being certain that the characteristics that he studied were constant in inheritance, and at the beginning of his experimentation he established **true-breeding** lines in which the plants produced only progeny like themselves when allowed to self-pollinate normally. These different lines—which bred true for flower color, pod shape, or one of the other well-defined characters that Mendel had selected for investigation (Figure 1-2)—provided the parents for hybridization.

Let us examine a few of Mendel's original experiments to see the methods that he used and how he interpreted his results. One pair of characters that he studied was round versus wrinkled seeds. When pollen from plants of a line with wrinkled seeds was used to cross-pollinate plants from a round-seeded line, all of the hybrid seeds produced were round. He also performed the **reciprocal**

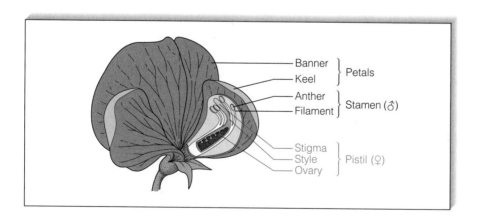

FIGURE 1-1 A pea flower with section removed to show the reproductive structures. Each flower has a single ovary that develops into the seed pod, and each ovary contains as many as ten ovules that give the seeds. The pistil is shown in green.

Banner } Petals
Keel

Anther } Stamen (♂)
Filament

Stigma
Style   } Pistil (♀)
Ovary

FIGURE 1-2 Six of the seven character differences in peas studied by Mendel (the seventh difference was long versus short stems). In each case, the characteristic shown at the left appears in the hybrid.

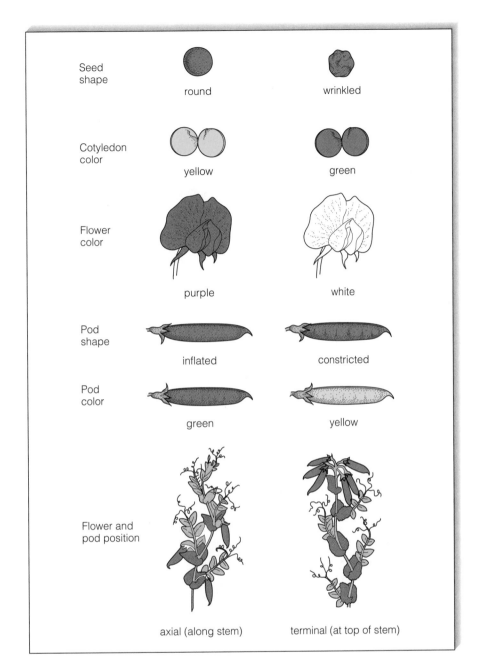

Seed shape	round	wrinkled
Cotyledon color	yellow	green
Flower color	purple	white
Pod shape	inflated	constricted
Pod color	green	yellow
Flower and pod position	axial (along stem)	terminal (at top of stem)

cross, in which plants from the round-seeded line were used as the pollen parents and those from the line with wrinkled seeds as female parents. As before, all of the hybrid seeds were round.

When the hybrid seeds from the round × wrinkled cross were allowed to undergo self-fertilization, the progeny were of two types—round and wrinkled—in definite numerical proportions. Mendel counted 5474 seeds that were round and 1850 that were wrinkled and noted that this ratio was approximately 3:1.

The hybrid progeny produced by crossing the lines with round and wrinkled seeds constitute the **F₁** generation, and all of the F₁ seeds were round. The progeny produced by self-fertilization of the F₁ constitute the **F₂** generation, and the F₂ progeny were in the proportions 3 round:1 wrinkled.

TABLE 1-1   Results of several of Mendel's experiments

Parental characteristics	F_1	Number of F_2 progeny	F_2 ratio
round × wrinkled (seeds)	round	5474 round, 1850 wrinkled	2.96 : 1
yellow × green (cotyledons)	yellow	6022 yellow, 2001 green	3.01 : 1
purple × white (flowers)	purple	705 purple, 224 white	3.15 : 1
inflated × constricted (pods)	inflated	882 inflated, 299 constricted	2.95 : 1
long × short (stems)	long	787 long, 277 short	2.84 : 1

(The *F* in the F_1, F_2 terminology stands for *filial*, referring to sons or daughters.) These results can be summarized in the following way:

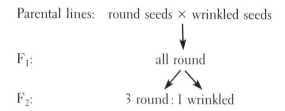

Parental lines: round seeds × wrinkled seeds

F_1:   all round

F_2:   3 round : 1 wrinkled

Similar results were obtained when Mendel made crosses between plants differing in six other pairs of alternative characteristics. The results of some of these experiments are summarized in Table 1-1. The principal observations were:

1. All F_1 hybrids possessed only one of the alternative parental traits.
2. In the F_2 generation, both parental traits were present.
3. The trait that appeared in the F_1 generation was always present in the F_2 about three times as frequently as the alternative trait.

In the succeeding sections, we will see how Mendel followed up these basic observations and performed experiments that led to his concept of discrete genetic units and to the principles governing their inheritance.

Particulate Hereditary Determinants

The prevailing concept of heredity in Mendel's time was that the traits of the parents were blended in the hybrid, as though the hereditary material consisted of fluids that became permanently mixed when combined. However, Mendel's observation that one of the parental characteristics was absent in F_1 hybrids and reappeared in unchanged form in the F_2 generation was inconsistent with the idea of blending. From this result, Mendel concluded that the traits from the parental lines were transmitted as two different elements *of a particulate nature* that retained their purity in the hybrids. The element associated with the trait seen in the F_1 hybrids (round seeds, in the example just used) he called **dominant,** and the other element, associated with the trait not seen in the hybrids but reappearing in their progeny (wrinkled seeds), he called **recessive.**

Mendel's conclusion was reinforced by observing the F_3 progeny produced by the self-pollination of individual F_2 plants. In the experiment with round versus wrinkled seeds, the plants grown from wrinkled F_2 seeds were true breeding; that is, they produced only wrinkled seeds in the F_3 generation. However, among 565 plants grown from round F_2 seeds, 193 were true breeding and produced only round seeds in the F_3 generation, but the other 372 plants produced both round and wrinkled seeds in a proportion very close to $3:1$. In quantitative terms,

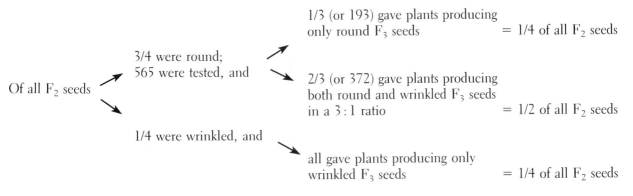

The same $1:2:1$ ratio was observed each time progeny from individual F_2 plants were tested.

The Principle of Segregation

From the arithmetical regularities observed in his experimental results and the deduction that the hereditary determinants are distinct particulate entities, Mendel formulated the following simple explanation of the $1:2:1$ ratio found in the F_2 generation:

1. A pea plant contains two hereditary determinants for each observed trait.
2. Each reproductive cell (**gamete**) of a plant contains only one of each pair of determinants, and the members of each pair of determinants are equally likely to be present in any individual reproductive cell.
3. The union of male and female reproductive cells in the formation of new **zygotes** (fertilized eggs) is a random process that reunites the hereditary determinants in pairs.

Figure 1-3 shows how these principles can explain the ratio of $1:2:1$ that Mendel observed in the F_2 generation from crosses between plants differing in any pair of alternative characters. The symbols A and *a* are used to represent the dominant and recessive determinants, respectively. The essential point in this explanation is the separation, or **segregation,** in unaltered form, of the two hereditary determinants in a hybrid plant during the formation of reproductive cells. This **principle of segregation** is sometimes called Mendel's first law.

> **The Principle of Segregation:** In the formation of gametes, the paired hereditary determinants separate (segregate) such that each gamete is equally likely to contain either one.

Note the assumption that the hereditary determinants are present in pairs in both parents and progeny but individually in reproductive cells.

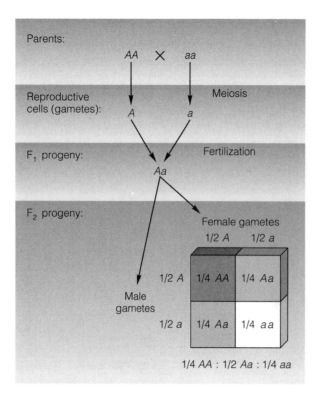

FIGURE 1-3 A diagrammatic explanation of the $1:2:1$ ratio observed by Mendel in the F_2 progeny from a cross between a dominant A and a recessive *a* pea plant.

Some Genetic Terminology

The following terminology is used in discussing genetic phenomena. These terms should be memorized.

1. Mendel's particulate hereditary elements are called **genes.**
2. The various forms of a given gene are called **alleles.**
3. Organisms in which the members of a pair of alleles are different, as in the Aa hybrids in Figure 1-3, are said to be **heterozygous,** and those in which the two alleles are alike are said to be **homozygous.** An individual organism may be homozygous for the dominant (AA) or for the recessive (aa) allele, and in both cases will be true breeding for the characteristic determined by the particular allele.
4. The **genotype** is the genetic constitution. A pea plant contains two alleles of each gene, and AA, Aa, and aa are examples of organism genotypes. Because gametes contain only one allele of a given gene, A and *a* are examples of gamete genotypes. Genotypes are usually designated incompletely in that only the alleles of genes of interest are specified.
5. The *observable* properties of an organism constitute its **phenotype.** Dominance results in the expression of the same phenotypic character—for example, round seeds—by both the homozygous dominant (AA) and the heterozygous (Aa) genotypes.

The Principle of Independent Assortment

In the experiments described so far, Mendel was concerned with the inheritance of single pairs of alleles. To determine whether the same pattern of inheritance applied to each pair of alleles when more than one pair was present in hybrids, he crossed parental plants that differed in two or three pairs of alleles affecting different characters. For example, plants from a true-breeding line having round and yellow seeds were crossed with plants from a line having wrinkled and green seeds. The F_1 seeds from this cross were round and yellow; that is, they were hybrid for both characteristics, or **dihybrid.** If we assume that seed shape and color are independent traits that do not affect one another, this phenotype can be expected from the results of the individual **monohybrid** crosses (Table 1-1), in which the alleles resulting in round seed shape and yellow seed color were dominant over their respective alternative alleles. The F_2 seeds obtained when F_1 plants were grown and allowed to self-pollinate had the four possible combinations of phenotypic characteristics with the following frequencies:

round, yellow	315
wrinkled, yellow	101
round, green	108
wrinkled, green	32
	556

Mendel saw that, when the pairs of alternative phenotypes were considered separately, the ratios 423 (315 + 108) round to 133 (101 + 32) wrinkled seeds and 416 (315 + 101) yellow to 140 (108 + 32) green seeds were in each case very close to the 3:1 ratio observed in the F_2 progenies from the monohybrid crosses. He also noticed that the four phenotypes in the F_2 from the dihybrid cross were present in approximately the proportions 9/16 round yellow, 3/16 wrinkled yellow, 3/16 round green, and 1/16 wrinkled green. Similar 9:3:3:1 ratios of F_2 phenotypes were found in the progeny of other dihybrid crosses.

Mendel recognized that a 9:3:3:1 ratio is expected to result when two independently occurring 3:1 ratios are combined, and formulated the **principle of independent assortment,** sometimes called his second law.

> **The Principle of Independent Assortment:** Segregation of the members of a pair of alleles is independent of the segregation of other pairs in the formation of reproductive cells.

To illustrate, using the dihybrid cross just considered, we can represent the dominant and recessive alleles of the pair affecting seed shape as W and w, respectively, and the allelic pair affecting seed color as G and g. Then, the genotype of the F_1 is the double heterozygote

$$\frac{W}{w} \frac{G}{g}$$

The phenotype of the F_1 seeds is round and yellow, indicating that the allele for round, W, is dominant to that for wrinkled, w, and the allele for yellow, G, is dominant to that for green, g.

The result of independent assortment is that W is as likely to be included in a gamete with G as it is with g, and w is equally likely to be included with G or g. Thus, when two pairs of alleles undergo independent assortment, the gametes produced by the double heterozygote are the following:

$$1/4 \ WG, \ 1/4 \ Wg, \ 1/4 \ wG, \ \text{and} \ 1/4 \ wg$$

In later chapters, we will see that there are important exceptions to the principle of independent assortment.

Mendel's hypothesis for the transmission of inherited characters—that is, alleles segregate and undergo independent assortment, and male and female gametes unite at random in the formation of progeny zygotes—explains the $9:3:3:1$ ratio of F_2 phenotypes in the dihybrid cross. This can be seen in Figure 1-4; the format used to show which combinations of F_1 female and male gametes produce which F_2 genotypes is called a **Punnett square.** The hypothesis accounted equally well for the more-complex ratio of phenotypes that occurred in the F_2 progeny from a trihybrid, in which three independent pairs of genes were segregating.

Mendel tested the predicted genotypes from the crosses by growing plants from the F_2 seeds and obtaining F_3 progeny by self-pollination. Note in Figure 1-4 that round green F_2 seeds are expected to have the genotype $WW\,gg$, or, twice as frequently, $Ww\,gg$. To test this prediction, Mendel grew 102 plants from such seeds and found that 35 of them produced only round green seeds (indicating that the parental genotype was $WW\,gg$), whereas the other 67 produced both round green and wrinkled green seeds (indicating that the

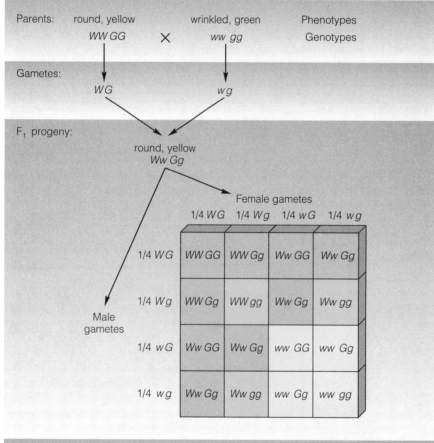

FIGURE 1-4 Diagram showing the basis for the $9:3:3:1$ ratio of F_2 phenotypes resulting from a cross between parents differing in two independently assorting characters.

parental plants must have been $Ww\,gg$.) The ratio $67:35$ is in good agreement with the expected $2:1$ ratio of genotypes. Similar good agreement with the predicted relative frequencies of the different genotypes was found when plants were grown from round yellow or from wrinkled yellow F_2 seeds. (As expected, plants grown from the wrinkled green seeds, which have the predicted homozygous recessive genotype $ww\,gg$, produced only wrinkled green seeds.)

Testcrosses

A second way that Mendel tested the hypothesis of independent assortment was by crossing the F_1 double heterozygotes ($Ww\,Gg$) with plants that were double homozygous recessive ($ww\,gg$.) As shown in Figure 1-5, the double heterozygotes will produce four types of gametes—$W\,G$, $W\,g$, $w\,G$, and $w\,g$—in equal frequencies, whereas the $ww\,gg$ plants will produce only $w\,g$ gametes. Thus, the progeny phenotypes are expected to consist of round yellow, round green, wrinkled yellow, and wrinkled green phenotypes in a $1:1:1:1$ ratio; this ratio is a direct reflection of the kinds of gametes produced by the double heterozygote because no dominant alleles are contributed by the $ww\,gg$ parent to obscure the results. In the actual cross, Mendel obtained 55 round yellow, 51 round green, 49 wrinkled yellow, and 53 wrinkled green, in good agreement with the predicted $1:1:1:1$ ratio. The results were the same in the reciprocal cross—that is, with the double heterozygote as female parent and the homozygous recessive as male parent. This observation confirmed Mendel's assumption that the gametes of both sexes included each possible genotype in approximately equal proportions.

A cross between a heterozygote for one or more genes, such as the genotype $Ww\,Gg$, and an individual organism that is homozygous for the recessive alleles is called a **testcross**. Testcrosses are extremely useful in genetic analysis for the following reason:

> In a testcross, the relative frequencies of the different gametes produced by the heterozygous parent can be observed directly in the phenotypes of the progeny, because the recessive parent contributes only recessive alleles.

Another valuable type of cross is a **backcross**, in which hybrid progeny are crossed with one of the parental genotypes. Backcrosses are commonly used by geneticists and by plant and animal breeders, as will be seen in later chapters. Note that the testcross used as an example in Figure 1-5 is also a backcross.

FIGURE 1-5 Genotypes and phenotypes resulting from a testcross of a $Ww\,Gg$ double heterozygote.

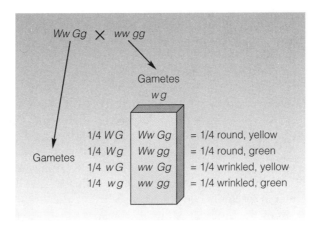

Multiple Alleles

So far, we have dealt with just two alleles of a gene—for example, A versus a, W versus w, or G versus g. Although each individual organism can contain no more than two alleles of any gene, some genes have more than two alleles that are present among a group of organisms. For example, the human blood groups designated A, B, O, or AB are determined by three types of alleles denoted I^A, I^B, and I^O, and the blood group of any particular person is determined by the particular pair of alleles present in his or her genotype. Genes with more than two alleles are often found in genetics, and they are said to have **multiple alleles.**

The record for multiple allelism, in terms of numbers of alleles, goes to certain flowering plants in which a single gene prevents self-pollination because pollen containing any particular allele will not fertilize flowers of genotypes containing the same allele. The species of red clover found in many pastures has this kind of self-sterility. Under these conditions, any plant with a new allele has an advantage because pollen containing the allele can fertilize all flowers except its own, and through evolution populations of red clover have accumulated hundreds of alleles of the self-sterility gene.

1-2 Mendelian Inheritance and Probability

A working knowledge of the rules of probability for predicting the outcome of chance events is basic to understanding the transmission of hereditary characteristics. In the first place, the proportions of the different types of offspring obtained from a cross are the cumulative result of numerous individual events of fertilization. Furthermore, in each fertilization the particular combination of dominant and recessive alleles that comes together occurs at random and is subject to chance variation.

For purposes of genetic analysis, the probability of an event may be considered equivalent to the proportion of times that the event is expected to occur in numerous repeated trials. Likewise, the proportion of times an event is expected to occur in numerous repeated trials is equivalent to the probability of occurrence in a single trial. For example, in the F_2 generation of the cross of round versus wrinkled seeds, Mendel observed 5474 round and 1850 wrinkled (Table 1-1); so the proportion of wrinkled was 1/3.96 [1850/(1850 + 5474)], or very nearly 1/4. We may therefore regard 1/4 as the approximate proportion of wrinkled seeds to be expected among a large number of progeny from this cross, or as the probability of obtaining a wrinkled seed from any particular fertilization.

Evaluating the probability of a genetic event usually requires an understanding of the mechanism of inheritance and knowledge of the particular cross. For example, in evaluating the probability of obtaining a round seed from a cross, you need to know that there are two alleles W and w, with W dominant to w; you also need to know the particular cross, because it makes a big difference to the probability of obtaining round seeds whether the cross is

WW × ww, in which all seeds are expected to be round,
Ww × Ww, in which 3/4 are expected to be round, or
Ww × ww, in which 1/2 are expected to be round.

In many genetic crosses, the possible outcomes of fertilization are equally likely. When there are n possible outcomes, all equally likely, and in m of these an event of interest occurs, then the probability of the event is m/n. For example, in progeny from the self-pollination of an Aa plant, or from the cross

Aa female × *Aa* male, four equally likely outcomes are possible—namely, *AA*, *Aa*, *aA*, and *aa*. The heterozygous genotype occurs in two of the four possible outcomes, and so the probability of a heterozygote is 2/4, or 1/2. Note that in this kind of tabulation it is important to regard the *Aa* and *aA* genotypes as different outcomes, which they in fact are because in one case the A allele comes from the mother and in the other case from the father.

Sometimes an event of interest can be expressed as consisting of two or more underlying events. For example, the phenotype of round seeds consists of either of two genotypes *WW* and *Ww*. A plant with round seeds must have either the genotype *WW* or the genotype *Ww* but cannot have both genotypes at the same time. Events like these, in which the occurrence of one precludes the occurrence of others so that only one at a time can occur, are said to be *mutually exclusive*. When events are mutually exclusive, their probabilities are combined according to the addition rule:

> **Addition Rule:** The probability *P* of occurrence of one or the other of two mutually exclusive events, A or B, is the sum of their individual probabilities.

In symbols, using *Prob* to mean *probability*, the addition rule is written

$$Prob(\text{A or B}) = Prob(\text{A}) + Prob(\text{B})$$

The addition rule can be applied to determine the proportion of round seeds expected from the cross *Ww* × *Ww*. The phenotype round seeds consists of the genotypes *WW* and *Ww*, which are mutually exclusive. In any individual progeny, the probability of genotype *WW* is 1/4 and that of *Ww* is 1/2; so the overall probability of either *WW* or *Ww* is

$$Prob(\text{WW or Ww}) = Prob(\text{WW}) + Prob(\text{Ww}) = 1/4 + 1/2 = 3/4$$

Because 3/4 is the probability of an individual seed being round, it is also the expected proportion of round seeds among a large number of progeny.

A second important situation in genetics is that in which events are *independent*, which means that the occurrence of one event has no influence on the possible occurrence of any others. For example, in Mendel's crosses for seed shape and color, the two traits are independent, and the ratio of phenotypes in the F_2 generation is expected to be 9/16 round yellow, 3/16 round green, 3/16 wrinkled yellow, and 1/16 wrinkled green. These proportions can be obtained by considering the traits individually, because they are independent. Considering only seed shape, we can expect the F_2 generation to consist of 3/4 round and 1/4 wrinkled seeds. Considering only seed color, we can expect the F_2 generation to consist of 3/4 yellow and 1/4 green. Because the traits are inherited independently, among the 3/4 of the seeds that are round, there should be 3/4 that are yellow, and so the overall proportion of round yellow seeds is expected to be $3/4 \times 3/4 = 9/16$. Likewise, among the 3/4 of the seeds that are round, there should be 1/4 green, giving $3/4 \times 1/4 = 3/16$ as the expected proportion of round green seeds. The proportions of the other phenotypic classes can be deduced in a similar way. The principle is that, when events are independent, the probability that they occur together is obtained by multiplication.

Successive offspring from a cross also are independent events, which means that the genotypes occurring in early births have no influence on the relative proportions of genotypes in later births. Widespread belief to the contrary, a couple is no more likely to have a girl on the next birth if they have already had five boys than if they have had five girls. The reason is that, though statistical averages tend to balance out across a large number of sibships, they do not need to balance within individual sibships. The term **sibship** refers to a group of offspring from the same parents. In the example of five children, the sibships with five boys balance those with five girls, for an overall sex ratio of $1 : 1$, but both types of sibship are unusual in their sex distribution.

When events are independent, such as independent traits or successive offspring from a cross, the probabilities are combined by means of the multiplication rule:

> **Multiplication Rule:** The probability of two independent events, A and B, occurring together is given by the product of their individual probabilities.

In symbols, the multiplication rule is

$$Prob(A \text{ and } B) = Prob(A) \cdot Prob(B)$$

The multiplication rule can be applied to answer questions like the following: Of two offspring from the mating $Aa \times Aa$, what is the probability that both have the dominant phenotype? Because the mating is $Aa \times Aa$, the probability that any particular offspring has the dominant phenotype equals 3/4. Using the multiplication rule, the probability that each of two offspring has the dominant phenotype is $3/4 \times 3/4 = 9/16$.

Here is another typical genetic question that can be answered by the use of the addition and multiplication rules: Of two offspring from the mating $Aa \times Aa$, what is the probability of one dominant phenotype and one recessive? Sibships of one dominant and one recessive can occur in two different ways—with the dominant born first or with the dominant born second—and these are mutually exclusive. The first case has probability $3/4 \times 1/4$ and the second has probability $1/4 \times 3/4$; because the events are mutually exclusive, the probabilities are added. The correct answer is therefore $(3/4 \times 1/4) + (1/4 \times 3/4) = 2(3/4)(1/4)$.

The addition and multiplication rules are very powerful tools for calculating the probabilities of genetic events. Figure 1-6 shows how the rules

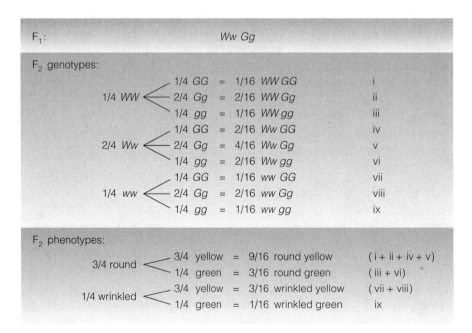

FIGURE 1-6 Example of the use of the addition and multiplication rules to determine the probabilities of the nine genotypes and four phenotypes in the F_2 produced by self-pollination of a dihybrid F_1. The roman numerals are arbitrary labels identifying the F_2 genotypes.

are applied to determine the expected proportions of the nine different genotypes occurring among the F_2 progeny produced by self-pollination of a $Ww\,Gg$ dihybrid.

In genetics, independence applies not only to the successive offspring formed by a mating, but also to the segregation of genes that undergo independent assortment. This means that the multiplication rule can be used to determine the probability of a specific genotype among the progeny from a cross in which numerous pairs of alleles undergo independent assortment. For example, the probability of the genotype $Aa\,Bb\,Cc\,Dd$ among the progeny from the cross $Aa\,Bb\,Cc\,Dd \times Aa\,Bb\,Cc\,Dd$ is $(1/2)(1/2)(1/2)(1/2) = (1/2)^4$, or 1/16.

1-3 Segregation in Pedigrees

Determining the genetic basis of a trait from the kinds of crosses that we have considered requires large numbers of offspring. The analysis of segregation by this method is not possible in human beings because matings cannot be controlled, and it is not usually feasible for some traits in large domestic animals. However, the mode of inheritance of a trait can sometimes be determined by examining the segregation of alleles in several generations of related individuals. This is typically done with a family tree that shows the phenotype of each individual member; such a diagram is called a **pedigree.** An important application of probability in genetics is its use in pedigree analysis.

In a pedigree diagram such as the one for the human family shown in Figure 1-7, females and males are represented by circles and squares, respectively. (A diamond is used if the sex of an individual member is unknown.) Individuals having the phenotype of interest are indicated by shaded symbols (in this case in color). The symbols of parents are joined by a horizontal line, which is connected vertically to a second horizontal line that extends above the symbols for their offspring. The offspring within a sibship, called **siblings** or **sibs** regardless of sex, are represented from left to right in order of their birth. Successive generations in a pedigree are designated by Roman numerals and the individuals in a generation by Arabic numbers.

The trait in Figure 1-7 is **Huntington disease,** a progressive nerve degeneration resulting in severe physical and mental disability, usually beginning about middle age. The pedigree shows that the trait affects both sexes, that it is transmitted from affected parent to affected offspring, and that about half of all the offspring of an affected parent are affected. These are characteristic features of simple Mendelian dominance. The dominant allele *HD* causing Huntington disease is very rare. Affected persons are heterozygous *HD / hd*, whereas unaffected persons are homozygous normal *hd / hd*.

A pedigree pattern for a recessive trait is shown in Figure 1-8. The trait is **albinism,** or absence of pigment in the skin, hair, and iris of the eyes. Both sexes can be affected, but the affected individuals need not have affected

FIGURE 1-7 Pedigree of a human family, showing the inheritance of the dominant gene for Huntington disease. Females and males are denoted by circles and squares, respectively. Red symbols indicate persons affected with the disease.

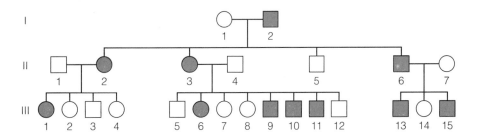

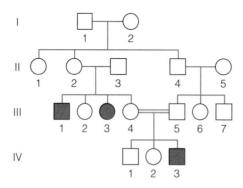

FIGURE 1-8 Pedigree of albinism. With recessive inheritance, affected persons (red) often have unaffected parents. The double horizontal line indicates a mating between relatives, in this case first cousins.

parents. The unaffected parents are called **carriers** because they are heterozygous for the recessive allele, and in a mating between carriers each offspring has a 1/4 chance of being affected. The pedigree also illustrates another feature often found with recessive traits, particularly rare traits, which is that the parents of affected individuals are often related. A mating between relatives, in this case first cousins, is indicated with a double line connecting the partners.

Matings between relatives are important for rare recessive alleles because a carrier may have multiple descendants who are carriers, and when the carriers mate the recessive allele can become homozygous. When a recessive allele is rare, it is more likely to become homozygous through inheritance from a common ancestor than if the parents were completely unrelated.

1-4 Genes and Cellular Products

The first suggestion of a relation between genes and specific cellular products was made by Archibald Garrod, an English physician interested in the rare disease **alkaptonuria** and several other human disorders. He proposed that these disorders result from inherited defects in body chemistry and called them **inborn errors of metabolism.** By examining pedigrees of human families in which alkaptonuria occurred, he correctly deduced that alkaptonuria is determined by a single recessive gene. The disease, which is quite harmless, is characterized by the accumulation and excretion of an innocuous substance that causes the urine of an affected person to turn black upon exposure to air. Noting that heterozygous and homozygous people have the same phenotype, he suggested (correctly) that the defect in alkaptonuria is the absence of an enzyme required for the breakdown of this substance. The active enzyme would be absent in homozygous recessive (aa) alkaptonurics and present in both homozygous dominant (AA) and heterozygous (Aa) persons. Because enzymes are catalytic proteins (they increase the rate of chemical reactions) that are unchanged in the reaction and can function repeatedly, a single enzyme molecule may be able to catalyze a particular reaction hundreds or thousands of times per second. Heterozygotes, with half the number of enzyme molecules possessed by homozygous dominants, will in most cases have sufficient enzyme for a normal phenotype. Although this is frequently the case, there are differences in phenotype between homozygous dominant and heterozygous individuals, as will be seen later in this chapter.

In Mendel's experiments, all traits had clear dominant-recessive patterns. This was fortunate because otherwise he might not have made his discoveries. However, lack of strict dominance is widespread in nature. In this section several alternative patterns will be described.

Absence of Dominance of Some Alleles

Absence of dominance of one member of a pair of alleles over the other is quite common in most organisms. For example, in crosses between snapdragon plants from a red-flowered variety and a variety with ivory-colored flowers, the F_1 plants produce only pink flowers intermediate in color between those of the parental varieties. In one experiment, the F_2 obtained by self-pollination of the F_1 hybrids consisted of 22 plants with red flowers, 52 with pink flowers, and 23 with ivory flowers. The numbers agree with the Mendelian ratio of 1 dominant homozygote : 2 heterozygotes : 1 recessive homozygote expected in the absence of dominance (Figure 1-9). In agreement with the predictions from this interpretation, the red-flowered F_2 plants produced only red-flowered progeny, the ivory-flowered plants produced only ivory-flowered progeny, and the pink-flowered plants produced red, pink, and ivory progeny in the proportions 1/4 red : 1/2 pink : 1/4 ivory.

FIGURE 1-9 Absence of dominance in the inheritance of flower color in snapdragons.

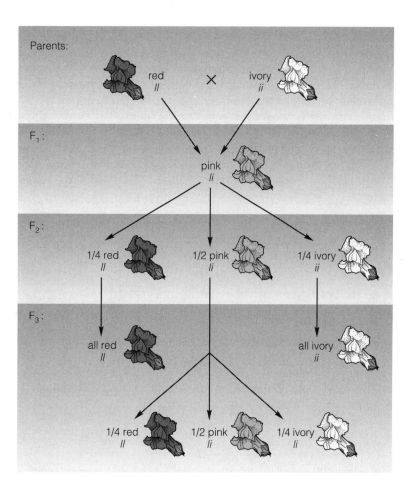

In this example of flower-color inheritance, the absence of dominance is due to the way in which the red pigment is formed. The pigment is formed by a complex sequence of enzymatic reactions. The *I* allele codes for an enzyme that is critical because the amount of enzyme controls how much red pigment can be formed. The alternative *i* allele codes for an inactive form of the enzyme. Because the amount of the critical enzyme is reduced in *Ii* heterozygotes, the amount of red pigment in the flowers is reduced also, and the effect of the dilution is pink flowers.

Intermediate heterozygote phenotypes are usually not observed with other genes that code for enzymes because, as mentioned earlier, enzymes are catalytic, and the heterozygotes usually contain enough enzyme to produce a normal phenotype. Most genes coding for enzymes exhibit complete dominance, which is the case for alkaptonuria.

On the other hand, even when dominance seems unmistakable, it is sometimes possible to find subtle features of the phenotype that are affected by the presence of the recessive allele in heterozygotes. One example is Mendel's familiar round versus wrinkled seeds. The genetic defect in wrinkled peas is the absence of an enzyme called starch-branching enzyme I (SBEI), which is needed for the synthesis of a branched-chain form of starch known as amylopectin. As Mendel found when he crossed round *WW* × wrinkled *ww*, the F_1 *Ww* seeds were all round, indicating complete dominance of the W allele. However, microscopic examination later revealed subtle differences in the form of the starch grains in seeds of the three genotypes. Homozygous *WW* peas contain large, well-rounded starch grains, with the result that the seeds retain water and shrink uniformly as they ripen and do not become wrinkled. The starch grains in wrinkled (*ww*) seeds lack amylopectin and are irregular in shape, and the ripening seeds lose water more rapidly and shrink unevenly. In heterozygous seeds, the starch grains are intermediate in shape, but their amylopectin content is high enough to result in uniform shrinking of the seeds and no wrinkling. Thus, the basic physiological determinant of round versus wrinkled seeds is the amount of amylopectin in the starch grains, and the heterozygotes have an amylopectin content intermediate between the two homozygotes. When the phenotypes are considered at the level of round versus wrinkled, W is dominant over *w*, but when the phenotypes are examined at the microscopic level, the heterozygotes are intermediate. The key point is the following:

> The phenotype consists of many different physical and biochemical attributes, and some of these attributes may be dominant, whereas others are not; thus, dominance is a property not only of a pair of alleles, but also of the attributes of phenotype that are observed.

For characters such as flower pigment in snapdragons and the starch content of pea seeds, the phenotype of heterozygotes appears to be almost exactly intermediate between the homozygotes. This is the expected condition when there is *no dominance* of one allele over the other (Figure 1-10). The expression of other phenotypic characters in heterozygotes, though also intermediate, may be more similar to one homozygote than to the other. Phenotypic expression of this type is the result of **partial dominance** of the effects of one of the alleles.

Codominance

Another exception to simple dominance is the full expression of two alternative alleles in a heterozygote, resulting in a phenotype in which the presence of both

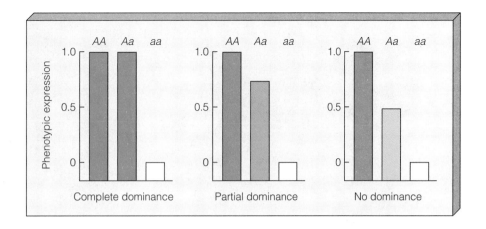

FIGURE 1-10 Levels of phenotypic expression in heterozygotes with varying degrees of dominance of one allele over the other.

alleles can be detected. This is called **codominance.** One of the best examples is the effect of the genes that determine the human A, B, AB, or O blood groups, which were discussed earlier in the context of multiple alleles. Blood type is determined by the types of polysaccharides (polymers of sugars) that are on the surface of red blood cells. Two different polysaccharides, A and B, can be present. The A type is synthesized by an enzyme coded by the I^A allele and the B type by an enzyme coded by the I^B allele. People of genotype I^A/I^A produce red cells having only the A polysaccharide and are said to have blood type A. Those of genotype I^B/I^B have red cells having only the B polysaccharide and have blood type B. Heterozygous I^A/I^B people have red cells with both A and B polysaccharides and have blood type AB. The I^A/I^B genotype illustrates codominance of these alleles, because both alleles are expressed.

A third allele I^O codes for a defective enzyme that produces neither A nor B polysaccharides. The I^O allele is recessive to both I^A and I^B, and so I^A/I^O heterozygotes have blood type A and I^B/I^O heterozygotes have blood type B. The homozygous recessives I^O/I^O, which lack both the A and B polysaccharides, are said to have blood type O.

Blood groups are important in medicine because of the frequent need for blood transfusions. An important feature of the ABO system is that most human blood contains antibodies to either the A or B polysaccharide. An **antibody** is a protein made by the immune system in response to a stimulating molecule called an **antigen** and capable of binding to the antigen. An antibody is usually specific in that it recognizes only one antigen. Some antibodies combine with antigen and form particulate clumps of large molecular aggregates. The ABO blood group system is summarized in Table 1-2.

TABLE 1-2   Relations between cellular antigens, serum antibodies, and genotypes of the four phenotypes in the ABO blood group

Phenotype	Red-cell antigen	Serum antibody	Genotype
A	A	Anti-B	$I^A I^A$ or $I^A I^O$
B	B	Anti-A	$I^B I^B$ or $I^B I^O$
AB	A and B	—	$I^A I^B$
O	—	Anti-A, anti-B	$I^O I^O$

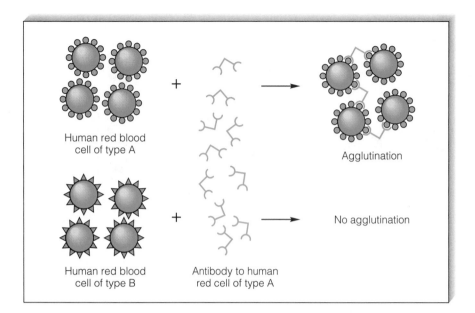

Antibodies are one line of the body's defense against invading viruses, and bacteria and other cells, and help in removing such invaders from the system. Although antibodies do not normally form without prior stimulation by the antigen, many common bacteria contain antigens similar to A and B, and so most people are repeatedly stimulated to produce anti-A and anti-B antibody. However, it would be disastrous for an organism to produce antibodies against its own antigens, and stimulation with A-antigen or B-antigen elicits antibody production only in people whose own red blood cells do not contain A or B, respectively. The end result is as follows:

> Individuals of blood type A and O make anti-B antibodies, those of blood type B and O make anti-A antibodies, and those of blood type AB make neither type of antibody.

The clinical significance of the ABO blood groups is that transfusion of blood containing A or B red-cell antigens into a person who makes antibodies against them results in an agglutination reaction in which the donor red blood cells are clumped (Figure 1-11). Many blood vessels are blocked, and the recipient of the transfusion goes into shock and may die. Incompatibility in the other direction, in which the donor blood contains antibodies against the recipient's red blood cells, is usually negligible because the donor's antibodies are diluted so rapidly that clumping does not occur. However, best medical practice is to use donor blood of the same ABO type as the recipient.

1-6 The Effects of Genes on the Expression of Other Genes

In the examples considered so far, distinct traits such as color and shape have been independent in their expression. However, all traits result ultimately from a series of complex biochemical reactions, and many phenotypes are determined by the combined action of the alleles of two or more genes. In these cases, the phenotype produced by the alleles of one gene is influenced by the genotype with respect to other genes. Such mutual influences on gene expression constitute **epistasis.** Epistatic interactions between genes can cause departures from the simple Mendelian ratios discussed so far.

Epistasis among genes affects eye color in *Drosophila*. The normal eye color is a dull red resulting from the combination of a bright scarlet pigment with a brown one. The normal color is called **wildtype**, a term used in *Drosophila* genetics for phenotypes or alleles that are normally found in flies in the wild. Many genes contribute to the formation of the scarlet and brown pigments, but among them are the genes for brown and scarlet. The recessive brown allele *bw* results in absence of the scarlet pigment, and homozygous *bw* / *bw* flies have brown eyes. The recessive scarlet allele *st* results in absence of the brown pigment, and homozygous *st* / *st* flies have scarlet eyes. The homozygous wildtype genotypes are designated bw^+ / bw^+ and st^+ / st^+, respectively. These symbols for alternative alleles are different from the upper-case-lowercase symbols used earlier in the book. They are used here because *Drosophila* geneticists often use the superscript plus sign to specify the wildtype allele. In the crosses of wildtype $bw^+ / bw^+ \times bw / bw$ and wildtype $st^+ / st^+ \times st / st$, the following results were obtained:

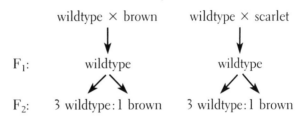

These results are typical of Mendelian segregation, with brown eyes determined by one recessive allele, *bw*, and scarlet eyes by a different recessive allele, *st*. However, the results of the cross brown × scarlet were not as straightforward. The F_1 hybrids were wildtype, but four phenotypes appeared among the F_2 progeny—wildtype, brown, scarlet, and white (Figure 1-12). These phenotypes were in the approximate proportions 9/16 wildtype, 3/16 brown, 3/16 scarlet, and 1/16 white, which suggests that white eye color results from the combined effects of homozygosity for *bw* and *st*. This model is summarized in Figure 1-13. In the grouping of F_2 genotypes into phenotypic classes in part A, the dash (–) indicates the presence of either a dominant or a recessive allele. This is a convenient abbreviation to use if the homozygous dominant and heterozygous genotypes are phenotypically indistinguishable. The reason that *bw* / *bw*; *st* / *st* double homozygotes have white eyes is that they lack both colored pigments: homozygosity for *bw* knocks out the scarlet pigment, and homozygosity for *st* knocks out the brown pigment.

Another type of interaction between different genes occurs in flower color in sweet peas. When a variety that breeds true for colored flowers is crossed with

FIGURE 1-12 Interaction between brown, *bw*, and scarlet, *st*, genotypes in determining eye color in *Drosophila*. The *bw* / *bw*; *st* / *st* double homozygote has white eyes.

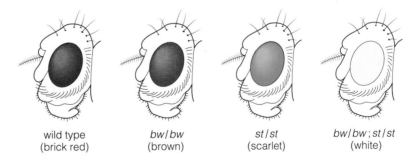

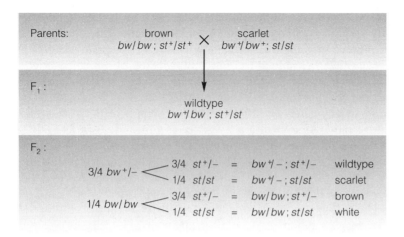

A

B

FIGURE 1-13 A. Independent assortment of the *bw* and *st* alleles in *Drosophila*. The dash indicates that the phenotype is unaffected by the identity of the second allele. B. Punnett square for a cross between the F$_1$ double heterozygotes in part A.

either of two different white-flowered varieties, the F$_1$ plants have colored flowers in both cases. Self-pollination of the F$_1$ hybrids from each cross results in an F$_2$ in which 3/4 of the plants have colored flowers and 1/4 are white flowered. The simplest explanation for these results is that the two white-flowered varieties are homozygous for a recessive allele of the same gene, in which case they would be expected to produce a white-flowered F$_1$ when crossed. However, when the two white varieties are crossed, the F$_1$ plants produce colored flowers (Figure 1-14). In the F$_2$ obtained in one experiment by self-pollination of the hybrids, 382 plants with colored flowers and 269 with white flowers were found—a ratio close to 9:7. Both the colored flowers in the F$_1$ hybrids and the occurrence of a 9:7 ratio in the F$_2$ can be explained by the effects of two pairs of genes undergoing independent assortment, with the

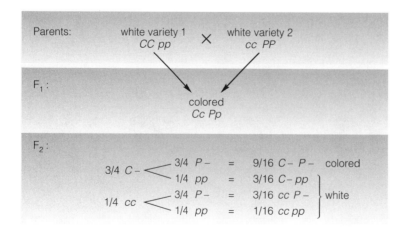

FIGURE 1-14 A cross showing epistasis in the determination of flower color in sweet peas. Color formation requires at least one dominant allele of each of two genes. Note the unusual ratio of 9 colored to 7 white (7 = 3 + 3 + 1) in the F$_2$ generation.

presence of at least one dominant allele of both genes required for the production of flower color (Figure 1-14). This interpretation could be tested by examining the self-pollinated progeny of individual F_2 plants with colored flowers. (To demonstrate the predictions for yourself, determine the actual genotypes of the F_2 plants grouped as $C–P–$ in Figure 1-14, the expected relative frequencies of these genotypes, and the relative proportions of colored and white-flowered progeny expected from the self-pollination of plants with each genotype.)

The biochemical explanation for this situation can be illustrated in the following way:

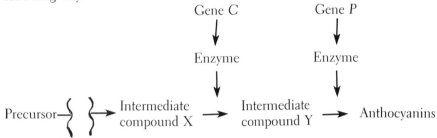

In the absence of gene C (that is, in cc homozygotes), little or no compound Y would be produced and the pathway would be blocked at that point. In turn, compound Y is the substrate for the enzyme coded by gene P, which converts it into an anthocyanin pigment. Therefore, the absence of gene P (in pp homozygotes) would block the final step in pigment synthesis. The genes are related in that pigment production requires both enzymes, and these are coded by the dominant alleles of the genes.

Penetrance, Expressivity, and Pleiotropy

The phenotypic expression of *some* genes is identical in all organisms with the same genotype. However, the phenotypes determined by *most* genes are more variable. The variation may result from the effects of other genes, as just seen, or because the biological processes that produce the particular phenotype are sensitive to environmental conditions. Moreover, most genes affect more than one trait. Metabolic pathways are complex and interconnected, and a defect in one enzyme is likely to affect not only the function of the pathway in which it is a participant, but also other pathways that interact with it.

The complexities of variable gene expression are usually considered under three categories:

1. **Variable expressivity** refers to genes that are expressed to different degrees in different individuals. The different degrees of expression often form a continuous series from full expression to almost no expression of the expected phenotypic characteristics. Many genetic diseases in human beings are caused by genes with variable expressivity, and the result is that some affected persons may be very severely affected, whereas others carrying the same gene are mildly affected.

2. **Incomplete penetrance** means that some individuals with a particular genotype do not express the phenotype associated with the genotype. Failure of the phenotype to be expressed may result from the effects of other genes or from environmental conditions. Incomplete penetrance is merely the extreme of variable expressivity, in which expression of the expected phenotype is so mild as to be undetectable. As an example of incomplete penetrance, there are identical human twins in which one twin has a genetically determined abnormality but the other does

not. The proportion of individuals whose phenotype matches their genotype for a given character is called the **penetrance** of the genotype. A genotype that is always expressed has a penetrance of 100 percent.

3. **Pleiotropy** refers to multiple phenotypic effects of a single mutant gene. Often the traits that are affected have no apparent physiological connection. However, all are secondary consequences of a single genetic defect. For example, several genetic human diseases result in anemia (insufficient numbers of red blood cells). The symptoms include lack of energy, easy fatigue, shortness of breath, rapid pulse, pounding heart, pallor in the hands and fingernails, and swelling of the ankles. All of these symptoms result from an inadequate supply of oxygen.

CHAPTER SUMMARY

Inherited traits are determined by particulate elements called genes. A gene has different forms, called alleles. In a higher plant or animal, the genes are present in pairs, one member of each pair having been transmitted from the maternal parent and the other member from the paternal parent. The specific allelic composition of an individual organism is called its genotype; the set of observable properties derived from its genotype is termed its phenotype. If the two alleles of a pair are the same (for example, AA or aa), the organism is homozygous for the A or a allele; if the alleles are different (Aa), the organism is heterozygous. When the phenotype of a heterozygote is the same as one of the homozygous genotypes, then the expressed allele is dominant and the hidden allele is recessive.

The organisms produced by a mating constitute the F_1 generation. Matings between members of the F_1 generation produce the F_2 generation. In a cross of the type $AA \times aa$, in which only one gene is considered, the ratio of the genotypes in the F_2 generation (obtained by self-fertilization of the Aa F_1 hybrids) is 1 dominant homozygote (AA):2 heterozygotes (Aa):1 recessive homozygote (aa). The phenotypes in the F_2 are in the ratio 3 dominant:1 recessive. This proves that in gamete formation the two members of an allelic pair are segregated randomly into different gametes, and the gametes participate in fertilization in random combinations. In crosses of the type $AA\,BB \times aa\,bb$, with two genes, the phenotypic ratios in the F_2 are 9:3:3:1, provided that the alleles undergo independent assortment. This ratio can be modified in various ways by interaction (epistasis) between the A,a and B,b alleles.

The random processes occurring in the formation of gametes and their union at fertilization follow the simple rules of probability, which form the basis for predicting outcomes of genetic crosses. With some organisms—for example, human beings—it is not possible to perform controlled crosses, and genetic analysis is carried out by the study of several generations of a family tree, called a pedigree. Pedigree analysis is the determination of both the genotypes of the family members and the probability of a particular genotype being associated with each member.

A pair of alleles does not always have a dominant member. For example, intermediate phenotypes occasionally arise because less of a gene product (which causes the phenotype) is made when only one wildtype allele is present than when two copies are present. This phenomenon is called partial dominance. When both of two different alleles in a heterozygote are expressed, they are codominant. A genotype that is expressed to different degrees in different individuals has variable expressivity; if it is not expressed at all in some individuals, it has incomplete penetrance. Most genes affect multiple traits, which is referred to as pleiotropy.

KEY TERMS

addition rule
albinism
alkaplonuria

allele
antibody
antigen

backcross
carrier
codominance

dihybrid
dominant
epistasis
F_1 generation
F_2 generation
gamete
gene
genotype
heterozygous
homozygous
Huntington disease
hybrid

inborn errors of metabolism
incomplete penetrance
independent assortment
monohybrid
multiple alleles
multiplication rule
partial dominance
pedigree
penetrance
phenotype
pleiotropy
Punnett square

recessive
reciprocal cross
segregation
sib
sibling
sibship
testcross
true breeding
variable expressivity
wildtype
zygote

EXAMPLES OF WORKED PROBLEMS

Problem 1: Both round fruit and elongated fruit are true breeding in tomatoes. The cross round × elongate produces F_1 progeny with round fruit, and the cross $F_1 \times F_1$ produces 3/4 progeny with round fruit and 1/4 progeny with elongate fruit. What kind of genetic hypothesis can explain these data?

Answer: In this kind of problem, a good strategy is to look for some evidence of Mendelian segregation. The 3 : 1 ratio in the F_2 generation is characteristic of Mendelian segregation when there is dominance. This observation suggests the genetic hypothesis of a dominant allele R for round fruit and a recessive allele r for elongate fruit. If the hypothesis were correct, then the true-breeding round and elongate genotypes would be RR and rr, respectively. The F_1 progeny of the cross RR (round) × rr (elongate) would be Rr, which has round fruit, as observed. The $F_1 \times F_1$ cross (Rr × Rr) gives 1/4 RR, 1/2 Rr, and 1/4 rr. Because both RR and Rr have round fruit, the expected F_2 ratio of round : elongate phenotypes is 3 : 1, providing additional evidence for the single-gene hypothesis.

Problem 2: Both red coat color and white coat color are true breeding in Shorthorn cattle. Crosses of red × white produce progeny that are roan—uniformly reddish brown but thickly sprinkled with white hairs. Crosses of roan × roan produce 1/4 red : 1/2 roan : 1/4 white. What kind of genetic hypothesis can explain these data?

Answer: In this case, the 1 : 2 : 1 ratio of phenotypes in the cross roan × roan suggests Mendelian segregation because this is the ratio expected from a mating between heterozygotes when dominance is incomplete. Supposing that roan is heterozygous, say Rr, we can then expect the cross roan × roan (Rr × Rr) to produce 1/4 RR, 1/2 Rr, and 1/4 rr genotypes. The observed result that 1/2 of the progeny are roan Rr fits this hypothesis, which implies that the RR and rr genotypes correspond to red and white. The problem states that red and white are true breeding, which is consistent with their being homozygous. Additional confirmation comes from the cross RR × rr, which gives Rr (roan) progeny, as expected. Note that the gene symbols R and r are assigned to red and white arbitrarily, and so it does

not matter whether RR stands for red and rr white, or the other way around.

Problem 3: The tailless trait in the mouse results from an allele of a gene in chromosome 17. In the cross tailless × tailless, both tailless and wildtype progeny are produced, in a ratio of 2 tailless : 1 wildtype. All tailless progeny from this cross, when mated with wildtype, produce a 1 : 1 ratio of tailless to wildtype progeny.
(a) Is the allele for the tailless trait dominant or recessive?
(b) What genetic hypothesis can account for the 2 : 1 ratio of tailless : wildtype and the results of the testcrosses with the tailless animals?

Answer:
(a) If the tailless phenotype were homozygous recessive, then the cross tailless × tailless should produce only tailless progeny. This is not the case, and so the tailless phenotype must result from a dominant allele, say T.
(b) Because the cross tailless × tailless produces both tailless and wildtype progeny, both parents must be heterozygous Tt. The expected ratio of genotypes among the zygotes is 1/4 TT, 1/2 Tt, and 1/4 tt. Because T is dominant, the Tt animals are tailless and the tt animals are wildtype. The 2 : 1 ratio can be explained if the TT zygotes do not survive (that is, are lethal). Because all surviving tailless animals must be Tt, this genetic hypothesis would also explain why *all* of the tailless animals from the cross, when mated with tt, give a 1 : 1 ratio of tailless (Tt) to wildtype (tt). (Developmental studies confirm that about 25 percent of the embryos do not survive.)

Problem 4: The illustration in the left-hand column on the next page shows four alternative types of combs in chickens, called rose, pea, single, and walnut. The following data summarize the results of crosses. The rose and pea strains used in crosses 1, 2, and 5 are true breeding.
1. rose × single → rose
2. pea × single → pea
3. (rose × single) F_1 × (rose × single) F_1 → 3 rose : 1 single

4. (pea × single) F$_1$ × (pea × single) F$_1$ → 3 pea : 1 single
5. rose × pea → walnut
6. (rose × pea) F$_1$ × (rose × pea) F$_1$ → 9 walnut : 3 rose : 3 pea : 1 single

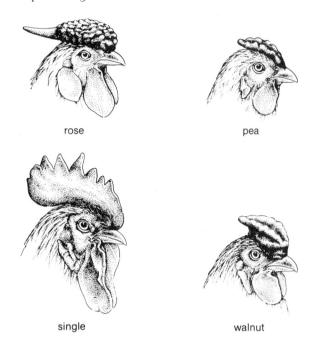

rose

pea

single

walnut

(a) What genetic hypothesis can explain these results?
(b) What are the genotypes of parents and progeny in each of the crosses?
(c) What are the genotypes of true-breeding strains of rose, pea, single, and walnut?

Answer:
(a) Cross 6 gives the Mendelian ratios expected when two genes are segregating, and so a genetic hypothesis with two genes is necessary. Crosses 1 and 3 give the results expected if the rose comb were due to a dominant allele, say *R*. Crosses 2 and 4 give the results expected if the pea comb were due to a dominant allele, say *P*. Cross 5 indicates that the walnut comb results from the interaction of *R* and *P*. The segregation in cross 6 means that *R* and *P* are not alleles of the same gene.
(b) 1. *RR pp* × *rr pp* → *Rr pp*.
   2. *rr PP* × *rr pp* → *rr Pp*.
   3. *Rr pp* × *Rr pp* → 3/4 *R−pp* : 1/4 *rr pp*.
   4. *rr Pp* × *rr Pp* → 3/4 *rr P−* : 1/4 *rr pp*.
   5. *RR pp* × *rr PP* → *Rr Pp*.

6. *Rr Pp* × *Rr Pp* → 9/16 *R−P−* : 3/16 *R−pp* : 3/16 *rr P−* : 1/16 *rr pp*.

(c) True-breeding genotypes are *RR pp* (rose), *rr PP* (pea), *rr pp* (single), and *RR PP* (walnut).

Problem 5: The pedigree in the accompanying illustration shows the inheritance of coat color in a group of cocker spaniels. The coat colors and genotypes are as follows:

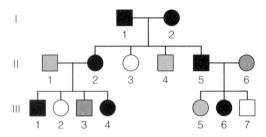

Black *A−B−* (black symbols)
Liver *aa B−* (pink symbols)
Red *A−bb* (red symbols)
Lemon *aa bb* (open symbols)

(a) Specify as completely as possible the genotype of each individual dog in the pedigree.
(b) What are the possible genotypes of individual III-4, and what is the probability of each genotype?
(c) If a single pup is produced from the mating of III-4 × III-7, what is the probability that the pup will be red?

Answer:
(a) All three matings (I-1 × I-2, II-1 × II-2, and II-5 × II-6) produce lemon-colored offspring, *aa bb*, and so each parent must carry at least one *a* allele and at least one *b* allele. Therefore, in consideration of the phenotypes, the genotypes are as follows: I-1 *Aa Bb*, I-2 *Aa Bb*, II-1 *aa Bb*, II-2 *Aa Bb*, II-5 *Aa Bb*, II-6 *Aa bb*. The genotypes of the offspring can be deduced from their own phenotypes and the genotypes of the parents. They are as follows: II-3 *aa bb*, II-4 *aa B−*, III-1 *Aa B−*, III-2 *aa bb*, III-3 *Aa bb*, III-4 *Aa B−*, III-5 *aa Bb*, III-6 *A−Bb*, III-7 *aa bb*.
(b) Individual III-4 is either *Aa BB* or *Aa Bb*, and the probabilities of these genotypes are 1/3 and 2/3, respectively.
(c) If individual III-4 is *Aa BB*, then the probability of a red pup is 0; if individual III-4 is *Aa Bb*, then the probability of a red pup is 1/2 × 1/2 = 1/4 (that is, the probability of an *A b* gamete from III-4). Overall, the probability of a red pup from the mating is 1/3 × 0 + 2/3 × 1/2 = 1/3.

PROBLEMS

1-1. With respect to homozygosity and heterozygosity, what can be said about the genotype of a strain or variety that breeds true for a particular trait?

1-2. What gametes can be formed by an individual organism of genotype *Aa*? Of genotype *Bb*? Of genotype *Aa Bb*?

1-3. How many different gametes can be formed by an organism with genotype $AA\,Bb\,Cc\,Dd\,Ee$, and in general by an organism that is heterozygous for m genes and homozygous for n genes?

1-4. If an allele R is dominant to r, how many different phenotypes are present in the progeny of a cross between Rr and Rr, and in what ratio? How many phenotypes are there, and in what ratio do they occur, if there is no dominance between R and r?

1-5. In a cross of a black-feathered Andalusian chicken with a strain having splashed-white feathers (in which there is an uneven sprinkling of black pigment through the feathers), the F_1 progeny are all slate blue. The cross $F_1 \times F_1$ produces black, slate blue, and splashed-white offspring in the ratio $1:2:1$, respectively. What genetic hypothesis can account for these data?

1-6. Assuming equal numbers of boys and girls, if a couple has a girl, what is the probability that the next child will be a boy? If a couple has two girls, what is the probability that the next child will be a boy? What type of probability argument do you base your answers on?

1-7. Assuming equal numbers of boys and girls, what is the probability that a family with two children will have two girls? One girl and one boy?

1-8. In the following questions you are asked to deduce the genotype of certain parents in a pedigree. The phenotypes are determined by dominant and recessive alleles of a single gene.
   (a) A homozygous recessive results from the mating of a heterozygote and a parent with the dominant phenotype. What does this tell you about the genotype of the parent with the dominant phenotype?
   (b) Two parents with the dominant phenotype produce nine offspring. Two have the recessive phenotype. What does this tell you about the genotype of the parents?
   (c) One parent has a dominant phenotype and the other has a recessive phenotype. Two offspring result and both have the dominant phenotype. What genotypes are possible for the parent with the dominant phenotype?

1-9. Pedigree analysis tells you that a particular parent may have the genotype $AA\,BB$ or $AA\,Bb$, each with the same probability. Assuming independent assortment, what is the probability of this parent's producing an $A\,b$ gamete? What is the probability of the parent's producing an $A\,B$ gamete?

1-10. Assume that the trihybrid cross $AA\,BB\,rr \times aa\,bb\,RR$ is made in a plant species in which A and B are dominant but there is no dominance between R and r. In the F_2 progeny from this cross, assuming independent assortment:
   (a) How many phenotypic classes are expected?

   (b) What is the probability of the parental $aa\,bb\,RR$ genotype?
   (c) What proportion would be expected to be homozygous for all three genes?

1-11. In the cross $Aa\,Bb\,Cc\,Dd \times Aa\,Bb\,Cc\,Dd$, in which all genes undergo independent assortment, what proportion of offspring are expected to be heterozygous for all four genes?

1-12. The pattern of coat coloration in dogs is determined by the alleles of a single gene, with S (solid) being dominant over s (spotted). Black coat color is determined by the dominant allele A of a second gene, and tan by homozygosity for the recessive allele a. A female having a solid tan coat is mated with a male having a solid black coat and produces a litter of six pups. The phenotypes of the pups are: 2 solid tan, 2 solid black, 1 spotted tan, and 1 spotted black. What are the genotypes of the parents?

1-13. In the human pedigree shown here, the daughter indicated by the solid circle has a form of deafness determined by a recessive allele. What is the probability that the phenotypically normal son (indicated by the open square) is heterozygous for the gene?

1-14. Huntington disease is a rare degenerative human disease determined by a dominant allele, H. The disorder is usually manifested after the age of forty-five. A young man has learned that his father has developed the disease.
   (a) What is the probability that the young man will later develop the disorder?
   (b) What is the probability that a child of the young man carries the H allele?

1-15. The Hopi, Zuni, and some other Southwest American Indians have a relatively high frequency of albinism (absence of skin pigment) resulting from homozygosity for a recessive allele, a. A normally pigmented man and woman, each having an albino parent, have two children. What is the probability that both children are albino? What is the probability that at least one of the children is albino?

1-16. Which combinations of donor and recipient ABO blood groups are compatible for transfusion? (Consider a combination to be compatible for transfusion if all the antigens in the donor red blood cells are also present in the recipient.)

1-17. Red kernel color in wheat results from the presence of at least one dominant allele of each of two independently segregating genes (that is, $R\!-\!B\!-$ genotypes have red kernels). Kernels on $rr\,bb$ plants are white, and the genotypes $R- bb$ and $rr B-$ result in brown kernel color. If plants of a variety that is true breeding for red kernels are crossed with plants true breeding for white kernels,

(a) What is the expected phenotype of the F_1 plants?

(b) What are the expected phenotypic classes in the F_2 progeny and their relative proportions?

1-18. Heterozygous $Cp\,cp$ chickens express a condition called creeper, in which the leg and wing bones are shorter than normal ($cp\,cp$). The dominant Cp allele is lethal when homozygous. Two alleles of an independently segregating gene determine white ($W-$) versus yellow ($w\,w$) skin color. From matings between chickens heterozygous for both of these genes, what phenotypic classes will be represented among the viable progeny, and what are their expected relative frequencies?

1-19. White Leghorn chickens are homozygous for a dominant allele, C, of a gene responsible for colored feathers and for a dominant allele, I, of an independently segregating gene that prevents the expression of C. The White Wyandotte breed is homozygous recessive for both genes ($cc\,ii$). What proportion of the F_2 progeny obtained from mating White Leghorn × White Wyandotte F_1 hybrids would be expected to have colored feathers?

1-20. The F_2 progeny from a particular cross exhibit a modified dihybrid ratio of $9:7$ (instead of $9:3:3:1$). What phenotypic ratio would be expected from a testcross of the F_1?

1-21. Phenylketonuria is a recessive inborn error in the metabolism of the amino acid phenylalanine that results in severe mental retardation of affected children. The female II-3 (solid symbol) in the pedigree shown here is affected. If persons III-1 and III-2—they are first cousins—mate, what is the probability that their offspring will be affected?

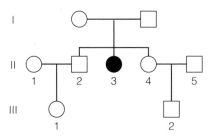

(Assume that persons II-1 and II-5 are homozygous for the normal allele.)

1-22. Black hair in rabbits is determined by a dominant allele, B, and white hair by homozygosity for a recessive allele, b. Two heterozygotes mate and produce a litter of three offspring.

(a) What is the probability that the offspring are born in the order white-black-white? What is the probability that the offspring are born in either the order white-black-white or the order black-white-black?

(b) What is the probability that exactly two of the three offspring will be white?

1-23. Assuming equal sex ratios, what is the probability that a sibship of four children consists entirely of boys? Of all boys or all girls? Of equal numbers of boys and girls?

1-24. Andalusian fowls are colored black, splashed white (resulting from an uneven sprinkling of black pigment through the feathers), or slate blue. Black and splashed white are true breeding, and slate blue is a hybrid that segregates in the ratio 1 black : 2 slate blue : 1 splashed white. If a pair of blue Andalusians is mated and the hen lays three eggs, what is the probability that one black chick, one blue, and one splashed white will be hatched from these eggs?

1-25. In the mating $Aa × Aa$, what is the smallest number of offspring, n, for which the probability of at least one aa offspring exceeds 95 percent?

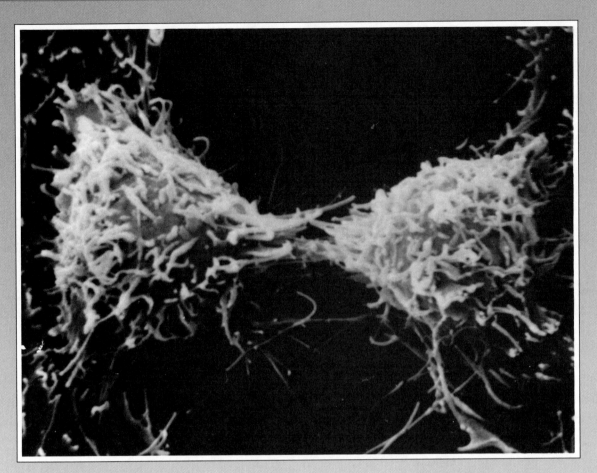

Genes and Chromosomes

Mendel's experiments made clear that the units of heredity are stable and particulate. However, at the time, the biological basis of the transmission of genes from one generation to the next was quite mysterious, and both the role of the nucleus in reproduction and the details of cell division were unknown. Once these phenomena became understood and chromosomes were seen by microscopy and recognized to be carriers of genes, new understanding came at a rapid pace. This chapter examines both the relation between chromosomes and genes and the mechanism of chromosome segregation in cell division.

Above: A cell in the act of division, as visualized by scanning electron microscopy. (From G. Shih and R. Kessel. 1982. *Living Images.* Jones and Bartlett.)

In the 1870s, the importance of the nucleus and its contents was recognized by the observation that the nuclei of two gametes fuse in the process of fertilization. The next major advance was the discovery of **chromosomes,** which had been made visible by light microscopy when stained with basic dyes. Then, chromosomes were found to segregate into both gametes and daughter cells by an orderly process before cell division. Finally, three important regularities were observed about the **chromosome complements** (the complete sets of chromosomes) of higher plants and animals:

1. The nucleus of each **somatic cell** (a cell of the body, in contrast with a **germ cell,** or gamete) contains a fixed number of chromosomes typical of the particular species. However, the numbers vary tremendously among species and have little relation to the complexity of the organism (Table 2-1).
2. The chromosomes in the nuclei of somatic cells are usually present in pairs. For example, the 46 chromosomes of human beings consist of 23 pairs, and the 14 chromosomes of peas consist of 7 pairs. Furthermore, one chromosome of each pair derives from the maternal parent and the other from the paternal parent of the organism. Cells with nuclei of this sort, containing two similar sets of chromosomes, are called **diploid.**
3. The germ cells or gametes that unite in fertilization to produce the diploid state of somatic cells have nuclei with only one set of chromosomes, consisting of one member of each pair—these nuclei are **haploid.**

The presence of a constant diploid chromosome number in cells of complex organisms that develop from single cells and the formation of gametes with the haploid chromosome number indicate that there are two processes of nuclear division, one that maintains chromosome number—mitosis—and another that halves the number—meiosis. These two processes are examined in the following sections.

TABLE 2-1 Somatic chromosome numbers of some plant and animal species

Plants		Animals	
Organism	Chromosome number	Organism	Chromosome number
Yeast (Saccharomyces cerevisiae)	32	Fruit fly (Drosophila)	8
Field horsetail	216	House fly	12
Bracken fern	116	Scorpion	4
Giant sequoia	22	Geometrid moth	224
Macaroni wheat	28	Common toad	22
Bread wheat	42	Chicken	78
Garden pea	14	Mouse	40
Corn (Zea mays)	20	Gibbon	44
Lily	12	Human being	46

2-2 Mitosis

Mitosis is a precise process of nuclear division that ensures that each of two daughter cells receives a complement of chromosomes identical with the complement of the parent cell. The essential details of the process are the same in all organisms. Moreover, the basic process is remarkably simple: each chromosome, present as a replicated structure at the beginning of nuclear division, divides into identical halves that are separated from each other, and one of them goes into each of the two daughter nuclei that are formed.

In a cell not ready for mitosis, the chromosomes are not visible. This stage of the cell cycle is called **interphase.** In preparation for mitosis, the genetic material, DNA, in the chromosomes is synthesized during a period of late interphase called **S** (Figure 2-1). DNA synthesis is accompanied by chromosome replication. Before and after S, there are periods, called G_1 and G_2, respectively, in which DNA synthesis does not occur. The **cell cycle,** or the life cycle of a cell, is commonly described as these three interphase periods followed by mitosis, **M**—that is, $G_1 \rightarrow S \rightarrow G_2 \rightarrow M$—as shown in Figure 2-1. This is a somewhat arbitrary representation in which the final event in cell reproduction—the division of the cytoplasm into two approximately equal parts containing the daughter nuclei—is included in the M period.

The length of time required for a complete life cycle varies with cell type. In higher organisms, the majority require between 18 and 24 hr. The relative duration of the different periods in the cycle also varies considerably with cell type. Mitosis is usually the shortest period, requiring between 1/2 and 2 hr.

The essential features of mitosis are illustrated in Figure 2-2. Mitosis is conventionally divided into four stages—**prophase, metaphase, anaphase,** and **telophase.** (If you have trouble remembering the order, you can jog your memory with **peas make awful tarts.**) The stages have the following characteristics:

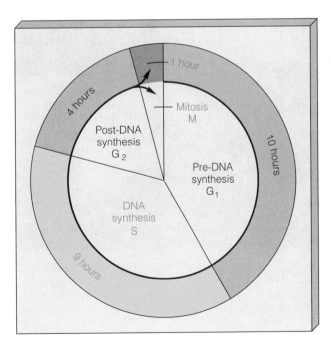

FIGURE 2-1 The cell cycle of a representative mammalian cell growing in tissue with a generation time of 24 hr.

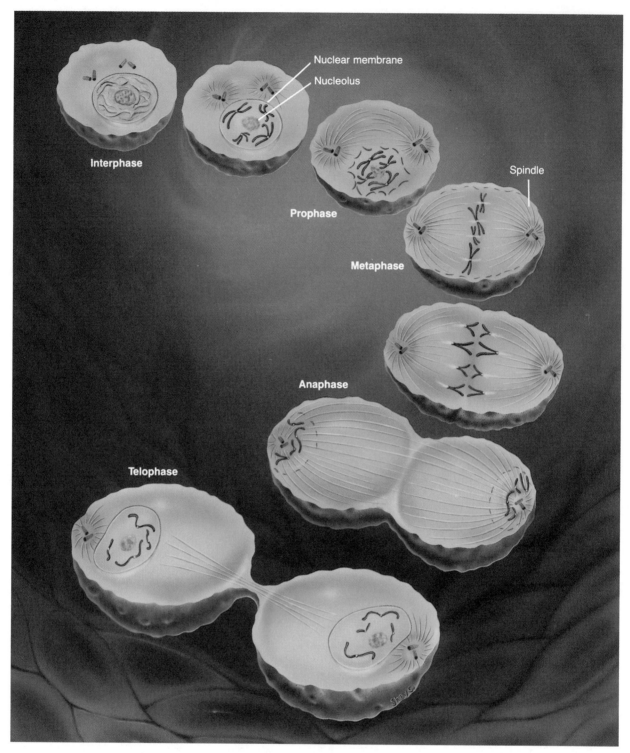

FIGURE 2-2 Diagram of mitosis in an organism having four chromosomes. For clarity, realistic dimensional changes in the chromosomes are not shown. Interphase, in which chromosomes are not visible, is not normally considered part of mitosis and is usually much longer than the rest of the cell cycle (see Figure 2-1). In early prophase, chromosomes become apparent as fine strands, and the nuclear membrane and one or more nucleoli are intact. As prophase progresses, chromosomes condense and each can be seen to consist of two chromatids; the nuclear membrane and nucleoli disappear. In metaphase, chromosomes are highly condensed and aligned on the central plane of the spindle, which forms at the conclusion of prophase. In anaphase, centromeres divide, and the halves of each chromosome move to opposite poles of the spindle. In telophase, the separation of sister chromosomes is complete, new nuclei are formed, the condensation process of prophase is reversed, the spindle breaks down, and nucleoli reorganize.

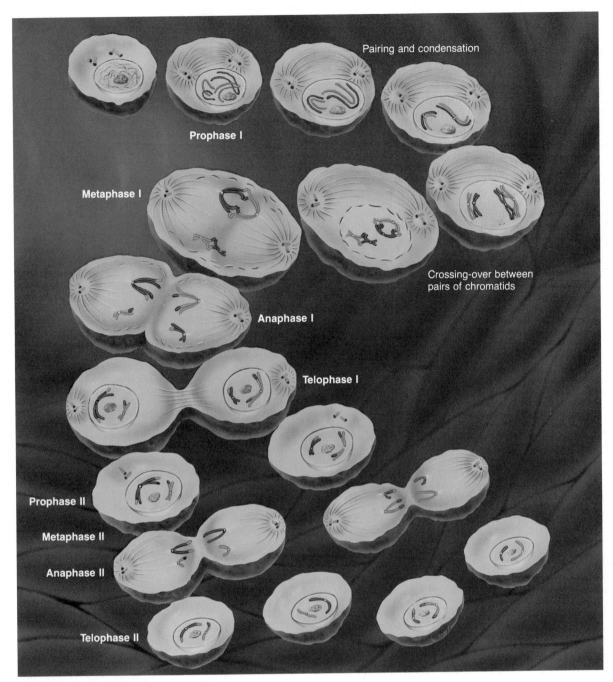

Pairing and condensation

Prophase I

Metaphase I

Crossing-over between pairs of chromatids

Anaphase I

Telophase I

Prophase II

Metaphase II

Anaphase II

Telophase II

FIGURE 2-3   Diagram illustrating the major features of meiosis in an organism having four chromosomes (two homologous pairs). In prophase I, chromosomes become apparent as elongated, fine strands. Homologous chromosomes pair and the individual chromosomes become more condensed. When pairing is completed, each chromosome can be seen to have two chromatids, and crossing-over occurs between chromatids. In metaphase I, the nuclear membrane disappears, the spindle apparatus forms, and homologous pairs of highly condensed chromosomes become oriented at the equator of the spindle. In anaphase I, members of each pair of homologous chromosomes move toward opposite poles of the spindle. In telophase I, chromosome movement is complete, the spindle breaks down, nuclear membranes may form, and the prophase condensation of chromosomes is partly reversed. In prophase II (not fully depicted), chromosomes, represented by one member of each homologous pair, condense; the nuclear membrane disappears; and the spindle apparatus forms. In metaphase II (not shown), chromosomes align in the central plane of the spindle. In anaphase II, centromeres divide, and the two chromatids of each pair move toward opposite poles of the spindle. In telophase II, chromosome movement is completed, the spindle breaks down, and the four haploid nuclei gradually revert to interphase morphology.

1. **Prophase** During interphase, the chromosomes have the form of extended filaments and cannot be seen as discrete bodies with a light microscope. Except for the presence of one or more conspicuous dark bodies (**nucleoli**), the nucleus has a diffuse, granular appearance. The beginning of prophase is marked by the condensation of chromosomes to form visibly distinct, thin threads within the nucleus. Occasionally, the chromosomes are seen rather early to be longitudinally double, consisting of two closely associated subunits called **chromatids.** Each pair of chromatids is the product of replication of one chromosome in the S period of interphase. The chromatids in a pair are held together at a specific region of the chromosome called the **centromere.** As prophase progresses, the chromosomes become shorter and thicker, as a result of intricate coiling (Chapter 5). In later prophase, three phenomena occur: (1) the nucleoli disappear; (2) the membrane enclosing the nucleus disintegrates; and (3) the **mitotic spindle** forms. The spindle is a bipolar structure consisting of fiberlike bundles that extend between the poles. Each chromosome becomes attached to several spindle fibers at its centromere.

2. **Metaphase** After the chromosomes are attached to spindle fibers, they move toward the center of the cell until all centromeres lie on a plane equidistant from the spindle poles. The period in which the chromosomes are located at the central plane of the spindle is called metaphase. At metaphase, the chromosomes reach their maximum contraction and are easiest to count and examine for differences in morphology.

3. **Anaphase** In anaphase, the centromeres divide, and the two **sister chromatids** of each chromosome move toward opposite poles of the spindle. The movement results in part from being pulled to the poles by the spindle fibers attached to the centromeres. Once the centromeres divide, each sister chromatid is regarded as a separate chromosome. At the completion of anaphase, the chromosomes lie in two groups near opposite poles of the spindle. Each group contains the same number of chromosomes that was present in the original interphase nucleus.

4. **Telophase** During telophase, a nuclear envelope forms around each compact group of chromosomes, nucleoli are formed, and the spindle disappears. The chromosomes undergo a reversal of the condensation that occurred in prophase until they are no longer visible as discrete entities. The two daughter nuclei slowly assume a typical interphase appearance as the cytoplasm of the cell divides into two by means of a gradually deepening furrow around the periphery. (In plants, a new cell wall is synthesized between the daughter cells and separates them.)

2-3 Meiosis

Meiosis is a mode of cell division in which cells are created with *half* the number of chromosomes present in the premeiotic cell. When a diploid cell with two sets of chromosomes undergoes meiosis, the result is four genetically different haploid products, each with one set of chromosomes.

Meiosis consists of two successive nuclear divisions, the essential details of which are shown in Figure 2-3. In the first division, the two members of each pair of chromosomes, which are said to be **homologous** to each other, become closely associated along their length. Because each homologous chromosome consists of a duplex of two sister chromatids joined at the centromere, the pairing of the homologous chromosomes produces a four-stranded structure. In the first meiotic division, the homologous chromosomes are separated from one another and distributed into two nuclei, each containing a haploid set of such duplex chromosomes. Without chromosome replication, a second division resembling a mitotic division occurs, in which the chromatids of each chromosome are separated and distributed into the daughter nuclei. The net effect of the two divisions is the creation of four haploid daughter nuclei, each containing the equivalent of a single sister chromatid from each pair of homologous chromosomes.

In animals, meiosis occurs in specific cells called **meiocytes,** a general term for the primary oocytes and spermatocytes in the gamete-forming tissues (Figure 2-4). Although the process of meiosis is similar in all sexually reproducing organisms, in the female of both animals and plants, only one of the four products develops into a functional cell (the other three disintegrate). In animals, the products of meiosis form gametes (sperm or eggs). In plants, the situation is slightly more complicated:

1. The products of meiosis typically form *spores*, which undergo one or more mitotic divisions to produce a haploid gametophyte organism. The gametophyte produces gametes by mitotic division of a haploid nucleus (Figure 2-5).
2. Fusion of haploid gametes creates a diploid zygote that develops into the sporophyte plant, which undergoes meiosis to produce spores and so restarts the cycle.

Meiosis is a more-complex and considerably longer process than mitosis and usually requires days or even weeks. The essence of meiosis, which we will

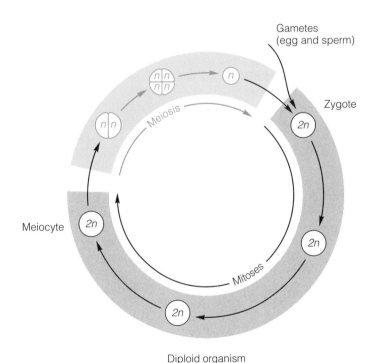

FIGURE 2-4 The life cycle of a higher animal. The number *n* is the number of chromosomes in the haploid chromosome complement. In males, the four products of meiosis develop into functional sperm, whereas, in females, only one of the four products develops into an egg.

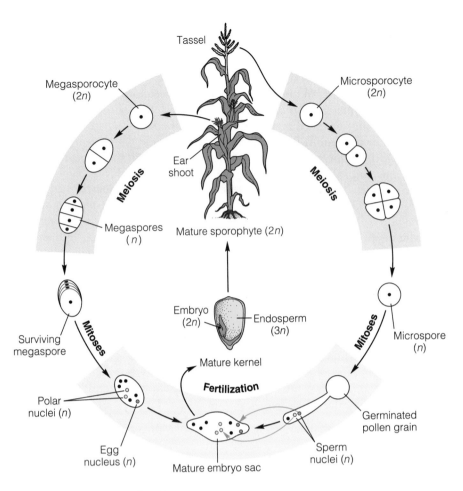

FIGURE 2-5 The life cycle of corn *Zea mays*. As is typical of higher plants, the diploid spore-producing (sporophyte) generation is conspicuous, whereas the gamete-producing (gametophyte) generation is microscopic. The egg-producing spore is the *megaspore* and the sperm-producing spore is the *microspore*. Nuclei participating in meiosis and fertilization are shown in yellow and green.

now examine in more detail, is that *it consists of two divisions of the nucleus but only one duplication of the chromosomes*. The nuclear divisions—called the **first meiotic division** and the **second meiotic division**—can be separated into a sequence of stages similar to those used to describe mitosis, as shown in Figure 2-2. The distinctive events of this important process take place in the first division of the nucleus; these events are described in the following section.

The First Meiotic Division

The four defined stages of the first meiotic division are **prophase I, metaphase I, anaphase I,** and **telophase I.** These stages are generally more complex than their counterparts in mitosis.

1. Prophase I This is a long stage, lasting several days in most higher organisms and commonly divided into five substages—**leptotene, zygotene, pachytene, diplotene,** and **diakinesis.**

In **leptotene,** the chromosomes become visible as long, threadlike structures. The pairs of sister chromatids can be distinguished by electron microscopy. In this initial phase of condensation of the chromosomes, numerous dense granules appear at irregular intervals along their length. These localized contractions, called **chromomeres,** have a characteristic number, size, and position in a given chromosome (Figure 2-6A).

The **zygotene** period is marked by the lateral pairing, or **synapsis,** of homologous chromosomes. This pairing begins at one or more points along the

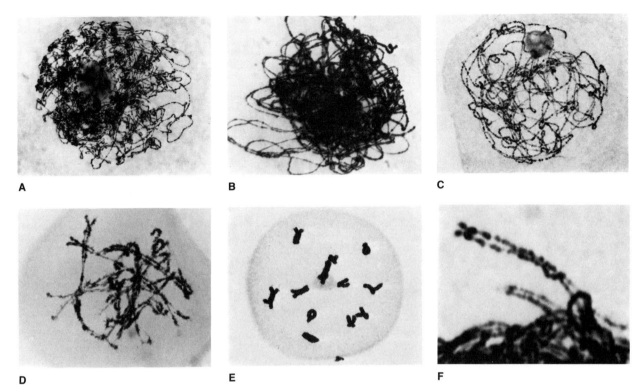

FIGURE 2-6   Substages of prophase of the first meiotic division in microsporocytes of a lily (*Lilium longiflorum*). A. Leptotene, in which condensation of the chromosomes is initiated and beadlike chromomeres are visible along the length of the chromosomes. B. Zygotene, in which pairing (synapsis) of homologous chromosomes occurs. Paired and unpaired regions can be seen particularly at the lower left in this photograph. C. Pachytene, in which crossing-over between homologous chromomes occurs. D. Diplotene, characterized by mutual repulsion of the paired homologous chromosomes, which remain held together at one or more cross points (chiasmata) along their length. E. Diakinesis, in which the chromosomes reach their maximum contraction. F. Zygotene (at higher magnification in another cell), showing paired homologs and matching of chromomeres during synapsis. [Courtesy of Marta Walters (A, B, C, E, F) and Herbert Stern (D).]

length of the chromosomes and results in a precise chromomere-by-chromomere association (Figure 2-6B). Each pair of synapsed homologous chromosomes is referred to as a **bivalent**.

During **pachytene** (Figure 2-6C) condensation of the chromosomes continues, and the chromosome complement is represented by the haploid number of bivalents. Each bivalent consists of a **tetrad** of four chromatids, but the two sister chromatids of each chromosome cannot usually be distinguished. A genetically important event called **crossing-over** takes place in pachytene, but it does not become apparent until the transition to diplotene.

At the onset of **diplotene,** the synapsed chromosomes begin to separate (Figure 2-6D). However, they remain held together at points along their length by cross connections called **chiasmata** (singular, **chiasma**). A chiasma is the result of breakage and joining between nonsister chromatids—that is, *a physical exchange between chromatids of homologous chromosomes* (Figure 2-7). In normal meiosis, each bivalent has at least one chiasma, and bivalents of long chromosomes often have three or more.

During the final period of prophase I—**diakinesis**—the chromosomes attain their maximum condensation (Figure 2-6E). The homologous chromosomes in a bivalent remain connected by one or two terminal chiasmata, which

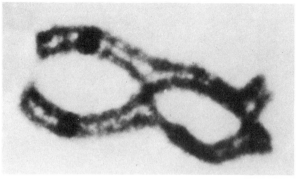
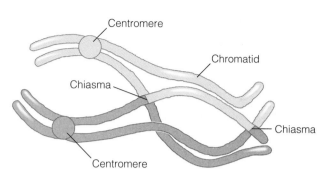

A   **B**

FIGURE 2-7   Light micrograph (A) and interpretive drawing (B) of a pair of homologous chromosomes of the salamander *Oedipina poelzi* at late diplotene in a spermatocyte, showing chiasmata where chromatids of the two chromosomes appear to exchange pairing partners. (From F. W. Stahl. 1964. *The Mechanics of Inheritance*. Prentice-Hall, Inc.; courtesy of James Kezer.)

persist until the first meiotic anaphase. Near the end of diakinesis, the formation of a spindle is initiated and the nuclear envelope breaks down.

2. Metaphase I   The bivalents become positioned with the centromeres of two homologous chromosomes on opposite sides of the plane through the middle of the spindle (Figure 2-8A). The co-orientation of the undivided centromeres of each bivalent relative to the two poles of the spindle is *random* and determines which member of each pair of chromosomes will subsequently move to each pole.

> The random alignment of nonhomologous chromosomes on the metaphase plate is the physical basis of Mendel's law of independent assortment.

3. Anaphase I   During this stage, homologous chromosomes, each composed of two chromatids joined at an undivided centromere, separate from one another and move to opposite poles of the spindle (Figure 2-8B).

> The physical separation of homologous chromosomes during anaphase is the physical basis of Mendel's law of segregation.

4. Telophase I   At the completion of anaphase I, a haploid set of chromosomes consisting of one homolog from each bivalent is located near each pole of the spindle. During telophase, the spindle breaks down and, depending on the species, either a nuclear envelope briefly forms around each group of chromosomes or the chromosomes enter the second meiotic division after only a limited uncoiling.

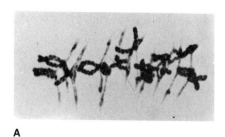
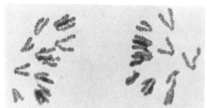

A   **B**

FIGURE 2-8   Metaphase I (A) and anaphase I (B) in microsporophytes of the lily *Lilium longiflorum*. (Courtesy of Herbert Stern.)

The Second Meiotic Division

In some species, the chromosomes pass directly from telophase I to **prophase II** without loss of condensation; in others, there is a brief pause between the two meiotic divisions. *Chromosome replication never occurs between the two divisions;* the chromosomes present at the beginning of the second division are identical with those present at the end of the first division. After a short prophase (prophase II) and the formation of second-division spindles, the centromeres of the chromosomes in each nucleus become aligned on the central plane of the spindle at **metaphase II** (Figure 2-9A). During **anaphase II,** the centromeres (replicated in the *first* division) separate and the chromatids of each chromosome move to opposite poles of the spindle (Figure 2-9B). **Telophase II** (Figure 2-9C) is marked by a transition to the interphase condition of the chromosomes in the four haploid nuclei accompanied by division of the cytoplasm. Thus, the second meiotic division superficially resembles a mitotic division. However, there is an important difference: *the chromatids of a chromosome are usually not identical sisters along their entire length because of crossing-over and the formation of chiasmata in prophase of the first division.*

2-4   Chromosomes and Heredity

The first clear proof that genes are parts of chromosomes was obtained in experiments concerned with the pattern of transmission of the **sex chromosomes,** the chromosomes responsible for the determination of the separate sexes in some plants and in almost all higher animals. The results are examined in this section.

Chromosomal Determination of Sex

The sex chromosomes are an exception to the rule that all chromosomes of diploid organisms are present in pairs of morphologically similar homologs. Early microscopic analysis showed that one of the chromosomes in males of some insect species does not have a homolog. This unpaired chromosome was called the **X chromosome,** and it was present in all somatic cells of the males

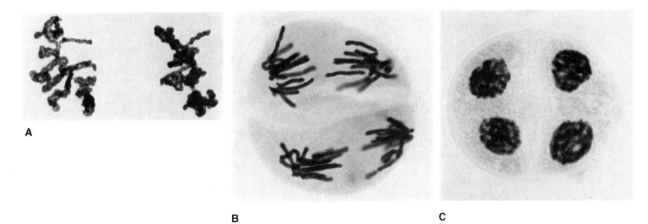

A

B   **C**

FIGURE 2-9   Metaphase II (A), anaphase II (B), and telophase II (C) in microsporocytes of the lily *Lilium longiflorum.* Cell walls have begun to form in telophase, whch will lead to the formation of four pollen grains. (Courtesy of Herbert Stern.)

but in only half the sperm cells. The biological significance of these observations became clear when females of the same species were shown to have two X chromosomes.

In other species in which the females have two X chromosomes, a morphologically different chromosome is present in males. This is referred to as the **Y chromosome**, and it pairs with the X chromosome during meiosis in males because the X and Y share a small region of homology. The difference in the chromosomal constitution of males and females is a chromosomal mechanism for determining sex at the time of fertilization; that is, whereas every egg cell contains an X chromosome, half the sperm cells contain an X chromosome and the rest contain a Y chromosome. Fertilization of an egg by an X-bearing sperm results in an XX zygote, which normally develops into a female; and fertilization by a Y-bearing sperm results in an XY zygote, which normally develops into a male (Figure 2-10). The result is a criss-cross pattern of inheritance of the X chromosome in which a male receives his X chromosome from his mother and transmits it only to his daughters.

The XX-XY type of chromosomal sex determination occurs in mammals, including human beings, many insects, and other animals, as well as in some flowering plants. The female is called the **homogametic** sex because only one type of gamete (X-bearing) is produced, and the male is called the **heterogametic** sex because two different types of gametes (X-bearing and Y-bearing) are produced. If the union of gametes in fertilization occurs at random, then a sex ratio at fertilization of 1:1 is expected because males produce equal numbers of X-bearing and Y-bearing sperm.

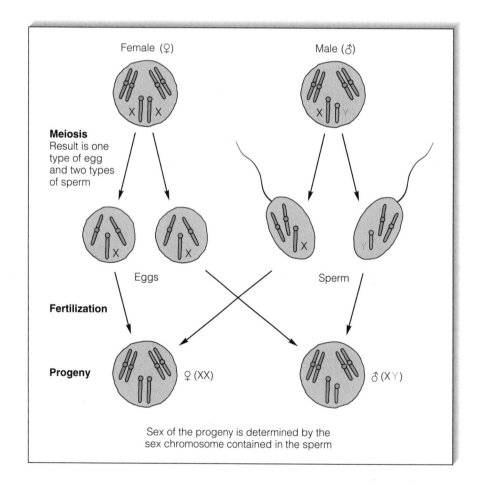

FIGURE 2-10 The chromosomal basis of sex determination found in mammals, many insects, and other animals.

The X and Y chromosomes are the **sex chromosomes.** Although they set the developmental switch that controls the earliest stages of female or male development, the developmental process itself requires many genes scattered throughout the chromosome complement, including genes on other chromosome pairs, which are called the **autosomes.** The X chromosome also contains many genes with functions unrelated to sexual differentiation, as will be seen in the next section. In most organisms, including human beings, the Y chromosome carries few genes other than those associated with male determination.

X-linked Inheritance

Crosses of the kind considered in Chapter 1 yield similar progeny when the genotypes of the male and female parents are reversed—that is, in *reciprocal crosses.* One of the earliest exceptions to this rule was found by Thomas Hunt Morgan in 1910, in an early study of white eyes in the fruit fly *Drosophila melanogaster* (Figure 2-11). As described in Chapter 1, the wildtype eye color is a dull red combination of scarlet and brown pigments. We know from Chapter 1 that white eyes can result from certain combinations of autosomal genes that eliminate the pigments individually. The white-eye allele that Morgan studied results in a metabolic block that knocks out both pigments simultaneously.

Morgan's study started with a single male with white eyes that appeared in a wildtype laboratory population that had been maintained for many generations. In a mating of this male with wildtype females, all of the F_1 progeny of both sexes had red eyes, showing that the allele for white eyes is recessive. In the F_2 progeny from the mating of F_1 males and females, he observed 2459 red-eyed females, 1011 red-eyed males, and 782 white-eyed males. The white-eyed phenotype was somehow connected with sex, because all of the white-eyed flies were males.

However, white eyes were not restricted to males. For example, when red-eyed F_1 females from the cross of wildtype ♀♀ × white ♂♂ were backcrossed with their white-eyed fathers, the progeny consisted of both red-eyed and white-eyed females and red-eyed and white-eyed males in approximately equal numbers.

A key observation came from the mating of white-eyed females with wildtype males. All of the female progeny had wildtype eyes, but all of the male

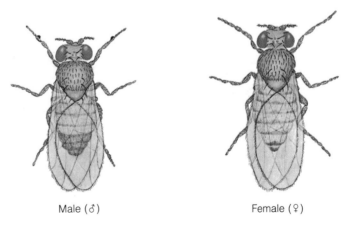

Male (♂)   Female (♀)

FIGURE 2-11   Drawings of a male and female fruit fly *Drosophila melanogaster.* (Courtesy of Carolina Biological Supply Company.)

progeny had white eyes. This is the reciprocal of the original cross of wildtype ♀♀ × white ♂♂, which had given only wildtype females and wildtype males, and so the reciprocal crosses gave different results.

Morgan realized that the reciprocal crosses had different progeny because the allele for white eyes is contained in the X chromosome. The gene is said to be **X-linked.** Figure 2-12 shows the normal chromosome complement of *Drosophila melanogaster*. Females have an XX chromosome complement, males are XY, and the Y chromosome does not contain a counterpart of the white gene. In terms of genotype, white-eyed males are w/Y and wildtype males are w^+/Y. The symbols w and w^+ denote the mutant and wildtype forms of the white gene present in the X chromosome. Because the w allele is recessive, white-eyed females are of genotype w/w, and wildtype females are either heterozygous, w^+/w, or homozygous, w^+/w^+.

Figure 2-13 illustrates the chromosomal interpretation of the reciprocal crosses wildtype ♀ × white ♂ (Cross 1) and white ♀ × wildtype ♂ (Cross 2). These diagrams account for the different phenotypic ratios observed in the F_1 and F_2 progeny from the crosses. Many other genes were later found in *Drosophila* that also follow the X-linked pattern of inheritance.

The characteristics of X-linked inheritance can be summarized as follows:

1. Reciprocal crosses resulting in different phenotypic ratios in the sexes often indicate X linkage.
2. Heterozygous females transmit each allele to approximately half their daughters and half their sons.
3. Males that inherit an X-linked recessive allele show the recessive trait because the Y chromosome does not contain a wildtype counterpart. Affected males transmit the recessive allele to all their daughters but none of their sons. Males that are not affected carry the wildtype allele in their X chromosome.

An example of a human trait with an X-linked pattern of inheritance is **hemophilia A,** a severe disorder of blood clotting determined by a recessive allele. Affected persons lack a blood protein needed for normal clotting, and they suffer excessive, often life-threatening bleeding after injury. A famous pedigree of hemophilia starts with Queen Victoria of England. One of her sons was hemophilic, and two of her daughters were heterozygous carriers of the gene who produced three hemophilic sons and four heterozygous daughters. Through two of these carrier granddaughters, the gene was introduced into the royal families of Russia and Spain. The present royal family of England is descended from a normal son of Victoria and free of the disease.

An opposite pattern of nonreciprocal inheritance can be found in organisms in which the male is homogametic and the female is heterogametic. This pattern is found in birds, some reptiles and fish, moths, and butterflies. For example, some breeds of chickens have feathers with alternating transverse bands of light and dark color, resulting in a phenotype referred to as barred. The feathers are uniformly colored in the nonbarred phenotypes of other breeds. Reciprocal crosses between true-breeding barred and nonbarred types give the following results

nonbarred ♀ × barred ♂   barred ♀ × nonbarred ♂

barred ♀♀ and ♂♂   nonbarred ♀♀ and barred ♂♂

indicating that the gene that determines barring is on the X chromosome and is dominant.

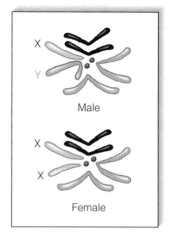

X

Y

Male

X

X

Female

FIGURE 2-12 The diploid chromosome complements of a male and female *Drosophila melanogaster.* The Y chromosome (red) contains a centromere that is not centrally located, giving the Y a characteristic J shape. The X chromosome (gray) has a nearly terminal centromere. The large autosomes (chromosomes 2 and 3) are not easily distinguishable in these types of cells. The tiny chromosome 4 appears as a dot.

FIGURE 2-13 A chromosomal interpretation of the results obtained in F_1 and F_2 progenies when a *Drosophila* female with wildtype red eyes is crossed with a white male (Cross 1) and when the reciprocal cross of white female with wildtype male (Cross 2) is made. In the X chromosome, the w^+ allele is shown in red and the w allele in black. The Y chromosome (gray) does not carry either allele of the gene.

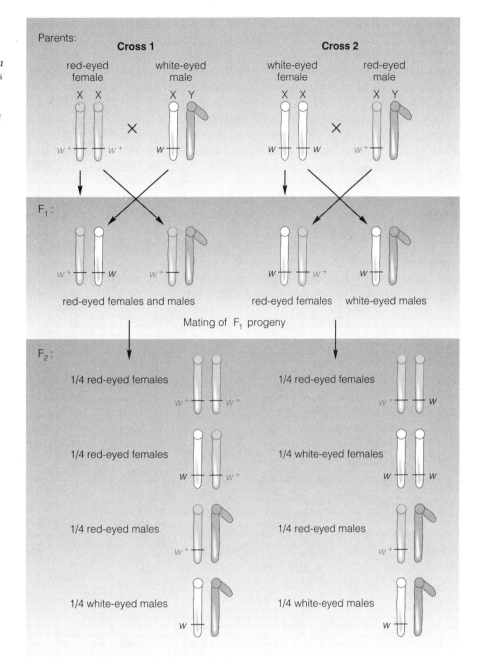

Nondisjunction as Proof of the Chromosome Theory of Heredity

The parallelism between the inheritance of the *Drosophila* white allele and the genetic transmission of the X chromosome containing it supported the chromosome theory of heredity that genes are parts of chromosomes. Other experiments with *Drosophila* provided the definitive proof.

One of Morgan's students, Calvin Bridges, discovered rare exceptions to the expected pattern of inheritance in crosses with several X-linked genes. For example, when white-eyed females were mated with red-eyed males, most of the progeny consisted of the expected red-eyed females and white-eyed males. However, about one in every 2000 F_1 flies was an exception, either a white-eyed female or a red-eyed male. Bridges showed that these rare exceptional offspring resulted from occasional failure of the two X chromosomes in the mother to separate from each other during meiosis—a phenomenon called **nondisjunction**. The consequence of nondisjunction of the X chromosome is the formation of some eggs with two X chromosomes and others with none. Four classes of zygotes are expected from the fertilization of these abnormal eggs (Figure 2-14). Offspring with no X chromosome are not detected because embryos lacking an X are not viable, and most progeny with three X chromosomes also die during early development. Microscopic examination of the chromosomes of the exceptional progeny from the cross white ♀♀ × wildtype ♂♂ showed that the exceptional white-eyed females had two X chromosomes *plus* a Y chromosome, and the exceptional red-eyed males had a single X but were *lacking* a Y. The latter, with a sex-chromosome constitution denoted XO, were sterile males.

These and related experiments demonstrated conclusively that genes are contained in chromosomes, because exceptional chromosomal behavior is precisely paralleled by exceptional inheritance of their genes. Bridges's proof of the chromosome theory ranks among the most-important and most-elegant experiments in genetics.

Sex Determination in Drosophila

Among organisms with an XX-XY type of sex determination, *Drosophila* is unusual in that the Y chromosome is not male determining. This is demonstrated by the finding that XXY embryos develop into morphologically normal, fertile females and XO embryos develop into morphologically normal,

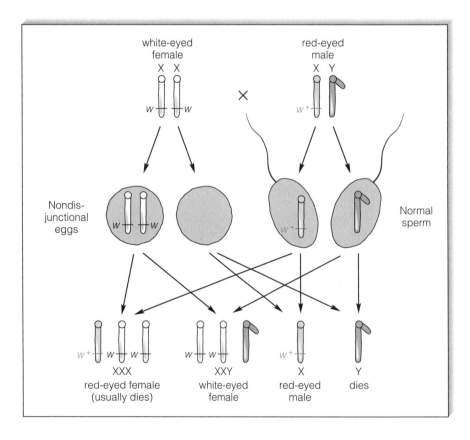

FIGURE 2-14 The results of meiotic nondisjunction of the X chromosomes in a female *Drosophila*.

but sterile, males. The sterility of XO males shows that the Y chromosome, though not necessary for male development, is essential for male fertility, and indeed the *Drosophila* Y chromosome contains a number of factors required for the formation of normal sperm.

The genetic determination of sex in *Drosophila* depends on the number of X chromosomes present in an individual fly compared with the number of sets of autosomes. Normal males have one X chromosome and two haploid sets of autosomes, giving a ratio of X chromosomes to sets of autosomes (X/A ratio) of 1/2. Normal females have two X chromosomes and two haploid sets of autosomes, for an X/A ratio of 1. Individual flies with X/A ratios smaller than 1/2 (for example, with one X and three sets of autosomes) are male, and those with X/A ratios greater than 1 (for example, with three X chromosomes and two sets of autosomes) are female. Intermediate X/A ratios such as 2/3 (for example, two X chromosomes and three sets of autosomes) develop as intersexes with some characteristics of each sex. The molecular basis of this type of sex determination is not understood.

2-5 Probability in Prediction and Analysis of Genetic Data

Genetic ratios result from the chance assortment of genes into gametes and the chance combination of gametes into zygotes. Although exact predictions are not possible for any particular event, it is possible to determine the probability that a particular event might occur, as described in Chapter 1. In this section, we consider some of the probability methods used in interpreting genetic data.

Use of the Binomial Distribution in Genetics

The probability that each of three children in a family will be of the same sex is an example of a simple problem that uses both the addition and multiplication rules of probability. The probability that all three will be girls is $(1/2)(1/2)(1/2) = 1/8$, and the probability that all three will be boys is also 1/8. Because these outcomes are mutually exclusive, the probability of one or the other is the sum of the two probabilities, which is 1/4. The remaining possible outcomes are that two of the children will be girls and the other a boy, or that two will be boys and the other a girl. For each of these outcomes only three different orders of birth are possible—for example, GGB, GBG, and BGG—each having a probability of 1/8. Thus, the probability of two girls and a boy, disregarding birth order, is the sum of the probabilities for the three possible orders, or 3/8; likewise, the probability of two boys and a girl is also 3/8. Therefore, the distribution of probabilities for the sex ratio in families with three children is:

GGG	GGB GBG BGG	GBB BGB BBG	BBB	
$(1/2)^3$	$3(1/2)^2(1/2)$	$3(1/2)(1/2)^2$	$(1/2)^3$	
1/8 +	3/8 +	3/8 +	1/8	= 1

This sex-ratio information can be obtained more directly by expanding the binomial expression $(p + q)^n$, in which p is the probability of the birth of a girl

(1/2), q the probability of the birth of a boy (1/2), and n is the number of children; in our example,

$$(p + q)^3 = 1p^3 + 3p^2q + 3pq^2 + 1q^3$$

in which the red numerals can be compared to the numerators in the fractions above. Similarly, the binomial distribution of probabilities for the sex ratios in families of five children is

$$(p + q)^5 = 1p^5 + 5p^4q + 10p^3q^2 + 10p^2q^3 + 5pq^4 + 1q^5$$

Each term tells us the probability of a particular combination. For example, the third term is the probability of three girls (p^3) and two boys (q^2) in a family having five children—namely,

$$10(1/2)^3(1/2)^2 = 10/32 = 5/16$$

There are $n + 1$ terms in a binomial expansion. The exponents of p decrease from n in the first term to 0 in the last term, and the exponents of q increase by one from 0 in the first term to n in the last term. The coefficients generated by successive values of n can be arranged in a regular triangle known as **Pascal's triangle** (Figure 2-15). Note that the horizontal rows of the triangle are symmetrical, and that each number is the sum of the two numbers on either side of it in the row above.

In general, if the probability of event A is p and that of event B is q, and the two events are independent and mutually exclusive (see Chapter 1), the probability that A will occur four times and B two times—in a specific order—is p^4q^2, by the multiplication rule. However, suppose that we are interested in the occurrence of this combination of events: four of A and two of B, irrespective of order. In that case, we multiply the probability that the combination 4A:2B will occur in any one specific order by the number of possible orders. The number of different combinations of six events, four of one kind and two of another, is

$$\frac{6!}{4!\ 2!} = \frac{1 \times 2 \times 3 \times 4 \times 5 \times 6}{1 \times 2 \times 3 \times 4 \times 1 \times 2} = 15$$

(The symbol ! is for factorial, or the product of all positive integers from one through a given number.) This calculation provides the coefficient of the p^4q^2 term in the expansion of the binomial $(p + q)^6$. Therefore, the probability that event A will occur four times and event B two times is $15p^4q^2$.

n	Coefficients							
0					1				
1					1	1			
2				1	2	1			
3			1	3	3	1			
4		1	4	6	4	1			
5	1	5	10	10	5	1			
6	1	6	15	20	15	6	1		

FIGURE 2-15 Pascal's triangle. The numbers are the coefficients of each term in the expansion of the polynomial $(p + q)^n$ for successive values of n from 0 through 6.

The general rule for repeated trials of events with constant probabilities is as follows:

If the probability of occurrence of event A is p, and the probability of the alternative event B is q, the probability that in n trials event A occurs s times and event B occurs t times is

$$\frac{n!}{s!\ t!}\ p^s q^t$$

in which $s + t = n$, and $p + q = 1$. To use this expression, remember that the symbol 0! is defined to equal 1 and that any number raised to the zero power (for example, 2^0) also equals 1. Each individual term in the expansion of the binomial $(p + q)^n$ is the same as the expression just given.

Let us consider a specific example, in which we calculate the probability that a mating between two heterozygous parents yields exactly the expected 3:1 ratio of the dominant and recessive traits among sibships of a particular size. The probability p of a child showing the dominant trait is 3/4, and the probability q of a child showing the recessive trait is 1/4. Suppose that we wanted to know how often families with eight children would contain exactly six with the dominant phenotype and two with the recessive. In this case, $n = 8$, $s = 6$, $t = 2$, and the probability of this combination of events is

$$\frac{8!}{6!\ 2!}\ (p)^6(q)^2 = \frac{6! \times 7 \times 8}{6! \times 2!}\ (3/4)^6(1/4)^2 = 0.31$$

That is, in only 31 percent of the families with eight children would the offspring exhibit the expected 3:1 phenotypic ratio; the other sibships would deviate in one direction or the other because of chance variation.

Evaluating the Fit of Observed Results to Theoretical Expectations: The Chi-square Method

Geneticists often need to decide whether an observed ratio is in satisfactory agreement with a theoretical prediction. Mere inspection of the data is unsatisfactory because different investigators may disagree. A better approach is to use a quantitative measure of how much the observed results deviate from the theoretical expectation and to base a judgement of goodness of fit on this criterion. **Goodness of fit** means how closely the observed and expected results agree.

A conventional measure of goodness of fit is a value called **chi-square** (symbol χ^2), which is calculated from the observed and expected numbers of progeny as follows:

1. For each class of progeny in turn, subtract the expected number from the observed number. Square this difference and divide the result by the expected number.
2. Sum the result of the numbers calculated in 1 for all classes of progeny. The summation is the value of χ^2 for these data.

In symbols, the calculation of χ^2 can be represented by the expression

$$\chi^2 = \sum \frac{(\text{Observed} - \text{Expected})^2}{\text{Expected}}$$

in which Σ means the summation over all the classes of progeny. Note that χ^2 is calculated using the observed and expected *numbers*, not the proportions, ratios, or percentages. Using something other than the actual numbers is the most-common beginner's mistake in applying the χ^2 method.

To illustrate the calculation of χ^2, suppose that the progeny of an $F_1 \times F_1$ cross includes two contrasting phenotypes observed in the numbers 99 and 45. We wish to know whether these data are in satisfactory agreement with the expected 3:1 ratio resulting from the segregation of a single pair of alleles. Calculation of the value of χ^2 is illustrated in Table 2-2. The total number of progeny is $99 + 45 = 144$. The *expected* numbers in the two classes, based on the hypothesis that the true ratio is 3:1, are calculated as $(3/4)144 = 108$ and $(1/4)144 = 36$. Because there are two classes of data, there are two terms in the χ^2, which are

$$\chi^2 = \frac{(99 - 108)^2}{108} + \frac{(45 - 36)^2}{36} = 0.75 + 2.25 = 3.00$$

The next step is to interpret the χ^2 value in terms of goodness of fit. This is done with the aid of the graphs in Figure 2-16. The x-axis gives χ^2 values measuring goodness of fit, and the y-axis gives the probability P that a worse fit would be obtained by chance, assuming that the expected theoretical numbers are correct. If the observed χ^2 is so large that the probability of a worse fit is very small, then the observed results do not fit the theoretical expectations. In this case, the genetic hypothesis that led to the expectations must be rejected.

In practice, the critical values of P are 0.05 (the 5 percent level) and 0.01 (the 1 percent level). For P values ranging from 0.01 to 0.05, the probability that chance alone would lead to a worse fit is between one in 20 experiments and one in 100. This is the pink region in Figure 2-16; if the P value falls in this range, the correctness of the genetic hypothesis is considered very doubtful. The result is said to be **significant** at the 5 percent level. For P values smaller than 0.01, the probability that chance alone would lead to a worse fit is less than one in 100 experiments. This is the green region in Figure 2-16; in this case, the result is said to be **highly significant** at the 1 percent level and the genetic hypothesis is rejected outright.

To determine the P value corresponding to a calculated χ^2, the number of **degrees of freedom** of the particular χ^2 test is needed. For the type of χ^2 test illustrated in Table 2-2, the number of degrees of freedom equals the number of classes of data minus 1. Table 2-2 contains two classes of data (wildtype and mutant), and so the degrees of freedom equals $2 - 1 = 1$. The reason for subtracting 1 is that the total number of progeny was set equal to the observed total in calculating the expected numbers. Analogous χ^2 tests with three classes of data have 2 degrees of freedom, and those with four classes of data have 3 degrees of freedom.

TABLE 2-2   Calculation of χ^2 for a monohybrid ratio

Phenotype (class)	Observed number	Expected number	Deviation from expected number	$\dfrac{(Deviation)^2}{Expected\ number}$
Wildtype	99	108	-9	0.75
Mutant	45	36	$+9$	2.25
Total	144	144	0	$\chi^2 = 3.00$

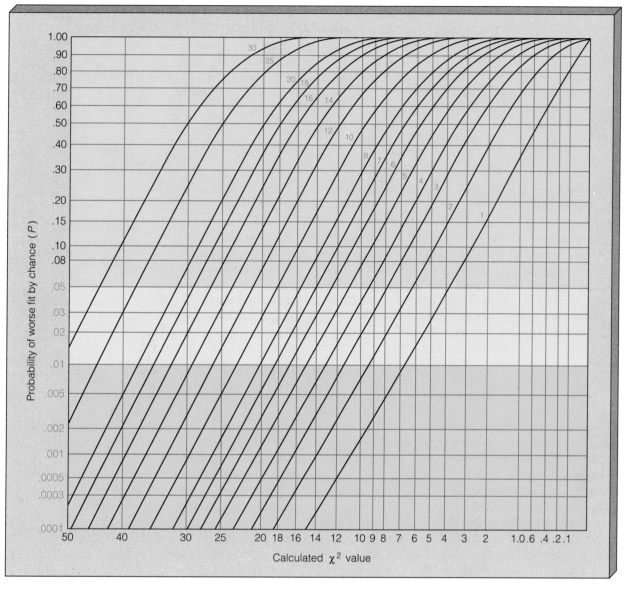

FIGURE 2-16 Graphs for interpreting goodness of fit of genetic predictions using the chi-square test. For any calculated value of χ^2 along the x-axis, the y-axis gives the probability P that chance alone would produce a result as bad or worse than that actually observed, when the genetic predictions are correct. Tests with P in the pink (less than 5 percent) or green (less than 1 percent) are regarded as statistically significant and normally require rejection of the genetic hypothesis leading to the prediction. Each χ^2 test has a number of degrees of freedom associated with it. In the type of tests illustrated in this chapter, the number of degrees of freedom equals the number of classes in the data minus 1.

Now we can interpret the χ^2 value in Table 2-2. Refer to Figure 2-16, and observe that each curve is labeled in red with its degrees of freedom. To determine the P value for the data in Table 2-2, first find the location of $\chi^2 = 3.00$ along the x-axis in Figure 2-16. Trace vertically until intersecting the curve with 1 degree of freedom (the straight diagonal line). Then trace horizontally until intersecting the y-axis and read the P value; in this case, $P = 0.08$. This means that chance alone would produce a χ^2 value as great or greater than 3.00 in about 8 percent of experiments of the type in Table 2-2; and, because the P value is above the pink region, the goodness of fit of the hypothesis of a 3:1 ratio of wildtype:mutant is judged to be satisfactory.

As a second illustration of the χ^2 test, we will determine the goodness of fit of Mendel's round versus wrinkled data to the expected 3:1 ratio. Among 7324 seeds, he observed 5474 round and 1850 wrinkled (Table 1-1 in Chapter 1). The expected numbers are (3/4)7324 = 5493 round and (1/4)7324 = 1831 wrinkled. The χ^2 value is calculated as

$$\chi^2 = \frac{(5474 - 5493)^2}{5493} + \frac{(1850 - 1831)^2}{1831} = 0.26$$

Because there are two classes of data (round and wrinkled), the degrees of freedom equals 1, and the P value for $\chi^2 = 0.26$ with 1 degree of freedom is approximately 0.60 (from Figure 2-16). This means that, in about 60 percent of all experiments of this type, a worse fit would be expected simply because of chance. Again and again Mendel's data are very close to the expected values. In fact, as assessed by χ^2 tests, they are often a little *too* close, and some statisticians have suggested that Mendel may have improved his data somewhat (perhaps unconsciously) to obtain rather better fits than chance alone would warrant.

Additional applications of the χ^2 test, including some more of Mendel's data, are included in the problems at the end of this chapter.

CHAPTER SUMMARY

The chromosomes in the somatic cells of higher plants and animals are in pairs. The members of each pair are homologous chromosomes, and each member is a homolog. Pairs of homologs are usually identical in appearance, whereas nonhomologous chromosomes often show differences in size and structural detail that make them visibly distinct from each other. Cells with nuclei containing two sets of homologous chromosomes are diploid. One set of chromosomes comes from the maternal parent and the other from the paternal parent. Gametes are haploid. They contain only one set of chromosomes, consisting of one member of each pair of homologs.

Mitosis is the process of nuclear division that maintains the chromosome number when a somatic cell divides. Before mitosis, chromosomes replicate, forming a two-part structure consisting of two sister chromatids joined at the centromere. At the onset of mitosis, chromosomes become visible and align along the midline of the spindle. The centromere of each chromosome divides, and the sister chromatids are pulled by spindle fibers to opposite poles of the cell. The separated sets of chromosomes form genetically identical daughter nuclei. Meiosis is the type of nuclear division that occurs in germ cells, and it reduces the diploid number of chromosomes to the haploid number. Replication of the genetic material occurs before the onset of meiosis; so each chromo-

some consists of two sister chromatids. In the first meiotic division, the homologous chromosomes first pair and then separate, and the resulting products contain chromosomes still consisting of two chromatids attached to a common centromere. As a result of crossing-over, which occurs in prophase I, the chromatids may not be genetically identical along their entire length. In the second meiotic division, the centromeres divide and the homologous chromatids separate. The end result of meiosis is the formation of four genetically different haploid nuclei. A distinctive feature of meiosis is the side-by-side pairing (synapsis) of homologous chromosomes in the zygotene substage of prophase I. In the pachytene substage, the paired chromosomes become connected by chiasmata (the physical manifestations of crossing-over) and do not separate until anaphase I. This separation is called disjunction (unjoining), and failure to separate is called nondisjunction (no unjoining, an often confusing double negative). Meiosis is the physical basis of the segregation and independent assortment of genes.

The X and Y sex chromosomes differ from other chromosome pairs in that they are visibly different and contain different genes. In mammals, insects, and many other organisms, females contain two X chromosomes (XX) and hence are homogametic, and males contain one X chromosome and one Y chromosome (XY) and hence are heterogametic. In birds, moths, butterflies, and some reptiles, the situation is the reverse, and females are the heterogametic sex, males the homogametic sex. The Y chromosome often contains only a few genes. In human beings and other mammals, these few genes include a male-determining factor. In *Drosophila*, presence of the Y chromosome does not determine male development. *Drosophila* sex determination depends on the ratio of the number of X chromosomes to the number of sets of autosomes. In most organisms, the X chromosome contains many genes unrelated to sexual differentiation. These are called X-linked genes, and they show a characteristic pattern of inheritance owing to their location in the X chromosome.

KEY TERMS

anaphase
anaphase I
anaphase II
autosome
bivalent
cell cycle
centromere
chiasma
chiasmata
chi-square
chromatid
chromomere
chromosome
chromosome complement
crossing-over
degrees of freedom
diakinesis
diploid
diplotene
first meiotic division
G_1

G_2
germ cell
goodness of fit
haploid
hemophilia A
heterogametic
homogametic
homologous
interphase
leptotene
M
meiocyte
meiosis
metaphase
metaphase I
metaphase II
mitosis
mitotic spindle
nondisjunction
nucleolus
pachytene

Pascal's triangle
prophase
prophase I
prophase II
S
second meiotic division
sex chromosome
somatic cell
spore
statistically significant
synapsis
telophase
telophase I
telophase II
tetrad
X chromosome
X-linked
Y chromosome
zygotene

EXAMPLES OF WORKED PROBLEMS

Problem 1: The black and yellow pigments in the coats of cats are determined by an X-linked pair of alleles, c^b (black) and c^y (yellow). Males are either black (c^b) or yellow (c^y), and females are either black ($c^b c^b$), calico with patches of black and patches of yellow ($c^b c^y$), or yellow ($c^y c^y$).

(a) What genotypes and phenotypes would be expected among the offspring of a cross between a black female and a yellow male?

(b) In a litter of eight kittens, there are two calico females, one yellow female, two black males, and three yellow males. What are the genotypes and phenotypes of the parents?

(c) Rare calico males are the result of nondisjunction. What is their sex-chromosome complement and their genotype?

Answer:

(a) The black female has genotype $c^b c^b$ and the yellow male has genotype c^y. The female offspring receive an X chromosome from each parent, and so their genotype is $c^b c^y$ and their phenotype is calico. The male offspring receive their X chromosome from their mother, and so their genotype is c^b and their phenotype is black.

(b) The male offspring provide information about the X chromosomes in the mother. Because some males are black (c^b) and some yellow (c^y), the mother must have the genotype $c^b c^y$ and have a calico coat. The fact that one of the offspring is a yellow female means that both parents carry an X chromosome with the c^y allele, and so the father must have the genotype c^y and have a yellow coat. The occurrence of the calico female offspring is also consistent with these parental genotypes.

(c) Nondisjunction occurring in a female of genotype $c^b c^y$ can produce an XX egg with the alleles c^b and c^y. When the XX egg is fertilized by a Y-bearing sperm, the result is an XXY male of genotype $c^b c^y$, which is the genotype of a male calico cat.

Problem 2: Certain breeds of chickens have reddish gold feathers due to a recessive allele, g, on the avian equivalent of the X chromosome; presence of the dominant allele results in silver plumage. An autosomal recessive gene, s, results in feathers called silkie that remain soft like chick down. What genotypes and phenotypes of each sex would be expected from a cross of a red rooster heterozygous for silkie and a silkie hen with silver plumage? (Remember that females are the heterogametic sex in birds.)

Answer: In this problem, you need to keep track of both the X-linked and the autosomal inheritance, and it has the additional complication that males are XX and females XY for the avian equivalents of the sex chromosomes. The parental red rooster that is heterozygous for silkie has the genotype gg for the X chromosome and Ss for the relevant autosome; and the silver, silkie hen has the genotype G for the X chromosome and ss for the autosome. The female offspring from the cross receive their X-chromosome from their father (the reverse of the situation in most animals), and so they have genotype g (reddish gold feathers) and half of them are Ss (wildtype) and half are ss (silkie). The male offspring, with respect to the X chromosome, have genotype Gg (silver feathers) and, again, half are Ss (wildtype) and half ss (silkie).

Problem 3: Ranch mink with the dark gray coat color known as aleutian are homozygous recessive, aa. Genotypes AA and Aa have the standard deep brown color. A mating of $Aa \times Aa$ produces eight pups.
(a) What is the probability that none of them have the aleutian coat color?
(b) What is the probability of a perfect 3 : 1 distribution of standard to aleutian?
(c) What is the probability of a 1 : 1 distribution in this particular litter?

Answer: This kind of problem demonstrates the effect of chance variation in segregation ratios that occurs in small families. In the mating $Aa \times Aa$, the probability that any particular pup has the standard coat color is 3/4 and that of aleutian is 1/4. Therefore, in a litter of size 8, the probability of 0, 1, 2, 3 . . . pups with the standard coat

color is given by successive terms in the binomial expansion of $[(3/4) + (1/4)]^8$. The specific probability of r standard and $8 - r$ aleutian pups is given by $8!/[r!(8 - r)!] \times (3/4)^r \times (1/4)^{(8 - r)}$.
(a) The probability that none are aleutian means that $r = 0$, and so the probability is $(3/4)^8 = 6561/65,536 = 0.10$.
(b) A perfect 3 : 1 ratio in the litter means that $r = 6$, and the probability is $[8!/(6!2!)](3/4)^6(1/4)^2 = 28(729/4096)(1/16) = 0.31$. (That is, a little less than a third of the litters would have the "expected" Mendelian ratio.)
(c) A 1 : 1 distribution in the litter means that $r = 4$, and the probability is $[8!/(4!4!)](3/4)^4(1/4)^4 = 70(81/256)(1/256) = 0.09$.

Problem 4: Certain varieties of maize are true breeding for either colored aleurone (the outer layer of the seed) or for colorless aleurone. A cross of a colored variety with a colorless variety gave an F_1 with colored seeds. Among 1000 seeds produced by the cross $F_1 \times$ colored, all were colored; and among 1000 seeds produced by the cross $F_1 \times$ colorless, 525 were colored and 475 were colorless.
(a) What genetic hypothesis can explain these data?
(b) Using the criterion of a χ^2 test, evaluate whether the data are in satisfactory agreement with the model.

Answer:
(a) Mendelian segregation in a heterozygote is suggested by the 525 : 475 ratio, and dominance is implied by the crosses of colored $\times$ colorless and $F_1 \times$ colored. Therefore, a hypothesis that seems to fit the data is that there is a dominant allele for colored aleurone—say, C—and the true-breeding colored and colorless varieties are CC and cc, respectively. The F_1 has the genotype Cc. The cross $F_1 \times$ colored (CC) produces 1 CC : 1 Cc progeny (all colored), and the cross $F_1 \times$ colorless (cc) is expected to produce a 1 : 1 ratio of colored (Cc) to colorless (cc).
(b) The expected numbers are 500 colored and 500 colorless, and therefore the χ^2 equals $(525 - 500)^2/500 + (475 - 500)^2/500 = 2.5$. Because there are two classes of data, there is one degree of freedom, and the P value is approximately 0.12 (see Figure 2-16). The P value is greater than 0.05, and so the goodness of fit is regarded as satisfactory.

PROBLEMS

2-1. At what stage in mitosis and meiosis does chromosome replication occur? When do the chromosomes first become visible in the light microscope?

2-2. If a cell contains 23 pairs of chromosomes just after completion of mitotic telophase, how many chromatids were present in metaphase?

2-3. The Greek roots of the terms *leptotene*, *zygotene*, *pachytene*, *diplotene*, and *diakinesis* literally mean

thin thread, *paired thread*, *thick thread*, *doubled thread*, and *moving apart*, respectively. How are these terms appropriate in describing the appearance and behavior of the chromosomes during prophase I? (Incidentally, the ending *-tene* denotes the adjective; the nouns are formed by dropping *-tene* and adding *-nema*, yielding the alternative terms *leptonema*, *zygonema*, *pachynema*, *diplonema*, and *diakinesis*, which are preferred by some authors.)

2-4. The first meiotic division is often called the *reductional* division and the second meiotic division the *equational* division. What feature of the chromosome complement is reduced in the first meiotic division and kept equal in the second?

2-5. Maize is a diploid organism with 10 pairs of chromosomes. How many chromatids and chromosomes are present in the following stages of cell division:
(a) Metaphase of mitosis?
(b) Metaphase I of meiosis?
(c) Metaphase II of meiosis?

2-6. Sweet peas have a somatic chromosome number of 14. If the centromeres of the 7 homologous pairs are designated as A/a, B/b, C/c, D/d, E/e, F/f, and G/g,
(a) How many different combinations of centromeres can be produced during meiosis?
(b) What is the probability that a gamete will contain only centromeres designated by capital letters?

2-7. Emmer wheat (*Triticum dicoccum*) has a somatic chromosome number of 28, and rye (*Secale cereale*) has a somatic chromosome number of 14. Hybrids produced by crossing these cereal grasses are highly sterile and have many characteristics intermediate between the parental species. How many chromosomes do the hybrids possess?

2-8. X-linked inheritance is occasionally called *crisscross* inheritance. In what sense does the X chromosome move back and forth between the sexes every generation? In what sense is the expression misleading?

2-9. The most-common form of color blindness in human beings results from an X-linked recessive gene. A phenotypically normal couple have a normal daughter and a son who is color-blind. What is the probability that the daughter is heterozygous?

2-10. The mutation for Bar-shaped eyes in *Drosophila* has the following characteristics of inheritance:
(a) Bar males × wildtype females produce wildtype sons and Bar daughters.
(b) The Bar females from the mating in part a, when mated with wildtype males, produce a 1 : 1 ratio of Bar : wildtype sons and Bar : wildtype daughters.
What mode of inheritance do these characteristics suggest?

2-11. Vermilion eye color in *Drosophila* is determined by the recessive allele *v* of an X-linked gene, and the wildtype eye color determined by the v^+ allele is brick red. What genotype and phenotype ratios would be expected from the following crosses?
(a) vermilion male × wildtype female
(b) vermilion female × wildtype male
(c) daughter from mating in part a × wildtype male
(d) daughter from mating in part a × vermilion male

2-12. The autosomal recessive allele *bw* for brown eyes in *Drosophila* interacts with the X-linked recessive allele *v* for vermilion to produce white eyes. What eye-color phenotypes, and in what proportions, would be expected from a cross of a white-eyed female (genotype *v* / *v*; *bw* / *bw*) with a brown-eyed male (genotype v^+ / Y; *bw* / *bw*)?

2-13. It is often advantageous to be able to determine the sex of newborn chickens from their plumage. How could this be done using the X-linked dominant gene S for silver plumage and the recessive allele *s* for gold plumage? (Remember that in chickens the homogametic and heterogametic sexes are reverses of those in mammals.)

2-14. A recessive mutation of an X-linked gene in human beings results in hemophilia, marked by a prolonged increase in the time needed for blood clotting. Suppose that a phenotypically normal couple produces two normal daughters and a son affected with hemophilia.
(a) What is the probability that both of the daughters are heterozygous carriers?
(b) If one of the daughters mates with a normal man and produces a son, what is the probability that the son will be affected?

2-15. In the pedigree shown here, the shaded symbols represent persons affected with an X-linked recessive form of mental retardation. What are the genotypes of all the persons in this pedigree?

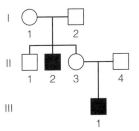

2-16. For an autosomal gene with five alleles, there are fifteen possible genotypes (five homozygotes and ten heterozygotes). How many genotypes are possible with five alleles of an X-linked gene?

2-17. Attached-X chromosomes in *Drosophila* are formed from two X chromosomes attached to a common centromere. Females of genotype C(1)RM / Y, in which C(1)RM denotes the attached-X chromosomes, produce C(1)RM-bearing and Y-bearing gametes in equal proportions. What progeny are expected to result from the mating between a male carrying the X-linked allele *w* for white eyes and an attached-X female with wildtype eyes? How does this result differ from the typical pattern of X-linked inheritance? (Note: *Drosophila* zygotes containing three X chromosomes or no X chromosomes do not survive.)

2-18. People who have the sex chromosome constitution XXY are male. A woman heterozygous for an X-linked mutation for color blindness mates with a normal male and produces an XXY son, who is color-blind. What kind of nondisjunction can explain this result?

2-19. Mice with a single X chromosome and no Y chromosome (an XO sex-chromosome constitution) are fertile females. If at least one X chromosome is required for viability, what sex ratio is expected among surviving progeny from the mating XO ♀ × XY ♂?

2-20. In the pedigree shown here, the shaded symbols represent persons affected with X-linked hemophilia, a blood-clotting disorder.
  (a) If the woman identified as II-2 has two more children, what is the probability that neither will be affected?
  (b) What is the probability that the first child of the mating II-4 × II-5 will be affected?

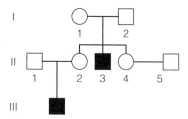

2-21. Assuming a sex ratio at birth of 1 : 1, consider two sibships, A and B, each with three children.

  (a) What is the probability that A consists only of girls and B only of boys?
  (b) What is the probability that either sibship consists only of girls and the other only of boys?

2-22. Assuming a sex ratio of 1 : 1, what is the probability that a sibship of size 4 will consist of 2 boys and 2 girls? What is the probability that a sibship of size 8 will consist of 4 boys and 4 girls?

2-23. In the mating Aa × Aa, what is the probability that a sibship of size 4 will contain no aa individuals? What is the probability that it will contain a perfect Mendelian ratio of 3 A– : 1 aa?

2-24. When 250 seedlings were grown from a sample of seeds produced by testcrossing a hybrid green corn plant with one having yellow-striped leaves, 140 of the seedlings had green leaves and 110 had yellow-striped leaves. Using the chi-square method, test this result for agreement with the expected 1 : 1 ratio.

2-25. A cross was made to produce D. *melanogaster* flies heterozygous for two pairs of alleles: dp^+ and dp, which determine long versus short wings, and e^+ and e, which determine gray versus ebony body color. The following F_2 data were obtained:

Long wing, gray body	462
Long wing, ebony body	167
Short wing, gray body	127
Short wing, ebony body	44

Test these data for agreement with the 9 : 3 : 3 : 1 ratio expected if the two pairs of alleles segregate independently.

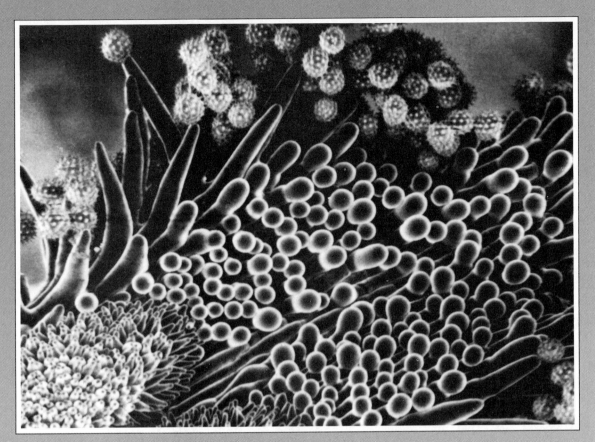

Gene Linkage and Chromosome Mapping

Recognition that homologous chromosomes segregate in meiosis led to the expectation that genes located in a particular chromosome would not undergo independent assortment but would be transmitted together with complete **linkage.** Linkage was indeed observed, but the observed linkages between genes were never complete. In this chapter, we will see that incomplete linkage between genes in the same chromosome is usual and results from the exchange

Above: The stigma of a flower covered with numerous pollen grains (spiked spheres), as visualized by scanning electron microscopy. (From G. Shih and R. Kessel. 1982. *Living Images.* Jones and Bartlett.)

of segments between homologous chromosomes. An exchange event between homologous chromosomes (crossing-over) results in the **recombination** of genes in the pair of chromosomes. The probability of crossing-over between any two genes serves as a measure of genetic distance between them and allows the construction of a genetic map, a diagram of a chromosome showing the relative positions of the genes.

3-1 Linkage and Recombination of Genes in a Chromosome

As described in Chapter 1, a simple and direct test of independent assortment is to testcross an F_1 double heterozygote (*Aa Bb*) with the double recessive homozygote (*aa bb*). If the two pairs of alleles assort independently, as expected when they are on different chromosomes,

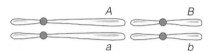

then the double heterozygote will produce the four possible types of gametes—*A B*, *A b*, *a B*, and *a b*—in equal proportions:

The result will be the same whether the *parents* of the double heterozygote had the genotypes *AA BB* and *aa bb* or the genotypes *AA bb* and *aa BB*. The four products of meiosis are expected in equal proportions, and thus the phenotypes of the testcross progeny are expected to be in the ratio of 1 : 1 : 1 : 1. An expected 50 percent of the testcross progeny consists of phenotypes with the same combination of traits present in the parents of the double heterozygote (**parental combinations**), and 50 percent consists of phenotypes with new combinations of the traits (**recombinants**). In this chapter, we will examine phenomena that cause deviations from the 1 : 1 : 1 : 1 ratio.

In early experiments with *Drosophila*, several X-linked genes were identified that provided ideal materials for studying the inheritance of genes in the same chromosome. One of these genes, whose w^+ and w alleles govern wildtype red versus white eye color, was discussed in Chapter 2; another such gene has the alleles m^+ and m, which determine normal versus miniature wings. The initial cross was between females with white eyes and normal wings and males with red eyes and miniature wings:

$$w\,m^+ / w\,m^+ \times w^+\,m / Y$$

The resulting F_1 progeny consisted of wildtype females and white-eyed, nonminiature males. When the F_1 progeny were crossed

$$w\,m^+ / w^+\,m \times w\,m^+ / Y$$

the female progeny consisted of a 1:1 ratio of red:white eyes (all were nonminiature), and the male progeny were as follows:

white eye, normal wing $(w\,m^+\,/\,\text{Y})$	226
red eye, miniature wing $(w^+\,m\,/\,\text{Y})$	202

66.5 percent are parental phenotyes

red eyed, normal wing $(w^+\,m^+\,/\,\text{Y})$	114
white eye, miniature wing $(w\,m\,/\,\text{Y})$	102

33.5 percent are recombinant phenotyes

644

The results of this experiment show a great departure from the 1:1:1:1 ratio of the four male phenotypes expected with independent assortment. The pattern of the deviation is what would be expected if genes in the same chromosome tended to remain together in inheritance but were not completely linked; that is, the combinations of phenotypic traits in the parents of the original cross (parental phenotypes) were present in 428/644 (66.5 percent) of the F_2 males, and nonparental combinations (recombinant phenotypes) of the traits were present in 216/644 (33.5 percent). The 33.5 percent recombinant gametes is the **frequency of recombination,** and it should be contrasted with the 50 percent recombination expected with independent assortment.

The recombinant gametes $w^+\,m^+$ and $w\,m$ result from crossing-over in meiosis in F_1 females. In this example, the frequency of recombination between the linked w and m genes was 33.5 percent, but with other linked genes it ranges from near 0 to 50 percent, depending on the genes. Note that genes that are in the same chromosome but sufficiently far apart do undergo independent assortment (frequency of recombination equal to 50 percent). This implies the following principle:

> Genes with recombination frequencies less than 50 percent are present in the same chromosome (linked). Two genes that undergo independent assortment, indicated by a recombination frequency of 50 percent, may be present in nonhomologous chromosomes or located far apart in a single chromosome.

The notation used in this book for linked genes has the general form $w^+\,m\,/\,w\,m^+$. This notation is a simplification of a more-descriptive, but cumbersome, form:

$$\frac{w^+ \qquad\qquad m}{w \qquad\qquad m^+}$$

In this form of notation, the horizontal lines represent the two homologous chromosomes in which the indicated alleles of the genes are located. The linked genes in a chromosome are always written in the same order for consistency. In the system of gene notation used for *Drosophila*, or other organisms for which a similar system is used, this convention makes it possible to indicate the wildtype allele of a gene with a plus sign in the appropriate position. For example, the genotype $w\,m^+/w^+\,m$ can be written without ambiguity as $w + / + m$.

A genotype like $w + / + m$, in which the mutant alleles are present in opposite homologous chromosomes, is said to have the **trans,** or **repulsion,** configuration. The alternative arrangement with the mutant alleles in the same homolog (for example, $w\,m / + +$) has the **cis,** or **coupling,** configuration.

The *cis* configuration of white and miniature results from the cross of white miniature females with red nonminiature males,

$$w\,m\,/\,w\,m \;\times\; + + /\,Y$$

In this case, the F_1 females were phenotypically wildtype double heterozygotes, and the males had white eyes and miniature wings. When these F_1 progeny were crossed

$$w\,m\,/ + + \;\times\; w\,m\,/\,Y$$

they produced the following progeny:

red eye, normal wing $(+ + /\,w\,m\;♀♀$ and $+ + /\,Y\;♂♂)$	395	62.4 percent are parental phenotyes
white eye, miniature wing $(w\,m\,/\,w\,m\;♀♀$ and $w\,m\,/\,Y\;♂♂)$	382	
white eye, normal wing $(w + /\,w\,m\;♀♀$ and $w + /\,Y\;♂♂)$	223	37.6 percent are recombinant phenotyes
red eye, miniature wing $(+ m\,/\,w\,m\;♀♀$ and $+ m\,/\,Y\;♂♂)$	247	
	1247	

Compared with the preceding experiment with w and m, the frequency of recombination between the genes is approximately the same, but the phenotypes constituting the parental and recombinant classes of offspring are reversed. The reversal is due to the fact that the original parents of the F_1 female were different. In the first cross, the F_1 female was the *trans* double heterozygote ($w + / + m$); in the second cross, the F_1 had the *cis* configuration ($w\,m / + +$). The repeated finding of equal recombination frequencies in experiments of this kind leads to the following conclusion:

> Recombination between linked genes occurs with the same frequency whether the alleles of the genes are in the *trans* or *cis* configuration.

The recessive allele y of another X-linked gene in *Drosophila* results in yellow body color instead of the usual gray color determined by the y^+ allele. When white-eyed females were mated with males having yellow bodies, and the wildtype F_1 females were testcrossed with yellow-bodied, white-eyed males,

$$+ w\,/ + w \;\times\; y + /\,Y$$
$$\downarrow$$
$$+ \;w\,/\,y\; + \;\times\; y\,w\,/\,Y$$

the progeny were

wildtype body, white eye (maternal gamete, $+w$)	4292	} 98.6 percent are parental phenotyes
yellow body, red eye (maternal gamete, $y+$)	4605	
wildtype body, red eye (maternal gamete, $++$)	86	} 1.4 percent are recombinant phenotyes
yellow body, white eye (maternal gamete, yw)	44	
	9027	

In a second experiment, yellow-bodied, white-eyed females were crossed with wildtype males, and the F_1 wildtype females and F_1 yellow-bodied, white-eyed males were intercrossed:

$$y\,w\,/\,y\,w \quad \times \quad +\,+\,/\,\mathrm{Y}$$

$$\downarrow$$

$$+\,+\,/\,y\,w \quad \times \quad y\,w\,/\,\mathrm{Y}$$

In this case, 98.7 percent of the F_2 progeny were parental phenotypes and 1.3 percent were recombinant phenotypes. The parental and recombinant phenotypes were reversed in the reciprocal crosses, but the recombination frequency was virtually the same. The recombination frequency was much lower than that obtained in crosses involving the genes determining red versus white eye color and normal versus miniature wing size. These and other experiments have led to the conclusions that (1) the recombination frequency is a characteristic of a particular pair of genes, and (2) recombination frequencies are the same in *cis* (coupling) and *trans* (repulsion) heterozygotes.

Drosophila is unusual in that recombination does not occur in males. The result is that all alleles located in a particular chromosome show complete linkage in *Drosophila* males. This feature is often made use of in experimental design. The phenomenon is atypical because recombination occurs in both sexes in most other animals and plants.

3-2   Crossing-over and Genetic Mapping

The linkage of the genes in a chromosome can be represented in the form of a **genetic map,** which shows the linear order of the genes along the chromosome with the distances between adjacent genes proportional to the frequency of recombination between them. A genetic map is also called a **linkage map** or a **chromosome map.** Early geneticists understood that recombination between genes occurs by an exchange of segments between homologous chromosomes in a process called crossing-over. This idea was based on an earlier theory about the origin of chiasmata, the cross-shaped configurations between homologous chromosomes seen during the first meiotic prophase (see Figure 2-7). The idea was that a chiasma results from the breaking and rejoining of chromatids during synapsis, with the result that there is an exchange of corresponding segments between them. The theory of crossing-over is that each chiasma results in new association of genetic markers. More will be said about the exchange process shortly.

The unit of distance in a genetic map is called a **map unit**; one map unit is equal to 1 percent recombination. For example, two genes that recombine with a frequency of 3.5 percent are said to be located 3.5 map units apart. One map unit corresponds physically to a length of the chromosome in which, on the average, an exchange occurs once in every 50 meioses. Because a crossover results in two recombinant chromatids and two nonrecombinant chromatids (Figure 3-1), this exchange frequency means that 2 of every 200 products of meiosis will have an exchange in the particular chromosome segment.

There is an important distinction between the distance between two genes in map units and the recombination frequency between the genes. Map units measure how much recombination actually occurs between the genes, and the recombination frequency says how much recombination is actually observed. These may differ because two recombination events occurring in the same region may undo each other's effects. In this case, two recombination events actually occur, but none is observed. If the region between the genes is short enough that no more than one crossover occurs in the region during any one meiosis, then map units and recombination frequencies are the same (because there are no multiple recombinations to undo each other). This is why, in defining map unit, we say that one map unit equals 1 percent recombination. Over a length of chromosome so short as to yield 1 percent observed recombination, multiple crossovers do not usually occur, and so the map distance equals the recombination frequency in this case.

When adjacent chromosome regions separating linked genes are sufficiently short that multiple crossovers do not occur, the recombination frequencies (and hence the map distances) between the genes are additive. This important feature of recombination and the logic used in genetic mapping is illustrated by the following example. Assume that the recombination frequency between genes *a* and *b* is 4 percent and the frequency between *b* and *c* is 7 percent. The order of the genes can be either *a b c* or *b a c*. Which of these alternative orders is correct can be determined by a separate measurement of the recombination frequency between *a* and *c*; that is, if *a* and *c* are found to recombine with a frequency of 11 percent, we can deduce that *b* is located between them:

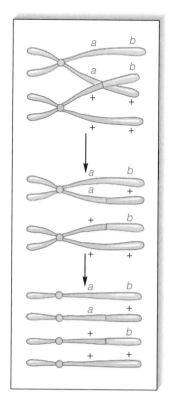

FIGURE 3-1   Diagram illustrating the occurrence of crossing-over between two genes. The exchange between two of the four chromatids results in two recombinant and two nonrecombinant products of meiosis.

On the other hand, if *a* and *c* recombine with a frequency of 3 percent, we would conclude that the order of the genes is *b a c*:

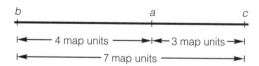

A genetic map can be expanded by this reasoning to include all of the known genes in a chromosome; these genes constitute a **linkage group.** The number of linkage groups is the same as the haploid number of chromosomes of the species. Genetic maps of the four linkage groups of *Drosophila melanogaster* are shown in Figure 3-2; only a small number of the genes that have been mapped are shown.

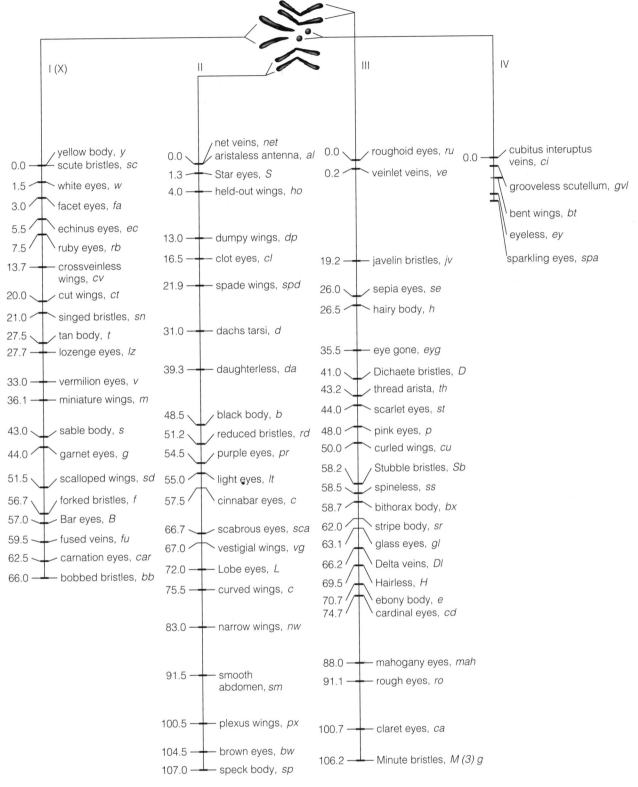

FIGURE 3-2 An abbreviated genetic map of *Drosophila melanogaster*, showing the correspondence between each of the four linkage groups and a pair of chromosomes. The map positions of the genes in a chromosome are in map units from the gene closest to one end of the chromosome. Only a few of the many genes that have been mapped are shown. Capitalized symbols indicate a dominant mutation.

Crossing-over

The orderly arrangement of genes represented by a genetic map is consistent with the conclusion that each gene occupies a well-defined site, or **locus**, in the chromosome, with the alleles of a gene in a heterozygote occupying corresponding locations in the pair of homologous chromosomes. Crossing-over, which occurs by a physical exchange of segments and results in a new association of genes in the same chromosome, has the following features:

1. The exchange of segments between parental chromatids occurs in eukaryotes during the first meiotic prophase, *after the chromosomes have replicated*. The group of four chromatids of a pair of homologous chromosomes are closely synapsed at this stage. Crossing-over is a physical exchange between chromatids in a pair of homologous chromosomes.

2. The exchange process consists of a breaking and rejoining of the two chromatids, resulting in the *reciprocal* exchange of equal and corresponding segments between them (see Figure 3-1).

3. Crossing-over occurs more or less at random along the length of a chromosome pair. Thus, the probability of its occurrence between two genes increases with increasing physical separation of the genes along the chromosome. This principle is the basis of genetic mapping.

One experimental proof that crossing-over occurs after the chromosomes have replicated came from a study of laboratory stocks of *D. melanogaster* in which the two X chromosomes in females are joined to a common centromere to form an aberrant chromosome called an **attached-X** or **compound-X** chromosome. The normal X chromosome in *Drosophila* has a centromere almost at the end of the chromosome, and the attachment of two of these chromosomes to a single centromere results in a chromosome with two equal arms—each consisting of a virtually complete X. Females with a compound-X chromosome usually contain a Y chromosome as well, and they produce two

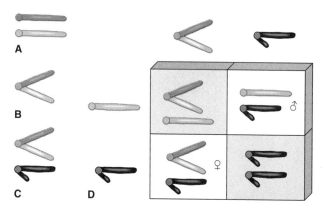

FIGURE 3-3 Attached-X (compound-X) chromosomes in *Drosophila*: (A) structurally normal X chromosomes in a female; (B) attached-X, with the long arms of two normal X chromosomes attached to a common centromere; (C) typical attached-X females also contain a Y chromosome; (D) outcome of a cross using an attached-X female and a normal male. In such a cross, the eggs contain either the attached-X or the Y chromosome, which combine at random with X-bearing or Y-bearing sperm. The shaded genotypes with three X chromosomes or no X chromosome are lethal. Note that the sons receive their X chromosome from their father and their Y chromosome from their mother—the opposite of the usual situation.

classes of viable offspring: females who receive the compound-X chromosome from their mothers and a Y from their fathers, and males with the maternal Y chromosome and a paternal X chromosome (Figure 3-3).

When females are heterozygous for genes in an attached-X chromosome, they produce some female progeny that are homozygous for the recessive alleles because of crossing-over between the homologous chromosome arms. The frequency with which homozygosity occurs increases with increasing map distance of the gene from the centromere. From the diagrams in Figure 3-4, it can be seen that homozygosity can occur only if the crossover between the gene and the centromere takes place after the chromosome has replicated—that is, at the *four-strand* stage of meiosis. Crossing-over before replication of the chromosome (at the *two-strand* stage) would result only in a swap of the alleles between the chromosome arms.

The demonstration that crossing-over (as detected by the recombination of two heterozygous markers) is associated with a physical exchange of segments between homologous chromosomes was made possible by the discovery of two structurally altered chromosomes that permitted the microscopic recognition of parental and recombinant chromosomes. In 1931, Curt Stern discovered two X chromosomes of *Drosophila* that had undergone structural changes that made

FIGURE 3-4   Diagram showing that crossing-over must occur at the four-strand stage in meiosis to produce a homozygous compound-X chromosome product from a compound-X chromosome that is heterozygous for an allele. The exchange must also occur between the centromere and the gene.

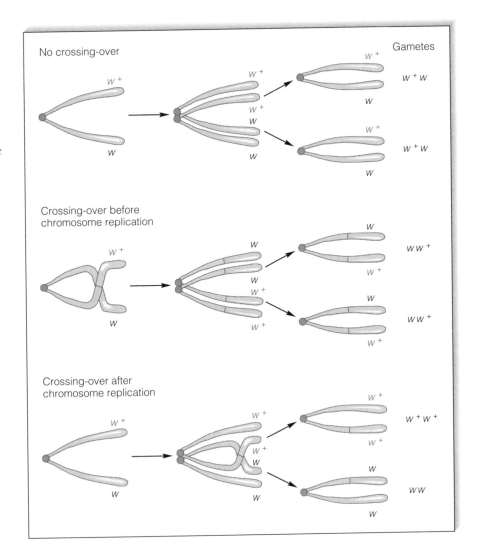

them distinguishable from each other and from a normal X chromosome. He used these X chromosomes in an experiment that provided one of the classical proofs of the physical basis of crossing-over (Figure 3-5). One of the X chromosomes had become separated into two parts by an exchange of fragments between the X chromosome and chromosome 4. The part remaining with the X centromere could be identified by its missing terminal segment. The second aberrant X chromosome had a small piece of a Y chromosome attached as a second arm. The mutant alleles *car* (a recessive allele resulting in carnation eye color instead of red) and *B* (a dominant allele resulting in bar-shaped eyes instead of round) were present in the first altered X chromosome, and the wildtype alleles of these genes were in the second altered X. Females with the two structurally and genetically marked X chromosomes were mated with males having a normal X that carried the recessive alleles of the genes (Figure 3-5). In the progeny from this cross, individual flies with parental or recombinant combinations of the phenotypic traits were recognized by their eye color and shape, and their chromosomal makeup could be determined by microscopic examination of tissue from some of their offspring. In the genetically recombinant progeny from the cross, the X chromosome had the morphology expected if recombination of the genes was accompanied by an exchange that recombined the chromosome markers; that is, the flies with red, bar-shaped eyes had an X chromosome with a missing terminal segment and the attached

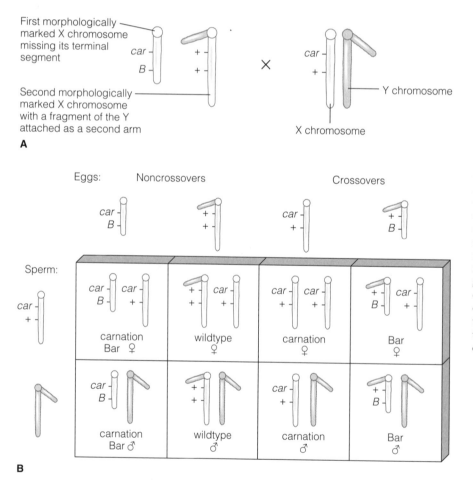

FIGURE 3-5   A. Diagram of a cross in which the two X chromosomes in a *Drosophila* female are morphologically distinguishable from each other and from a normal X. One X chromosome has a missing terminal segment, the other has a second arm consisting of a fragment of the Y chromosome. B. Result of the cross. The carnation offspring contain a structurally normal X chromosome, and the Bar offspring contain an X chromosome with both morphological markers. The result demonstrates that genetic recombination between marker genes is associated with physical exchange between homologous chromosomes. Segregation of the missing terminal segment of the X, which is attached to chromosome 4, is not shown.

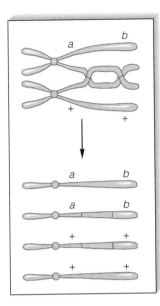

FIGURE 3-6 Diagram showing that the occurrence of two exchanges between the same two chromatids in the region between two genes has no genetically detectable effect.

Y arm, and those with carnation-colored, round eyes had a structurally normal X with no missing terminus and no Y arm. As expected, the nonrecombinant progeny were found to have an X chromosome morphologically identical with one in their mothers.

Multiple Crossing-over

When two genes are far apart along a chromosome, more than one crossover can occur between them in a single meiosis, and this complicates the interpretation of recombination data. The probability of multiple crossovers increases with the distance between the genes. Multiple crossing-over complicates genetic mapping because map distance is based on the number of physical exchanges that occur, and some of the multiple exchanges occurring between two genes will not result in recombination of the genes and hence will not be detected. The effect of one crossover can be cancelled by another crossover occurring further along the way. For example, if two exchanges between the same two chromatids occur between genes *a* and *b*, then their net effect will be that shown in Figure 3-6. Two of the products of this meiosis are double-crossover chromosomes, but the chromosomes are not recombinant for *a* and *b* and so are genetically indistinguishable from noncrossover chromosomes. The occurrence of such cancelling events means that the observed recombination value is an *underestimate* of the true exchange frequency, and, consequently, of the map distance between the genes. In higher organisms double crossing-over effectively never occurs in chromosome segments less than about 10 map units long; so the difficulty is avoided by the use of recombination data for closely linked genes to build up genetic linkage maps.

The maximum frequency of recombination between any two genes is 50 percent—the same value that would be observed if the genes were on nonhomologous chromosomes and assorted independently. Fifty percent recombination occurs when the genes are so far apart in the chromosome that at least one crossover almost always occurs between them. From Figure 3-1, it is easy to see that the occurrence of a single exchange in every meiosis would result in half of the products having parental combinations and the other half having recombinant combinations of the genes. The occurrence of two exchanges between two genes has the same effect, as shown in Figure 3-7. Again, no recombination of the marker genes is detectable if the same chromatids (strands) participate in the two exchanges (two-strand double crossover). When the two exchanges have one chromatid in common (three-strand double crossover), the result is indistinguishable from that of a single exchange—two products with parental combinations and two with recombinants are produced. Note that there are two types of three-strand doubles, depending on which three chromatids participate. The final possibility is that the second exchange occurs between the chromatids that did not participate in the first exchange (four-strand double crossover), in which case all four products are recombinant. If the chromatids taking part in the two exchange events are chosen at random (which appears to be the case), the expected proportions of the three types of double exchanges are 1/4, 1/2, and 1/4, respectively. This means that, on the average, there will be $(1/4)(0) + (1/2)(2) + (1/4)(4) = 2$ recombinant chromatids among the 4 chromatids that result from meioses in which two exchanges occur between a pair of genes. This is the same proportion obtained when a single exchange always occurs between

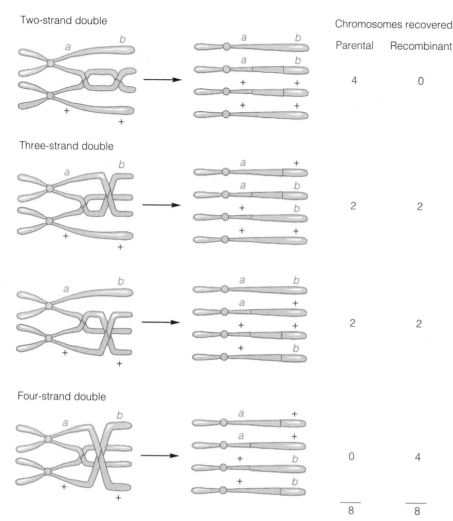

Two-strand double

Three-strand double

Four-strand double

Chromosomes recovered	
Parental	Recombinant
4	0
2	2
2	2
0	4
8	8

FIGURE 3-7 Diagram showing that when two exchanges always occur in the region between two genes, and the chromatids participate at random in the exchanges, the result is indistinguishable from independent assortment of the genes.

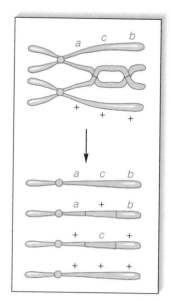

FIGURE 3-8 Diagram showing that two exchanges occurring between the same chromatids and spanning the middle pair of alleles in a triple heterozygote result in a reciprocal exchange of that pair of alleles between the two chromatids.

the genes. Moreover, this result (a maximum of 50 percent recombination) is obtained for any number of exchanges.

The occurrence of double crossing-over is detectable in recombination experiments that employ **three-point crosses**—those using three pairs of alleles. If a third pair of alleles, c^+ and c, is located between the two with which we have been concerned (the outermost markers), double exchanges in the region can be detected when the crossovers flank the c gene (Figure 3-8). The two crossovers, occurring in this example between the same pair of chromatids, result in a reciprocal exchange of the c^+ and c alleles between the chromatids. A three-point cross is an efficient way to obtain recombination data; it is also a simple method for determining the order of the three genes, as will be seen in the next section.

3-3 Gene Mapping from Three-Point Testcrosses

Table 3-1, the results of a testcross in corn with three genes in a single chromosome, illustrates the analysis of a three-point cross. The recessive alleles of the genes in this cross are *lz* (for lazy, or prostrate, growth habit), *gl* (for glossy leaf), and *su* (for sugary endosperm), and the multiple heterozygote parent in the cross had the genotype *Lz Gl Su / lz gl su*. Therefore, the two classes of segregants among the progeny that represent noncrossover (parental-type) gametes are the normal plants and those with the lazy glossy sugary phenotype. These classes are far larger than any of the crossover classes. Thus, if the combination of dominant and recessive alleles in the chromosomes of the heterozygous parent were unknown, we would deduce from their relative frequency in the progeny that the noncrossover gametes were *Lz Gl Su* and *lz gl su*. This is a point important enough to state as a general principle:

> In any genetic cross, no matter how complex, the two most-frequent classes of chromosomes identify the *nonrecombinants*; these classes provide the linkage phase (*cis* versus *trans*) of the alleles in the multiply heterozygous parent.

In mapping experiments, the gene sequence is usually not known—the order in which the three genes are shown in this example is entirely arbitrary. However, there is an easy way to determine the correct order from three-point data—namely, by identifying the genotypes of the double-crossover gametes produced by the heterozygous parent and then comparing them with the parental gametes. Because the probability of the simultaneous occurrence of two exchanges is considerably smaller than that of either single exchange, the double-crossover gametes will be the least frequent types. It can be seen from Table 3-1 that the classes composed of four plants with the sugary phenotype and two plants with the lazy and glossy phenotype—products of the *Lz Gl su* and *lz gl Su* gametes, respectively—are the least frequent and therefore constitute the double-crossover progeny.

> The effect of double crossovers, as Figure 3-8 shows, is to interchange the members of the *middle* pair of alleles between the chromosomes.

This means that if the parental chromosomes are

$$Lz\,Gl\,Su \quad \text{and} \quad lz\,gl\,su$$

and the double-crossover chromosomes are

$$Lz\,Gl\,su \quad \text{and} \quad lz\,gl\,Su$$

TABLE 3-1 Progeny from a three-point testcross in corn

Phenotype of testcross progeny	Genotype of gamete from hybrid parent			Number
normal (wildtype)	*Lz*	*Gl*	*Su*	286
lazy	*lz*	*Gl*	*Su*	33
glossy	*Lz*	*gl*	*Su*	59
sugary	*Lz*	*Gl*	*su*	4
lazy, glossy	*lz*	*gl*	*Su*	2
lazy, sugary	*lz*	*Gl*	*su*	44
glossy, sugary	*Lz*	*gl*	*su*	40
lazy, glossy, sugary	*lz*	*gl*	*su*	272

Su and *su* are interchanged by double crossing-over and must be the middle pair of alleles. Therefore, the genotype of the heterozygous parent in the cross should be written as

$$\frac{Lz \qquad Su \qquad Gl}{lz \qquad su \qquad gl}$$

which is now diagrammed correctly with respect to both the order of the genes and the array of alleles in the homologous chromosomes. A two-strand double crossover between chromatids of these parental types is diagrammed below, and the products can be seen to correspond to the two types of gametes identified in the data as the double crossovers:

From this diagram, it can also be seen that the reciprocal products of a single crossover between *lz* and *su* would be *Lz su gl* and *lz Su Gl, and the products of a single exchange between su and gl* would be *Lz Su gl* and *lz su Gl*.

We can now summarize the data in a more-informative way, writing the genes in correct order and identifying the numbers of the different chromosome types produced by the heterozygous parent that are present in the progeny:

Lz	Su	Gl	286 ⎫	
lz	su	gl	272 ⎭	Parental types
Lz	su	gl	40 ⎫	
lz	Su	Gl	33 ⎭	Single crossovers between *lz* and *su*
Lz	Su	gl	59 ⎫	
lz	su	Gl	44 ⎭	Single crossovers between *su* and *gl*
Lz	su	Gl	4 ⎫	
lz	Su	gl	2 ⎭	Double-crossover types
			740	

Note that each class of recombinants consists of two reciprocal products (grouped together in braces) and that the reciprocal products occur in approximately equal frequencies (286 versus 272, 40 versus 33, and so forth). This observation illustrates an important principle:

The two reciprocal products resulting from any crossover, or any combination of crossovers, are expected to occur among the progeny in approximately equal frequencies.

In calculating the frequency of crossing-over from the data, remember that the double-recombinant chromosomes result from *two* exchanges, one in each of the chromosome regions defined by the three genes. Therefore, crossovers between *lz* and *su* are represented by the following chromosome types:

Lz	su	gl	40
lz	Su	Gl	33
Lz	su	Gl	4
lz	Su	gl	2
			79

That is, 79/740, or 10.7 percent, of the chromosomes recovered in the progeny are crossovers between the *lz* and *su* genes, and so the map distance between these genes is 10.7 units. Similarly, crossovers between *su* and *gl* are represented by

$$
\begin{array}{cccr}
Lz & Su & gl & 59 \\
lz & su & Gl & 44 \\
Lz & su & Gl & 4 \\
lz & Su & gl & \underline{2} \\
 & & & 109
\end{array}
$$

The crossover frequency between this second pair of genes is 109/740, or 14.8 percent, and so the map distance between them indicated by these data is 14.8 units. The genetic map of the chromosome segment in which the three genes are located is therefore

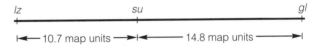

Interference in the Occurrence of Double Crossing-over

The detection of double crossing-over makes it possible to determine whether exchanges in two different regions of a pair of chromosomes occur independently of each other. Using the information from the example with corn, we know from the recombination frequencies that the probability of exchange is 0.107 between *lz* and *su* and 0.148 between *su* and *gl*. If crossing-over occurs independently in the two regions (that is, if the occurrence of one exchange does not alter the probability of the second exchange), the probability of the occurrence of an exchange in both regions is the product of these individual probabilities: that is, $(0.107)(0.148) = 0.0158$, or 1.58 percent. This means that in a sample of 740 the expected number of double crossovers would be 740×0.0158, or 12, whereas the number observed was only 6. Such deficiencies in the observed number of double crossovers are common and identify a phenomenon called **interference**: the occurrence of crossing-over in one region of a chromosome reduces the probability of a second crossover in a nearby region.

The **coefficient of coincidence** is the observed number of double recombinants divided by the expected number; its value provides a simple measure of the degree of interference, defined as

$$
\text{Interference} = 1 - (\text{Coefficient of coincidence})
$$

From the data in our example, the coefficient of coincidence is $6/12 = 0.50$, meaning that the number of double crossovers that occurred was only 50 percent of the number expected if crossing-over in the two regions were independent. It has been found experimentally that interference usually increases as the distance between the two outside markers becomes smaller, until a point is reached at which double crossing-over does not occur; that is, no double crossovers are found, and the coefficient of coincidence equals 0 (or, to say the same thing, the interference equals 1). The distance is about 10 map units in a variety of organisms. Conversely, when the total distance between the gene loci is greater than about 45 map units, interference disappears and the coefficient of coincidence becomes 1.

In some species of fungi and unicellular algae, each meiotic tetrad is contained in a saclike structure called an **ascus** and can be recovered as an intact group. The advantage of these organisms for the study of recombination is the potential for analyzing all of the products of an individual meiosis. Two other features of the organisms are especially useful for genetic analysis: (1) they are haploid, and so dominance is not a complicating factor because the genotype is expressed directly in the phenotype; and (2) they produce very large numbers of progeny, making it possible to detect rare events and to estimate their frequencies accurately.

The life cycles of these organisms tend to be short. The only diploid stage is the zygote, which undergoes meiosis only a few cell divisions after it is formed; the resulting haploid meiotic products, called *spores*, germinate to regenerate the vegetative stage (Figure 3-9). In some species, each of the four products of meiosis subsequently undergoes a mitotic division, with the result that each member of the tetrad yields a *pair* of genetically identical spores. In most of the organisms, the meiotic products, or their derivatives, are not arranged in any particular order in the spore sac. However, bread molds of the genus *Neurospora* and related organisms have the useful characteristic that the meiotic products are arranged in a definite order directly related to the planes of the meiotic divisions. We will examine the ordered system after first looking at unordered tetrads.

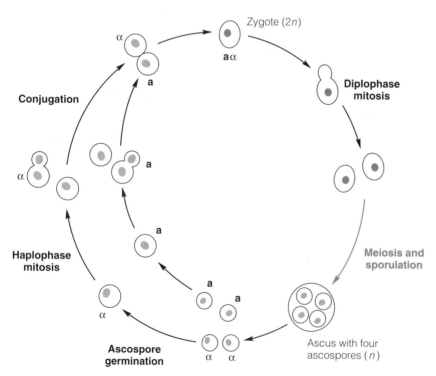

FIGURE 3-9   Life cycle of the yeast *Saccharomyces cerevisiae*. Mating type is determined by the alleles **a** and α. Both haploid and diploid cells normally multiply by mitosis (budding). Depletion of nutrients in the growth medium induces meiosis and sporulation of cells in the diploid state. Diploid nuclei are solid red; haploid nuclei are gray.

Analysis of Unordered Tetrads

Three patterns of segregation are possible in the tetrads when two pairs of alleles are segregating. For example, in a cross $AB \times ab$, the three types of tetrads are:

AB AB ab ab, referred to as **parental ditype,** or PD. Only two genotypes are represented, and their alleles have the same combinations found in the parents.

Ab Ab aB aB, referred to as **nonparental ditype,** or NPD. Only two genotypes are represented, but their alleles have nonparental combinations.

AB Ab aB ab, referred to as **tetratype,** or TT. All four of the possible genotypes are present.

Tetrad analysis is an effective way of determining whether two genes are linked, because of the following principle:

When genes are *unlinked,* the parental ditype tetrads and nonparental ditype tetrads occur with equal frequencies.

This is shown in Figure 3-10A for two pairs of alleles, Aa and Bb, located in different chromosomes—that is, unlinked. In the absence of crossing-over between either gene and its centromere, the two chromosomal configurations are equally likely at metaphase I, and so PD = NPD. If crossing-over does occur (Figure 3-10B), a tetratype tetrad results, but this does not change the fact that PD = NPD.

In contrast, when genes are linked, parental ditypes are far more frequent that nonparental ditypes. To see why, assume that the genes are linked and consider the events required for production of the three types of tetrads. Figure 3-11 shows that, when no crossing-over occurs between the genes, a PD tetrad is formed. Single crossing-over between the genes results in a TT tetrad. The occurrence of a two-strand, three-strand, or four-strand double crossover results in a PD, TT, or NPD tetrad, respectively. With linked genes, meiotic cells with no crossovers will always outnumber those with four-strand double crossovers. Therefore, *an observation that nonparental ditype tetrads occur with lower frequency than parental ditype tetrads is a sensitive indicator of linkage.*

The relative frequencies of the different types of tetrads can be used to determine the map distance between two linked genes. To do this, the number of NPD tetrads is used to estimate the frequency of double crossovers. The four types of double crossovers in Figure 3-11 are expected to occur with equal frequency. Thus, the number of TT tetrads resulting from three-strand double crossing-over (parts D and E) should equal twice the number of NPD tetrads (part F), and so the number of TT owing to single crossing-over is [TT] − 2[NPD], in which the brackets denote observed number. The number of tetrads resulting from double crossing-over (parts C, D, E, and F) should equal four times the number of NPD tetrads (part F), or 4[NPD]. The map distance between two genes equals one-half the frequency of single-crossover tetrads (part B), plus the frequency of double-crossover tetrads. Thus,

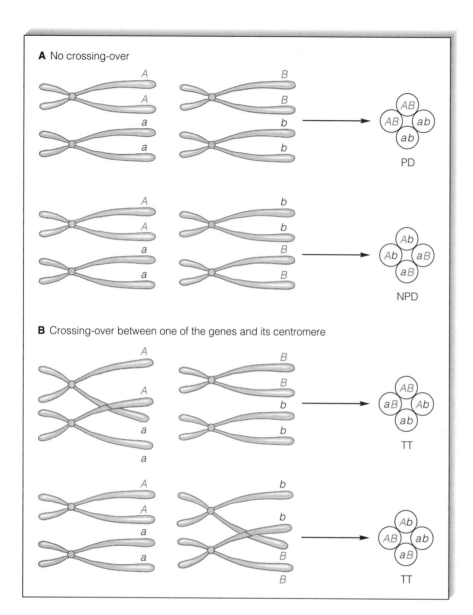

A No crossing-over

A ... A ... a ... a   *B ... B ... b ... b*

AB
AB ab
ab
PD

A ... A ... a ... a   *b ... b ... B ... B*

Ab
Ab aB
aB
NPD

B Crossing-over between one of the genes and its centromere

A ... A ... a ... a   *B ... B ... b ... b*

AB
aB Ab
ab
TT

A ... A ... a ... a   *b ... b ... B ... B*

Ab
AB ab
aB
TT

FIGURE 3-10 Types of unordered asci produced with two genes in different chromosomes. A. In the absence of crossing-over, random arrangement of chromosome pairs at metaphase I yields two different combinations of chromatids, one yielding PD tetrads and the other NPD tetrads. B. When crossing-over occurs between one gene and its centromere, the two chromosome arrangements yield TT tetrads. If both genes are closely linked to their centromeres (so crossing-over is rare), few TT tetrads are produced.

$$\text{Map distance} = \frac{(1/2)([TT] - 2[NPD]) + 4[NPD]}{\text{Total number of tetrads}} \times 100$$

$$= \frac{(1/2)[TT] + 3[NPD]}{\text{Total number of tetrads}} \times 100$$

$$= \frac{(1/2)([TT] + 6[NPD])}{\text{Total number of tetrads}} \times 100 \tag{1}$$

This equation takes into account the fact that only half of the meiotic products in a TT tetrad produced by single crossing-over are crossovers. When the linked genes are close enough that [NPD] = 0 (no double crossovers), the map distance

FIGURE 3-11   Types of
tetrads produced with two
linked genes: (A) in the absence
of crossing-over (PD tetrad);
(B) with a single crossover
between the genes (TT tetrad);
(C through F) with the four
possible types of double
crossovers between the genes
(only the four-strand double in
part F gives an NPD tetrad).

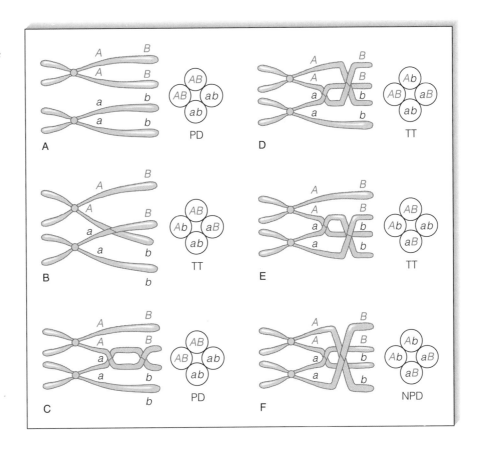

equals the percentage of TT tetrads divided by 2. (You divide by 2 because only half the spores in TT tetrads are recombinant for the alleles.)

As an example, consider a two-factor cross that yields 112 PD, 4 NPD, and 24 TT. The fact that NPD < PD (that is, 4 < 112) indicates that the two genes are linked. Substitution of the values into Equation 1 yields a map distance of $(1/2)[24 + 6(4)](100)/(112 + 4 + 24) = 17.1$ map units. Note that this mapping procedure differs from that presented earlier in the chapter in that recombination frequencies are not calculated directly from the number of recombinant and nonrecombinant chromatids.

Analysis of Ordered Tetrads

In *Neurospora crassa*, a species used extensively in genetic investigations, the products of meiosis are contained in an *ordered* array of spores (Figure 3-12). A zygote nucleus, contained in a saclike structure (an ascus), undergoes meiosis almost immediately after it is formed. The four nuclei produced by meiosis are in a linear, ordered sequence in the ascus and each of them undergoes a mitotic division to form two genetically identical and adjacent **ascospores**. Thus, each mature ascus contains eight ascospores in four pairs, each pair representing one of the products of meiosis. The ascospores can be removed in sequence from an ascus and germinated in individual culture tubes to determine their genotypes.

If two genes have segregated, it is possible to classify ordered asci as PD, NPD, or TT, and linkage analysis is carried out as described in the preceding section. However, the ordered arrangement of meiotic products also makes it possible to map each gene with respect to its centromere; that is, to determine

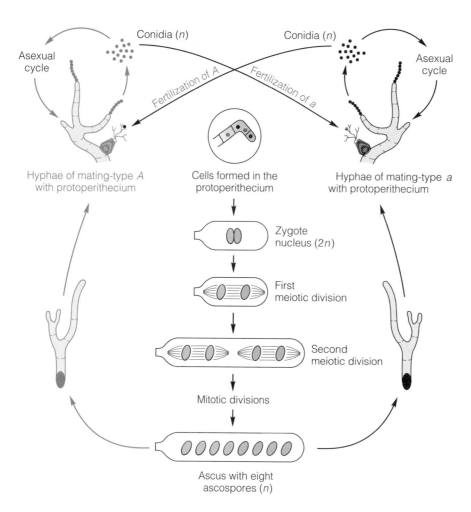

FIGURE 3-12 The life cycle of *Neurospora crassa*. The vegetative body consists of partly segmented filaments called hyphae. Conidia are asexual spores that also function in the fertilization of organisms of the opposite mating type. A protoperithecium develops into a stucture in which numerous cells undergo meiosis.

the recombination frequency between a gene and its centromere. This mapping technique is based on a feature of meiosis shown in Figure 3-13:

> Homologous centromeres of parental chromosomes separate at the first meiotic division; the centromeres of sister chromatids separate at the second meiotic division.

Thus, in the absence of crossing-over between a gene and its centromere, the alleles of the gene (for example, A and *a*) must separate in the first meiotic division; this separation is called **first-division segregation.** If, instead, a crossover does occur between the gene and its centromere, the A and *a* alleles do not become separated until the second meiotic division; this separation is called **second-division segregation.** The distinction between first-division and second-division segregation is shown in Figure 3-13. As shown in part A, only two possible arrangements of the products of meiosis are possible when first-division segregation occurs, namely, AA*aa* or *aa*AA. However, four patterns of second-division segregation are possible because of the random arrangement of homologous chromosomes at metaphase I and of the chromatids at metaphase II. These four arrangements, shown in part B, are A*a*A*a*, *a*AA*a*, A*aa*A, and *a*AA*a*.

The percentage of asci with second-division segregation patterns for a gene can be used to map the gene with respect to its centromere. For example, let us assume that 30 percent of a sample of asci from a cross have a second-division

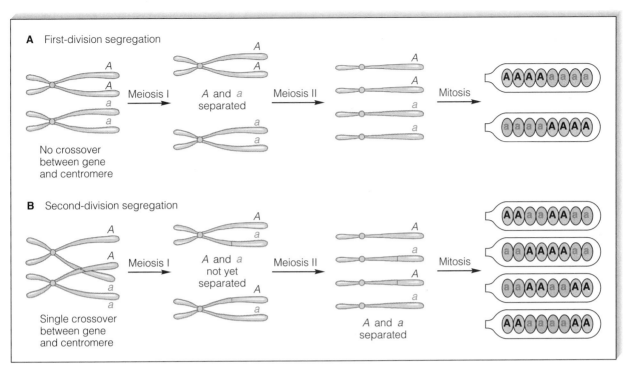

A First-division segregation

A and a
separated

Meiosis I ⟶ Meiosis II ⟶ Mitosis

No crossover
between gene
and centromere

B Second-division segregation

A and a
not yet
separated

Meiosis I ⟶ Meiosis II ⟶ Mitosis

Single crossover
between gene
and centromere

A and a
separated

FIGURE 3-13 First- and second-division segregation in *Neurospora*. A. First-division segregation patterns are found in the ascus when crossing-over between the gene and centromere does not occur. The alleles separate (segregate) in meiosis I. Two spore patterns are possible, depending on the orientation of the pair of chromosomes on the first-division spindle. The orientation shown results in the pattern in the upper ascus. B. Second-division segregation patterns are found in the ascus when crossing-over between the gene and the centromere delays separation of A from *a* until meiosis II. Four spore patterns are possible, depending on the orientation of the pair of chromosomes on the first-division spindle and of the chromatids of each chromosome on the second-division spindle. The orientation shown results in the pattern in the top ascus.

segregation pattern for the A and *a* alleles. This means that 30 percent of the cells undergoing meiosis had a crossover between the A gene and its centromere. Furthermore, in each cell in which crossing-over occurs, two of the chromatids are recombinant and two are nonrecombinant. In other words, a single crossover results in four meiotic products, half of which are recombinant and half nonrecombinant. Thus, a frequency of crossing-over of 30 percent corresponds to a recombination frequency of 15 percent. By convention, map distance refers to the frequency of recombinant meiotic products rather than the frequency of cells in which the crossover occurred. Therefore, the map distance between a gene and its centromere is given by the equation

$$\frac{1/2 \ (\text{Asci with second-division segregation patterns})}{\text{Total number of asci}} \times 100 \qquad (2)$$

This equation is valid as long as the gene is close enough to the centromere that few multiple crossovers occur. Thus, reliable linkage values are best determined for genes that are fairly near the centromere. The location of more-distant genes can then be determined by mapping them with respect to genes nearer the centromere.

Mitotic Recombination

Genetic exchange also occurs in mitosis, although its frequency is about 1000-fold lower than that in meiosis. The first evidence for mitotic recombination was obtained from experiments with *Drosophila*, but the phenomenon has been studied most carefully in fungi such as yeast and *Aspergillus*, in which the frequencies are higher than in most other organisms. Genetic maps can be constructed from mitotic recombination frequencies. The relative map distances between particular genes sometimes correspond to those based on meiotic recombination frequencies, but for unknown reasons the distances are often markedly different. The discrepancies may be a result either of different mechanisms of exchange or from a nonrandom distribution of potential sites of exchange.

3-5 Recombination Within Genes

Genetic analyses and microscopic observations made before 1940 led to the view that a chromosome was a linear array of particulate units—the genes—joined in some way resembling a string of beads. The gene was believed to be the smallest unit of genetic material capable of alteration by mutation and the smallest unit of inheritance. Crossing-over had been seen in all regions of the chromosome but not within genes, and thus the idea of indivisibility of the gene had developed. This classical concept of a gene began to change when it was realized that mutant alleles of a gene might be the result of alterations at different mutable sites within the gene. The evidence that led to this idea was the finding of rare recombination events within several genes in *Drosophila*. The initial observation came from investigations of the lozenge gene (lz) in the X chromosome of *D. melanogaster*. Mutant alleles of lozenge are recessive, and their phenotypic effects include disturbed arrangement of the facets of the compound eye and a reduction in the red pigment. Numerous lz alleles with distinguishable phenotypes are known.

    Females heterozygous for two different lz alleles, one in each homologous chromosome, have lozenge eyes. When such heterozygous females are crossed with males having one of the lz mutant alleles in the X chromosome and large numbers of progeny are examined, flies having normal eyes are occasionally found. For example, one cross carried out was

$$
\begin{array}{c}
\mathrm{X} \\
\mathrm{X}
\end{array}
\quad
\frac{+ \quad lz^{\mathrm{BS}}+ \quad +}{ct \quad + lz^{\mathrm{g}} \quad v}
\quad \times \quad
\frac{+ \quad + lz^{\mathrm{g}} \quad v}{\qquad\qquad\qquad}
\quad
\begin{array}{c}
\mathrm{X} \\
\mathrm{Y}
\end{array}
$$

in which lz^{BS} and lz^{g} are mutant alleles of lozenge, and ct (cut wing) and v (vermilion eye color) are genetic markers that map 7.7 units to the left and 5.3 units to the right, respectively, of the lozenge locus. From this cross, 134 males and females with wildtype eyes were found among more than 16,000 progeny, for a frequency of 8×10^{-3}. These rare individuals might have resulted from a reverse mutation of one or the other of the mutant lozenge alleles to lz^{+}, but the observed frequency, though small, is much greater than the known frequencies of reverse mutations. The genetic constitution of the maternally derived X chromosomes in the normal-eyed offspring was also inconsistent with such an explanation: the male offspring had cut wings and the females were

$ct / +$ heterozygotes. That is, all of the rare nonlozenge progeny had an X chromosome with the constitution

$$\underline{\qquad ct \qquad ++ \qquad + \qquad}$$

which could be accounted for by crossing-over between the two lozenge alleles. Proof of this conclusion came from the detection of the reciprocal crossover chromosome

$$\underline{\qquad + \qquad lz^{BS}lz^{g} \qquad v \qquad}$$

in five male progeny having a lozenge phenotype distinctly different from the phenotype resulting from the presence of either an lz^{BS} or lz^{g} allele alone. The exceptional males also had vermilion eyes. The observation of intragenic crossing-over indicated that genes indeed have fine structure and that the multiplicity of allelic forms of some genes might be due to mutations at different sites in the gene. In the next chapter, we will see that these sites are the nucleotide constituents that make up DNA.

3-6   Complementation

A particular phenotype often requires the activities of a number of genes. A genetic investigation of a system determining, for example, the synthesis of a pigment or the breakdown of a sugar commonly requires the isolation of numerous mutations that produce a particular phenotype. Invariably, it becomes important to know whether mutations causing similar phenotypes are located in the same or different genes—that is, if they are alleles. The genetic test used to determine both allelism among a group of mutations and the number of different genes represented is called **complementation.** It can be used only with recessive mutations.

    Complementation tests are done with the mutations in the *trans* configuration

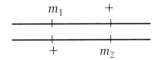

in which m_1 and m_2 represent two mutations. This configuration is obtained in *Drosophila* and other diploid organisms simply by crossing individuals homozygous for the respective mutations. How it is done with viruses and with bacteria is explained in Chapter 10. A cell or organism with two mutations in the *trans* configuration (one in each chromosome) is called a **trans-heterozygote.**

    Consider a *trans*-heterozygote in which the two mutations are in different genes and both genes must be active for an organism to have a wildtype phenotype. The two mutations impair different functional units but, because each chromosome has one wildtype allele of each gene and both m_1 and m_2 are recessive, the organism has the wildtype phenotype. When a *trans*-heterozygote has the wildtype phenotype, the mutations are said to *complement* one another, because the phenotype is that associated with the normal expression of both

A *Trans*-heterozygote for two mutations in different genes

B *Trans*-heterozygote for two mutations in the same gene

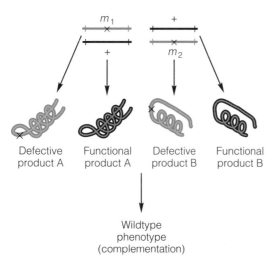

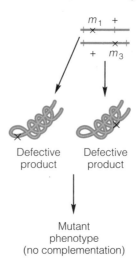

FIGURE 3-14 The basis for interpretation of a complementation test used to determine whether two mutations are in different genes (A) or are alleles of the same gene (B).

genes. The genetic constitution of such a heterozygote and its relation to the synthesis of functional products of each gene is shown in Figure 3-14A. However, if the two mutations are in the same gene, the state of the organism is the same as that with homozygosity for a particular recessive mutation—that is, synthesis of a normal, functional product of a gene is prevented because both members of the pair of genes are mutant. Figure 3–14B shows that no functional product is made and the *trans*-heterozygote has a mutant phenotype—complementation is absent. In summary, in the absence of complicating interactions,

> A *trans*-heterozygote for two recessive mutations has the wildtype phenotype if the mutations are in different genes and a mutant phenotype if the mutations are allelic.

Mutations that complement each other are said to belong to different **complementation groups,** whereas alleles are members of the same complementation group.

Complementation and recombination must not be confused. Complementation is a property of *trans*-heterozygotes that have mutations in two different genes, whereas recombination is detected by genetic analysis of the progeny of heterozygotes.

As an example of complementation analysis, consider the eight different mutations analyzed pair-by-pair in *trans*-heterozygotes giving the results shown in Table 3-2. Each + or − entry designates the result of a single test of a pair of mutations. A + entry denotes complementation, and a − entry indicates noncomplementation. To identify the complementation groups, find the sets of − entries that appear down one column or across one row. These indicate noncomplementing (allelic) mutations. For example, in column 1, mutations 1 and 5 fail to complement, and so they are alleles. Altogether, three genes can be identified by the lack of complementation: the first contains mutations 1 and 5; the second contains mutations 2, 4, and 8; and the third contains mutations 3, 6, and 7. Individual mutations that intersect at a + sign are complementing, and so, to confirm the three genes just identified, tests of pairs of mutations in different genes should complement one another. This is confirmed—for

TABLE 3-2   An example of complementation data

		Mutation number							
		1	*2*	*3*	*4*	*5*	*6*	*7*	*8*
Mutation	1	−							
number	2	+	−						
	3	+	+	−					
	4	+	−	+	−				
	5	−	+	+	+	−			
	6	+	+	−	+	+	−		
	7	+	+	−	+	+	−	−	
	8	+	−	+	−	+	+	+	−

Note: + = complementation; − = no complementation. An entry at the intersection of a horizontal row and a vertical column represents the result of one complementation test between two mutations. For example, the + entry at the intersection of row 3 and column 2 indicates that mutations 2 and 3 complement; the corresponding entry for row 2 and column 3 is not given because it is the same as the one just stated—that is, the table is symmetric about the diagonal.

example, the combination of mutations 1 and 8 yields a + entry. Thus, the three complementation groups, denoted A, B, and C, are the following:

Group	Mutation
A	1, 5
B	2, 4, 8
C	3, 6, 7

The simplest explanation of the data is that the phenotype being studied can result from mutations in at least three different genes. One cannot say that there are only three genes, because mutations in other genes may not have been isolated.

CHAPTER SUMMARY

Nonallelic genes located in the same chromosome tend to remain together in meiosis rather than undergoing independent assortment. This phenomenon is called linkage. The indication of linkage is deviation from the 1:1:1:1 ratio of phenotypes in the progeny of a cross of the form *Aa Bb* × *aa bb*. When genes in a cross with two linked genes segregate, more than 50 percent of the gametes produced have parental combinations of the segregating alleles and less than 50 percent have nonparental (recombinant) combinations of the alleles. The recombination of linked genes occurs by crossing-over, a process in which nonsister chromatids of the homologous chromosomes exchange corresponding segments during the first meiotic prophase.

The frequencies of crossing-over between different genes can be used to determine the relative order and locations of the genes in chromosomes. This is called genetic mapping. Distance between adjacent genes in such a map (a genetic or linkage map) is defined to be proportional to the frequency of recombination between them; the unit of map distance (the map unit) is defined as 1 percent recombination. One map unit corresponds to a physical length of the chromosome in which a crossover event will occur, on the average, once in every fifty meioses. For short distances, map units are additive. (For example, for three genes with order *a b c*, if the map distances *a–b* and *b–c* are 2 and 3 map units, respectively, the map distance *a–c* is 2 + 3 = 5 map units.) The recombination frequency underestimates actual genetic distance if the region between the genes being considered is too great. This discrepancy results from the occurrence of multiple crossover events, which yield either no recombinants or the same number produced by a single event. For example, two crossovers in the region between two genes may yield no recombinants, and three crossover events may yield recombinants of the same type as that from a single crossover.

When many genes are mapped in a particular species, they are found in linkage groups equal in number to the haploid chromosome number of the species. The maximum frequency of recombination between any two genes in a cross is 50 percent; this frequency is obtained when the genes are in nonhomologous chromosomes and assort independently or when the genes are sufficiently far apart in the same chromosome that at least one crossover occurs between them in every meiosis.

The four haploid products of individual meiotic divisions can be used to analyze linkage and recombination in some species of fungi and unicellular algae. The method is called tetrad analysis. In *Neurospora* and related fungi, the meiotic tetrads are contained in a tubular sac, or ascus, in a linear order, making it possible to determine whether a pair of alleles segregated in the first or the second meiotic division. With such asci, it is possible to use the centromere as a genetic marker, and in fact the centromere serves as a reference point from which all genes in the same chromosome can be mapped. Linkage analysis in unordered tetrads uses the frequencies of parental ditype (PD), nonparental ditype (NPD), and tetratype (TT) tetrads. The occurrence of NPD < PD is a sensitive indicator of leakage.

Complementation tests tell whether two recessive mutations are in the same gene or in different ones. If two mutations, *a* and *b*, are present in a cell in the *trans* configuration and the phenotype is wildtype (complementation), the mutations are in different genes. If the phenotype is mutant (noncomplementation), the mutations are in the same gene. Collections of mutations with similar phenotypes can be assigned to complementation groups that identify the genes affecting the trait.

K E Y   T E R M S

ascospore
ascus
attached-X chromosome
chromosome map
cis-heterozygote
coefficient of coincidence
complementation
complementation group
compound-X chromosome
coupling

first-division segregation
frequency of recombination
genetic map
interference
linkage
linkage group
linkage map
locus
map unit
nonparental ditype

parental combinations
parental ditype
recombinant
recombination
repulsion
second-division segregation
tetratype
three-point cross
trans-heterozygote

E X A M P L E S   O F   W O R K E D   P R O B L E M S

Problem 1: The genes *cn* (cinnabar eyes) and *bw* (brown eyes) are located in the same chromosome of *Drosophila*, but they are so far apart that they are unlinked. Compare the possible offspring produced by the cross *cn bw* / + + female × *cn bw* / *cn bw* male and the cross *cn bw* / + + male × *cn bw* / *cn bw* female. The + symbols denote the wildtype alleles. Both *cn* and *bw* are recessive. Flies with the genotype *cn* / *cn* have cinnabar (bright red) eyes, *bw* / *bw* flies have brown eyes, and the double mutant *cn bw* / *cn bw* has white eyes.

Answer: The key to this problem is to remember that crossing-over does not occur in *Drosophila* males. When the male is *cn bw* / + +, his gametes are *cn bw* and + +, and so the progeny are *cn bw* / *cn bw* (white eyes) and

+ + / *cn bw* (wildtype eyes) in equal proportions. Crossing-over does occur in female *Drosophila*, but *cn* and *bw* are so far apart that they undergo independent assortment. When the female is *cn bw* / + +, her gametes are *cn bw*, *cn* +, + *bw*, and + +, and so the progeny are *cn bw* / *cn bw* (white eyes), *cn* + / *cn bw* (bright red), + *bw* / *cn bw* (brown), and + + / *cn bw* (wildtype) in equal proportions.

Problem 2: One of the earliest reported cases of linkage was in peas in the year 1905, between the genes for purple versus red flowers and for elongate versus round pollen, which are separated by about 12 map units. Purple flowers result from the genotypes *BB* and *Bb*, red flowers from *bb*; elongate pollen results from the genotypes *EE* and *Ee*, round pollen from *ee*. What genotypes and phenotypes among the progeny would be expected from crosses of F_1 hybrids × red, round when the F_1 hybrids are obtained as follows: **(a)** purple, elongate × red, round → purple, elongate F_1. **(b)** purple, round × red, elongate → purple, elongate F_1.

Answer: The red, round parents have genotype $b\,e\,/\,b\,e$. You must deduce the genotype of the F_1 parent. Although both F_1 hybrids have the dominant purple, elongate phenotype, they are different kinds of double heterozygotes. In mating **(a)**, both the b and the e alleles come from the red, round parent, and so the F_1 genotype is the *cis*, or coupling, heterozygote $B\,E\,/\,b\,e$. In mating **(b)**, the b allele comes from one parent and the e allele from the other, and so the F_1 genotype is the *trans*, or repulsion, heterozygote $B\,e\,/\,b\,E$.

To calculate the types of progeny expected, use the fact that 1 map unit corresponds to 1 percent recombination: because the genes are 12 map units apart, 12 percent of the gametes will be recombinant (6 percent of each of the reciprocal classes) and 88 percent will be nonrecombinant (44 percent of each of the reciprocal classes). Therefore, the two types of heterozygotes produce gametes with the following expected percentages:

Gametes	$B\,E\,/\,b\,e$	$B\,e\,/\,b\,E$
$B\,E$	44	6
$B\,e$	6	44
$b\,E$	6	44
$b\,e$	44	6

The answers to the specific questions are: **(a)** $B\,E\,/\,b\,e$ (purple, elongate) 44 percent; $B\,e\,/\,b\,e$ (purple, round) 6 percent; $b\,E\,/\,b\,e$ (red, elongate) 6 percent; $b\,b\,/\,b\,e$ (red, round) 44 percent. **(b)** $B\,E\,/\,b\,e$ (purple, elongate) 6 percent; $B\,e\,/\,b\,e$ (purple, round) 44 percent; $b\,E\,/\,b\,e$ (red, elongate) 44 percent; $b\,e\,/\,b\,e$ (red, round) 6 percent.

Problem 3: In *Drosophila*, the genes *ct* (cut wing), *y* (yellow body), and *v* (vermilion eye color) are all X-linked. Females heterozygous for all three markers were mated with wildtype males, and the following male progeny were obtained. As is conventional in *Drosophila* genetics, the wildtype allele of each gene is designated by a + sign in the appropriate column.

ct	y	v	4
ct	y	+	93
ct	+	v	54
ct	+	+	349
+	y	v	331
+	y	+	66
+	+	v	97
+	+	+	6
		Total	1000

(a) What genotype did the female parents have?
(b) What is the order of the genes?
(c) Calculate the recombination frequencies and the interference between the genes.
(d) Draw a genetic map of the genes.

Answer: This is a conventional kind of three-point linkage problem. The way to start is to deduce the genotype of the triply heterozygous parent. This is done by noting that *the most-common classes of offspring are the nonrecombinants*, in this case $ct + +$ and $+ y\,v$. Then deduce the order of the genes. This is done by noting that *the least-common classes of offspring are the double recombinants*, in this case $ct\,y\,v$ and $+ + +$. Moreover, *each class of double recombinants will be identical with one of the nonrecombinants except for the marker that is in the middle*. In this case, comparing $ct\,y\,v$ with $+ y\,v$ and $+ + +$ with $ct + +$ shows that ct is in the middle. Therefore, the answer to part **a** is that the mothers' genotype was $+ ct + /\,y + v$ and the answer to part **b** is that ct is in the middle. (It is not possible to determine from these data whether the gene order is $y\,ct\,v$ or $v\,ct\,y$.)

The next step is to rearrange the data with the genes in the correct order, with reciprocal chromosomes adjacent, and with nonrecombinants at the top, single recombinants for region 1 and 2 in the middle, and nonrecombinants at the bottom, as follows:

+	ct	+	349	Nonrecombinant
y	+	v	331	Nonrecombinant
y	ct	+	93	Region 1
+	+	v	97	Region 1
+	ct	v	54	Region 2
y	+	+	66	Region 2
y	ct	v	4	Region 1 + Region 2
+	+	+	6	Region 1 + Region 2

Now, to answer part **c**, calculate the recombination frequencies. The most-common mistake is to ignore the double recombinants. In fact, they are recombinant in both regions and so must be counted twice—once in calculating the recombination frequency in region 1 and again in calculating the recombination frequency in region 2. The frequency of recombination in region 1 (y–ct) is $(93 + 97 + 4 + 6)/1000 = 0.20$, or 20 percent. The frequency of recombination in region 2 (ct–v) is $(54 + 66 + 4 + 6)/1000 = 0.13$, or 13 percent. To calculate the interference, note that the expected number of double recombinants is $0.20 \times 0.13 \times 1000 = 26$, whereas the actual number is 10. The coincidence is the ratio $10/26 = 0.385$, and therefore the interference across the region is $1 - 0.385 = 0.615$. **(d)** The genetic map based on these data is shown below.

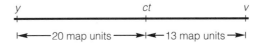

Problem 4: The gene *met14* in the yeast *Saccharomyces cerevisiae* codes for an enzyme used in the synthesis of the

amino acid methionine and is very close to the centromere of chromosome XI. Suppose that *met14* is used in mapping a new mutation, *new*, whose location in the genome is unknown. Diploids are made by mating *met14 NEW* × *MET14 new*, in which the uppercase symbols denote the wildtype alleles, and asci are examined. In this mating, what kinds of asci correspond to parental ditype (PD), nonparental ditype (NPD), and tetratype (TT)? What relative proportions of PD, NPD, and TT asci would be expected in the following cases: **(a)** *new* closely linked with *met14*? **(b)** *new* on a different chromosome from that of *met14* but close to the centromere? **(c)** *new* on a different chromosome from that of *met14* and far from the centromere?

Answer: The genotype of the diploid is the *trans*, or repulsion, heterozygote *met14 NEW / MET14 new*, and so the parental ditype (PD) tetrads contain two *met14 NEW* spores and two *MET14 new* spores; the nonparental ditype (NPD) tetrads contain two *met14 new* spores and two *MET14 NEW* spores; and the tetratype (TT) tetrads contain one spore each of *met14 NEW*, *MET14 new*, *met14 new*, and *MET14 NEW*.

The specific problems illustrate the principle that *any* centromere-linked marker in a fungus with unordered tetrads serves to identify first-division and second-division segregation of every other gene, which enables mapping other genes with respect to their centromeres. Reasoning through problems of tetrad analysis is best approached by considering the possible chromosome configurations during the meiotic divisions.

The possibilities are illustrated in the figure below, in which for simplicity A and a represent *MET14* and *met14*, respectively, and B and b represent *NEW* and *new*, respectively. Parts A through C correspond to questions **a** through **c**. Part A shows the genes closely linked to each other (and therefore to the same centromere). They are shown in different arms, but they could just as well be in the same arm. With no crossing-over between either gene and the centromere, all tetrads are PD. In part B, the genes are closely linked to different centromeres. With no crossing-over between either gene and its centromere, half the tetrads are PD and half NPD. In part C, the *new* (b) gene is not centromere linked. Again PD = NPD (indicating that A and B are in different chromosomes), but crossing-over between the gene and the centromere gives TT tetrads. *The frequency of TT tetrads increases with the recombination frequency between the gene and the centromere, to a maximum of 2/3 when the gene and centromere undergo free recombination.*

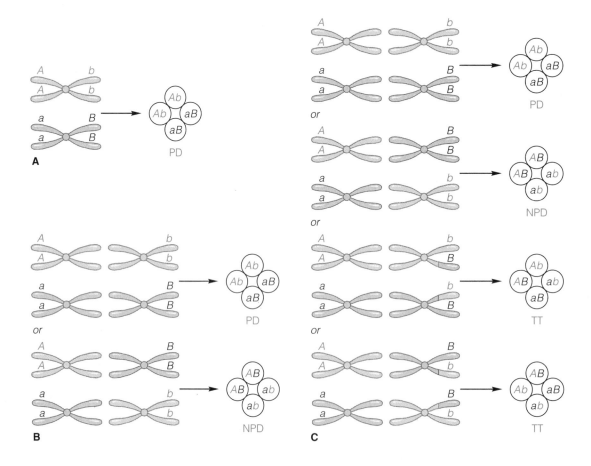

3-1. Each of two recessive mutations, m_1 and m_2, can produce albino guinea pigs. The genes are close together in one chromosome and may be allelic. The cross $m_1/m_1 \times m_2/m_2$ produces progeny with wildtype pigmentation. Are the mutations in the same gene?

3-2. What gametes are produced by an individual of genotype Ab/aB if the genes are (a) in different chromosomes or (b) in the same chromosome with no recombination between them?

3-3. What gametes are produced by an individual of genotype AB/ab if recombination can occur between the genes? Which gametes are the most frequent?

3-4. In the absence of recombination, what genotypes of progeny are expected from the cross $AB/ab \times Ab/aB$? If A and B are dominant, what ratio of phenotypes is expected?

3-5. Which of the following arrays represents the *cis* configuration of the alleles A and B: ab/AB or aB/Ab?

3-6. What gametes are produced by male and female *Drosophila* of genotype AB/ab when the genes are present in the same chromosome and the frequency of recombination between them is 5 percent?

3-7. If a large number of genes were mapped, how many linkage groups would be found in (a) a haploid organism with 17 chromosomes per somatic cell; (b) a bacterial cell with a single, circular DNA molecule, and (c) the rat, with 42 chromosomes per diploid somatic cell?

3-8. If the recombination frequency between genes A and B is 6.2 percent, what is the distance between the genes in map units in the linkage map?

3-9. Consider an organism heterozygous for two genes located in the same chromosome, AB/ab. If a single crossover occurs between the genes in every cell undergoing meiosis, and no multiple crossovers occur between the genes, what is the recombination frequency between the genes? What effect would multiple crossing-over have on the recombination frequency?

3-10. Normal plant height in wheat requires the presence of either or both of two dominant alleles, A or B. Plants homozygous for both recessive alleles (ab/ab) are dwarfed but otherwise normal. The two genes are in the same chromosome and recombine with a frequency of 16 percent. From the cross $Ab/aB \times AB/ab$, what is the expected frequency of dwarfed plants among the progeny?

3-11. In the yellow-fever mosquito *Aedes aegypti*, sex is determined by a single pair of alleles, M and m. Heterozygous Mm mosquitos are male, homozygous mm are female. The recessive gene *bz* for bronze body color ($bz^+/-$ is black) is linked to the sex-determining gene and recombines with it at a frequency of 6 percent. In the cross of a heterozygous $bz^+ M/bz\,m$ male with a bronze female, what genotypes and phenotypes of progeny are expected?

3-12. Two genes in chromosome 7 of corn are identified by the recessive alleles *gl* (glossy), determining glossy leaves, and *ra* (ramosa), determining branching of ears. When a plant heterozygous for each of these alleles was crossed with a homozygous recessive plant, the progeny consisted of the following genotypes with the numbers of each indicated:

$Gl\,ra/gl\,ra$ 88 $gl\,Ra/gl\,ra$ 103

$Gl\,Ra/gl\,ra$ 6 $gl\,ra/gl\,ra$ 3

Calculate the frequency of recombination between these genes.

3-13. The recessive alleles *b* and *cn* of two genes in chromosome 2 of *Drosophila* determine black body color and cinnabar eye color, respectively. If the genes are 8 map units apart, what are the expected genotypes and phenotypes and their relative frequencies among the progeny of the cross $+ +/b\,cn \times + +/b\,cn$? (Remember that crossing-over does not occur in *Drosophila* males.)

3-14. In a testcross of an individual heterozygous for each of three linked genes, the most-frequent classes of progeny were ABc/abc and abC/abc, and the least-frequent classes were ABC/abc and abc/abc. What was the genotype of the triple heterozygote parent and what is the order of the genes?

3-15. Construct a map of a chromosome from the following recombination frequencies between individual pairs of genes: $r-c$ 10, $c-p$ 12; $p-r$ 3; $s-c$ 16; $s-r$ 8. You will discover that the distances are not strictly additive. Why aren't they?

3-16. Yellow versus gray body color in *D. melanogaster* is determined by the alleles y and y^+, vermilion versus dark eyes by the alleles v and v^+, and singed versus straight bristles by the alleles sn and sn^+. When females heterozygous for each of these X-linked genes were testcrossed with yellow, vermilion, singed males, the following classes and numbers of progeny were obtained:

yellow, vermilion, singed	53
yellow, vermilion	108
yellow, singed	331
yellow	5
vermilion, singed	3
vermilion	342
singed	95
wildtype	63

(a) What is the order of the three genes? Construct a linkage map with the genes in their correct order and indicate the map distances between the genes.

(b) How does the frequency of double crossovers observed in this experiment compare with the frequency expected if crossing-over occurs independently in the two chromosome regions? Determine the coefficient of coincidence and the interference.

3-17. In corn, the alleles C and c result in colored versus colorless seeds, Wx and wx in nonwaxy versus waxy endosperm, and Sh and sh in plump versus shrunken endosperm. When plants grown freom seeds heterozygous for each of these pairs of alleles were testcrossed with plants from colorless, waxy, shrunken seeds, the progeny seeds were as follows:

colorless, nonwaxy, shrunken	84
colorless, nonwaxy, plump	974
colorless, waxy, shrunken	20
colorless, waxy, plump	2349
colored, waxy, shrunken	951
colored, waxy, plump	99
colored, nonwaxy, shrunken	2216
colored, nonwaxy, plump	15
	6708

Determine the order of the three genes and construct a linkage map showing the genetic distances between adjacent genes.

3-18. The following diagram summarizes the recombination frequencies observed in a large experiment to study three linked genes:

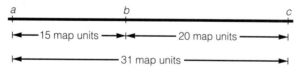

What was the observed frequency of double crossing-over in this experiment? Calculate the interference.

3-19. Three genes in chromosome 9 of corn determine shrunken sh versus plump Sh kernels, waxy wx versus nonwaxy Wx endosperm, and glossy gl versus non-

glossy Gl leaves. The genetic map of this chromosomal region is $sh-30-wx-10-gl$. From a plant of genotype $Sh\,wx\,Gl\,/\,sh\,Wx\,gl$, what is the expected frequency of $sh\,wx\,gl$ gametes: (a) in the absence of interference, and (b) assuming 60 percent interference?

3-20. The recessive mutations b (black body color), st (scarlet eye color), and hk (hooked bristles) identify three autosomal genes in D. melanogaster. The following progeny were obtained from a testcross of females heterozygous for all three genes:

black, scarlet	243
black	241
black, hook	15
black, hook, scarlet	10
hook	235
hook, scarlet	226
scarlet	12
wildtype	18

What conclusions are possible concerning the linkage relations of these three genes? Calculate any appropriate map distances.

3-21. In the nematode Caenorhabditis elegans, the mutations dpy-21 (dumpy) and unc-34 (uncoordinated) identify linked genes affecting body conformation and coordination of movement. The frequency of recombination between the genes is 24 percent. If the heterozygote dpy-21 + / + unc-34 undergoes self-fertilization (the normal mode of reproduction in this organism), what fraction of the progeny is expected to be both dumpy and uncoordinated?

3-22. Dark eye color in rats requires the presence of a dominant allele of each of two genes r and p. Animals homozygous for either or both of the recessive alleles have light-colored eyes. In one experiment, homozygous dark-eyed rats were crossed with doubly recessive light-eyed rats, and the resulting F_1 rats were testcrossed with rats of the homozygous light-eyed strain. The progeny from the testcross consisted of 628 dark-eyed and 889 light-eyed rats. In another experiment, $R\,p\,/\,R\,p$ animals were crossed with $r\,P\,/\,r\,P$ animals, and the F_1 rats were crossed with animals from an $r\,p\,/\,r\,p$ strain. The progeny consisted of 86 dark-eyed and 771 light-eyed rats. Determine whether the genes are linked and, if they are, estimate the frequency of recombination between the genes from each of the experiments.

3-23. A Drosophila geneticist exposes flies to a mutagenic chemical and obtains nine mutations in the X chromosome that are lethal when homozygous. The mutations are tested in pairs in complementation tests, with the results shown in the adjoining table. A + indicates complementation (that is, flies carrying both mutations survive), and a − indicates noncomplementation (flies carrying both mutations die). How many genes (complementation groups) are represented by the mutations, and which mutations belong to each complementation group?

	1	2	3	4	5	6	7	8	9
1	–	+	–	+	+	+	–	+	+
2		–	+	+	+	+	+	+	+
3			–	+	+	+	–	+	+
4				–	+	–	+	+	+
5					–	+	+	+	–
6						–	+	+	+
7							–	+	+
8								–	+
9									–

3-24. The following classes and frequencies of ordered tetrads were obtained from the cross $a^+b^+ \times a\,b$ in Neurospora. (Only one member of each pair of spores is shown.)

Spore pair				Number of asci
1–2	3–4	5–6	7–8	
$a^+ b^+$	$a^+ b^+$	$a\,b$	$a\,b$	1766
$a^+ b^+$	$a\,b$	$a^+ b^+$	$a\,b$	220
$a^+ b^+$	$a\,b^+$	$a^+ b$	$a\,b$	14

What is the order of the genes in relation to the centromere?

3-25. The following spore arrangements were obtained in the indicated frequencies from ordered tetrads in a cross between a Neurospora strain, com val, which exhibits a compact growth form and is unable to synthesize the amino acid valine, and a wildtype strain, + / +. (Only one member of each pair of spores is shown.)

Spore pair	Ascus composition									
1–2	c	v	c	+	c	v	+	v	c	v
3–4	c	v	c	+	c	+	c	+	+	v
5–6	+	+	+	v	+	v	c	v	c	+
7–8	+	+	+	v	+	+	+	+	+	+
Number:	34		36		20		1		9	

What can you conclude about the linkage and location of the genes with respect to each other and the centromere?

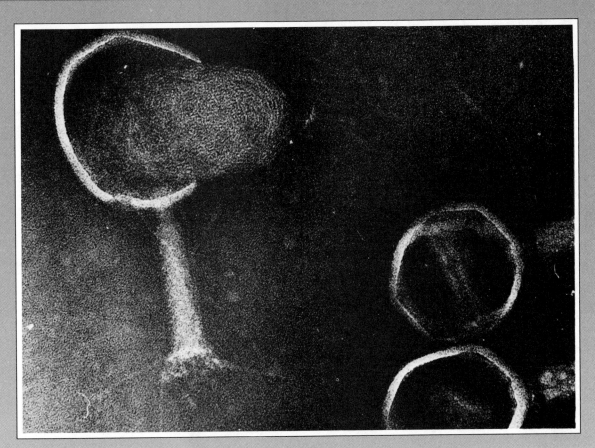

The Chemical Nature
and Replication
of the Genetic Material

Analysis of the phenotypic expression and patterns of inheritance of genes reveals nothing about gene structure, how genes are copied to yield exact replicas, or how they determine cellular characteristics. Understanding these basic features of heredity requires identification of the chemical nature of the

Above: An electron micrograph of *E. coli* phage T4 with a broken head, releasing its single DNA molecule as a tightly tangled mass. The two particles at the right are phages lacking DNA. (Courtesy of Robley Williams.)

genetic material and the processes involved in its replication. These topics are the subjects of this chapter.

4-1 The Importance of Bacteria and Viruses in Genetics

Most cells are organized in either of two fundamentally different ways (Figure 4-1). In bacteria and blue-green algae, which constitute the group of unicellular organisms called **prokaryotes**, the genetic material is located in a region that

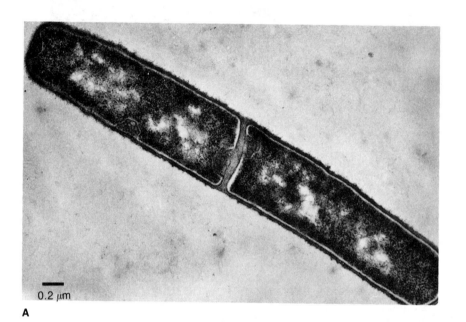

0.2 μm

A

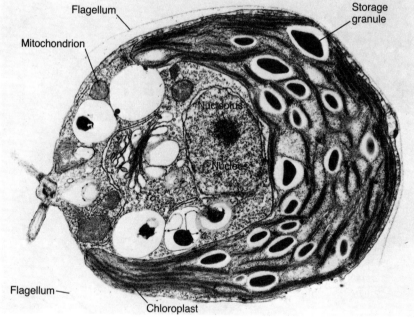

Flagellum

Mitochondrion

Storage granule

Nucleolus

Nucleus

Flagellum

Chloroplast

B

FIGURE 4-1   The organization of a prokaryotic cell and a eukaryotic cell. A. An electron micrograph of a dividing bacterial cell (a prokaryote), showing the dispersed genetic material (light areas). B. An electron micrograph of a section through a cell of the alga *Tetraspora* (a eukaryote), showing the membrane-bound nucleus, the nucleolus (dark central body), and other membrane systems (chloroplasts and mitochondria) that subdivide the cytoplasm into regions of specialized function. The dark sharply bounded regions are starch storage granules, which store carbohydrate. Note the complexity of this structure compared with the simpler organization of the bacterial cell shown in part A. (Courtesy of (A) A. Benichou-Ryter and (B) Jeremy Pickett-Heaps.)

lacks clear boundaries and is called the **nucleoid.** In most other single-celled organisms and in all cells of multicellular organisms, the genetic material is enclosed in the nucleus by a membrane envelope that separates it from the cytoplasm. Organisms whose cells have nuclei are called **eukaryotes.** In eukaryotic cells, other membrane systems subdivide the cytoplasm into regions of specialized function. Meiosis and nuclear fusion in the course of fertilization also are properties of eukaryotes.

Viruses are small particles, considerably smaller than cells, able to infect susceptible cells and multiply within them to form large numbers of progeny virus particles. Few, if any, organisms are not subject to viral infection. Many human diseases are caused by different viruses—for example, influenza, measles, acquired immune deficiency syndrome (AIDS), and the common cold. Viruses that infect bacteria are called **bacteriophages,** or simply **phages.** Most viruses consist of a single molecule of genetic material enclosed in a protective coat composed of one or more kinds of protein molecules; however, their size, molecular constituents, and structural complexity vary greatly (Figure 4-2). Isolated viruses possess no metabolic systems, and so a virus can multiply only within a cell; it is "living" only in the sense that its genetic material directs its own multiplication; outside its host cell a virus is an inert particle.

Bacteria and viruses bring to traditional types of genetic experiments four important advantages over multicellular plants and animals. First, they are haploid, and so dominance or recessiveness of alleles is not a complication in identifying genotype. Second, a new generation is produced in minutes rather

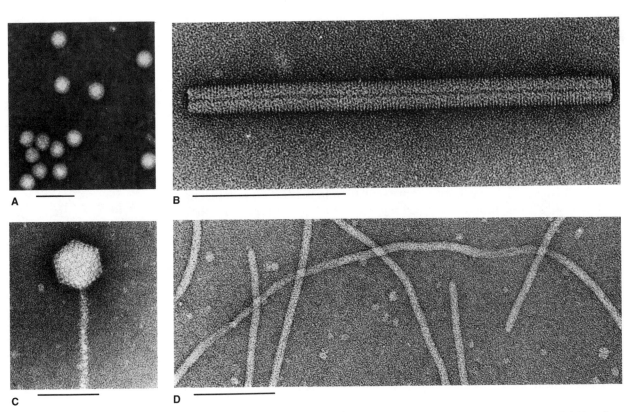

FIGURE 4-2 Electron micrographs of four different viruses: (A) poliovirus; (B) *E. coli* phage λ; (C) *E. coli* phage M13; (D) tobacco mosaic virus. The length of the bar is 1000 Å in each case. (One Å unit equals 10^{-8} cm.) (Courtesy of Robley Williams.)

The Importance of Bacteria and Viruses in Genetics 87

than weeks or months, which vastly increases the rate of accumulation of data. Third, they are easy to grow in enormous numbers under controlled laboratory conditions, which facilitates biochemical studies and the analysis of rare genetic events. Fourth, the individual members of these large populations are genetically identical; that is, each laboratory population is a **clone.** Later chapters will show the extent to which work with these organisms has contributed to genetics.

4-2 Evidence That the Genetic Material Is DNA

A weakly acidic substance of unknown function, later called **deoxyribonucleic acid (DNA),** was discovered by Johann Friedrich Miescher in 1869 in the nuclei of human white blood cells. The substance, called **nucleic acid,** could be isolated from the nuclei of many cell types. Chemical analysis done in 1910 showed that there were two classes of nucleic acids—DNA and ribonucleic acid (RNA). In 1924, microscopic studies using stains for DNA and protein showed that both substances are present in chromosomes. Indirect evidence suggested a close relation between DNA and the genetic material. For example, almost all somatic cells of a given species contain a constant amount of DNA, whereas the RNA content and the amount and kinds of proteins differ greatly in different cell types. Furthermore, nuclei resulting from meiosis in both plants and animals have only half the DNA content of nuclei in their somatic cells. However, DNA was not considered to be identical with the genetic material, mainly because crude chemical analyses had suggested that DNA lacks the chemical diversity needed for a genetic substance. In contrast, proteins are an exceedingly diverse collection of molecules; so it was widely believed that proteins were the genetic material and that DNA was merely the structural framework of chromosomes. The identity between DNA and the genetic material was shown directly in two experiments that are described in this section.

Bacterial pneumonia in mammals is caused by strains of *Streptococcus pneumoniae* able to synthesize a polysaccharide (complex carbohydrate) capsule around itself; this capsule protects the bacterium from the defense mechanisms of the infected animal and enables the bacterium to cause disease. When a bacterial cell is grown on solid medium, by repeated division it forms a visible clone called a **colony.** The enveloping capsule gives the colony a glistening or smooth (S) appearance. Some strains have mutations in the enzyme required to synthesize the capsular polysaccharide, and these bacteria form colonies that have a rough (R) surface. R strains do not cause pneumonia, because without their capsules the bacteria are inactivated by the immune system of the host.

Both R and S strains of the bacteria breed true, except for rare mutations in R strains that give rise to the S phenotype and rare mutations in S strains that give rise to the R phenotype. Because simple mutations can cause conversion between the forms, the S and R phenotypes are likely to be determined by different allelic forms of a simple genetic unit.

When mice are injected with either living R cells or with heat-killed S cells, they remain healthy. However, in 1928 it was discovered that mice injected with a mixture containing a small number of R bacteria and a large number of heat-killed S cells often died of pneumonia. Bacteria isolated from blood samples of the dead mice produced *pure* S cultures having a capsule typical of the heat-killed S cells. The purity is significant because it showed that the dead S cells could in some way restore to a living R bacterial cell the ability to resist the immunological system of the mouse, multiply, and cause pneumonia. Furthermore, the R bacteria were changed, or **transformed**, into S bacteria, and the new characteristics were inherited by descendants of the transformed bacteria.

In 1944, in a milestone experiment, Oswald Avery, Colin MacLeod, and Maclyn McCarty added DNA from S cells to growing cultures of R cells and observed the production of a few type-S cells. The DNA preparations contained traces of protein, but the transforming activity was not altered by treatment with enzymes that degrade proteins or by treatment with an enzyme that degrades RNA. However, transforming activity was completely destroyed by treatment with deoxyribonuclease, an enzyme that degrades DNA (Figure 4-3). Such experiments implied that the substance responsible for genetic transformation was the DNA of the cell and, hence, that DNA is the genetic material.

A second pivotal experiment was reported by Alfred Hershey and Martha Chase in 1952. In their experiments, reproduction of bacteriophage T2 in the bacterium *Escherichia coli* was studied. Infection by T2 (Figure 4-4) had earlier been shown to proceed by sequential attachment of a phage particle by the tip of its tail to the bacterial cell wall, entry of phage material into the cell, multiplication of this material to form a hundred or more progeny phage, and release of progeny by disruption of the host cell (see Chapter 10). T2 particles were also known to be composed of DNA and protein in approximately equal amounts.

Because DNA contains phosphorus but no sulfur, and proteins generally contain some sulfur but no phosphorus, it is possible to label DNA and proteins differentially using radioactive isotopes of the two elements. Hershey and Chase produced particles with radioactive DNA by infecting *E. coli* cells that had been grown for several generations in a medium containing ^{32}P (a radioactive isotope of phosphorus) and then collecting the phage progeny. Other particles with

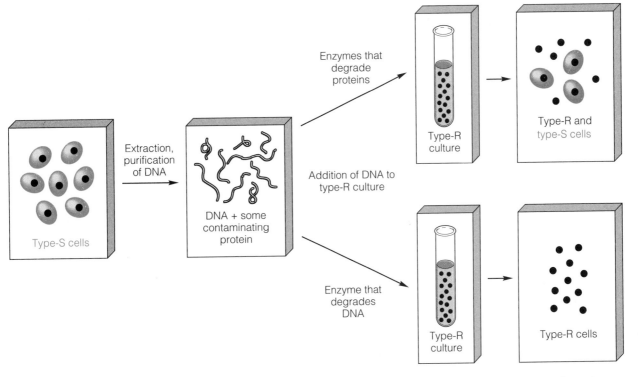

FIGURE 4-3   A diagram of the experiment that demonstrated that DNA is the active material in bacterial transformation.

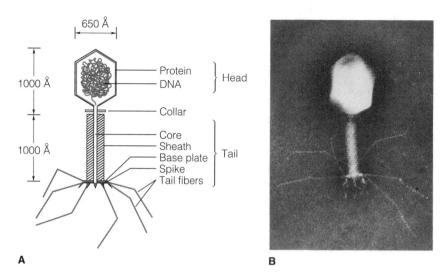

650 Å

1000 Å

Protein ⎤
DNA ⎦ Head

Collar

Core ⎤
Sheath ⎥
Base plate ⎥ Tail
Spike ⎥
Tail fibers ⎦

1000 Å

A

B

FIGURE 4-4   A. Drawing of *E. coli* phage T2, showing various components. B. An electron micrograph of phage T4, a closely related phage. (Courtesy of Robley Williams.)

labeled proteins were obtained in the same way, using medium containing ^{35}S (a radioactive isotope of sulfur).

In the experiment summarized in Figure 4-5, nonradioactive *E. coli* cells were infected with phage labeled with *either* ^{35}S or ^{32}P in order to follow the proteins and DNA during infection. Infected cells were separated from unattached particles by centrifugation, resuspended in fresh medium, and then agitated in a kitchen blender, which tore attached phage material from the cell surfaces. This treatment was found to have no effect on the subsequent course of the infection in that the number of progeny phage was the same with and without the blending step. Thus, the genetic material rapidly entered the infected cells. When intact bacteria were separated from the material removed during the blending by a second centrifugation, most of the radioactivity from ^{32}P-labeled phage was found to be associated with the bacteria; however, when the infecting phage was labeled with ^{35}S, only a tiny fraction of the radioactivity was associated with the bacterial cells. From these results, it appears that a T2 phage transfers most of its DNA, but very little of its protein, to the cell that it infects. A critical finding was that about 50 percent of the transferred ^{32}P-labeled DNA, but less than 1 percent of the transferred ^{35}S-labeled protein, was inherited by the *progeny* phage particles, which indicated that the chemical identity of the genetic material in T2 phage is DNA.

These two experiments—the transformation experiment and the "blender" experiment—were widely accepted as proof that DNA is the genetic material in all organisms. In later years, a few exceptions were found in certain viruses that lack DNA and utilize RNA as their genetic material. However, DNA *is exclusively the genetic material in cells.*

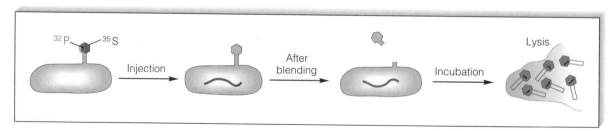

^{32}P ^{35}S

Injection

After blending

Incubation

Lysis

FIGURE 4-5   A diagrammatic summary of the Hershey-Chase ("blender") experiment, which demonstrated that DNA and not protein is responsible for directing the reproduction of phage T2 in infected *E. coli* cells.

DNA is a polymer (a large molecule containing repeating units) composed of a five-carbon sugar (2'-deoxyribose), phosphoric acid, and four nitrogen-containing **bases.** Two of these nitrogenous bases are **purines,** which have a double-ring structure; the other two are **pyrimidines,** which contain a single ring.

1. The purine bases are **adenine (A)** and **guanine (G)** (Figure 4-6).
2. The pyrimidine bases are **thymine (T)** and **cytosine (C)** (Figure 4-6).

In DNA, each base is chemically linked to a deoxyribose, forming a compound called a **nucleoside.** A phosphate group also is attached to the sugar of each nucleoside, yielding a **nucleotide** (Figure 4-7); the terminology is that a nucleotide is a nucleoside phosphate. In the conventional numbering of the carbon atoms in the sugar in Figure 4-7, the carbon atom to which the base is attached is the 1' carbon. (The atoms in the sugar are given primed numbers to distinguish them from atoms in the bases.) The nomenclature of the nucleoside and nucleotide derivatives of the DNA bases is somewhat complicated and is summarized in Table 4-1. Most of these terms are not needed in this book; they are included because they may be encountered in further reading.

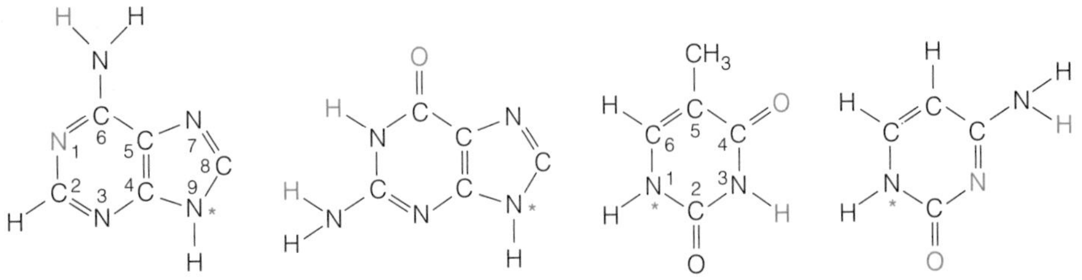

Adenine   **Guanine**   **Thymine**   **Cytosine**

FIGURE 4-6   The four nitrogen-containing bases in DNA. The N linked to deoxyribose is denoted by a red asterisk. The red atoms are engaged in hydrogen bonds in the DNA base pairs, as explained in Section 4-4.

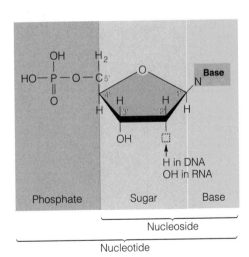

FIGURE 4-7   A typical nucleotide showing the three major components (phosphate, sugar, and base), the difference between DNA and RNA, and the distinction between a nucleoside (no phosphate group) and a nucleotide (with phosphate). Nucleotides may contain one phosphate unit (monophosphate), two (diphosphate), or three (triphosphate).

TABLE 4-1   DNA nomenclature

Base	Nucleoside	Nucleotide
Adenine (A)	Deoxyadenosine	Deoxyadenosine-5'- monophosphate (dAMP) diphosphate (dADP) triphosphate (dATP)
Guanine (G)	Deoxyguanosine	Deoxyguanosine-5'- monophosphate (dGMP) diphosphate (dGDP) triphosphate (dGTP)
Thymine (T)	Deoxythymidine	Deoxythymidine-5'- monophosphate (dTMP) diphosphate (dTDP) triphosphate (dTTP)
Cytosine (C)	Deoxycytidine	Deoxycytidine-5'- monophosphate (dCMP) diphosphate (dCDP) triphosphate (dCTP)

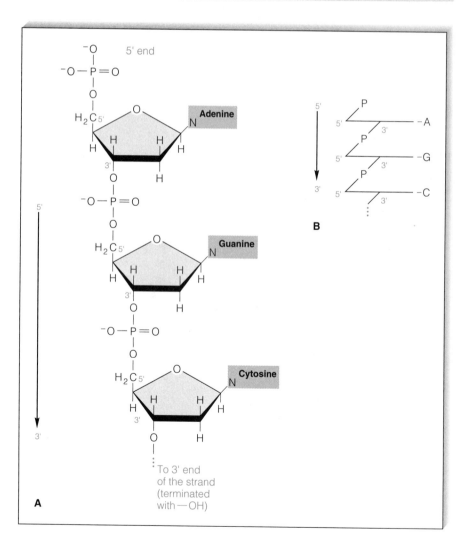

FIGURE 4-8   Three nucleotides at the 5' end of a single polynucleotide strand: (A) the chemical structure of the sugar-phosphate linkages showing the 5'-to-3' orientation of the strand (the red numbers are the carbon-atom numbers); (B) a common schematic way to depict a polynucleotide strand.

In nucleic acids, the nucleotides are joined to form a **polynucleotide chain,** in which the phosphate attached to the 5′ carbon of one sugar is linked to the hydroxyl attached to the 3′ carbon of the next sugar (Figure 4-8). The chemical bonds by which the sugar components of adjacent nucleotides are linked through the phosphate groups are called **phosphodiester bonds.** The 5′-3′-5′-3′ orientation of these linkages continues throughout the chain, which typically consists of millions of nuclotides. Note that the terminal groups of each polynucleotide chain are a 5′-phosphate (**5′-P**) group at one end and a 3′-hydroxyl (**3′-OH**) group at the other.

The molar concentrations (denoted by square brackets [ ]) of the bases in DNA exhibit two important features:

1. The concentration of purine bases equals that of the pyrimidine bases; that is,

[total purines] = [A] + [G] = [total pyrimidines] = [T] + [C].

2. The concentrations of adenine and thymine are equal, as are the concentrations of guanine and cytosine; that is,

[A] = [T]   and   [G] = [C]

The **base composition,**

$$([G] + [C])/([G] + [C] + [A] + [T])$$

sometimes called the **percent GC,** varies among species but is constant in all cells of an organism and within a species. Data on the base composition of DNA from a variety of organisms are given in Table 4-2.

TABLE 4-2   Base composition of DNA from different organisms

Organism	Base (and percentage of total bases)				Base composition (percent GC)
	Adenine	Thymine	Guanine	Cytosine	
Bacteriophage T7	26.0	26.0	24.0	24.0	48.0
Bacteria					
Clostridium perfringens	36.9	36.3	14.0	12.8	26.8
Streptococcus pneumoniae	30.3	29.5	21.6	18.7	40.3
Escherichia coli	24.7	23.6	26.0	25.7	51.7
Sarcina lutea	13.4	12.4	37.1	37.1	74.5
Fungi					
Saccharomyces cerevisiae	31.7	32.6	18.3	17.4	35.7
Neurospora crassa	23.0	23.3	27.1	27.6	54.7
Higher plants					
Wheat	27.3	27.1	22.7	22.8*	45.5
Maize	26.8	27.2	22.8	23.2*	46.0
Animals					
Drosophila melanogaster	30.7	29.4	19.6	20.2	39.8
Pig	29.4	29.7	20.5	20.5	41.0
Salmon	29.7	29.1	20.8	20.4	41.2
Human being	29.8	31.8	20.2	18.2	39.0

*Includes one-fourth 5-methylcytosine, a modified form of cytosine found in most plants more complex than algae and in many animals.

4-4 Physical Structure of DNA: The Double Helix

An essentially correct three-dimensional structure of the DNA molecule was proposed in 1953 by James Watson and Francis Crick. DNA consists of two polynucleotide chains twisted around one another forming a double-stranded helix in which adenine and thymine, and guanine and cytosine, are paired (Figure 4-9). Each chain makes one complete turn every 34 Å. The helix is right-handed, meaning that, as you look down the barrel, each chain follows a clockwise path as it progresses. The bases are spaced at 3.4 Å, and so there are ten bases per helical turn in each strand, or ten base pairs per turn of the double helix. Each base is paired to a base in the other strand by hydrogen bonds, which are the main force holding the strands together. (A hydrogen bond is a weak bond in which two negatively charged atoms share a hydrogen atom between them.) The paired bases are planar, parallel to one another, and perpendicular to the long axis of the double helix. When discussing a DNA molecule, one frequently refers to the individual strands as single strands or single-stranded DNA, and to the double helix as double-stranded DNA or a **duplex** molecule.

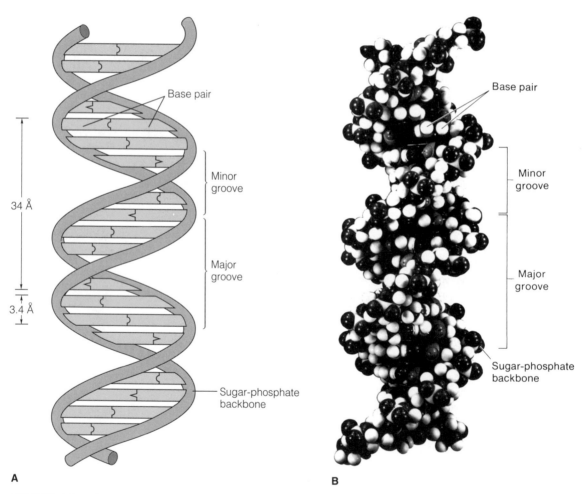

FIGURE 4-9 The DNA molecule: (A) a diagrammatic model of the double helix; (B) space-filling model of the DNA double helix. The sugar-phosphate backbone winds around the outside of the paired strands, and the base pairs bridge between the strands.

A central feature of DNA structure is the pairing between specific bases:

The purine adenine pairs with the pyrimidine thymine (forming an AT pair) and the purine guanine pairs with the pyrimidine cytosine (forming a GC pair).

The adenine-thymine base pair and the guanine-cytosine pair are illustrated in Figure 4-10. Note that an AT pair has two hydrogen bonds and a GC pair has three hydrogen bonds. This means that bonding between G and C is stronger in the sense that it requires more energy to break.

The two polynucleotide strands of the double helix are oriented in opposite directions in the sense that the bases that are paired are attached to sugars lying above and below the plane of pairing, respectively. Recall that the backbone of a chain (Figure 4-9) consists of deoxyribose molecules alternating with phosphate groups that link the 3' carbon atom of one sugar to the 5' carbon of the next in line. The two strands of a double helix are **antiparallel** with respect to this linkage: that is, if the strands are followed from the 5' end to the 3' end (the 5' → 3' direction), the structure . . .-P-5'-sugar-3'-P-. . . lies in one direction in one strand and in the opposite direction in the other strand. Note that this means that each terminus of the double helix possesses one 5'-P group (on one strand) and one 3'-OH group (on the other strand).

The specificity of base-pairing means that each base along one polynucleotide strand of the DNA is matched with the base in the opposite position on the other strand; the sequences of bases along the two strands are said to be **complementary**. Nothing restricts the sequence of bases in a single strand, and so any sequence could be present along one strand.

The principal features of the double helix are summarized diagrammatically in Figure 4-11. The diagram is static and so somewhat misleading. DNA is actually a very dynamic molecule, constantly in motion. In some regions, the strands can separate briefly and then come together again in the same or a different conformation. Although the right-handed double helix in Figure 4-9

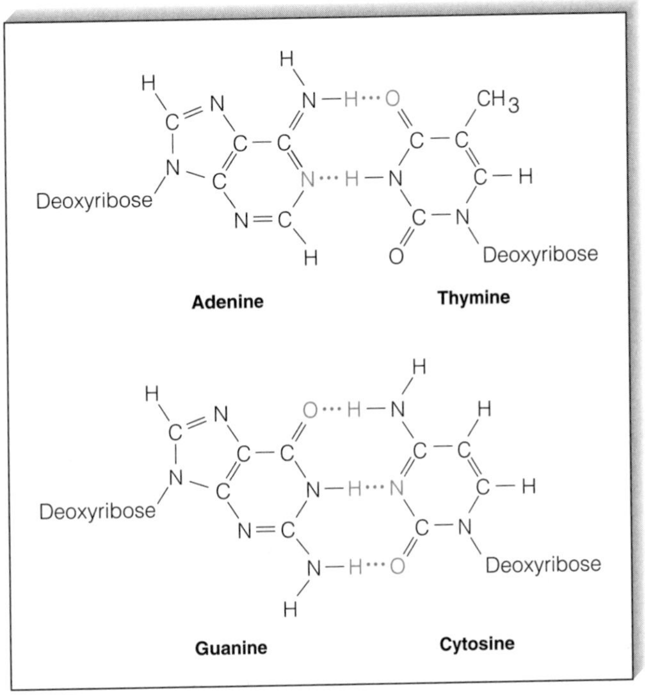

FIGURE 4-10 Normal base pairs in DNA. Hydrogen bonds (dotted lines) and the joined atoms are shown in red.

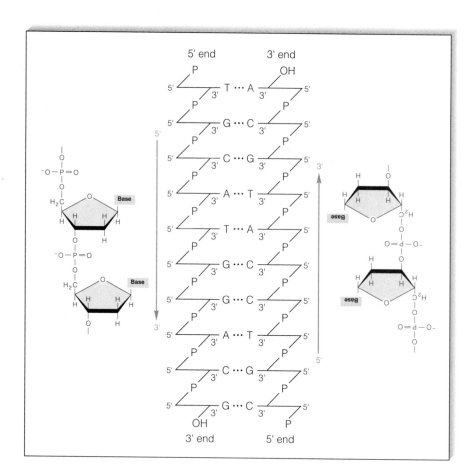

FIGURE 4-11 A segment of a DNA molecule showing the antiparallel orientation of the complementary strands. The red arrows indicate the 5'-to-3' direction of each strand. The phosphates (P) join the 3' carbon atom of one deoxyribose (horizontal line) to the 5' carbon atom of the adjacent deoxyribose. The orientations of the chemical structures are shown on either side of the sugar-phosphate chain.

is the standard form, DNA can form more than twenty slightly different variants of right-handed helices, and some regions can even form helices in which the strands twist to the left. If there are complementary stretches of nucleotides in the same strand, a single strand, separated from its partner, can fold back upon itself like a hairpin. Even triple helices formed of three strands can form in regions of DNA containing suitable base sequences.

4-5   What a Genetic Material Needs That DNA Supplies

Not every polymer would be useful as genetic material. However, DNA is admirably suited to a genetic function because it satisfies the following three essential requirements:

1. *A genetic material must carry all of the information needed to direct the specific organization and metabolic activities of the cell.* The product of most genes is a protein molecule—a polymer composed of hundreds of different kinds (twenty in all) of molecular units called amino acids (Chapter 11). The sequence of amino acids in the protein determines its chemical and physical properties. A gene is expressed when its protein product is synthesized, and one requirement of the genetic material must be to direct the order in which amino acid units are added to the end of a growing protein molecule. In DNA, this is done

FIGURE 4-12 Replication of
DNA according to the
mechanism proposed by
Watson and Crick. The two
double-stranded replicas consist
of one parental strand (black)
and one daughter strand (red).

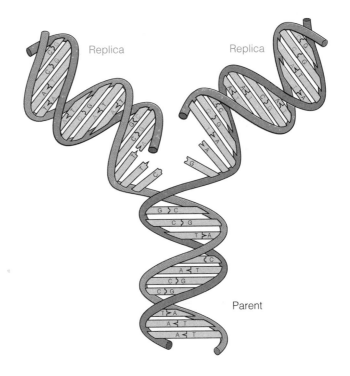

by means of a genetic code in which groups of three bases specify amino acids. Because the four bases in a DNA molecule can be arranged in any sequence, and because the sequence can vary from one part of the molecule to another and from organism to organism, DNA can contain a great many unique regions, each of which can be a distinct gene. Thus, synthesis of a variety of different protein molecules can be directed by a long DNA chain.

2. *The genetic material must replicate accurately, so that the information it contains is precisely inherited by daughter cells.* The basis for exact duplication of a DNA molecule is the complementarity of the AT and GC pairs in the two polynucleotide chains. Unwinding and separation of the chains, with each free chain being copied, results in the formation of two identical double helices (Figure 4-12).

3. *The genetic material must be capable of undergoing occasional mutation, such that the information it carries is altered in a heritable way.* Watson and Crick suggested that this would be possible in DNA by rare rearrangements of atoms in the bases. For example, adenine and cytosine do not usually pair, but a shift in the hydrogen atom at the 6-amino position in adenine to the N-1 position permits hydrogen-bonding between these bases (Figure 4-13). If this rare pairing occurred during replication of the DNA, one of the polynucleotide strands would have cytosine instead of thymine at that point in its base sequence. The normal pairing of this cytosine with guanine in the next round of replication would result in replacement of the original AT base pair by a GC pair, and the information in the molecule would be altered by a change in one letter of the genetic message. Although such reversible changes in location of the hydrogen atoms in bases are a theoretical possibility for mutations, and do occur occasionally, other mechanisms are far more important and are discussed in Chapter 12.

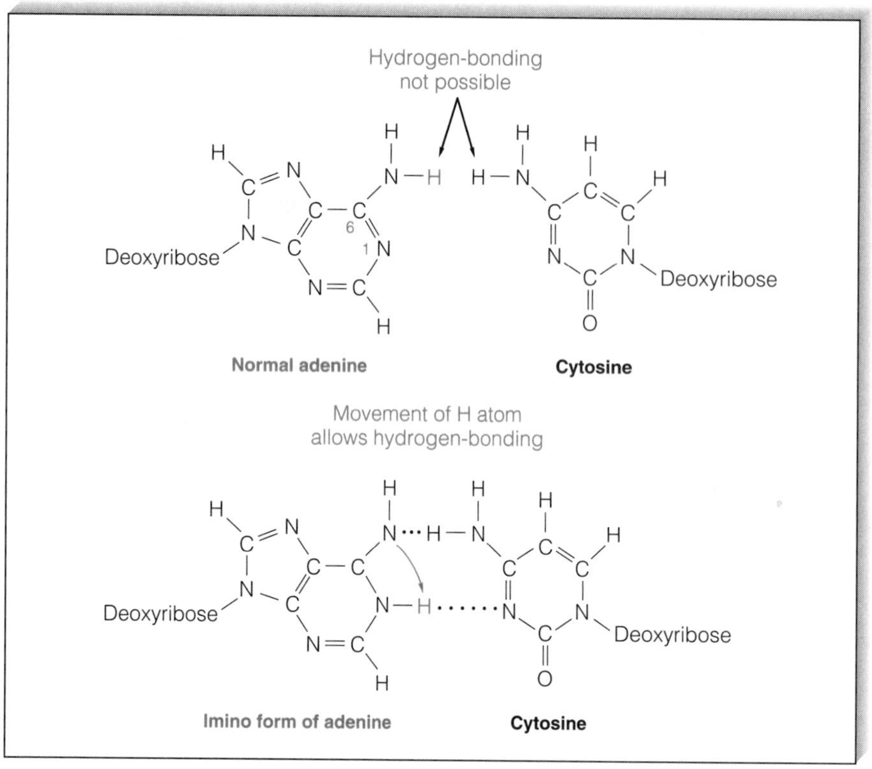

FIGURE 4-13 Formation of an adenine-cytosine pair when a hydrogen atom in adenine moves from the 6-amino position to the 1-nitrogen position.

4-6 The Replication of DNA

The process of replication, in which each strand of the double helix serves as a template for synthesis of a new strand (see Figure 4-12), is simple in principle. It requires only that the hydrogen bonds joining the bases break to allow separation of the chains and that appropriate free nucleotides of the four types pair with the newly accessible bases in each strand. However, it is a complex process with geometric problems requiring a variety of enzymes and other proteins. These processes will be examined in the following sections.

The Basic Rule for Replication of Nucleic Acids

The primary function of any mode of replication is to duplicate the base sequence of the parent molecule. The specificity of base-pairing—adenine with thymine (replaced by uracil in RNA) and guanine with cytosine—provides the mechanism used by all replication systems. Furthermore,

1. Nucleotide monomers are added one by one to the end of a growing strand by an enzyme called a *DNA polymerase.*
2. The sequence of bases in each new strand, or **daughter strand,** is complementary to the base sequence in the old strand, or **parent strand,** being copied—that is, if there is an adenine in the parent strand, a thymine nucleotide will be added to the end of the growing daughter strand when the adenine is being copied.

The following section explains how the two strands of a daughter molecule are physically related to the two strands of the parent molecule.

The Geometry of DNA Replication

The production of daughter DNA molecules from a single parental molecule gives rise to several topological problems, which result from the helical structure and great length of typical DNA molecules and the circularity of many DNA molecules. These problems and their solutions are described in this section.

Semiconservative replication of double-stranded DNA   In the **semiconservative** mode of replication, each parental DNA strand serves as a template for one new, or daughter, strand and, as each new strand is formed, it is hydrogen-bonded to its parental template (see Figure 4-12). Thus, as replication proceeds, the parental double helix unwinds and then rewinds again into two new double helices, each of which contains one originally parental strand and one newly formed daughter strand.

Semiconservative replication was demonstrated in 1958 by Matthew Meselson and Franklin Stahl by an experiment in which the heavy ^{15}N isotope of nitrogen was used for the physical separation of parental and daughter DNA molecules. DNA isolated from the bacterium *E. coli* grown in a medium containing ^{15}N as the only available source of nitrogen is denser than DNA from bacteria grown in media with the normal ^{14}N isotope. ^{15}N- and ^{14}N-containing DNA can be separated by means of **equilibrium centrifugation in a density gradient**. DNA molecules have about the same density as very concentrated cesium chloride (CsCl) solutions; for example, 5.6 molar CsCl has a density of 1.700 g/cm^3, and *E. coli* DNA containing ^{14}N in the purine and pyrimidine rings has a density of 1.708 g/cm^3. Total replacement of the ^{14}N with ^{15}N increases the density of *E. coli* DNA to 1.722 g/cm^3.

When a CsCl solution containing DNA is centrifuged at high speed, the Cs^+ ions gradually sediment toward the bottom of the centrifuge tube. This movement is counteracted by diffusion (the random movement of molecules), which prevents complete sedimentation. At equilibrium, a linear gradient of increasing CsCl concentration—and, thus, of density—is present from the top to the bottom of the centrifuge tube. The DNA also moves—upward or downward—to a position in the gradient at which the density of the solution is precisely equal to its own density. At equilibrium, a mixture of ^{14}N-containing ("light") and ^{15}N-containing ("heavy") *E. coli* DNA will separate into two distinct zones in a density gradient.

In the Meselson-Stahl experiment summarized in Figure 4-14, bacteria were grown for many generations in a ^{15}N-containing medium. At the beginning of the experiment, essentially all DNA was uniformly labeled with ^{15}N and therefore had a heavy density. The cells were then transferred to a ^{14}N-containing medium, and DNA was isolated from samples of cells taken from the culture at intervals after this density transfer. After one generation of growth (one round of replication of the DNA molecules and a doubling of the number of cells), all of the DNA had a density exactly intermediate between the densities of [^{15}N]DNA and [^{14}N]DNA, indicating that the replicated molecules contained equal amounts of the two nitrogen isotopes. After a second generation of growth in the ^{14}N medium, half of the DNA had the density of DNA with ^{14}N in both strands ("light" DNA) and the other half had the intermediate ("hybrid") density. This distribution of ^{15}N atoms is precisely the result predicted from semiconservative replication of the Watson-Crick structure. Subsequent experiments with replicating DNA from numerous viruses, bacteria, and higher organisms have indicated that semiconservative replication is universal.

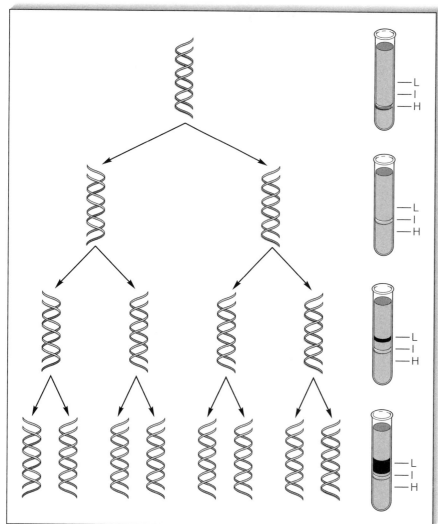

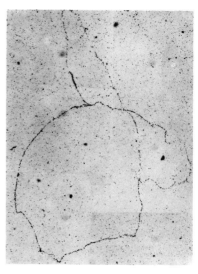

FIGURE 4-15 Autoradiogram of the intact replicating chromosome of an *E. coli* cell that has grown in a medium containing [³H]thymine for slightly less than two generations. The continuous lines of dark grains were produced by electrons emitted by decaying ³H atoms in the DNA molecule. The pattern is seen by light microscopy. (From J. Cairns, *Cold Spring Harbor Symp. Quant. Biol.* 1963. 28:44.)

FIGURE 4-14 The DNA replication experiment of Meselson and Stahl. The experiment begins with cells grown in ¹⁵N-containing medium. The state of DNA molecules after various periods of growth in ¹⁴N-containing medium is shown by the wavy lines. The density of the DNA is determined by mixing it with a CsCl solution having a density approximately equal to that of the DNA molecules. The DNA-CsCl solution is centrifuged. During this time, the Cs⁺ and Cl⁻ ions are distributed, and at equilibrium a stable density gradient is formed. The DNA molecules also move during centrifugation and come to rest at positions in the centrifuge tube at which their density equals the density of the CsCl. Diagrams of centifuge tubes are shown at the right, with DNA molecules of different densities forming bands at different positions. The letters L, I, and H stand for light, intermediate, and heavy density, respectively.

    Initially all molecules have a heavy density (two red strands). After one round of replication, the molecules are intermediate in density (one red and one black strand). After two rounds of replication, half the molecules have intermediate density (one red and one black strand) and half have a light density (two black strands). After an additional round of replication, three-quarters of the molecules are light and one-quarter are intermediate, indicated by the difference in the width of the bands in the centrifuge tube.

In the Meselson-Stahl experiment, *E. coli* DNA was extensively fragmented when isolated, and so the form of the molecule was unknown. Later, isolation of unbroken molecules and examination of them by two techniques—autoradiography and electron microscopy—showed that the DNA in *E. coli* cells is circular.

FIGURE 4-16 Electron micrograph of a small circular DNA molecule replicating by the θ mode. The parental and daughter segments are shown in the drawing. (Courtesy of Donald Helinski.)

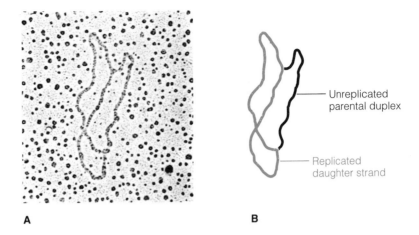

Unreplicated parental duplex

Replicated daughter strand

A   **B**

Geometry of the replication of circular DNA molecules   The first proof that *E. coli* DNA replicates as a circle came from an autoradiographic experiment. (Genetic-mapping experiments to be described in Chapter 10 already had suggested that the bacterial chromosome is circular.) Cells were grown in a medium containing [³H]thymine so that all DNA synthesized would be radioactive. The DNA was isolated without fragmentation and placed on photographic film. Each ³H-decay exposed one grain in the film, and after several months there were enough grains to visualize the DNA with a microscope; the pattern of black grains on the film located the molecule. One of the well-known autoradiograms from this experiment is shown in Figure 4-15. On a smaller scale, Figure 4-16 shows a replicating circular molecule of a plasmid, which is found inside certain bacterial cells. A replicating circle is schematically like the Greek letter θ (theta); so this mode of replication is usually called **θ replication**.

The circularity of the replicating molecules in Figures 4-15 and 4-16 brings out an important geometric feature of semiconservative replication. There are about 400,000 turns in an *E. coli* double helix and, because the two chains of a replicating molecule must make a full rotation to unwind each of these gyres, some kind of swivel must exist to avoid tangling the entire structure (Figure 4-17). Current evidence suggests that the axis of rotation for unwinding

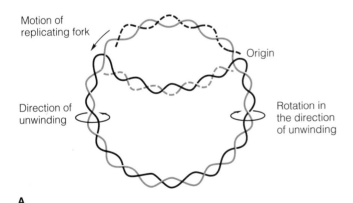

Motion of replicating fork

Origin

Direction of unwinding

Rotation in the direction of unwinding

A

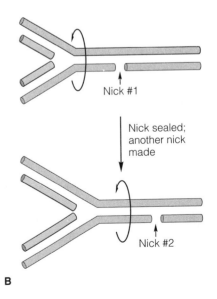

Nick #1

Nick sealed; another nick made

Nick #2

B

FIGURE 4-17   Replication of a circular DNA molecule: (A) the unwinding motion of the daughter branches of a replicating circle, without positions at which free rotation can occur, causes overwinding of the unreplicated portion; (B) mechanism by which a single-strand break (a nick) ahead of a replication fork allows rotation.

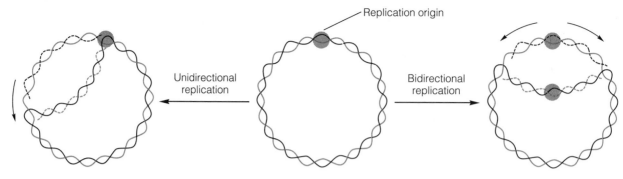

FIGURE 4-18   The distinction between unidirectional and bidirectional DNA replication. In unidirectional replication, there is only one replication fork; bidirectional replication requires two replication forks. The short arrows indicate the direction of movement of the forks. Most DNA replicates bidirectionally. For clarity, double-stranded DNA is drawn as a single line.

is provided by cuts made in the backbone of *one* strand of the double helix during replication; these cuts are rapidly repaired after unwinding. Enzymes capable of making such cuts and then rapidly repairing them have been isolated from both bacterial and mammalian cells; they are called **topoisomerases.**

The position along a molecule at which DNA replication begins is called a **replication origin,** and the region in which parental strands are separating and new strands are being synthesized is called a **replication fork.** The process of generating a new replication fork is **initiation.** At one extreme, initiation may occur at random positions; alternatively, the origin may be a unique site. In most bacteria, bacteriophages, and viruses, *DNA replication is initiated at a unique origin.* Furthermore, with only a few exceptions, two replication forks move in opposite directions from the origin (Figure 4-18); that is, *DNA almost always replicates bidirectionally in both prokaryotes and eukaryotes.*

Multiple origins of replication in eukaryotes   The single DNA molecule in a eukaryotic chromosome is linear and replicates **bidirectionally** as a linear molecule. Furthermore, replication is initiated at many sites in the DNA. The structures resulting from the numerous initiations are seen in electron micrographs as multiple loops along a DNA molecule (Figure 4-19). Multiple initiation is a means of reducing the total replication time of a large molecule.

Replication of an *E. coli* DNA molecule, which contains about 4×10^6 nucleotide pairs, requires 40 minutes. Thus, each of the two replication forks generated at the single origin moves at a rate of approximately 65,000 nucleotide pairs per minute; similar rates are found in the replication of all phage DNA molecules studied. Movement of a replication fork in eukaryotic DNA molecules is much slower—in *Drosophila melanogaster*, about 2600 nucleotide pairs per minute at 24°C. The DNA molecule in the largest chromosome of this organism contains about 7×10^7 nucleotide pairs; so, with a single replication origin, replication would take about 15 days. Developing *Drosophila* embryos use about 8500 replication origins per chromosome, which reduces the replication time to a few minutes. In a typical eukaryotic cell, origins are spaced about 40,000 nucleotide pairs apart, which allows each chromosome to replicate in 1/4 to 1/2 hr. Because all chromosomes do not replicate simultaneously, complete replication of all chromosomes in eukaryotes usually takes from 5 to 10 hr.

Rolling-circle replication   Some circular DNA molecules, including those of a number of bacterial and eukaryotic viruses, replicate by a process that does not

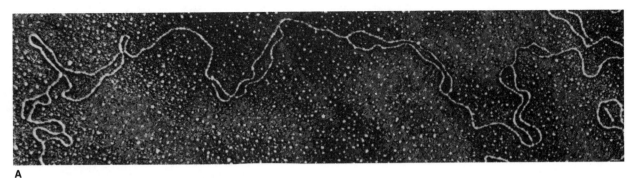

A

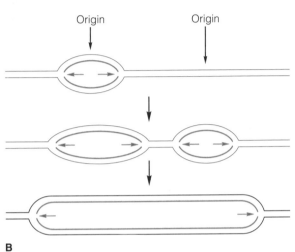

FIGURE 4-19 Replicating DNA of *Drosophila melanogaster*: (A) an electron micrograph of a 30,000-nucleotide-pair segment showing seven replication loops; (B) an interpretive drawing showing how loops merge. Two replication origins are shown in the drawing. The red arrows indicate the direction of movement of the replication forks. (Electron micrograph courtesy of David Hogness.)

B

include θ-shaped intermediates. This replication mode is called **rolling-circle replication**. In this process, replication starts with a cut at a specific sugar–phosphate bond in a double-stranded circle (Figure 4-20). This cut produces two chemically distinct ends—a 3′ end (at which the nucleotide has a free 3′-OH group) and a 5′ end (at which the nucleotide has a free 5′-P group). New DNA is synthesized by the addition of deoxynucleotides to the 3′ end with simultaneous displacement of the 5′ end from the circle. As replication proceeds around the circle, the 5′ end rolls out as a tail of increasing length.

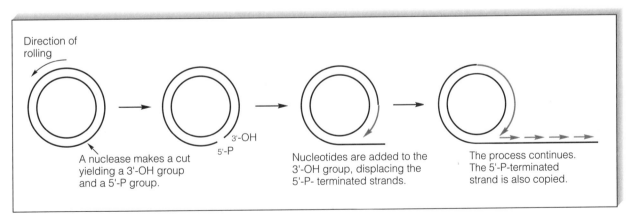

FIGURE 4-20 Rolling-circle replication. Newly synthesized DNA is in red. The displaced strand is replicated in short fragments, as explained in Section 4-7.

The Replication of DNA 103

In most cases, as the tail is extended, a complementary chain is synthesized, which results in a double-stranded DNA tail. Because the displaced strand is chemically linked to the newly synthesized DNA in the circle, replication does not terminate, and extension proceeds without interruption, forming a tail that may be many times as long as the circumference of the circle. Rolling-circle replication is a common feature in late stages of replication of double-stranded DNA phages having circular intermediates. An important example of rolling-circle replication will be given in Chapter 10, where matings between donor and recipient *E. coli* are described.

So far, we have considered only certain geometrical features of DNA replication. In the next section, the enzymes and protein factors used in DNA replication are described.

4-7 DNA Synthesis

Nucleic acids are synthesized in chemical reactions controlled by enzymes, as is the case for most metabolic reactions in living cells. The enzymes that form the sugar–phosphate bond (the phosphodiester bond) between adjacent nucleotides in a nucleic acid chain are called **DNA polymerases.** A variety of DNA polymerases have been purified and DNA synthesis has been carried out **in vitro** in a cell-free system prepared by disrupting cells and combining purified components in a test tube under precisely defined conditions. The best-understood polymerases are those isolated from *E. coli.*

Three principal requirements must be fulfilled for DNA polymerases to catalyze the synthesis of DNA:

1. *The 5′-triphosphates of the four deoxynucleosides must be present.* These are the compounds denoted in Table 4-1 as dATP, dGTP, dTTP, and dCTP. They contain the bases adenine, guanine, thymine, and cytosine, respectively. Details of the structures of dCTP and dGTP are shown in Figure 4-21, in which the phosphate groups cleaved off in DNA synthesis are indicated in red. DNA synthesis requires all four nucleoside 5′-triphosphates and does not occur if any of them are omitted.

2. *A preexisting single strand of DNA to be copied must be present.* Such a strand is called a **template** strand.

3. *A nucleic acid segment, which may be very short and either DNA or RNA, must be present and hydrogen-bonded to the template strand.* This segment is called a **primer** (Figure 4-22). There is need for a primer because *no known DNA polymerase is able to initiate chains.* Thus, presence of a primer chain with a free 3′-OH group is absolutely necessary for initiation of replication.

The reaction catalyzed by the DNA polymerases is the formation of a phosphodiester bond between the free 3′-OH group of the primer and the innermost phosphorus atom of the nucleoside triphosphate being incorporated at the new primer terminus (Figure 4-22). The result is as follows:

DNA synthesis occurs by the elongation of primer chains, *always in the 5′ → 3′ direction.*

Recognition of the appropriate incoming nucleoside triphosphate in the growth of the primer chain depends on base-pairing with the opposite nucleotide in the template chain. DNA polymerase will usually catalyze the polymerization

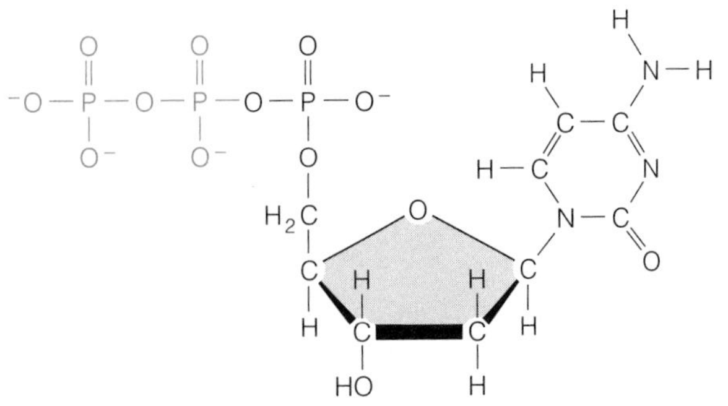

Deoxycytidine 5'- triphosphate (dCTP)

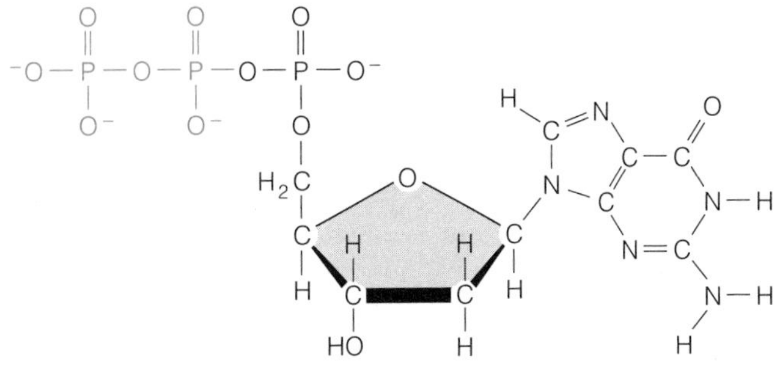

Deoxyguanosine 5'- triphosphate (dGTP)

FIGURE 4-21 Two deoxynucleoside triphosphates used in DNA synthesis. The red part of each molecule is removed in synthesis.

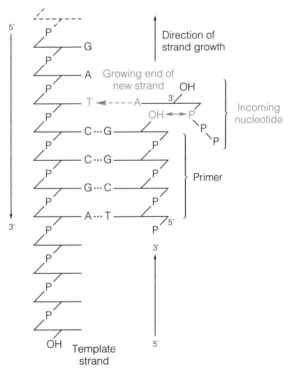

FIGURE 4-22 Addition of nucleotides to the 3'-OH terminus of a primer. The recognition step is shown as the formation of hydrogen bonds between the red A and the red T. The chemical reaction is between the red 3'-OH group and the red phosphate of the triphosphate.

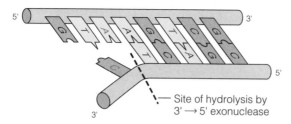

FIGURE 4-23 The 3'-to-5' exonuclease activity of the proofreading function; the growing strand is cleaved to release a nucleotide containing base C (red), which does not pair with base A (red) in the template strand.

reaction that incorporates the new nucleotide at the primer terminus only when the correct base pair is present. The same DNA polymerase is used to add each of the four deoxynucleoside phosphates to the 3'-OH terminus of the growing strand.

Two DNA polymerases are needed for DNA replication in *E. coli*—**DNA polymerase I**, often written **Pol I**, and **DNA polymerase III (Pol III)**. Pol III is the major replication enzyme. Polymerase I plays a secondary role in replication that will be described in a later section. The eukaryotic DNA polymerase responsible for replication of chromosomal DNA is called **polymerase α**.

In addition to their ability to polymerize nucleotides, most DNA polymerases are capable of **nuclease** activities that break phosphodiester bonds in the sugar-phosphate backbones of nucleic acid chains. Many other enzymes are capable of nuclease activity, and these are of two types: (1) **exonucleases**, which can remove a nucleotide only from the end of a chain, and (2) **endonucleases**, which break bonds within the chains. DNA polymerases I and III of *E. coli* engage in an exonuclease activity only at the 3' terminus (a 3' → 5' exonuclease activity), which serves as a built-in mechanism for correcting rare errors in polymerization. Occasionally, a polymerase adds an incorrect nucleotide—one that cannot base-pair—to the end of the growing chain. The presence of an unpaired nucleotide activates the 3'–5'-exonuclease, also called the **proofreading** or **editing function**; this activity excises the unpaired nucleotide from the 3'-OH end of the growing chain (Figure 4-23). The proofreading function can "look back" only one base (the one added last.) Nevertheless:

> The genetic significance of the proofreading function is that it is an error)correcting mechanism that serves to reduce the frequency of mutation resulting from the incorporation of incorrect nucleotides in DNA replication.

Two properties of all known DNA polymerases—(1) synthesis along the template strand only in the 5' → 3' direction, and (2) the inability to initiate new chains—create problems in replication. These problems and their solutions are described in the next section.

4-8 Discontinuous Replication

In the model of replication shown in Figure 4-12, both daughter strands are drawn as if replicating continuously. However, no known DNA molecule replicates in this way—instead, *one of the daughter strands is made in short fragments, which are then joined together*. The reason for this mechanism and the properties of these fragments are described in the following subsections.

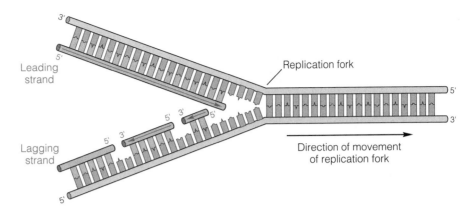

FIGURE 4-24 Short fragments in the replication fork. For each tract of base pairs, the lagging strand is synthesized later than the leading strand.

Leading strand

Lagging strand

Replication fork

Direction of movement of replication fork

Fragments in the Replication Fork

All known DNA polymerases can add nucleotides only to a 3′-OH group. Thus, if both daughter strands grew in the same overall direction, each growing strand would need a 3′-OH terminus. However, the two strands of DNA are antiparallel, and so only one of the growing strands can terminate in a free 3′-OH group; the other must terminate in a free 5′ end. The solution to this topological problem is that, within a single replication fork, both strands grow in the 5′ → 3′ orientation, which requires that they grow in opposite directions along the parental strands. One strand of the newly made DNA consists of small **precursor fragments,** as shown in Figure 4-24. Synthesis of the discontinuous strand is initiated only at intervals, which causes a single-stranded region of the parental strand to be present on one side of the replication fork. The 3′-OH terminus of the continuously replicating strand is always ahead of the discontinuous strand, which has led to the use of the convenient terms **leading strand** and **lagging strand** for the continuously and discontinuously replicating strands, respectively. Single-stranded regions have been seen in high-resolution electron micrographs of replicating DNA molecules (Figure 4-25).

In the next subsection, we examine how synthesis of a precursor fragment is initiated.

FIGURE 4-25 A. A replicating θ molecule of phage λ DNA. The arrows show the two replication forks. The segment between each pair of thick lines at the arrows is single-stranded; note that it appears thinner and lighter. B. An interpretive drawing. (Courtesy of Manuel Valenzuela.)

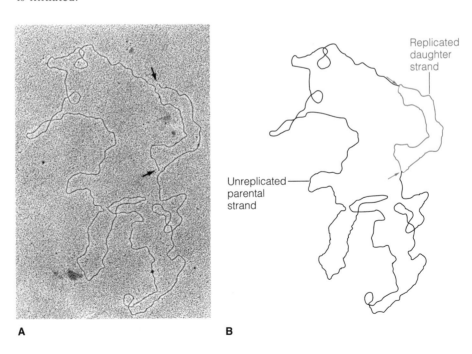

Replicated daughter strand

Unreplicated parental strand

A

B

Initiation by an RNA Primer

As mentioned earlier, DNA polymerases cannot initiate synthesis of a new strand; so a free 3'-OH is needed. In most organisms, initiation is accomplished by a special type of RNA polymerase. RNA is usually a single-stranded nucleic acid consisting of four types of nucleotides joined together by $3' \rightarrow 5'$ phosphodiester bonds—the same chemical bonds as those in DNA. Two chemical differences distinguish newly synthesized RNA from DNA (Figure 4-26). The first difference is in the sugar component. RNA contains **ribose,** which is identical with the deoxyribose of DNA except for the presence of an –OH group on the 2' carbon atom. The second difference is in one of the four bases: the thymine found in DNA is replaced by the closely related pyrimidine **uracil (U)** in RNA. An RNA strand is synthesized by using a DNA strand as a template and forming a complementary strand in which the bases in the DNA are paired with those in the RNA (Chapter 11). Synthesis is catalyzed by enzymes called **RNA polymerases.** These enzymes differ from DNA polymerases in that they can initiate synthesis of RNA chains without need for a primer.

In the initiation of DNA synthesis, a short stretch of RNA, usually ranging from fifty to seventy-five nucleotides in length, remains bound to the DNA template after the RNA polymerase dissociates from the DNA. This segment of RNA serves as a primer to which a DNA polymerase can add deoxynucleotides (Figure 4-27). The RNA polymerase that produces the primer for DNA synthesis is called **primase.** Thus, a precursor fragment in the lagging strand has the following structure while it is being synthesized:

$$\text{PPP-5'} \underline{\hspace{3cm}} \mathbf{\underline{\hspace{3cm}}} \underline{\hspace{2cm}} \text{3'-OH}$$

$$\text{RNA} \qquad\qquad \text{DNA}$$

The Joining of Precursor Fragments

The precursor fragments are ultimately joined to yield a continuous strand of DNA. This strand contains no RNA sequences, and so assembly of the lagging strand must require removal of the RNA primer, replacement with a DNA sequence, and then joining. In *E. coli*, the first two processes are accomplished by DNA polymerase I, and joining is catalyzed by the enzyme **DNA ligase,** which can link adjacent 3'-OH and 5'-P groups at a nick. How this is done is shown in Figure 4-28. Pol III extends the growing strand until the RNA of the primer of the previously synthesized precursor fragment is reached. Where the DNA and RNA strands meet there is a single-strand interruption, or **nick.** The *E. coli* DNA ligase cannot seal the nick because a triphosphate is present (it can only link a 3'-OH and a 5'-*mono*phosphate). However, polymerase I has an exonuclease activity that can remove nucleotides from the 5' end of a

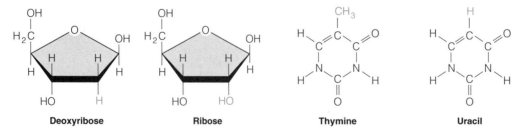

| Deoxyribose | Ribose | Thymine | Uracil |

FIGURE 4-26 Differences between DNA and RNA. The red groups are the distinguishing features of deoxyribose and ribose and of thymine and uracil.

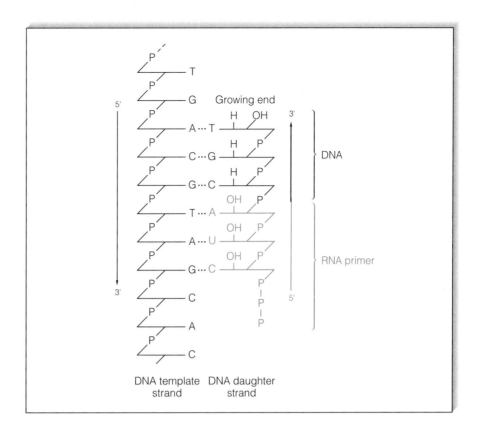

FIGURE 4-27 Priming of DNA synthesis with an RNA segment. The DNA template strand and newly synthesized DNA are shown in black. The RNA segment, which has been made by an RNA-polymerase enzyme, is shown in red.

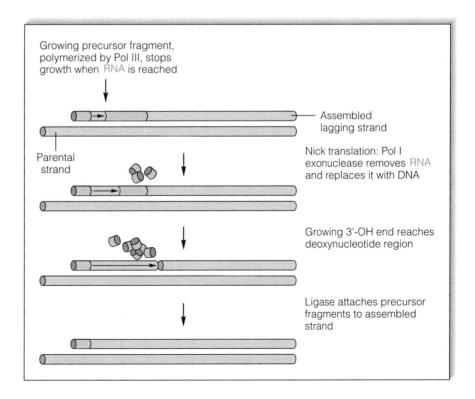

FIGURE 4-28 Sequence of events in the assembly of precursor fragments. RNA is indicated in red. The replication fork (not shown) is at the left.

base-paired fragment. It is effective with both DNA and RNA. This activity is a **5′ → 3′ exonuclease** activity. Thus, Pol I acts at the nick and moves the nick in the 5′ → 3′ direction, removing RNA nucleotides one by one and adding DNA nucleotides to the 3′ end of the DNA strand. When all of the RNA nucleotides have been removed, DNA ligase joins the 3′-OH group to the terminal 5′-P of the precursor fragment. By this sequence of events, the precursor fragment is assimilated into the lagging strand. When the next precursor fragment reaches the RNA primer of the fragment just joined, the sequence begins again.

4-9 Determination of the Sequence of Bases in DNA

A great deal of information about gene structure and gene expression can be obtained by direct determination of the sequence of bases in a DNA molecule. Several techniques are available for base sequencing; one is described in this section. No technique can determine the sequence of bases in an entire chromosome in a single experiment; so chromosomes are first cut into fragments a few hundred base pairs long, a size that can be sequenced easily. (Procedures for fragmentation are described in Section 4-10.) To obtain the sequence of a long stretch of DNA, a set of overlapping fragments is prepared, the sequence of each is determined, and all sequences are then combined. The procedures are straightforward, and DNA sequences have accumulated at such a rapid rate that large computer databases are necessary to manage the more than 50×10^6 nucleotides of DNA sequences already determined for a variety of genes from many organisms. A still larger database will be required to handle the DNA sequence of the human chromosome complement, which consists of 3×10^9 nucleotide pairs in the haploid chromosome set.

The **dideoxy sequencing method,** also called the **Sanger method,** employs DNA synthesis in the presence of small amounts of abnormal nucleotides that contain the sugar **dideoxyribose** instead of deoxyribose (Figure 4-29). Dideoxyribose lacks the 3′-OH group, which is essential for attachment of the next nucleotide in a growing DNA strand, and so incorporation of a dideoxynucleotide instead of a deoxynucleotide immediately terminates further synthesis of the strand. To sequence a DNA strand, four DNA synthesis reactions are carried out. Each reaction contains the single stranded DNA template to be sequenced, a small primer fragment complementary to a stretch of the template strand, all four deoxyribonucleoside triphosphates, and *one* of the nucleoside triphosphates in the dideoxy form. Each reaction produces a set of fragments that terminate where a dideoxynucleotide was randomly incorporated in place of the normal deoxynucleotide. That is, in each of the four reactions, the lengths of the fragments are determined by the positions in the daughter strand where the particular dideoxynucleotide present in that reaction was incorporated. The sizes of the fragments produced by chain termination are determined by gel electrophoresis, and the base sequence is then determined by the following rule:

> If a fragment containing *n* nucleotides is generated in the reaction containing a particular dideoxynucleotide, then position *n* in the *daughter strand* is occupied by the base present in the dideoxynucleotide. The numbering is from the 5′ nucleotide of the primer.

For example, if a 93-base fragment is present in the reaction containing the dideoxy form of dATP, then the 93rd base in the daughter strand produced by

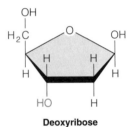

Deoxyribose

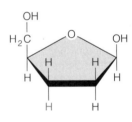

Dideoxyribose

FIGURE 4-29 Structure of normal deoxyribose and the dideoxyribose sugar used in DNA sequencing. The dideoxyribose has a hydrogen atom (red) attached to the 3′ carbon, in contrast with the hydroxyl group (red) at this position in deoxyribose. Because the 3′ hydroxyl is essential for attachment of the next nucleotide in line in a growing DNA strand, incorporation of a dideoxynucleotide immediately terminates synthesis.

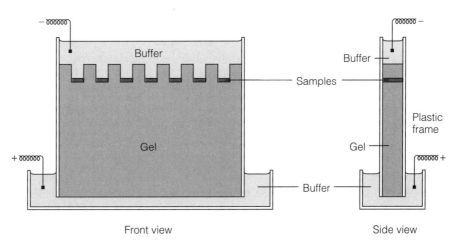

Front view   Side view

FIGURE 4-30 Apparatus for gel electrophoresis capable of handling seven samples simultaneously. Liquid gel is allowed to harden in place, with an appropriately shaped mold placed on top of the gel during hardening in order to make "wells" for the sample (red). After electrophoresis, the sample is made visible by removing the plastic frame and immersing the gel in a solution containing a reagent that binds to or reacts with the molecules that were electrophoresed, which are located at various positions in the gel. The separated components of the sample appear as bands, which may be either visibly colored or fluorescent when illuminated with fluorescent light, depending on the particular reagent used. The region of a gel in which the components of one sample can move is called a *lane*. Thus, this gel has seven lanes.

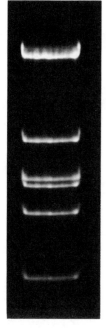

FIGURE 4-31 Gel electrophoresis of DNA. Molecules of different sizes were mixed and placed in a well. Electrophoresis was in the vertical direction. The DNA has been made visible by the addition of a dye (ethidium bromide) that binds only to DNA and that fluoresces when the gel is illuminated with short-wavelength ultraviolet light.

DNA synthesis must be an A. Because most native DNA molecules are double stranded, it does not matter whether the sequence of the template strand or the daughter strand is determined. The sequence of the template strand can be deduced from the daughter strand because their nucleotide sequences are complementary. However, in practice, both strands of a molecule are usually sequenced independently and compared in order to eliminate most of the errors made in misreading the gels.

Gel Electrophoresis

DNA molecules are negatively charged and thus can move in an electric field. For example, if the terminals of a battery are connected to the opposite ends of a horizontal tube containing a DNA solution, the molecules will move toward the positive end of the tube, at a rate depending on the electrical field strength and on the shape and size of the molecules.

The most-common type of electrophoresis used in genetics is **gel electrophoresis.** In this procedure, the rate of movement depends primarily on the molecular weight of the molecule, as long as all molecules are linear and not too large.

An experimental arrangement for gel electrophoresis of DNA is shown in Figure 4-30. A thin slab of a gel is prepared containing small slots (called wells) into which samples are placed. An electric field is applied and the negatively charged DNA molecules penetrate and move through the gel. A gel is a complex network of molecules, containing narrow, tortuous passages, and so smaller molecules pass through more easily; hence, the rate of movement increases as the molecular weight decreases. Figure 4-31 shows the result of electrophoresis of a collection of double-stranded DNA molecules. Each discrete region containing DNA is called a **band.**

Gel electrophoresis can be carried out with both double-stranded and single-stranded DNA. With short, single-stranded fragments, molecules differing in size by a single base can be separated. For example, if a reaction mixture contains a set of DNA fragments consisting of 20, 21, 22, . . . , 100 nucleotides, electrophoresis will yield 81 bands—one containing the fragment of size 20, the next the fragment of size 21, and so forth. This extraordinary sensitivity to size is the basis of the DNA sequencing procedure.

The Sequencing Procedure

The procedure for sequencing a DNA fragment is diagrammed in Figure 4-32. In this example the primer length is 20 nucleotides. Termination at G produces fragments of size 21, 24, 27, and 29 nucleotides, termination at A produces fragments of size 22 and 30 nucleotides, termination at T produces a fragment of 26 nucleotides, and termination at C produces fragments of 23, 25, and 28 nucleotides. Typically, either the primer or one of the deoxynucleotides used contains a radioactive atom, so that after electrophoresis to separate the fragments by size, the fragments can be visualized as bands on photographic film placed over the gel. After electrophoresis and autoradiography, the

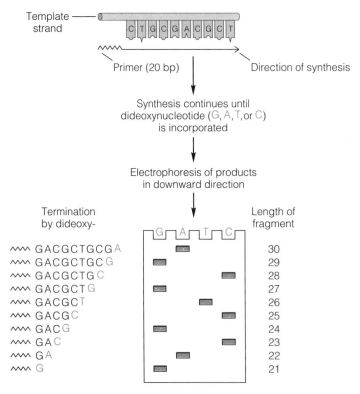

FIGURE 4-32   Dideoxy method of DNA sequencing. Four DNA synthesis reactions are carried out in the presence of all normal nucleotides plus *one* of the dideoxy nucleotides containing G, A, T, or C. Synthesis continues along the template strand until a dideoxynucleotide is incorporated. The products resulting from termination at each dideoxy nucleotide are indicated at the left. The fragments are separated by size by electrophoresis and the positions of the nucleotides determined directly from the gel. In this example, the length of the primer needed to initiate DNA synthesis is 20 nucleotides. Lengths of the DNA fragments are shown at the right of the gel. The sequence of the daughter strand is read from the bottom of the gel as 5'-GACGCTGCGA-3'.

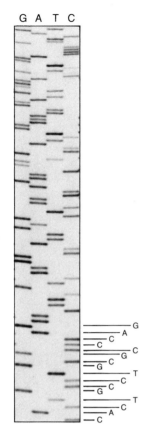

FIGURE 4-33 A section of a dideoxy sequencing gel. The sequence is read from the bottom to the top. Each horizontal row represents a single nucleotide position in the DNA strand synthesized from the template. The vertical columns result from termination by the dideoxy forms of G, A, T, or C. The sequence from the lower part of the gel is indicated. (Courtesy of Kyoko Maruyama.)

positions of the bases in the daughter strand can be read directly from the gel, starting with position 21, as GACGCTGCGA. This corresponds to the template-strand sequence CTGCGACGCT.

Figure 4-33 is an autoradiogram of a sequencing gel. The shortest fragments are those that move the fastest and farthest. Each fragment contains the primer fragment at the 5′ end of the daughter strand. The sequence can be read directly from the bottom to the top of the gel. Thus, the initial sequence of the segment in this gel, read from bottom to top, is CACTGCCTGCGC-CCAG, and so forth.

4-10 Isolation and Characterization of Particular DNA Fragments

DNA sequencing and other applications often require the isolation and identification of *particular* DNA fragments. This is accomplished by techniques for breaking DNA molecules at specific sites and for isolating particular DNA fragments. These techniques are described in this section.

DNA fragments are obtained by treatment of DNA samples with a specific class of nucleases. Many nucleases have been isolated from a variety of organisms, and most produce breaks at random sites within a DNA sequence. However, the class of nucleases called **restriction endonucleases,** or, more simply, **restriction enzymes,** consists of sequence-specific enzymes. Most restriction enzymes recognize only one short base sequence, usually four or six nucleotide pairs. The enzyme binds with the DNA at these sites and makes a break in each strand of the DNA molecule, resulting in 3′-OH and 5′-P groups at each position. The nucleotide sequence recognized for cleavage by a restriction enzyme is called the **restriction site** of the enzyme. Several hundred restriction enzymes specific for different restriction sites have been isolated from many species of prokaryotic microorganisms. Some examples are given in Table 4-3.

Restriction enzymes have the following important characteristics:

1. The restriction-site sequences recognized by most restriction enzymes are **palindromes**—that is, the sequence is the same when read from either direction (Table 4-3).
2. Most restriction enzymes recognize a single restriction site.
3. The restriction site is recognized without regard to the source of the DNA.
4. Because most restriction enzymes recognize a unique restriction-site sequence, *the number of cuts made in the DNA from a particular organism by a particular enzyme is fixed.*

The DNA fragment produced by a pair of adjacent cuts in a DNA molecule is called a **restriction fragment.** A large DNA molecule will typically be cut into many restriction fragments of different sizes. For example, a typical bacterial DNA molecule, which contains roughly 4×10^6 base pairs, is cut into several hundred to several thousand fragments, and mammalian nuclear DNA is cut into more than a million fragments. Although these numbers are large, they are actually quite small relative to the number of sugar–phosphate bonds in the DNA of an organism. Restriction fragments are usually short enough that they can be separated by electrophoresis and manipulated in various ways—for example, by ligating them into self-replicating molecules such as bacteriophages, plasmids, or even small artificial chromosomes. These procedures

TABLE 4-3   Some restriction endonucleases, their sources, and their cleavage sites

Enzyme	Microorganism	Target sequence and cleavage sites
EcoRI	*Escherichia coli*	G A A T T C C T T A A G
BamHI	*Bacillus amyloliquefaciens* H	G G A T C C C C T A G G
HaeII	*Haemophilus aegyptius*	Pu G C G C Py Py C G C G Pu
HindIII	*Haemophilus influenza*	A A G C T T T T C G A A
PstI	*Providencia stuartii*	C T G C A G G A C G T C
TaqI	*Thermus aquaticus*	T C G A A G C T

Note: The vertical dashed line indicates the axis of symmetry in each sequence. Arrows indicate the sites of cutting. The enzyme TaqI yields cohesive ends consisting of two nucleotides, whereas the cohesive ends produced by the other enzymes contain four nucleotides. Pu and Py refer to any purine and pyrimidine, respectively.

constitute **DNA cloning** and are the basis of one form of *genetic engineering*, discussed further in Chapter 16.

Because of the sequence specificity, *a particular restriction enzyme produces a unique set of fragments for a particular DNA molecule.* Another enzyme will produce a different set of fragments from the same DNA molecule. Figure 4-34A shows the sites of cutting of *E. coli* phage λ DNA by the enzymes EcoRI and BamHI. A map showing the unique sites of cutting of the DNA of a particular organism by a single enzyme is called a **restriction map.** The family of fragments generated by a single enzyme can be detected easily by gel electrophoresis of enzyme-treated DNA (Figure 4-34B), and particular DNA fragments can be isolated by cutting out the part of the gel containing the fragment and removing the DNA from the gel.

Several techniques enable one to locate particular DNA fragments in a gel. One of the most generally applicable procedures is the **Southern blot.** In this procedure, a gel in which DNA molecules have been separated by electrophoresis is treated with alkali to render the DNA single stranded (**denaturation** of the DNA), and then the DNA is transferred to a sheet of nitrocellulose in such a way that the relative positions of the DNA bands are maintained (Figure 4-35). The nitrocellulose, to which the single-stranded DNA tightly binds, is then exposed to radioactive complementary RNA or DNA (the **probe**) in a way that leads to the complementary strands coming together to form duplex molecules (**renaturation**). Radioactivity becomes stably bound (resistant to

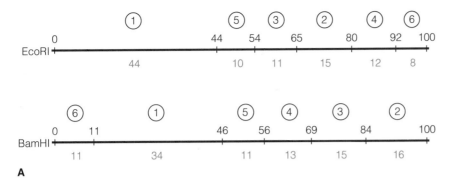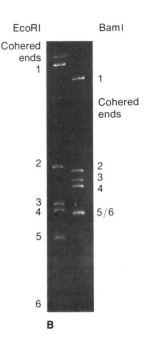

A

FIGURE 4-34 A. Restriction maps of λ DNA for the restriction enzymes EcoRI and BamHI. The vertical bars indicate the sites of cutting. The black numbers indicate the approximate percentage of the total length of λ DNA measured from the end of the molecule arbitrarily designated the left end. The red numbers are the lengths of each fragment, again expressed as percentages of the total length. B. An electrophoretic gel of EcoRI and BamHI enzyme digests of λ DNA. The band at the top consists of the two terminal fragments joined together. Numbers indicate fragments in order from largest (1) to smallest (6); the circled numbers on the maps correspond to the numbers beside the gel. The DNA has not been electrophoresed long enough to separate bands 5 and 6 of the BamHI digest.

removal by washing) to the DNA only at positions at which base sequences complementary to the radioactive molecules are present, so that duplex molecules can form. The radioactivity is located by placing the paper in contact with x-ray film; after development of the film, blackened regions indicate positions of radioactivity. For example, a cloned DNA fragment from one species may be used as probe DNA in a Southern blot with DNA from another species; the probe will hybridize only with restriction fragments containing DNA sequences that are sufficiently homologous to allow stable duplexes to form.

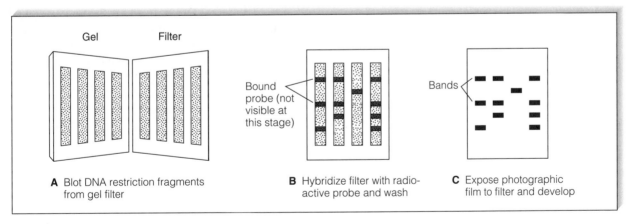

FIGURE 4-35 Southern blot. A. DNA restriction fragments are separated by electrophoresis, blotted from the gel onto a special filter material, and chemically attached using ultraviolet light. B. The strands are denatured and mixed with radioactive probe DNA, which binds with complementary sequences present on the filter. The bound probe remains, whereas unbound probe washes off. C. Bound probe is revealed by darkening of photographic film placed over the filter. The positions of the bands indicate which restriction fragments contain DNA sequences homologous with those in the probe.

Except for a few viruses, the genetic material of all organisms is DNA. In eukaryotes, the DNA is enclosed in the nucleus, except for small DNA molecules present in mitochondria and chloroplasts. DNA rather than proteins was shown to be the genetic material through (1) the genotypic transformation of bacteria by isolated DNA and (2) the injection of phage DNA, but not phage protein, into bacteria that subsequently produce progeny phages.

DNA is a double-stranded polymer consisting of nucleotides. A nucleotide has three components—a base, a sugar (which is deoxyribose in DNA), and a phosphate. Sugars and phosphates alternate in forming a single polynucleotide chain with one terminal 3'-OH group and one 5'-P group. In double-stranded (duplex) DNA, the two strands are antiparallel: each end of the double helix carries one 3'-OH group and one 5'-P group. Four bases are found in DNA—adenine (A) and guanine (G), which are purines, and cytosine (C) and thymine (T), which are pyrimidines. Equal numbers of purines and pyrimidines are found in double-stranded DNA, because the bases are paired as AT pairs and GC pairs. This pairing holds the two polynucleotide strands together in a double helix. The base composition of DNA varies from one organism to the next. The information content of a DNA molecule resides in the sequence of bases along the chain, and each gene consists of a unique sequence.

The double helix replicates using enzymes called DNA polymerases, but other proteins are also needed. Replication is semiconservative in that each parental single strand, called a template strand, is found in one of the double-stranded progeny molecules. Replication proceeds by a DNA polymerase (1) bringing in a nucleotide with a base capable of hydrogen-bonding with the corresponding base in the template strand, and (2) joining the 5'-P group of the nucleotide to the free 3'-OH group of the growing strand. Because double-stranded DNA is antiparallel, only one strand—the leading strand—grows in the direction of movement of the replication fork. The other strand—the lagging strand—is synthesized in the opposite direction as short fragments that are subsequently joined together. DNA polymerases cannot initiate synthesis, and so a primer is always needed. The primer is an RNA fragment made by an RNA-polymerizing enzyme; the RNA primer is removed at later stages of replication. DNA molecules of prokaryotes usually have a single replication origin; eukaryotic DNA molecules usually have many origins.

The base sequence of a DNA molecule can be determined by one of several methods. With each method, the DNA is isolated in discrete fragments containing several hundred nucleotide pairs. Complementary strands of each fragment are sequenced in order to generate the complete sequence. The dideoxy sequencing method uses dideoxynucleotides to terminate daughter-strand synthesis and reveal the identity of the base present in the daughter strand at the site of termination.

Restriction enzymes cleave DNA molecules at the positions of specific sequences (restriction sites) of usually four or six nucleotides. Each restriction enzyme produces a unique set of fragments for any particular DNA molecule. These fragments can be separated by electrophoresis and used for purposes such as DNA cloning or sequencing. The positions of particular restriction fragments in a gel can be visualized by means of a Southern blot, in which radioactive probe DNA is mixed with single-stranded (denatured) restriction fragments that have been transferred to a filter membrane after electrophoresis. The probe DNA will form stable duplexes with whatever fragments contain sufficiently homologous sequences, and the positions of these duplexes can be determined by autoradiography of the filter.

KEY TERMS

adenine (A)	colony	dideoxy sequencing method
antiparallel	complementary	DNA cloning
bacteriophage	cytosine (C)	DNA ligase
base	daughter strand	DNA polymerase
base composition	denaturation	duplex DNA
bidirectional replication	deoxyribonucleic acid (DNA)	editing function
clone	dideoxyribose	endonuclease

EXAMPLES OF WORKED PROBLEMS

Problem 1: A technique is used for determining the base composition of double-stranded DNA. Rather than giving the relative amounts of each of the four bases—[A], [T], [G], and [C]—it yields the value of the ratio [A]/[C]. If this ratio is 1/3, what are the relative amounts of each of the four bases?

Answer: In double-stranded DNA, [A] = [T] and [G] = [C] because of the base-pairing between complementary strands. Therefore, if [A]/[C] = 1/3, then [C] = [G] = 3[A]. Because [A] + [T] + [G] + [C] = 1, everything can be put in terms of [A] as [A] + [A] + 3[A] + 3[A] = 1, or [A] = [T] = 0.125. This makes [C] = [G] = (1 − 0.25)/2 = 0.375. In other words, the DNA is 12.5 percent A, 12.5 percent T, 37.5 percent G, and 37.5 percent C.

Problem 2: The restriction enzyme EcoRI cleaves double-stranded DNA at the sequence 5'-GAATTC-3', and the restriction enzyme HindIII cleaves at 5'-AAGCTT-3'. A 20 kilobase (kb) circular plasmid is digested with each enzyme individually and then in combination, and the resulting fragment sizes are determined by means of electrophoresis. The results are as follows:

EcoRI alone: fragments of 6 kb and 14 kb
HindIII alone: fragments of 7 kb and 13 kb
EcoRI and HindIII: fragments of 2 kb, 4 kb, 5 kb, and 9 kb

How many possible restriction maps are compatible with these data? For each possible restriction map, make a diagram of the circular molecule and indicate the relative positions of the EcoRI and HindIII restriction sites.

Answer: Because the single-enzyme digests give two bands each, there must be two restriction sites for each enzyme in the molecule. Furthermore, because digestion with HindIII makes both the 6-kb and 14-kb restriction fragments disappear, each of these fragments must contain one

HindIII site. Considering the sizes of the fragments in the double digest, the 6-kb EcoRI fragment must be cleaved into 2-kb and 4-kb fragments, and the 14-kb EcoRI fragment must be cleaved into 5-kb and 9-kb fragments. Two restriction maps are compatible with the data, depending on which end of the 6-kb EcoRI fragment the HindIII site is nearest. The position of the remaining HindIII site is determined by the fact that the 2-kb and 5-kb fragments in the double digest must be adjacent in the intact molecule in order that a 13-kb fragment can be produced by HindIII digestion alone.

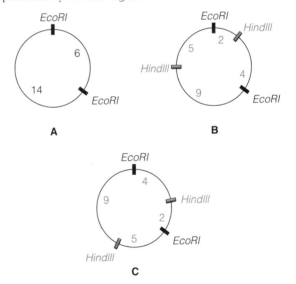

The accompanying figure shows the relative positions of the EcoRI sites (part A). Parts B and C are the two possible restriction maps, which differ according to whether the EcoRI site at the top yields the 2-kb or the 4-kb fragment in the double digest.

Problem 3: The DNA sequencing gels diagrammed in parts A and B of the adjoining illustration were obtained from human DNA by the use of the dideoxy sequencing method. Part A comes from a person homozygous for the normal allele of the cystic fibrosis gene, part B from an affected person homozygous for a mutant allele. Use the bands in the gel to answer the following questions:

(a) Why are there discrete bands in the gel?
(b) Which end of the gel—top or bottom—corresponds to the 5′ end of the DNA sequence?
(c) What nucleotide sequence is indicated by the bands in the gels?
(d) What is the sequence of the complementary strand in duplex DNA?
(e) Comparing the sequences in parts A and B, what is special about the three nucleotides that give the bands indicated in red?

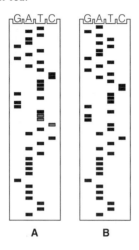

A   **B**

Answer:

(a) Each band results from a fragment of DNA whose replication was terminated by the incorporation of a dideoxy nucleotide. The fragments differ by one nucleotide in length, and they are arranged by size, with the smallest fragments at the bottom, the largest at the top.

(b) Because DNA replication elongates at the 3′ end, each fragment of increasing length has an additional 3′ nucleotide. Therefore, the DNA sequence is oriented with the 5′ end at the bottom of the gel and the 3′ end at the top.

(c) The sequences, read directly from the gel from bottom to top, are (part A)

5′-ATTAAAGAAAATATCATCTTTG-GTGTTTCCTATGATGAATAT-3′

and (part B)

5′-ATTAAAGAAAATATCATTGGT-GTTTCCTATGATGAATATAGA-3′.

(d) The complementary strands in duplex DNA are antiparallel and have A paired with T and G paired with C. For part A , this is

3′-TAATTTCTTTTATAGTAGAAA-CCACAAAGGATACTACTTATA-5′

and, for part B,

3′-TAATTTCTTTTATAGTAACCA-CAAAGGATACTACTTATATCT-5′.

(e) The three nucleotides in red in the wildtype gene are missing in the mutant gene. This three-base deletion is found in about 70 percent of the mutations in cystic fibrosis patients, and it results in a missing amino acid at position 508 in the corresponding protein (see Chapter 11).

PROBLEMS

4-1. What are the four bases commonly found in DNA? Which form base pairs? What five-carbon sugar is found in DNA? What is the difference between a nucleoside and a nucleotide?

4-2. How many phosphate groups are there per base in DNA, and how many phosphates are there in each precursor for DNA synthesis?

4-3. Which chemical groups are at the ends of a single polynucleotide strand?

4-4. What is the relation between the amount of DNA in a somatic cell and in a gamete?

4-5. Name four requirements for initiation of DNA synthesis. To what chemical group in a DNA chain is an incoming nucleotide added, and what group in the nucleotide reacts with the DNA terminus?

4-6. In what sense are the two strands of DNA antiparallel?

4-7. What is the base sequence of a DNA strand that is complementary to the hexanucleotide 3′-A-G-G-C-T-C-5′? Label the termini as to 5′ or 3′.

4-8. What is a nuclease enzyme, and how do endonucleases and exonucleases differ?

4-9. In what direction (5′ → 3′ or 3′ → 5′) does a DNA polymerase move along the template strand? How do organisms solve the problem that all DNA polymerases move in the same direction along a template strand, yet double-stranded DNA is antiparallel?

4-10. What is the chemical difference between the groups joined by DNA polymerase and DNA ligase?

4-11. What are three enzymatic activities of DNA polymerase I?

4-12. Why can RNA polymerases and primases initiate DNA replication, whereas DNA polymerases cannot?

4-13. In gel electrophoresis, do smaller double-stranded molecules move more slowly or more rapidly than larger molecules?

4-14. Consider a hypothetical phage whose DNA replicates exclusively by rolling-circle replication. A phage with radioactive DNA in both strands infects a bacterium and is allowed to replicate in nonradioactive medium. Assuming that only daughter DNA from the elongating branch ever gets packaged into progeny phage particles,
(a) What fraction of the parental radioactivity will appear in progeny phage?
(b) How many progeny phage will contain radioactive DNA, and how will the number be affected by the occurrence of crossing-over?

4-15. What is the fundamental difference between the initiation of θ replication and of rolling-circle replication?

4-16. When the base composition of DNA from the bacterium *Mycobacterium tuberculosis* was determined, 18 percent of the bases were found to be adenine.
(a) What is the percentage of thymine?
(b) What is the entire base composition of the DNA and the [G] + [C] content?

4-17. The double-stranded DNA molecule of a newly discovered virus was found by electron microscopy to have a length of 34 μm.
(a) How many nucleotide pairs are present in one of these molecules?
(b) How many complete turns of the two polynucleotide chains are present in such a double helix?

4-18. Chemical analysis of the DNA extracted from a formerly unknown virus indicates the following composition: 40 percent adenine, 20 percent thymine, 15 percent guanine, and 25 percent cytosine. What does this result imply about the structure of the viral DNA?

4-19. An elegant combined chemical and enzymatic technique enables one to identify "nearest neighbors" of bases—that is, nucleotides that are adjacent in a DNA strand. For example, if the single-stranded tetranucleotide 5'-AGTC-3' were treated in this way, the nearest neighbors would be AG, GT, and TC

(they are always written with the 5' terminus at the left). Before techniques were available for determining the complete base sequence of DNA, nearest-neighbor analysis was used to determine sequence relationships. Nearest-neighbor analysis also indicated that complementary DNA strands are antiparallel, which you are asked to examine in this problem by predicting some nearest-neighbor frequencies. Assume that you have determined the frequencies of the following nearest neighbors: AG, 0.15; GT, 0.03; GA, 0.08; TT, 0.10.
(a) What are the nearest-neighbor frequencies of CT, AC, TC, and AA?
(b) If DNA had a parallel, rather than an antiparallel, structure, what nearest-neighbor frequencies could you deduce from the observed values?

4-20. What is the chemical group (3'-P, 5'-P, 3'-OH, or 5'-OH) at the sites indicated by the dots labeled a, b, and c in the adjoining figure?

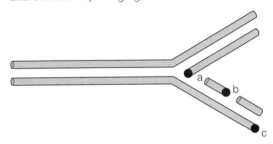

4-21. Identify the chemical group that is at the indicated terminus of the daughter strand of the extended branch of the rolling circle in the accompanying diagram?

4-22. The following sequence of bases is present in one chain of a DNA duplex that has opened up at a replication fork, and synthesis of an RNA primer on this template begins by copying the base in red.

3'- . . . TCTGATATCAGTACG . . . -5'

(a) If the RNA primer consists of eight nucleotides, what is its base sequence?
(b) In the intact RNA primer, which nucleotide has a free hydroxyl (—OH) terminus and what is the chemical group on the nucleotide at the other end of the primer?
(c) If replication of the other strand of the original DNA duplex proceeds continuously (with few or no intervening RNA primers), does the replication fork most likely move in a left-to-right or right-to-left direction?

4-23. A DNA fragment produced by the restriction enzyme SalI is inserted into a unique SalI cloning site in a vector molecule. Digestion with restriction enzymes produces the following fragment sizes that originate from the inserted DNA:

(a) SalI: 20 kb
(b) SalI + EcoRI: 7 kb, 13 kb
(c) SalI + HindIII: 4 kb, 5 kb, 11 kb
(d) SalI + EcoRI + HindIII: 3 kb, 4kb, 5 kb, 8 kb

What restriction map of the insert is consistent with these fragment sizes?

4-24. The DNA sequencing gel shown here was obtained by the dideoxy sequencing method. What is the nucleotide sequence of the strand synthesized by steps in the sequencing reactions? What is the nucleotide sequence of the complementary template strand? Label each end 3′ or 5′.

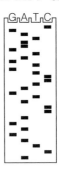

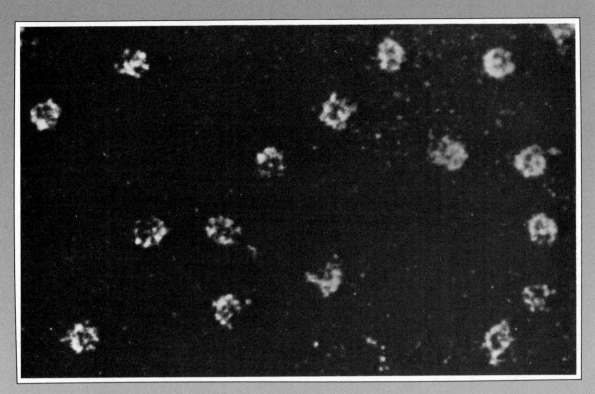

The Molecular Organization of Chromosomes

Understanding genetic processes requires knowledge of the organization of the genetic material. In this chapter, we will see that chromosomes are diverse in size and structural properties and that their DNA differs in the composition and arrangement of nucleotide sequences. The most-pronounced differences in structure and genetic organization are between the chromosomes of eukaryotes and those of prokaryotes. Some viral chromosomes are especially noteworthy in that they consist of one single-stranded (rather than double-stranded) DNA molecule and, in a small number of viruses, of one or more molecules of RNA instead of DNA. Eukaryotic cells contain several chromosomes, each of which contains one intricately coiled DNA molecule, whereas prokaryotes contain a

Above: A dark-field electron micrograph of individual nucleosomes at high magnification. (Courtesy of Ada Olins.)

single major chromosome (plus, occasionally, several copies of one or more small circular DNA molecules called plasmids).

The genetic complement of a cell or virus is referred to as a **genome.** In eukaryotes, this term is commonly used to refer to one complete haploid set of chromosomes, such as that found in a sperm or egg.

5-1 Genome Size and Evolutionary Complexity

Measurement of the nucleic acid content of the genome of viruses, bacteria, and lower and higher eukaryotes has led to the following generalization: *genome size increases roughly with evolutionary complexity.* That is, the single nucleic acid molecule of a typical virus is smaller than the DNA molecule in a bacterial chromosome; unicellular eukaryotes, such as the yeasts, contain more DNA than a typical bacterium and the DNA is organized in several chromosomes; and multicellular eukaryotes have the greatest amount of DNA. However, among the higher eukaryotes, no correlation exists between evolutionary complexity and the amount of DNA because, in higher eukaryotes, DNA content is not directly proportional to the number of genes (Table 5-1).

Bacteriophage MS2 is one of the smallest viruses. It has only four genes in a single-stranded RNA molecule containing 3569 nucleotides. SV40 virus, which infects monkey and human cells, has a genetic complement of five genes

TABLE 5-1   DNA content of some representative viral, bacterial, and eukaryotic genomes

Genome	Approximate number of nucleotide pairs (kb)*	Form
Virus		
SV40	5	Circular double-stranded
ØX174	5	Circular single-stranded;
M13	6	double-stranded replicative form
λ	50	
Herpes simplex	152	
T2, T4, T6	165	Linear double-stranded
Smallpox	267	
Bacteria		
Mycoplasma hominis	760	
Escherichia coli	4700	Circular double-stranded
Eukaryotes		Haploid chromosome number
Saccharomyces cerevisiae (yeast)	14,000	16
Caenorhabditis elegans (nematode)	100,000	6
Drosophila melanogaster (fruit fly)	165,000	4
Homo sapiens (human being)	3,000,000	23
Zea mays (maize)	4,500,000	10
Amphiuma sp. (salamander)	76,500,000	14

*kb = kilobase, or thousands of base pairs. The approximate molecular length in μm can be calculated dy dividing the length in base pairs by 3000. The approximate molecular mass can be calculated by multiplying the length in base pairs by 660.

in a circular double-stranded DNA molecule consisting of about 5000 nucleotide pairs (Table 5-1). Large DNA molecules are measured in **kilobase pairs (kb)**, or thousands of base pairs. The genome of SV40 is thus about 5 kb. The more-complex phages and animal viruses have as many as 250 genes and DNA molecules ranging from 50 to 300 kb. Bacterial genomes are substantially larger. For example, the chromosome of *E. coli* illustrated in Figure 4-15 contains from about 1500 to 3000 genes in a DNA molecule composed of about 4700 kb.

Eukaryotes have a more-complex genetic apparatus. Their genomes are packaged in chromosomes. The number of chromosomes is characteristic of the particular species, as noted in Chapter 2. In moving up the evolutionary scale of animals or plants, the DNA content per haploid genome generally tends to increase, but there are many individual exceptions. The number of chromosomes shows no pattern. One of the smallest genomes in a multicellular animal is that of the nematode worm *Caenorhabditis elegans*, with a DNA content about 20 times that of the *E. coli* genome. The *D. melanogaster* and the human genomes have about 40 and 700 times as much DNA, respectively, as the *E. coli* genome. However, exceptions to the general correspondence between genome size and evolutionary complexity exist; for example, among the amphibia and fish, several species have genomes many times the size of mammalian genomes.

The genomes of higher organisms are extremely large—big enough for perhaps 10^6 genes in human beings to perhaps 25×10^6 in *Amphiuma* (Table 5-1). However, for a variety of reasons, the actual number of genes in these organisms is much less than the theoretical maximum. For example, some closely related species of salamanders differ by 30-fold in their DNA content per cell, yet they must have about the same number of genes. In higher organisms, it appears that most of the DNA has functions other than carrying genetic information (discussed in Chapter 12).

A remarkable feature of the genetic apparatus of eukaryotes is that the enormous amount of genetic material contained in the nucleus of each cell is precisely divided in each cell division. A haploid human genome contained in a gamete has a DNA content equivalent to a linear DNA molecule 1 meter (10^6 μm) in length. The largest of the 23 chromosomes in the human genome contains a DNA molecule that is 82 mm (8.2×10^4 μm) long. However, at the metaphase of a mitotic division, the DNA molecule is condensed into a compact structure about 10 μm long and less than 1 μm in diameter. The genomes of prokaryotes and viruses, though much smaller than eukaryotic genomes, also are very compact. For example, an *E. coli* chromosome, which contains a DNA molecule about 1500 μm long, is contained in a cell about 2 μm long and 1 μm in diameter. How this compactness is attained will be described in Section 5-4.

5-2 Supercoiling of DNA

The DNA of prokaryotic and eukaryotic chromosomes is **supercoiled**—that is, double-stranded segments are twisted around one another. The geometry of supercoiling can be illustrated by a simple example. Consider first a linear duplex DNA molecule whose ends are joined in such a way that each strand forms a continuous circle. Such a DNA molecule is called a **covalent circle,** and it is said to be **relaxed** if no further twisting is present (Figure 5-1A). The individual polynucleotide strands of a relaxed circle form the usual right-

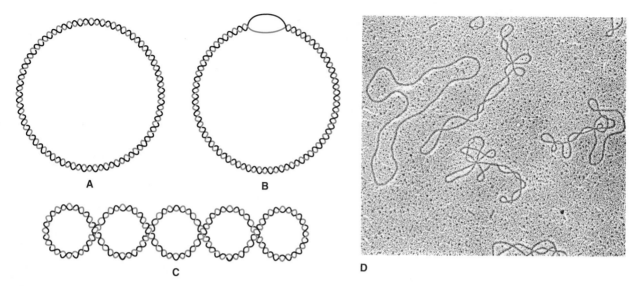

FIGURE 5-1   Different states of a covalent circle: (A) a nonsupercoiled (relaxed) covalent circle with 36 helical turns; (B) an underwound covalent circle with only 32 helical turns; (C) the molecule in part B, but with four twists to eliminate the underwinding; (D) electron micrograph showing nicked circular and supercoiled DNA of phage PM2. Note that no bases are unpaired in part C. In solution, parts B and C would be in equilibrium. (Electron micrograph courtesy of K. G. Murti.)

handed (positive) helical structure with ten nucleotide pairs per turn of the helix. Before the ends of the linear molecule are joined, if one end is rotated one or more times through 360° with respect to the other end in a direction that produces unwinding of the double helix, then joining the ends results in an underwound circular helix. A DNA molecule has a strong tendency to maintain its standard helical form with ten nucleotide pairs per turn and therefore will respond to the underwinding in one of two ways: (1) by forming regions in which the bases are unpaired (Figure 5-1B), or (2) by twisting the circular molecule in the opposite sense from the direction of underwinding (Figure 5-1C). This twisting is called **supercoiling,** and a molecule with this sense of twisting is **negatively supercoiled.** Examples of supercoiled molecules are shown in Figure 5-1C and D. The two responses to underwinding are not independent, and underwinding is usually accommodated by a combination of the two processes—namely, an underwound molecule contains some unpaired bases and some supercoiling, with the supercoiling predominating. If, instead, the molecule were overwound, supercoiling in the opposite (positive) sense would result. The supercoiling of naturally occurring molecules is always of the negative type.

The supercoiling of natural DNA molecules is not produced by the unwinding of a linear molecule before it is joined into a circle. In bacteria, it is the result of the activity of a DNA topoisomerase enzyme, mentioned in Chapter 4. The DNA of eukaryotic chromosomes from which the proteins have been chemically removed is also supercoiled. Although supercoiling occasionally plays a role in the expression of some genes, the overall biological function of supercoiling is unknown.

The introduction of one single-strand break (a nick)—for example, by a **deoxyribonuclease (DNase)** enzyme that breaks sugar–phosphate bonds in DNA strands—in a typical supercoiled DNA molecule eliminates all supercoiling because the constraint of underwinding can be removed by a free rotation of the intact strand about the sugar–phosphate bond opposite the break.

The chromosome of *E. coli* is a condensed unit—called a **nucleoid** or **folded chromosome**—containing a single circular DNA molecule. The most-striking feature of the nucleoid is that the DNA is organized into a set of looped domains (Figure 5-2), a feature that is also characteristic of eukaryotic chromosomes. As isolated, the nucleoid contains, in addition to DNA, small amounts of several proteins, which are thought to be responsible in some way for the multiply looped arrangement of the DNA. The degree of condensation of the isolated nucleoid (that is, its physical dimensions) is affected by a variety of factors, and some controversy exists about the state of the nucleoid within a cell.

Figure 5-2 also shows that loops of the DNA of the *E. coli* chromosome are supercoiled. Notice that some loops are not supercoiled; this is a result of the action of several DNases during isolation and indicates that the loops are in some way independent of one another. In the preceding section, it was stated that supercoiling is generally eliminated in a DNA molecule by one single-strand break. However, such a break in the *E. coli* chromosome does not eliminate all supercoiling. If nucleoids, all of whose loops are supercoiled, are treated with a DNase and examined at various times after single-strands have been nicked, it is observed that a single-strand break removes the supercoiling of only one loop, not all loops (Figure 5-3). Thus, the loops must be isolated from one another in such a way that rotation in one loop is not transmitted to other loops. How this occurs is unknown.

It is possible experimentally to eliminate all supercoiling (by introduction of many single-strand breaks or by treatment with certain types of topoi-

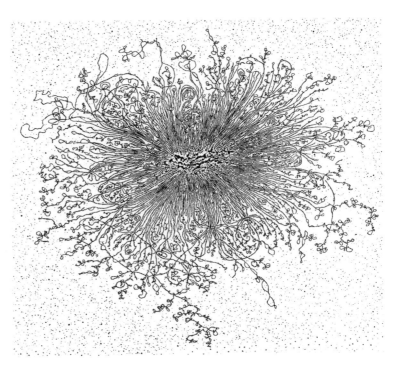

FIGURE 5-2   An electron micrograph of an *E. coli* chromosome showing the multiple loops emerging from a central region. (Courtesy of Ruth Kavenoff. Bluegenes #1. © 1983. All rights reserved by Designergenes Posters Ltd. Posters and shirts are available from Carolina Biological Supply, 2700 York Road, Burlington, North Carolina 27215. Telephone: (800)334-5551.)

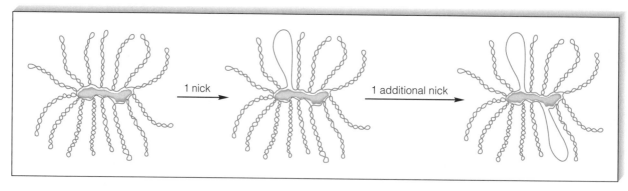

FIGURE 5-3   A schematic drawing of the folded supercoiled *E. coli* chromosome, showing only 15 of the 40-to-50 loops attached to a putative protein core (shaded area) and the opening of loops by nicks.

somerases). When this is done, the overall looped structure of the chromosome is not lost, which indicates that folding and supercoiling of the DNA are independent phenomena.

5-4   Structure of Eukaryotic Chromosomes

A eukaryotic chromosome contains a single DNA molecule of enormous length. For example, the largest chromosome in the *D. melanogaster* genome has a DNA content of about 65,000 kb (6.5×10^7 nucleotide pairs), which is equivalent to a continuous linear duplex about 22 mm long. These long molecules usually fracture in isolation, but some fragments that are recovered are still very long. Figure 5-4 is an autoradiograph of radioactively labeled *Drosophila* DNA more than 36,000 kb in length.

The DNA of all eukaryotic chromosomes is associated with numerous protein molecules in a stable ordered aggregate called **chromatin**. Some of the proteins present in chromatin determine chromosome structure and the changes in structure that occur in the division cycle of the cell. Other chromatin proteins appear to have important roles in regulating chromosome functions.

The Nucleosome Is the Basic Structural Unit of Chromatin

The simplest form of chromatin is present in nondividing eukaryotic cells, when chromosomes are not sufficiently condensed to be visible by light microscopy. Chromatin isolated from such cells is a complex aggregate of DNA and proteins, the major class of which consists of the **histone** proteins.

Histones are largely responsible for the structure of chromatin. There are five major types—**H1, H2A, H2B, H3, and H4**—in the chromatin of almost all eukaryotes, and they are present in amounts about equal in mass to that of the DNA. Histones are small proteins that contain between 100 and 200 amino acids and differ from most other proteins in that from 20 to 30 percent of the amino acids are lysine and arginine, both of which have a positive charge. (Only a few percent of the amino acids of a typical protein are lysine and arginine.) The positive charges enable histone molecules to bind to DNA, primarily by electrostatic attraction to the negatively charged phosphate groups

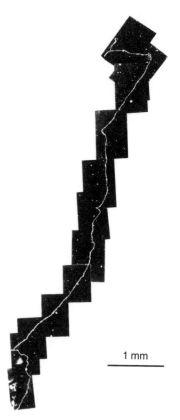

1 mm

FIGURE 5-4   Autoradioagram of a DNA molecule from *D. melanogaster*. The molecule is 12 mm long (approximately 36,000 kb). (From R. Kavenoff, L. C. Klotz, and B. H. Zimm. 1974. *Cold Spring Harbor Symp. Quant. Biol.*, 38: 4.)

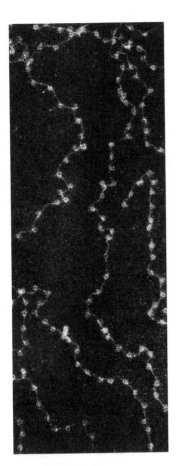

FIGURE 5-5 Dark-field electron micrograph of chromatin showing the beaded structure at low salt concentration. The beads have diameters of about 100 Å. (Courtesy of Ada Olins.)

in the sugar-phosphate backbone of DNA. Placing chromatin in a solution with a high salt concentration (for example, 2 molar NaCl) to eliminate the electrostatic attraction causes the histones to dissociate from the DNA. Histones also bind tightly to each other; both DNA–histone and histone–histone binding are important for chromatin structure.

The histones from different organisms are remarkably similar to one another, with the exception of H1. In fact, the amino acid sequences of H3 molecules from widely different species are almost identical. For example, the sequences of H3 of cow chromatin and pea chromatin differ by only 4 of 135 amino acids. The H4 proteins of all organisms are also quite similar; again, cow and pea H4 differ by only 2 of their 102 amino acids. There are few other proteins whose amino acid sequences vary so little from one species to the next. When the variation is very small between organisms, one says that the sequence is highly **conserved.** The extraordinary conservation in histone composition through hundreds of millions of years of evolutionary divergence is consistent with the important role of these proteins in the structural organization of eukaryotic chromosomes.

By electron microscopy, chromatin looks like a regularly beaded thread (Figure 5-5). Brief treatment of chromatin with certain DNases yields a collection of small particles of quite uniform size consisting only of histones and DNA (Figure 5-6). When the histones are removed from these particles, the DNA fragments are found to be of lengths equal to about 200 nucleotide pairs or small multiples of that unit size (the precise size varies with species and tissue).

The beadlike units in chromatin are called **nucleosomes.** Each unit has a definite composition—namely, one molecule of H1, two molecules each of H2A, H2B, H3, and H4, and one segment of DNA containing about 200 nucleotide pairs. Extensive digestion of these units with a nuclease removes some of the DNA and causes the loss of H1. The resulting structure, called a **core particle,** consists of an octamer of pairs of H2A, H2B, H3, and H4, around which the remaining 145-nucleotide-pair length of DNA is wound in about one and three-fourths turns (Figure 5-7). Thus, a nucleosome is composed of a core particle, additional DNA that links adjacent core particles (the DNA that is

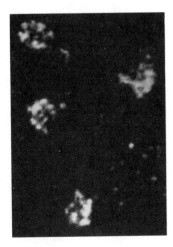

FIGURE 5-6 Electron micrograph of nucleosome monomers. (Courtesy of Ada Olins.)

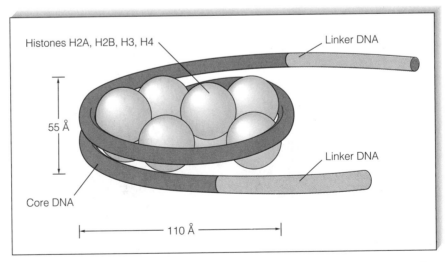

FIGURE 5-7 Diagram of a nucleosome core particle. The DNA molecule is wound one and three-fourths turns around a histone octamer. If H1 were present, it would bind to the octamer surface and to the linkers, causing the linkers to cross.

removed by nuclease digestion), and one molecule of H1; the H1 binds to the histone octamer and to the linker DNA, causing the linkers extending from both sides of the core particle to cross and draw nearer to the octamer, though some of the linker DNA does not come into contact with any histones. The size of the linker ranges from 20 to 100 nucleotide pairs for different species and even in different cell types in the same organism ($200 - 145 = 55$ nucleotide pairs is usually considered an average size). Little is known about the structure of the linker DNA or whether it has a special genetic function, and the cause of the variation in its length is also unknown.

Arrangement of Chromatin Fibers in a Chromosome

The DNA molecule of a chromosome is folded and refolded in such a way that it is convenient to think of chromosomes as having several levels of organization, each responsible for a particular degree of shortening of the enormously long strand (Figure 5-8). Assembly of DNA and histones represents the first level—namely, a sevenfold reduction in length of the DNA and the formation of a beaded flexible fiber 110 Å (11 nm) wide (Figure 5-8B), roughly five times the width of free DNA (Figure 5-8A). The structure of chromatin varies with the concentration of salts, and the 110-Å fiber is present only when the salt concentration is quite low. If the salt concentration is increased slightly, the fiber becomes shortened somewhat by forming a zigzag arrangement of closely spaced beads between which the linking DNA is no longer visible in electron micrographs. If the salt concentration is further increased to that present in living cells, a second level of compaction is reached—namely, the organization of the 110-Å nucleosome fiber into a shorter thicker fiber with an average diameter ranging from 300 to 350 Å, called the **30-nm fiber** (Figure 5-8C). In forming this structure, the 110-Å fiber apparently coils in a somewhat irregular left-handed superhelix or solenoidal supercoil with six nucleosomes per turn (Figure 5-9). It is believed that most intracellular chromatin has the solenoidal supercoiled configuration.

The final level of organization is that in which the 30-nm fiber condenses into a chromatid of the compact metaphase chromosome (Figure 5-8D through

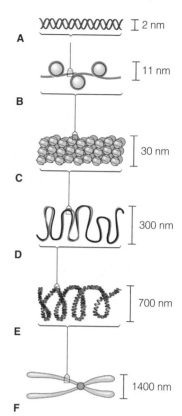

FIGURE 5-8 Various stages in the condensation of (A) DNA and (B through E) chromatin in forming (F) a metaphase chromosome. The dimensions indicate known sizes of intermediates, but the detailed structures are hypothetical.

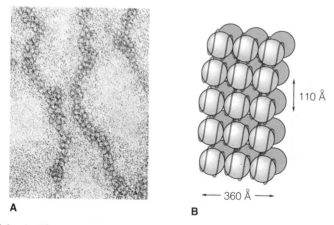

FIGURE 5-9 A. Electron micrograph of the 30-nm component of mouse metaphase chromosomes. (Courtesy of Barbara Hamkalo.) B. A proposed solenoidal model of chromatin. The DNA (red) is wound around each nucleosome. It is unlikely that the real structure is so regular. (After J. T. Finch and A. Klug. 1976. *Proc. Nat. Acad. Sci.*, 73: 1900.)

FIGURE 5-10 Electron micrograph of a partly disrupted anaphase chromosome of the milkweed bug *Oncopeltus fasciatus*, showing multiple loops of 30-nm chromatin at the periphery. (From V. Foe, H. Forrest, L. Wilkinson, and C. Laird. 1982. *Insect Ultrastructure*, 1: 222.)

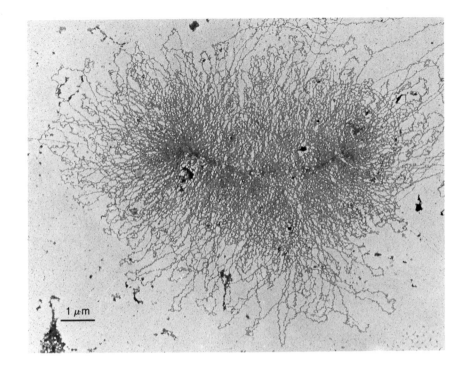

F). Little is known about this process other than that it seems to proceed in stages. In electron micrographs of isolated metaphase chromosomes from which histones have been removed, the partly unfolded DNA has the form of an enormous number of loops that seem to extend from a central core or **scaffold** composed of nonhistone chromosomal proteins (Figure 5-10). Electron microscopic studies of chromosome condensation in mitosis and meiosis suggest that the scaffold extends along the chromatid and that the 30-nm fiber becomes arranged into a helix of loops radiating from the scaffold. Details are not known about the additional folding that is required of the fiber in each loop to produce the fully condensed metaphase chromosome.

The genetic significance of the compaction of DNA and protein into chromatin and ultimately into the chromosome is that it greatly facilitates the movement of the genetic material during nuclear division. Without chromosome condensation there would be many more abnormalities in the distribution of genetic material into daughter cells.

5-5 Polytene Chromosomes

A typical eukaryotic chromosome contains only a single DNA molecule. However, in the nuclei of cells of the salivary glands and certain other tissues of the larvae of *Drosophila* and other two-winged (dipteran) flies, there are giant chromosomes, called **polytene chromosomes**, which contain about 1000 DNA molecules laterally aligned (Figure 5-11). Each of these chromosomes has a volume many times greater than that of the corresponding chromosome at mitotic metaphase in ordinary somatic cells, and a constant and distinctive pattern of transverse banding (Figure 5-12). The polytene structures are formed by repeated replication of the DNA in a closely synapsed pair of homologous chromosomes without separation of the replicated chromatin strands or of the two chromosomes. Polytene chromosomes are atypical chromosomes and are formed in "terminal" cells; that is, the larval cells containing them do not divide

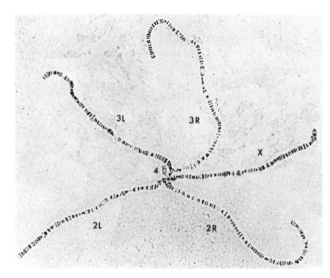

FIGURE 5-11 Polytene chromosomes from a larval salivary-gland cell of *Drosophila melanogaster*. The centromeric regions of all chromosomes are united in the common chromocenter. (Courtesy of George Lefevre.)

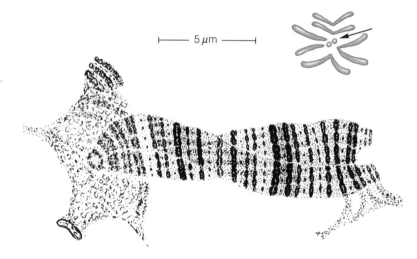

FIGURE 5-12 The polytene fourth chromosome of *Drosophila melanogaster* adhering to the chromocenter, shown at the left. At the upper right, drawn to the same scale, are the somatic chromosomes as they appear in mitotic prophase, with the pair of dotlike fourth chromosomes indicated by the arrow. (From C. Bridges. 1935. *J. Heredity*, 26: 60.)

further during development of the fly and are later eliminated in the formation of the pupa. However, they have been especially valuable in the genetics of *Drosophila*, as will become apparent in Chapter 6.

In polytene nuclei of *D. melanogaster* and other species, large blocks of heterochromatin (a particular type of chromatin described in the following section) adjacent to the centromeres are aggregated into a single compact mass called the **chromocenter**. Because the two largest chromosomes (numbers 2 and 3) have centrally located centromeres, the chromosomes appear in the configuration shown in Figure 5-11: the paired X chromosomes (in a female), the left and right arms of chromosomes 2 and 3, and a short chromosome (chromosome 4) project from the chromocenter. In a male, the Y chromosome, which consists almost entirely of heterochromatin, is incorporated in the chromocenter.

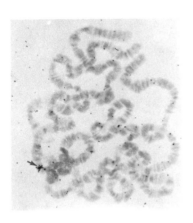

FIGURE 5-13 Autoradiogram of *Drosophila melanogaster* polytene chromosomes hybridized *in situ* with radioactively labeled RNA copied from the histone genes, showing hybridization to a particular region (arrow). This region identifies the position of the histone genes in the chromosomes. (Courtesy of Mary Lou Pardue.)

The darkly staining transverse bands in polytene chromosomes have about a tenfold range in width. These bands result from side-by-side alignment of tightly folded regions of the individual chromatin strands that are often visible in mitotic and meiotic prophase chromosomes as chromomeres (see Figure 2-6). More DNA is present within the bands than in the interband (lightly stained) regions. About 5000 bands have been identified in the *D. melanogaster* polytene chromosomes. This linear array of bands, which has a pattern that is constant and characteristic for each species, provides a finely detailed **cytological map** of the chromosomes. The banding pattern is such that short regions in any of the chromosomes can be identified, as can be seen in Figure 5-12.

Because of their large size and finely detailed morphology, polytene chromosomes are exceedingly useful for *in situ* nucleic acid hybridization. In the ***in situ* hybridization** procedure, labeled probe DNA or RNA is added to squashed polytene nuclei after denaturation of the chromosomal DNA, under conditions that favor renaturation. After washing, the only probe that remains in the chromosomes has formed hybrid duplexes with chromosomal DNA and its position can be identified cytologically (Figure 5-13).

5-6 The Organization of Nucleotide Sequences in Eukaryotic Genomes

In bacteria, the variation of average base composition from one part of the genome to another is quite small. However, in eukaryotes, some components of the genome can be detected because their base composition is quite different from the average of the rest of the genome (for example, one component of crab DNA is only 3 percent GC). These components are called **satellite DNA.** In the mouse, satellite DNA accounts for 10 percent of the genome. A striking feature of satellite DNA is that it consists of fairly short nucleotide sequences that may be repeated *as many as a million times in a haploid genome.* Other **repetitive sequences** also are present in eukaryotic DNA. Some of the special features of repetitive DNA are considered in the next section.

5-7 Nucleotide Sequence Composition

Eukaryotic organisms differ widely in the proportion of the genome consisting of repetitive DNA sequences and in the types of these sequences that are present. A eukaryotic genome typically consists of three components:

1. **Unique,** or **single-copy, sequences.** This is usually the major component and is typically from 30 to 75 percent of the chromosomal DNA in most organisms.
2. **Highly repetitive sequences.** This component constitutes from 5 to 45 percent of the genome. Some of these sequences are the satellite DNA referred to earlier. The sequences in this class are typically from 5 to 300 base pairs per repeat and are duplicated as many as 10^5 times per genome.
3. **Middle-repetitive sequences.** This component is from 1 to 30 percent of a eukaryotic genome and includes sequences that are repeated from a few times to 10^5 times per genome.

These different components can be identified according to the number of bands that appear in Southern blots with the use of appropriate probes or by other methods. It should be noted that the dividing line between many middle-repetitive sequences and highly repetitive sequences is arbitrary.

Unique Sequences

Most gene sequences and the adjacent nucleotide sequences required for their expression are contained in the unique-sequence component. With minor exceptions (for example, the repetition of one or a few genes), the genomes of viruses and prokaryotes are composed entirely of single-copy sequences; in contrast, such sequences constitute only 38 percent of the total genome in some sea urchin species, a little more than 50 percent of the human genome, and about 70 percent of the *D. melanogaster* genome.

Highly Repetitive Sequences

Many highly repetitive sequences are localized in blocks of tandem repeats, whereas others are dispersed throughout the genome. An example of the dispersed type is a family of related sequences in the human genome called the **Alu** family because the sequences contain a characteristic restriction site for the enzyme AluI (Section 4-10). Alu sequences are 300 base pairs in length and are present in about 500,000 copies in the human genome; this repetitive DNA family alone accounts for about 5 percent of human DNA.

Among the localized highly repetitive sequences, most are fairly short. Sequences of this type make up about 6 percent of the human genome and 18 percent of the *D. melanogaster* genome, but 45 percent of the DNA of *D. virilis*. One of the simplest possible repetitive sequences is composed of an alternating . . . ATAT . . . sequence with about 3 percent GC interspersed, which makes up 25 percent of the genomes of certain species of land crabs. In the *D. virilis* genome, the major components of the highly repetitive class are three related sequences of seven base pairs, which have the following compositions in one of the complementary strands:

$$5'\text{-ACAAACT-}3'$$
$$5'\text{-ATAAACT-}3'$$
$$5'\text{-ACAAATT-}3'$$

Blocks of satellite (highly repetitive) sequences in the genomes of several organisms have been located by *in situ* hybridization with metaphase chromosomes (Figure 5-14). The satellite sequences located by this method have been found to be in the regions of the chromosomes called **heterochromatin.** These are regions that condense earlier in prophase than the rest of the chromosome and are darkly stainable by many standard dyes used to make chromosomes visible (Figure 5-15); sometimes the heterochromatin remains highly condensed throughout the cell cycle. The **euchromatin,** which makes up most of the genome, is visible only in the mitotic cycle. The major heterochromatic regions are adjacent to the centromere; smaller blocks are at the ends of the chromosome arms (the telomeres) and interspersed with the euchromatin. In many species, an entire chromosome, such as the Y chromosome in *D. melanogaster,* is almost completely heterochromatic. Different highly repetitive sequences have been purified from *D. melanogaster* and *in situ* hybridization has shown that each chromosome has its own distinctive types and distribution of these sequences.

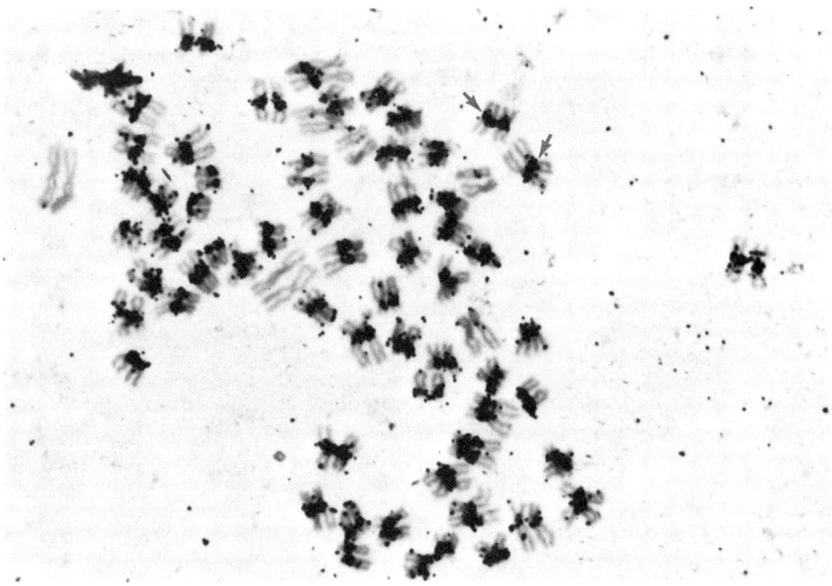

FIGURE 5-14 Autoradiogram of metaphase chromosomes of the kangaroo rat *Dipodomys ordii*; radioactive RNA copied from purified satellite DNA sequences has been hybridized to the chromosomes to show the localization of the satellite DNA. Hybridization is principally in the regions adjacent to the centromeres (arrows). Note that some chromosomes are apparently free of this satellite DNA. They contain a different satellite DNA not examined in this experiment. (Courtesy of David Prescott.)

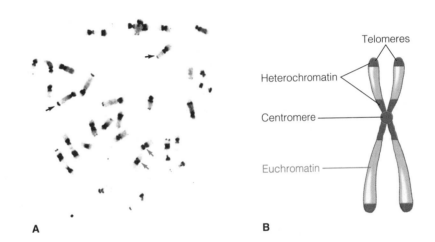

FIGURE 5-15 A. Metaphase chromosomes of the ground squirrel *Ammospermophilus harrissi*, stained to show the heterochromatic regions near the centromere of most chromosomes (red arrows) and the telomeres of some chromosomes (black arrows). (Courtesy of T. C. Hsu.) B. An interpretive drawing.

The number of genes located in heterochromatin is small relative to the number in euchromatin. The relatively small number of genes means that large blocks of heterochromatin are genetically almost inert, or devoid of function. Indeed, such heterochromatic blocks can often be rearranged in the genome, duplicated, or even deleted without major phenotypic consequences.

Middle-repetitive Sequences

Middle-repetitive sequences constitute about 12 percent of the *D. melanogaster* genome and 40 percent or more of the human and other eukaryotic genomes. These sequences differ greatly in the number of copies and their distribution within a genome. They represent many families of related sequences and include several groups of genes. For example, the genes for the RNA components of the ribosomes—the particles on which proteins are synthesized (Chapter 11)—and the genes for tRNA molecules, which also participate in protein synthesis, are repeated in the genomes of all organisms. Genes for the two major ribosomal RNA molecules are a tandem pair that is repeated several hundred times in most eukaryotic genomes. The genomes of all eukaryotes also contain multiple copies of the histone genes. Each histone gene is repeated about ten times per genome in chickens, twenty times in mammals, about one hundred times in *Drosophila*, and as many as six hundred times in certain sea urchin species.

The dispersed middle-repetitive-DNA component of the *D. melanogaster* genome consists of about fifty families of related sequences, and from twenty to sixty copies of each family are widely scattered throughout the chromosomes. The positions of these sequences differ from one individual fly to the next except in completely homozygous laboratory strains. The variability in position occurs because many of these sequences are able to move from one location to another in a chromosome and between chromosomes; they are said to be **transposable elements.** Analogous types of sequences are found in the genomes of yeast, maize, and bacteria (Chapter 10) and probably exist in all organisms. An important dimension has been added to our understanding of the genome as a structural and functional unit by the discovery of these mobile genetic elements, because they can in some cases cause chromosome breakage, chromosome rearrangements, modification of the expression of genes, and novel types of mutations.

5-8 Transposable Elements

In the 1940s, in a study of the genetics of kernel mottling in maize (Figure 5-16), Barbara McClintock discovered an element that regulated the mottling and caused the chromosome carrying the genes for color and consistency of the kernels to break. The element was called Dissociation (*Ds*). Mapping data showed that the chromosome breakage is always at or very near the location of *Ds*. McClintock's critical observation was that *Ds* does not have a constant location but occasionally moves to another (**transposition**), causing chromosome breakage at a new site. Furthermore, *Ds* move only if a second element, called Activator (*Ac*), is present. In addition, *Ac* itself moves within the genome and can cause alterations in the expression of genes at or near its insertion site similar to the modifications resulting from the presence of *Ds*.

Additional transposable elements with characteristics and genetic effects similar to those of *Ac* and *Ds* are known in maize. Much of the color variation seen in kernels of the varieties used for decorative purposes are attributable to the presence of one or more of these elements.

Since McClintock's discovery, transposable nucleotide sequences have been observed to be widespread in eukaryotes and prokaryotes. In *D. melanogaster,* they constitute from 5 to 10 percent of the genome and represent about fifty distinct families of sequences. One well-studied family of closely related, but not identical, sequences is called *copia.* This element is present in about thirty copies per genome. The *copia* element (Figure 5-17) contains about 5000 base pairs with two identical sequences of 267 base pairs located terminally and in the same orientation (the sequences are called **direct repeats**). The ends of each of these terminal repeats contain two segments of 17 base pairs, whose sequences are also nearly identical. These shorter segments have opposite orientations and are called **inverted repeats.** Other transposable elements have a similar organization of direct or inverted terminal repeats, as do many such elements in other organisms—for example, the transposable elements in *E. coli* described in Chapter 10.

The molecular processes responsible for the movement of transposable elements are not well understood (some information will be presented in

FIGURE 5-16 Sectors of colored and waxy tissue in the endosperm of maize kernels resulting from the presence of the transposable elements *Ds* and *Ac*. (Courtesy of Barbara McClintock and the Cold Spring Harbor Laboratory Library Archives. Photograph by David Greene.)

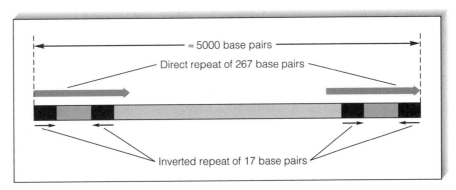

FIGURE 5-17 Sequence organization of a *copia* transposable element of *Drosophila melanogaster*.

Chapter 10). A common feature is that transposition of the element is usually accompanied by the duplication of a small number of base pairs originally present at the insertion site, with the result that a copy of this short chromosomal sequence is found immediately adjacent to both ends of the inserted element (Figure 5-18). The length of the duplicated segment ranges from 2 to 12 base pairs, depending on the particular transposable element. Insertion of a transposable element is not a sequence-specific process, in that at each location the element is flanked by a different duplicated sequence; however, the *number* of duplicated base pairs is usually the same at all locations and is characteristic of a particular transposable element. Experimental deletion or mutation of part of the base sequences of several different elements has shown that the short terminal inverted repeats are essential for transposition, probably because they are necessary for binding an enzyme called a **transposase** that is required for transposition. Many transposable elements code for their own transposase by means of a gene located in the central region between the terminal repeats, and hence these elements are able to promote their own transposition. Elements in which the transposase gene has been lost or inactivated by mutation are transposable only if a related element is present in

FIGURE 5-18 The sequence arrangement of one type of transposable element—in this case, *Ds* of maize—and the changes that occur during insertion. *Ds* is inserted into the maize *sh* gene at the position indicated. In the insertion process, a sequence of eight base pairs next to the site of insertion is duplicated and flanks the *Ds* element.

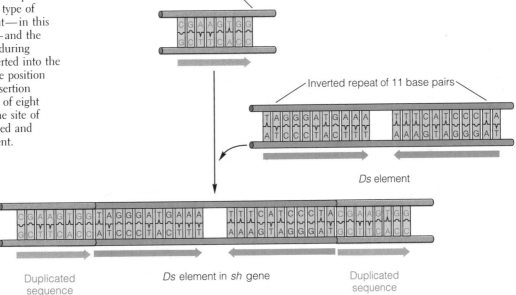

the genome to provide this activity. Thus, the inability of the maize *Ds* element to transpose without *Ac* results from the absence of a functional transposase gene in *Ds*.

Transposable elements are responsible for many visible mutations. For example, a transposable element in peas is responsible for the wrinkled-seed mutation studied by Gregor Mendel. The wildtype allele of the gene codes for starch-branching enzyme I (SBEI), which is used in the synthesis of amylopectin (starch with branched chains). In the wrinkled mutation, a transposable element inserted into the gene renders the enzyme nonfunctional. The pea transposable element has terminal inverted repeats that are very similar to those in the maize *Ac* element, and the insertion site in the *wrinkled* allele is flanked by a duplication of eight base pairs of the SBEI coding sequence. This particular insertion appears to be genetically quite stable, because the transposable element does not seem to have been excised in the long history of wrinkled peas.

Recognition in recent years that transposable DNA sequences exist in most genomes and are quite numerous has greatly altered our perception of the organization and stability of the genetic material. Transposable elements are examples of **selfish DNA,** which maintains itself in the genome because of its ability to replicate and transpose. Selfish DNA does not appear to have a specific function in the genome, nor does its presence appear to benefit the host organism. Transposable elements do have genetic consequences, however, and in later chapters some of the important genetic consequences of the insertion and excision of these elements will be discussed.

5-9 Centromere and Telomere Structure

The centromere is a specific region of the eukaryotic chromosome that becomes visible as a distinct morphological entity along the chromosome during condensation. It is responsible for chromosome movement in both mitosis and meiosis, functioning, at least in part, by serving as an attachment site for one or more spindle fibers. It is also the site at which the spindle fibers shorten, causing the chromosomes to move toward the poles. Electron microscopic analysis has shown that in some organisms—for example, the yeast *Saccharomyces cerevisiae*—a single spindle-protein fiber is attached to centromeric chromatin. Most other organisms have multiple spindle fibers attached to each centromeric region.

The chromatin segment of the centromeres of *Saccharomyces cerevisiae* has a unique structure in that it is exceedingly resistant to the action of various DNases and has been isolated as a protein-DNA complex containing from 220 to 250 base pairs. The nucleosomal constitution and DNA base sequences of many of the yeast chromosomes have been determined. Several common features of the base sequences are shown in Figure 5-19A. There are four regions, labeled I through IV. All yeast centromeres have the sequence characteristics indicated for regions I, II, and III, but the sequence of region IV varies from one centromere to another. Region II is noteworthy in that more than 90 percent of the base pairs are AT pairs. The centromeric DNA is contained in a structure (the centromeric core particle) that contains more DNA than a typical yeast nucleosome core particle (160 base pairs) and is larger. This structure is responsible for the resistance of centromeric DNA to DNase. The spindle fiber is believed to be attached directly to this particle (Figure 5-19B).

The base-sequence arrangement of the yeast centromeres is not typical of other eukaryotic centromeres. In higher eukaryotes, the chromosomes are about

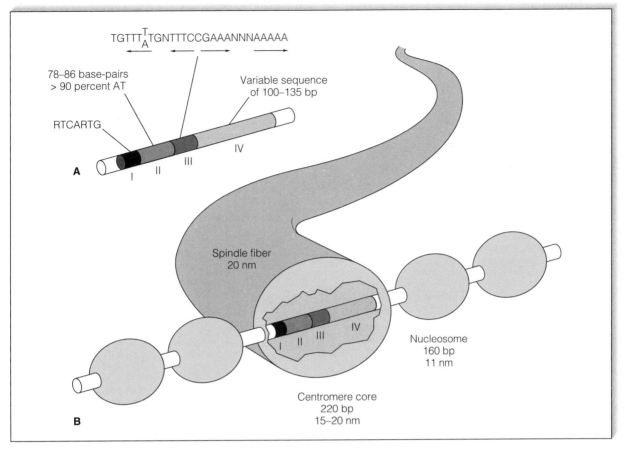

FIGURE 5-19 A yeast centromere. A. Diagram of centromere DNA showing the major regions (I–IV) shared by all yeast centromeres. The symbol R indicates any purine (A or G), and the symbol N indicates any nucleotide. Note the inverted repeat segments in region III (arrows). The sequence of region IV varies from one centromere to the next. B. Positions of the centromere core and the nucleosomes on the DNA. The DNA is wrapped around histones in the nucleosomes, but the detailed organization and composition of the centromere core are unknown. (After K. S. Bloom, M. Fitzgerald-Hayes, and J. Carbon. 1982. *Cold Spring Harbor Symp. Quant. Biol.*, 47: 1175.)

one hundred times as large as yeast chromosomes, and several spindle fibers are usually attached to each chromosome. Furthermore, the centromeric regions of the chromosomes of many higher eukaryotes contain large amounts of heterochromatin, consisting of repetitive satellite DNA, as described in Section 5-6. For example, the centromeric regions of human chromosomes contain a tandemly repeated DNA sequence of about 170 base pairs called the alpha satellite. The number of copies in the centromeric region ranges from 5000 to 15,000, depending on the chromosome. The DNA sequences needed for spindle-fiber attachment may be interspersed among the alpha-satellite sequences. Whether the alpha-satellite sequences themselves contribute to centromere activity is unknown.

A **telomere** is a special DNA sequence at the tip of a eukaryotic chromosome essential for chromosome stability. The evidence that telomeres are special is derived from genetic and microscopic observation. For example, in both maize and *Drosophila*, the ends of broken chromosomes that lack telomeres are not genetically stable. The broken ends tend to fuse together whenever possible. Furthermore, if one end of a single metaphase chromosome is broken away and lost, the sister chromatids often fuse, forming a chromosome with two centromeres.

The special telomere DNA sequences are *added* to the ends of eukaryotic chromosomes by an enzyme called **telomerase.** The substrate for the telomerase is a telomere addition sequence consisting of a short repetitive sequence that is highly conserved in most organisms. Among vertebrates (and the ciliate *Tetrahymena,*) the telomere sequence consists of 300 to 600 tandem repeats of the simple sequence 5'-TTAGGG-3'. Relatively few copies of this sequence at the end of a chromosome are necessary to prime the telomerase to add additional copies and form a telomere. Most telomere regions also contain longer, moderately repetitive DNA sequences located subterminally in the chromosomes. These sequences differ among organisms and even among different chromosomes in the same organism.

CHAPTER SUMMARY

The DNA content of organisms varies widely. Small viruses exist whose DNA contains only a few thousand nucleotides, and among the higher animals and plants the DNA content can be as large as 1.5×10^{11} nucleotides. Generally DNA content increases with the complexity of the organism, but within particular orders and genera the DNA content varies as much as tenfold.

DNA molecules come in a variety of forms. Except for a few of the smallest viruses, whose DNA is single stranded, and for some viruses in which RNA is the genetic material, the DNA of all organisms is double stranded. The chromosomal DNA of higher organisms is always linear. Bacterial DNA is circular, as is the DNA of many animal viruses and of some bacteriophages. Circular DNA molecules are invariably supercoiled. The bacterial chromosome consists of independently supercoiled domains: the independence is probably the result of proteins that bind to the DNA in a way that prevents rotation of the helix.

The DNA of both prokaryotic and eukaryotic cells and of viruses is never in a fully extended state but is folded in an intricate way, thereby reducing its effective volume. In viruses, the DNA is tightly folded but without bound protein molecules. In bacteria, the DNA is folded to form a multiply looped structure, called a nucleoid, which includes several proteins that are essential for folding. In eukaryotes, the DNA is compacted into chromosomes, which contain several proteins and which are thick enough to be visible by light microscopy in the mitotic phase of the cell cycle. The DNA-protein complex of eukaryotic chromosomes is called chromatin. The protein component of chromatin consists primarily of five distinct proteins: the histones H1, H2A, H2B, H3, and H4. The last four types aggregate to form an octameric protein

containing two copies of each type of histone. DNA is wrapped one and three-fourths turns around the histone octamer, forming a particlelike structure called a nucleosome. This wrapping is the first level of compaction of the DNA in chromosomes. Each nucleosome unit contains about 200 nucleotide pairs of which about 145 are in contact with the protein. The remaining 55 nucleotide pairs link adjacent nucleosomes. Histone H1 binds to the linker segment and draws the nucleosomes nearer to one another. The DNA in its nucleosome form is further compacted to a helical fiber, the 30-nm fiber. In forming a visible chromosome, this unit undergoes several additional levels of folding, producing a highly compact visible chromosome. The result is that a eukaryotic DNA molecule, whose length and width are about 50,000 and 0.002 μm, respectively, is folded in many ways to form a chromosome with a length of tens of micrometers and a width of about 0.5 μm.

Polytene chromosomes are found in certain organs in insects. These gigantic chromosomes consist of about 1000 molecules of partly folded chromatin aligned side by side. Seen by microscopy, they have about 5000 transverse bands. Polytene chromosomes do not replicate further, and cells containing them do not divide. They are useful to geneticists primarily as morphological markers for particular genes and chromosome segments.

The number of copies of individual base sequences in a DNA molecule can vary tremendously. In prokaryotic DNA, most sequences are unique. However, in eukaryotic DNA, only a fraction of the DNA consists of unique sequences present once per haploid genome. Many sequences are present in hundreds to millions of copies. Some highly repetitive sequences are primarily located in the centromeric regions of the chromosomes, whereas others are dispersed. A significant fraction of the DNA, the middle-repetitive DNA, consists of sequences of which from ten to a thousand copies per cell are present. Much of middle-repetitive DNA in the higher eukaryotes consists of transposable elements, sequences able to move from one part of the genome to another. A typical transposable element is a sequence of one to several thousand nucleotide pairs terminating in short sequences that are either identical (direct repeats) or inverted, compared with one another (inverted repeat). The terminal repeats plus a transposase enzyme are necessary for the movement of these elements, a process known as transposition. Many transposable elements contain a gene coding for their own transposase. Insertion of most transposable elements causes a duplication of a short chromosomal nucleotide sequence flanking the point of insertion. Transposable elements are examples of selfish DNA, which maintains itself in the genome without benefitting the host; such elements do have genetic consequences.

Centromeres and telomeres are regions of eukaryotic chromosomes specialized for spindle-fiber attachment and stabilization of the tips, respectively. The centromeres of most higher eukaryotes are associated with localized, highly repeated, satellite DNA sequences. Telomeres are formed by a telomerase enzyme adding nucleotides to the end of a telomerase addition site. In vertebrates, the telomere sequence consists of 300 to 600 tandem repeats of the simple sequence 5'-TTAGGG-3'. Relatively few copies of this sequence are needed to prime the telomerase.

KEY TERMS

chromatin
chromocenter
conserved sequence
copia element
core particle
covalently closed circle
cytological map
deoxyribonuclease
direct repeat
DNase
euchromatin
folded chromosome

genome
heterochromatin
histone
in situ hybridization
inverted repeat
negatively supercoiled
nucleosome
polytene chromosome
relaxed DNA
repetitive sequence
satellite DNA
scaffold

selfish DNA
single-copy sequence
supercoiling
telomerase
telomere
30-nm fiber
transposable element
transposase
transposition
unique sequence

EXAMPLES OF WORKED PROBLEMS

Problem 1: The smallest known chromosome in *Drosophila* was induced by x-rays and is about 3 megabases in length. What is the approximate physical length of this chromosome in millimeters? (A 1-megabase (Mb) molecule of duplex DNA contains one million nucleotide pairs, and $1 \text{ Å} = 10^{-7}$ mm.)

Answer: The standard form of duplex DNA contains 10 nucleotide pairs per 34Å, or 3.4 Å per nucleotide pair. Therefore, a 3-Mb molecule has a length of $3 \times 10^6 \times 3.4$ Å $= 1.02 \times 10^7$ Å $= 1.02$ mm; that is, a 3-megabase molecule of DNA has a physical size of approximately 1 mm.

Problem 2: The genome size of *Drosophila melanogaster* is 165,000 kilobases (kb), approximately 2/3 of which is euchromatic and 1/3 heterochromatic. In salivary-gland chromosomes, the euchromatic part of the genome becomes polytene, and the giant chromosomes exhibit approximately 5000 transverse bands used as landmarks for the locations of genes, chromosome breakpoints, and other cytogenetic features. What is the approximate DNA content of an average band in the salivary-gland chromosomes?

Answer: The euchromatic part of the *Drosophila* genome consists of approximately 165,000 kb × 2/3 = 110,000 kb of nucleotide pairs. This is distributed over 5000 bands, for an average of 110,000 kb/5000 = 22 kb per band.

Problem 3: Suppose that the DNA sequence repeated at the ends of a transposable element undergoes pairing and recombination. What are the genetic consequences if: (a) the sequences are direct repeats? or (b) the sequences are inverted repeats? Draw diagrams to support your answers.

Answer: The diagrams show the consequences.

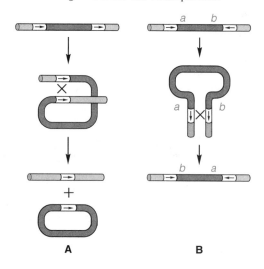

A   B

(a) When direct repeats pair and undergo crossing-over, the result is the deletion of one of the repeats and the DNA between them.

(b) When the repeats are inverted repeats, the result is that the transposable element is inverted in the chromosome, as indicated by the red letters.

PROBLEMS

5-1. Histones are basic proteins, which means that they interact with acids. What is the acid with which they interact? How many individual protein molecules are in a nucleosome? Which type of histone molecule is not present in pairs in nucleosomes?

5-2. Are the ratios of the different histone types the same in all cells of a eukaryotic organism? In all eukaryotic organisms?

5-3. Are all repetitive sequences transposable elements? If you were determining the base sequence of a long DNA sequence, what features of the sequence would make you suspect that a transposable element was present?

5-4. The w^{pch} mutant allele of the *white* gene in *Drosophila* contains an insertion of 1.3 kb within the controlling region of the gene. The w^{pch} allele undergoes further mutation at a high rate, including reverse mutations to w^+. The w^+ derivatives are found to lack the 1.3-kb insert. What kind of DNA sequence is the 1.3-kb insert likely to be?

5-5. Some transposable elements in animals are active only in the germ line and not in somatic cells. How might this restriction of activity be advantageous to the persistence of the transposable element?

5-6. Most transposable elements create a short duplication of host sequence when they insert. The duplication is always in a direct, rather than an inverted, orientation. What does this tell you about the process of insertion?

5-7. The *E. coli* chromosome is 4700 kb in length. What is its length in millimeters? The haploid human genome contains 3×10^9 base pairs of DNA. What is its length in millimeters?

5-8. If you had never seen a chromosome in a microscope but had seen nuclei, what feature of DNA structure would tell you that DNA must exist in a highly coiled state within cells?

5-9. What causes the bands in a polytene chromosome? How do polytene chromosomes undergo division?

5-10. Are genes found in heterochromatin?

5-11. Where are the telomeres located in chromosomes? Is it possible to have a chromosome with the centromere exactly at the tip? What type of chromosome has no telomeres?

5-12. Is there an analog of a telomere in the *E. coli* chromosome?

5-13. Many transposable elements contain a pair of repeated sequences. What is their position in the element? Are they in direct or inverted orientation?

5-14. Consider a long linear DNA molecule, one end of which is rotated four times with respect to the other end in the unwinding direction.
(a) If the two ends are joined to keep the molecule in the underwound state, how many base pairs will be broken?
(b) If the underwound molecule is allowed to form a supercoil, how many twists will be present?

5-15. Endonuclease S1 can break only single-stranded DNA but does not break double-stranded linear DNA. However, S1 can cleave supercoiled DNA, usually making a single break. Why does this occur?

5-16. Circular DNA molecules of the same size migrate at different rates in electrophoretic gels, depending on whether they are supercoiled or relaxed circles. Why?

5-17. Denaturation of DNA refers to breakdown of the double helix and ultimate separation of the individual strands, and it can be induced by heat and other treatments. Because GC base pairs have three hydrogen bonds and AT base pairs have only two, the temperature required for denaturation increases with the GC content and with the length of continuous GC tracts in the molecule. Renaturation of DNA refers to the formation of double-stranded DNA from complementary single strands.

(a) Which of the two—denaturation or renaturation—is dependent on the concentration of DNA?

(b) Which of the following two DNA molecules would have the lower temperature for strand separation? Why?

(1) AGTTGCGACCATGATCTG
    TCAACGCTGGTACTAGAC

(2) ATTGGCCCCGAATATCTG
    TAACCGGGGCTTATAGAC

5-18. DNA from species A, labeled with ^{14}N and randomly fragmented, is renatured with an equal concentration of DNA from species B, labeled with ^{15}N and randomly fragmented, and then centrifuged to equilibrium in CsCl. Five percent of the total renatured DNA has a hybrid density. What fraction of the base sequences are common to the two species?

5-19. What is meant by the terms "direct repeat" and "inverted repeat"? Use the base sequence shown here as an example.

5-20. Mutations have not been observed that result in nonfunctional histones. Why should this be expected?

5-21. A fraction of middle-repetitive DNA from *Drosophila* is isolated and purified. It is used as a template in a polymerization reaction with DNA polymerase and radioactive substrates, and highly radioactive probe DNA is prepared. The probe DNA is then used in an *in situ* hybridization experiment with cells containing polytene chromosomes obtained from ten different flies of the same species. Autoradiography indicates that the radioactive material is localized to about twenty sites in the genome, but they are in different sites in each fly examined. What does this observation suggest about the DNA sequence being studied?

5-22. A sample of identical DNA molecules, each containing about 3000 base pairs per molecule, is mixed with histone octamers under conditions that allow the formation of chromatin. The reconstituted chromatin is then treated with a nuclease and enzymatic digestion is allowed to take place. The histones are removed and the positions of the cuts in the DNA are identified by sequencing the fragments. It is found that the breaks have been made at *random* positions, and, as expected, at about 200-base-pair intervals. The experiment is then repeated with a single variation. A protein, X, known to bind to DNA is added to the DNA sample before the addition of the histone octamers. Again, reconstituted chromatin is formed and digested with nuclease. In this experiment, it is found that the breaks are again at intervals of about 200 base pairs, but they are localized at particular positions in the base sequence. At each position, the site of breakage can vary over only a 2-to-3-base range. However, examination of the base sequences in which the breaks have occurred does not indicate that breakage occurs in a particular sequence—that is, each 2-to-3-base region in which cutting occurs has a different sequence. Explain the difference between the two experiments.

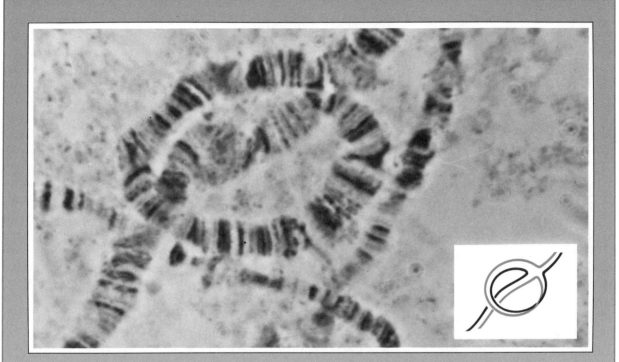

Variation in Chromosome Number and Structure

In all species, individual organisms are occasionally found that have extra chromosomes or that lack a particular chromosome. These conditions are due to abnormalities in chromosome *number*. Other rare individuals are found to have an alteration in the arrangement of genes in the genome, such as by having a chromosome with a particular segment missing, reversed in orientation, or attached to a different chromosome. These alterations are due to abnormalities in chromosome *structure*. This chapter will deal with the genetic effects of both numerical and structural chromosome abnormalities. It will be seen that plants are much more tolerant of such changes than are animals and that, in animals, numerical alterations often produce stronger effects on phenotype than do structural alterations.

Above: An inversion of the distal half of the X chromosome of *Drosophila melanogaster* in heterozygous combination with a normal X. Inset: An interpretive drawing. (Courtesy of Dan Lindsley.)

6-1 Forms of Chromosomes

In both mitosis and meiosis, spindle fibers attach to the centromere of each chromosome in cell division and cause the sister chromatids to be separated and moved to opposite poles. Occasionally, chromosomes arise having an abnormal number of centromeres, as diagrammed in Figure 6-1A. The upper chromosome has two centromeres and is said to be **dicentric.** Such aberrant chromosomes are unstable because they are frequently lost from cells when the two centromeres proceed to opposite poles in cell division; in this case, the chromosome is stretched and forms a *bridge* between the daughter cells, which may not be included in either daughter nucleus or may break, with the result that each daughter nucleus receives a broken chromosome. The lower chromosome in the figure is an **acentric** chromosome, which lacks a centromere. Acentric chromosomes also are unstable because they cannot be maneuvered properly during cell division and tend to be lost. Only chromosomes that have a single centromere are regularly transmitted from parents to offspring.

Chromosomes are conveniently described by their form during anaphase movement. Three distinct shapes are seen, resembling a V, or a J, or an I. The shape is determined by the position of the centromere, which determines the relative length of the lagging chromosome arms (Figure 6-1B). A V-shaped chromosome has its centromere approximately in the middle, forming arms of about equal length, and is called a **metacentric** chromosome. A J-shaped chromosome has an off-center centromere, forming arms of unequal length; such chromosomes are **submetacentric.** When the centromere is very close to one end, the chromosome appears I-shaped at anaphase because the arms are grossly unequal in length; such a chromosome is **acrocentric.**

The distinctions between metacentric, submetacentric, and acrocentric chromosomes are somewhat arbitrary, but the terms are useful because they supply a physical image of the chromosome. More important, chromosome evolution often tends to conserve the number of chromosome *arms* without conserving the number of *chromosomes.* For example, *Drosophila melanogaster* has two large metacentric autosomes, but many other *Drosophila* species have four acrocentric autosomes instead. Detailed comparison of the genetic maps of these species reveals that the acrocentric chromosomes in the other species correspond, arm for arm, with the large metacentrics in *D. melanogaster* (Figure 6-2). Among higher primates, chimpanzees and human beings have 22 pairs of chromosomes that are morphologically similar, but chimpanzees have two pairs of acrocentrics not found in human beings, and human beings have one pair of metacentrics not found in chimpanzees. In this case, each arm of the human metacentric is homologous to one of the chimpanzee acrocentrics.

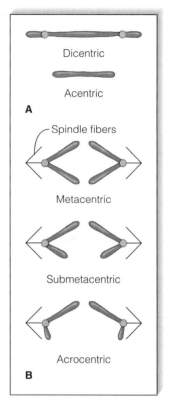

FIGURE 6-1   A. Diagram of a dicentric (two centromeres) chromosome and an acentric (no centromere) one. Dicentric and acentric chromosomes are frequently lost in cell division, the former because the two centromeres may bridge between the daughter cells and the latter because the chromosome cannot attach to the spindle fibers. B. The three possible shapes of chromosomes in anaphase. The centromeres are outlined in red.

6-2   Polyploidy

The genus *Chrysanthemum* illustrates an important phenomenon frequently found in higher plants. One *Chrysanthemum* species has 18 chromosomes, whereas a closely related species has 36. However, comparison of chromosome morphology indicates that the 36-chromosome species has two complete sets of the chromosomes found in the 18-chromosome species. This phenomenon is known as **polyploidy.** The basic haploid chromosome number in the group is 9, which is the chromosome number found in gametes of the 18-chromosome species; that is, the 18-chromosome species has 2 copies of each of the 9

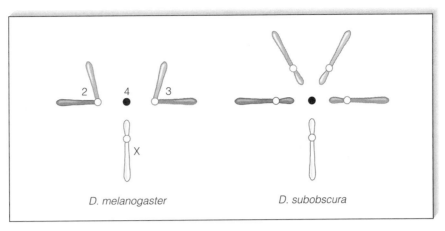

FIGURE 6-2   The haploid chromosome complement of two species of *Drosophila*. Shading indicates homology of chromosome arms. The large metacentric chromosomes of *Drosophila melanogaster* (chromosomes 2 and 3) correspond arm for arm with the four large acrocentric autosomes of *Drosophila subobscura*.

chromosomes of the haploid set and so is a normal diploid. The 36-chromosome species has 4 copies of each of the 9 basic chromosomes ($4 \times 9 = 36$) and is a **tetraploid**. Other species of *Chrysanthemum* have 54 chromosomes (6×9), 72 chromosomes (8×9), and 90 (10×9).

In meiosis, the chromosomes of all *Chrysanthemum* species synapse normally in pairs to form bivalents (Section 2-3). The 18-chromosome species forms 9 bivalents, the 36-chromosome species forms 18 bivalents, the 54-chromosome species forms 27 bivalents, and so forth. Gametes receive one chromosome from each bivalent; so the number of chromosomes in the gametes of any species is exactly half the number of chromosomes in its somatic cells. For example, the 90-chromosome species forms 45 bivalents; so the gametes contain 45 chromosomes. When two 45-chromosome gametes come together during fertilization, the complete set of 90 chromosomes in the species is restored. Thus, *the gametes of a polyploid organism are not always haploid*, as they are in a diploid; for example, a tetraploid organism has diploid gametes.

Polyploidy is widespread in certain plant groups, and it is found in many valuable crop plants, such as wheat, oats, cotton, potatoes, bananas, coffee, and sugar cane. Among flowering plants, at least one-third of existing species originated as some form of polyploid. Polyploidy often leads to an increase in the size of individual cells, and polyploid plants are often larger and more vigorous than their diploid ancestors; however, there are many exceptions to these generalizations. Polyploidy is rare in vertebrate animals, but it is found in a few groups of invertebrates.

Polyploid plants occurring in nature almost always have an even number of sets of chromosomes because organisms having an odd number have low fertility. Organisms with three complete sets of chromosomes are known as **triploids**. As far as growth is concerned, a triploid is quite normal because the triploid condition does not interfere with mitosis; in mitosis in triploids (or any other type of polyploid), each chromosome replicates and divides just as in a diploid. However, because each chromosome has more than one pairing partner, chromosome segregation is severely upset in meiosis, and most gametes are defective. Unless the organism can perpetuate itself by means of asexual reproduction, it will eventually become extinct.

The infertility of triploids is sometimes of commercial benefit. For example, the seeds in commercial bananas are small and edible because the plant is triploid and most of the seeds fail to develop to full size. In oysters, triploids are produced by treating fertilized diploid eggs with a chemical that causes the second polar body of the egg to be retained. The triploid oysters are sterile and do not spawn, and so they remain edible through the hot summer months of June, July, and August (months lacking the letter *r*), when the spawning of normal oysters renders them inedible. In Florida and in certain other states, weed control in waterways is aided by release of weed-eating fish (the grass carp), which do not become overpopulated and a problem themselves, because the released fish are sterile triploids.

Tetraploid organisms can be produced in several ways. The simplest mechanism is a failure of chromosome separation in mitosis, which instantly doubles the chromosome number. In a plant species that can undergo self-fertilization, such an occurrence creates a new, genetically stable species because the chromosomes in the tetraploid can pair two by two in meiosis and therefore segregate regularly, each gamete receiving a full diploid set of chromosomes. Self-fertilization of the tetraploid restores the chromosome number; so the tetraploid condition can be perpetuated.

An **octoploid** (eight sets of chromosomes) can be produced by failure of chromosome separation in mitosis in a tetraploid. If only bivalents form in meiosis, an octoploid organism can be perpetuated sexually by self-fertilization or through crosses with other octoploids. Furthermore, cross-fertilization between an octoploid and a tetraploid results in a **hexaploid** (six sets of chromosomes). Repeated episodes of polyploidization and cross-fertilization may ultimately produce an entire polyploid series of closely related organisms differing in chromosome number, as exemplified in *Chrysanthemum*.

Chrysanthemum represents just one type of polyploidy. In this case, all chromosomes in the polyploid species derive from a single diploid ancestral species. Polyploidy derived from the multiplication of a single ancestral set of chromosomes is known as **autopolyploidy.** However, in many cases of polyploidy, the polyploid species have complete sets of chromosomes from two or more different ancestral species. Such polyploids are known as **allopolyploids,** and they originate from occasional hybrids that occur between distinct diploid species.

Hybridization between species occurs when pollen from one species germinates on the stigma of another species and sexually fertilizes the ovule. The pollen may be carried to the wrong flower by wind, insects, or other pollinators. Figure 6-3 illustrates hybridization between species A and B leading

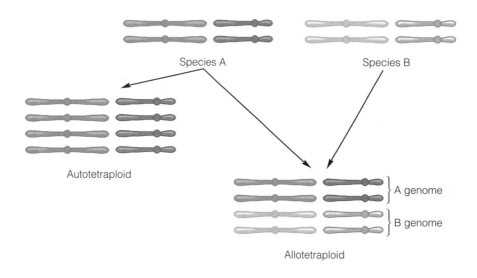

Species A

Species B

Autotetraploid

} A genome

} B genome

Allotetraploid

FIGURE 6-3 Formation of an autotetraploid by doubling of a complete diploid set of chromosomes and of an allotetraploid by union of two different diploid sets.

FIGURE 6-4 Hybridizations that occurred in the ancestry of cultivated bread wheat (*Triticum aestivum*), which is an allohexaploid containing complete diploid genomes (AA, BB, DD) from three ancestral species. (Data from Chris Chapman, University of Missouri.)

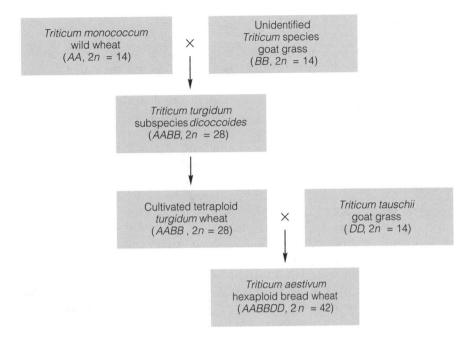

to the formation of an allopolyploid (in this case, an *allotetraploid*), which carries a complete diploid genome from each of its two ancestral species. Hybridization and the formation of allopolyploids is an extremely important process in plant evolution and plant breeding. At least half of all naturally occurring polyploids are allopolyploids. Cultivated wheat is an excellent example of allopolyploidy. Cultivated bread wheat is a hexaploid with 42 chromosomes constituting a complete diploid genome of 14 chromosomes from each of three ancestral species. The 42-chromosome allopolyploid is thought to have originated by the hybridizations outlined in Figure 6-4.

The genetics of polyploid species is more complex than that of diploid species because a polyploid individual carries more than two alleles of any gene. With two alleles in a diploid, only three genotypes are possible—AA, Aa, and aa,—whereas, in a tetraploid, five possible genotypes are possible—AAAA, AAAa, AAaa, Aaaa, and aaaa—the middle three of which represent different types of tetraploid heterozygotes.

6-3 Monoploidy

Monoploids are individual organisms containing a single gametic chromosome set. For example, if the parental species is diploid, the monoploids contain a haploid chromosome set; if the parent is a tetraploid, the monoploid contains a diploid set. Monoploids are quite rare but occur naturally in certain insect species (ants, bees) in which males are derived from unfertilized eggs. Meiosis cannot occur normally in the germ cells of a monoploid, and hence most monoploids are sterile. However, some species, for example, male honeybees, produce gametes by a modified meiosis in which chromosome separation in meiosis I does not occur. In many plants, the production of monoploids can be stimulated by creating conditions that yield aberrant cell divisions. Monoploids are important in modern plant breeding. In selecting organisms with desired properties, diploidy is always a problem because favorable recessive alleles may be masked by being heterozygous. This problem can be avoided by studying monoploids—provided the sterility can be overcome. Two techniques make this possible.

With some plants, monoploids can be derived from cells in the anthers (the pollen-bearing structures). Extreme chilling of the anthers causes some of the haploid cells destined to become pollen grains to begin to divide. If these cold-shocked cells are placed on an agar surface containing suitable nutrients and certain plant hormones, a small dividing mass of cells forms, called an **embryoid.** A subsequent change in plant hormones causes the embryoid to form a small plant with roots and leaves that can be potted in soil and allowed to grow normally. In diploid species, monoploid derivatives are haploid, which enables their genotypes to be identified without regard to the dominance or recessiveness of individual alleles. A monoploid plant with the traits desired by the plant breeder is then selected. In some cases, the desired genes are present in the original diploid (or polyploid) plant and merely sorted out and selected in the monoploids. In other cases, the anthers are treated with mutagenic agents in the hope of producing the desired traits.

The procedure is not yet complete, for at this point the monoploid plant is sterile, and so seed is not produced. What is necessary is to convert the monoploid into a homozygous diploid. This is possible by treatment of meristematic tissue (the growing point of a stem or branch) with the substance **colchicine.** This chemical is an inhibitor of the formation of the mitotic spindle. When the treated haploid cells in the meristem begin mitosis, chromosomes double in number by the normal replication process but metaphase and anaphase never occur. Many of the cells are killed by colchicine, but fortunately for the plant breeder some of the haploid cells are converted into the diploid state (Figure 6-5). The colchicine is removed to allow continued cell multiplication, and many of the now-diploid cells multiply to form a small sector of tissue that can be recognized microscopically. This tissue can be placed on a nutrient agar surface and will develop into a complete plant. These plants, which are completely homozygous, are fertile and produce normal seeds.

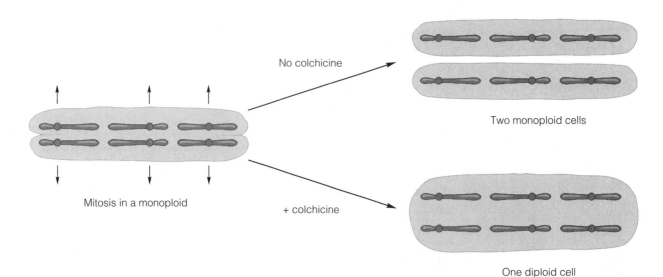

No colchicine

Two monoploid cells

Mitosis in a monoploid

+ colchicine

One diploid cell

FIGURE 6-5   Production of a diploid from a monoploid by treatment with colchicine. The colchicine disrupts the spindle and thereby prevents separation of the chromatids after the centromeres (circles) divide.

Occasionally, organisms arise that have extra copies of individual chromosomes, rather than extra entire sets of chromosomes. This situation is called **polysomy.** In contrast with polyploids, which in plants are often healthy and in some cases more vigorous than the diploid, polysomics are usually less vigorous than the diploid and have abnormal phenotypes. Thus, whereas the presence of complete extra sets of chromosomes in polyploids is not necessarily harmful to the organism, the occurrence of a single extra chromosome (or a missing chromosome) may have large effects. For example, Figure 6-6 shows the seed capsule of the Jimson weed, *Datura stramonium,* beneath which is a series of capsules of strains, each having an extra copy of a different chromosome. An otherwise diploid organism having an extra copy of an individual chromosome is called a **trisomic.** Note in Figure 6-6 that the seed capsule of each of the trisomics has distinctive abnormalities.

Because polysomy generally results in more-severe phenotypic effects than polyploidy does, the harmful phenotypic effects in trisomics must be related to the imbalance in the number of copies of different genes. A polyploid organism has a balanced genome in the sense that the ratio of the number of copies of any pair of genes is the same as in the diploid. For example, in a tetraploid, each gene is present in twice as many copies as in a diploid; so no gene or group of genes is out of balance with the others. Balanced chromosome abnormalities, which retain equality in the number of copies of each gene, are said to be **euploid.** In contrast, gene equality is upset in a trisomic because three copies of the genes located in the trisomic chromosome are present, whereas two copies of the genes in the other chromosomes are present. Such unbalanced chromosome complements are said to be **aneuploid.** In general, aneuploid abnormalities are usually more severe than euploid abnormalities. For example, in *Drosophila,* triploid females are viable, fertile, and nearly normal in morphology, whereas trisomy for either of the two large autosomes is invariably lethal.

Diploid

Trisomics

FIGURE 6-6 Seed capsules of the normal diploid *Datura stramonium* (Jimson weed), which has a haploid number of 12 chromosomes, and 4 of the 12 possible trisomics. The phenotype of the seed capsule in trisomics differs according to the chromosome that is trisomic.

Just as an occasional individual may have an extra chromosome, a chromosome may also be missing. Such an individual is said to be **monosomic** for the missing chromosome. In general, a missing copy of a chromosome results in more-harmful effects than those produced by an extra copy of the same chromosome, and monosomy is often lethal.

6-5 Human Chromosomes

The chromosome complement of a normal male human being is illustrated in Figure 6-7. The chromosomes have been treated with a staining reagent, called Giemsa, which causes the chromosomes to exhibit transverse bands that are specific for each pair of homologs. These bands permit the chromosome pairs to be identified individually. By convention, the autosome pairs are arranged and numbered from longest to shortest and separated into seven groups designated by the letters A through G; this conventional representation of chromosomes is called a **karyotype,** and it is obtained by cutting individual chromosomes out of photographs taken during metaphase and pasting them into place. In a karyotype of a normal female human being, the autosomes would not differ from those of a male and hence would be identical with those in Figure 6-7; however, there would be two X chromosomes instead of an X and a Y. Abnormalities in chromosome number and morphology are made evident by a karyotype.

Trisomy in Human Beings

Monosomy or trisomy of most human autosomes is usually incompatible with life, although there are a few exceptions. One exception is **Down syndrome,** which is caused by trisomy of chromosome 21. Down syndrome affects about

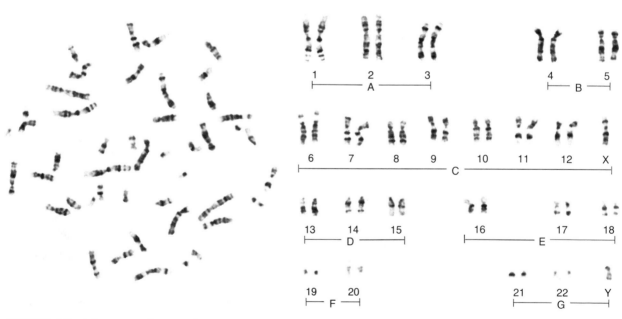

FIGURE 6-7 A karyotype of a normal human male. Blood cells arrested in metaphase were stained with Giemsa and photographed with a microscope. Left: The chromosomes as seen in the cell by microscopy. Right: The chromosomes have been cut out of the photograph and paired with their homologs. (Courtesy of Patricia Jacobs.)

1 in 750 live-born children. Its major symptom is mental retardation, but there can be multiple physical abnormalities as well, such as major heart defects in many cases.

Students familiar with trisomy 21 may know it as *Down's* syndrome, with an apostrophe *s*. However, this text follows current practice in human genetics in not using the possessive form of proper names to designate syndromes. For example, Down's syndrome, Huntington's disease, and Fanconi's anemia become Down syndrome, Huntington disease, and Fanconi anemia. Use of the definite article is also recommended (the Down syndrome, the Huntington disease, the Fanconi anemia), but this seems a bit fussy, if not pedantic. The nonpossessive form without the definite article is also consistent with such usages as La Guardia airport, Hoover dam, Lincoln Memorial, Johnson Space Center, and St. Louis Cardinals.

The great majority of cases of Down syndrome are caused by nondisjunction—failure in the separation of homologous chromosomes in meiosis (as explained in Chapter 2)—resulting in a gamete containing two copies of chromosome 21. For unknown reasons, nondisjunction of chromosome 21 is more likely to occur during oogenesis than during spermatogenesis, and so the abnormal gamete in Down syndrome is usually the egg. Moreover, nondisjunction of chromosome 21 increases dramatically with the age of the mother, with the risk of Down syndrome reaching 6 percent in mothers of age forty-five and older. Thus, many physicians recommend that women older than thirty-five who are pregnant have cells from the fetus tested to detect Down syndrome prenatally. This can be done from 15 to 16 weeks after fertilization by **amniocentesis,** in which cells of a developing fetus are obtained by insertion of a fine needle through the wall of the uterus and into the sac of fluid (the *amnion*) containing the fetus, or even earlier in pregnancy by sampling cells from an embryonic membrane called the *chorion*. In about 3 percent of families with a Down syndrome child, the risk of another affected child is very high—as much as 20 percent of births. This high risk is caused by a chromosome abnormality called a *translocation* in one of the parents, which will be discussed later in this chapter.

Sex-Chromosome Abnormalities and Dosage Compensation

Abnormalities in the number of sex chromosomes usually produce less-severe phenotypic effects than do abnormalities in the number of autosomes. For example, the effects of extra Y chromosomes are relatively mild. This is in part because the Y chromosome in mammals is largely heterochromatic. In human beings, there is a region at the tip of the short arm of the Y that is homologous with a corresponding region at the tip of the short arm of the X chromosome. It is in this homologous region that the X and Y synapse in spermatogenesis, and an obligatory crossover in the region holds the chromosomes together and ensures proper separation during anaphase I. The crossover is said to be obligatory because it occurs somewhere in this region in every meiotic division. Near the region of XY homology, but not within it, the Y chromosome contains a **testis-determining factor (TDF)** that triggers male embryonic development. Beyond the XY pairing region and the TDF, the Y chromosome appears to contain very few genes, and so extra Y chromosomes are milder in their effects on phenotype than are extra autosomes.

Extra X chromosomes have milder effects than extra autosomes because, in mammals, all X chromosomes except one are genetically inactivated very early in embryonic development. The inactivation tends to minimize the phenotypic effects of extra X chromosomes, but there are still some effects due to a block of genes near the tip of the short arm that are *not* inactivated.

In female mammals, X-chromosome inactivation is a normal process in embryonic development. In human beings, at an early stage of embryonic development, one of the two X chromosomes is inactivated in each somatic cell; different tissues undergo X inactivation at different times. The X chromosome that is inactivated in a particular somatic cell is selected at random, but once the decision is made, the same X chromosome remains inactive in all of the descendants of the cell.

X-chromosome inactivation has two consequences. First, it equalizes the number of active copies of X-linked genes in females and males. Although a female has two X chromosomes and a male has only one, because of inactivation of one X chromosome in each of the somatic cells of the female, the number of *active* X chromosomes in both sexes is one. In effect, gene dosage is equalized except for the block of genes in the short arm of the X that escapes inactivation. The equalization in dosage of active genes is called **dosage compensation.**

Second, a normal female is a **mosaic** for X-linked genes (Figure 6-8). That is, each somatic cell expresses the genes in only one X chromosome, but the X chromosome that is genetically active differs from one cell to the next. This mosaicism has been observed directly in females that are heterozygous for X-linked alleles that determine different forms of an enzyme, A and B; when cells from the heterozygous female are individually cultured in the laboratory, half of the clones are found to produce only the A form of the enzyme and the other half to produce only the B form. Mosaicism can be observed directly in women who are heterozygous for an X-linked recessive mutation resulting in the absence of sweat glands; these women exhibit patches of skin in which sweat glands are present (these patches are derived from embryonic cells in which the normal X chromosome remained active and the mutant X was inactivated), and other patches of skin in which sweat glands are absent (these patches are derived

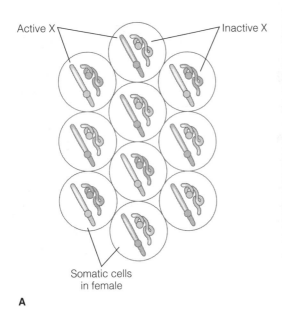

Active X   Inactive X

Somatic cells
in female

A

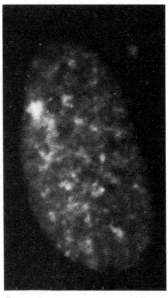

B

FIGURE 6-8 A. Schematic diagram of cells of a normal female showing that the female is a mosaic for X-linked genes. The two X chromosomes are indicated in red and black. An active X is depicted as a straight line, and an inactive X as a tangle. Each cell has just one active X, but the particular X that remains active is a matter of chance. In human beings, the inactivation includes all but a few genes in the tip of the short arm. B. A fluorescence micrograph of a human cell showing a Barr body (bright spot at the upper left—see arrow). This cell is from a normal human female, and it has one Barr body. (Courtesy of A. J. R. de Jonge.)

from embryonic cells in which the normal X chromosome was inactivated and the mutant X remained active.)

In certain cell types, the inactive X chromosome in females can be observed microscopically as a densely staining body in the nucleus of interphase cells. This is called a **Barr body** (see arrow in Figure 6-8B). Although cells of normal females have one Barr body, cells of normal males have none. Individuals with two or more X chromosomes have all but one X chromosome per cell inactivated through dosage compensation, and the number of Barr bodies equals the number of inactivated X chromosomes.

Many types of sex-chromosomal abnormalities have been observed. As noted, they are are usually less severe in their phenotypic effects than are abnormal numbers of autosomes. The four most-common types are:

1. **47,XXX.** This condition is often called the **trisomy-X syndrome.** The number 47 in the chromosome designation refers to the total number of chromosomes, and XXX indicates that the person has three X chromosomes. People having 47,XXX are female. Many are phenotypically normal or nearly normal, though the frequency of mild mental retardation is somewhat greater than it is among 46,XX females.

2. **47,XYY.** This condition is often called the **double-Y syndrome.** People having 47,XYY are male and tend to be tall, but they are otherwise phenotypically normal. At one time it was thought that 47,XYY males developed severe personality disorders and were at a high risk of committing crimes of violence, a belief based on an elevated incidence of 47,XYY among violent criminals. More-careful study indicates that most 47,XYY males have moderately impaired mental function and, although their rate of criminality is higher than that of normal males, the crimes are mainly nonviolent petty crimes such as theft. The majority of 47,XYY males are phenotypically and psychologically normal and have no criminal convictions.

3. **47,XXY.** This condition is called the **Klinefelter syndrome.** Affected persons are male. They tend to be tall, do not undergo normal sexual maturation, are sterile, and in some cases have enlargement of the breasts. Mild mental impairment is common.

4. **45,X.** Monosomy of the X chromosome in females is called the **Turner syndrome.** Affected persons are phenotypically female but short in stature and without sexual maturation. Mental abilities are typically within the normal range.

Chromosomal Abnormalities in Spontaneous Abortion

Approximately 15 percent of all recognized human pregnancies terminate in spontaneous abortion, and in about half of them the fetus has a major chromosomal abnormality. Table 6-1 summarizes the average rates of chromosomal abnormality found per 100,000 recognized pregnancies in several studies. Although many autosomal trisomies are found in spontaneous abortions, autosomal monosomies are not found. Monosomic embryos undoubtedly exist, but abortion probably occurs so early in development that the pregnancy goes unrecognized. Triploids and tetraploids are also common in spontaneous abortions. The majority of trisomy-21 fetuses, and the vast majority of 45,X fetuses, are spontaneously aborted; this serves the biological function of eliminating many fetuses that are grossly abnormal in their development because of major chromosomal abnormalities.

TABLE 6-1 Chromosome abnormalities per 100,000 recognized human pregnancies

Chromosome constitution	Number among spontaneously aborted fetuses	Number among live births
Normal	7500	84,450
Trisomy		
13	128	17
18	223	13
21	350	113
Other autosomes	3176	0
Sex chromosomes		
47,XYY	4	46
47,XXY	4	44
45,X	1350	8
47,XXX	21	44
Translocations		
Balanced (euploid)	14	164
Balanced (aneuploid)	225	52
Polyploid		
Triploid	1275	0
Tetraploid	450	0
Others (mosaics, etc.)	280	49
Total	15,000	85,000

6-6 Abnormalities in Chromosomal Structure

So far, abnormalities in chromosome *number* have been described. The remainder of this chapter deals with abnormalities in chromosomal *structure*. There are several principal types of structural aberrations, each of which has characteristic genetic effects. Chromosomal aberrations were initially discovered through their genetic effects, which, though confusing at first, were eventually understood as resulting from abnormal chromosomal structure and later confirmed directly by microscopic observations.

Deletions

Chromosomes sometimes arise in which a segment is missing. Such chromosomes are said to have a **deletion** or a **deficiency**. Deletions are generally harmful to the organism, and the usual rule is the larger the deletion the greater the harm. Very large deletions are usually lethal, even when heterozygous with a normal chromosome. Small deletions are often viable when they are heterozygous with a normal chromosome; however, they are usually lethal when homozygous, because the deletion frequently eliminates one or more genes that are essential to survival and must be supplied by the normal homologous chromosome.

Deletions can be detected genetically by making use of the fact that a chromosome with a deletion no longer carries the wildtype alleles of the genes that have been eliminated. For example, in *Drosophila*, many Notch deletions are large enough to remove the nearby wildtype allele of white, also. When these chromosomes having deletions are heterozygous with a structurally

FIGURE 6-9   Mapping of a deletion by testcrosses. The F₁ heterozygotes with the deletion express the recessive phenotype of all deleted genes. The expressed recessive alleles are said to be uncovered by the deletion.

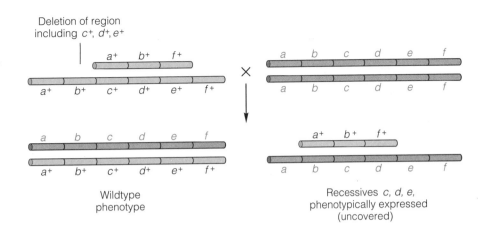

normal chromosome carrying the recessive w allele, the fly has white eyes because the wildtype w^+ allele is no longer present in the deleted-Notch chromosome. This "uncovering" of a recessive allele implies that the corresponding wildtype allele of white has been deleted. Once a deletion has been identified, its size can be assessed genetically by determining which recessive mutations in the region are uncovered by the deletion. This method is illustrated in Figure 6-9.

With the banded polytene chromosomes in *Drosophila* salivary glands, it is possible to study deletions and other chromosomal aberrations physically. For example, all the Notch deletions cause particular bands to be missing in the salivary chromosomes. Physical mapping of deletions also allows individual genes, otherwise known only from genetic studies, to be assigned to specific bands or regions in the salivary chromosomes.

Physical mapping of genes in a part of the *Drosophila* X chromosome is illustrated in Figure 6-10. The banded chromosome is shown near the top, along with the numbering system used to refer to specific bands. Genes are

FIGURE 6-10   Part of the X chromosome in the salivary glands of *Drosophila melanogaster* and the extent of six deletions (I–VI) in a set of chromosomes. Any recessive allele that is uncovered by a deletion must be located inside the boundaries of the deletion. This principle can be used to assign genes to specific bands in the chromosome.

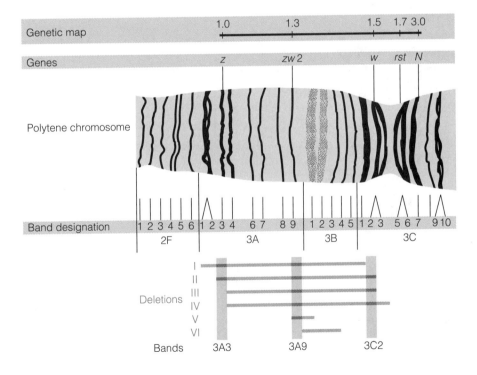

assigned to particular bands on the basis of their presence or absence in each of a set of chromosomes containing overlapping deletions. For example, the mutant X chromosomes I through VI have deletions, and the deleted part is shown in red. These deletions define regions along the chromosome, some of them corresponding to specific bands. For example, the deleted region in both chromosome I and II that is present in all the other chromosomes consists of band 3A3. In crosses, only deletions I and II uncover the mutation zeste (*z*), and so the *z* gene must be in band 3A3, as indicated at the top. Similarly, mutations in *zw2* are uncovered by all deletions except VI; therefore, the *zw2* gene must be in band 3A9. As a final example, the *w* mutation is uncovered only by deletions II, III, and IV; thus, the *w* gene must be in band 3C2.

Duplications

Some abnormal chromosomes have a region that is present twice. These chromosomes are said to have a **duplication.** Certain duplications have phenotypic effects of their own. An example is the *Bar* duplication in *Drosophila*, which is a tandem duplication of a small group of bands in the X chromosome that produces a dominant phenotype of bar-shaped eyes. (A **tandem duplication** is one in which the duplicated segment is directly adjacent to the normal region in the chromosome.)

Tandem duplications are able to produce even more copies of the duplicated region by means of a process called **unequal crossing-over,** outlined in Figure 6-11. Part A of Figure 6-11 illustrates the chromosomes in meiosis of an individual that is homozygous for a tandem duplication (red region). During synapsis, these chromosomes can mispair with each other, as illustrated in part B. A crossover within the mispaired part of the duplication (part C) will thereby produce a chromatid carrying a **triplication** and a reciprocal product (labeled "single copy" in part D) that has lost the duplication. For the *Bar* gene, the

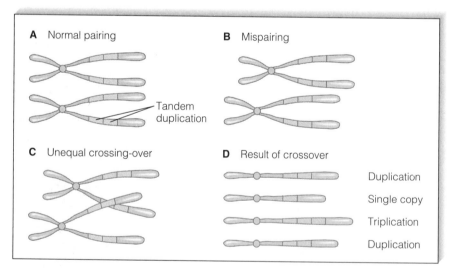

FIGURE 6-11   An increase in the number of copies of a chromosome segment resulting from unequal crossing-over of tandem duplications (red): (A) normal synapsis of chromosomes with a tandem duplication; (B) mispairing—the right element of the lower chromosome is paired with the left element of the upper chromosome; (C) crossing-over within the mispaired duplication, which is called unequal crossing-over; (D) the outcome of unequal crossing-over—one product contains a single copy of the duplicated region, another chromosome contains a triplication, and the two strands not participating in the crossover retain the duplication.

triplication can be recognized because it produces an even greater reduction in eye size than the duplication.

The most-frequent effect of a duplication is a reduction in viability (probability of survival); in general, survival decreases with increasing size of the duplication. However, deletions are usually more harmful than duplications of comparable size.

Inversions

Another abnormality is an **inversion,** a segment of a chromosome in which the order of the genes is the reverse of the normal order. An example is shown in Figure 6-12.

Inversions have a great effect on the recovery of recombinants for topological reasons, as shown in Figure 6-13. The initial problem occurs during

FIGURE 6-12 Partial genetic map of chromosome 2 of *D. melanogaster:* (A) normal order; (B) genetic map of same region in a chromosome that has undergone an inversion. Note that the order of *vg* and *L* are reversed in the inversion.

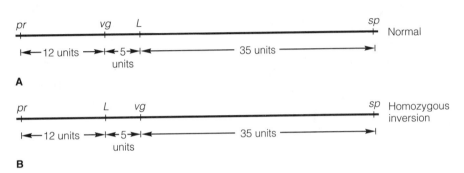

FIGURE 6-13 Result of pairing and crossing-over within an inversion: (A) normal and inverted chromosomes; (B) synapsis requires one of the chromosomes to form a loop in the inverted region; (C) crossing-over within the inversion loop (only the two chromatids participating in the crossover are shown); (D) results of the crossover are chromosomes with a deficiency of one terminal region and a duplication of the other. The centromeres of the chromosomes are not indicated.

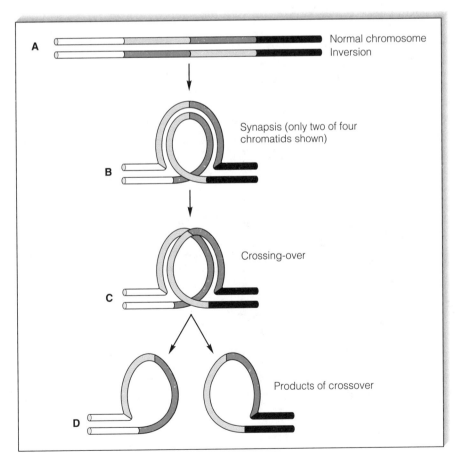

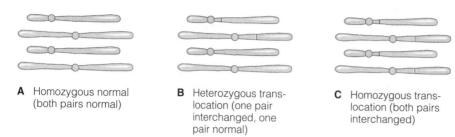

A Homozygous normal (both pairs normal)

B Heterozygous translocation (one pair interchanged, one pair normal)

C Homozygous translocation (both pairs interchanged)

FIGURE 6-14   A. Two pairs of nonhomologous chromosomes in a diploid individual. B. Heterozygous translocation, in which two nonhomologous chromosomes (the two at the top) have interchanged terminal segments. C. Homozygous translocation.

synapsis, because a loop must form in the inverted region in either the normal homolog or the homolog with the inversion (Figure 6-13B). (Part A shows the normal and inverted chromosomes.) Such looping is observed and apparently forms without difficulty. However, the main problem occurs when the paired chromosomes undergo crossing-over in the region of the inversion (Figure 6-13C). When crossing-over occurs within the inverted region, the result is the formation of chromosomes containing large duplications and deletions (Figure 6-13D), which are usually lethal in the zygote. If the inverted region includes the centromere, the abnormal chromosomes resulting from crossing-over in the inverted region are acentric or dicentric.

Translocations

Another chromosomal aberration results from the interchange of parts between nonhomologous chromosomes. This is called a **translocation.** Figure 6-14B illustrates two pairs of homologous chromosomes in an individual that is heterozygous for a translocation (part A shows the normal array). The top two chromosomes (red and black) have undergone an interchange of terminal parts. Compared with the normal individual in part A, the individual with a heterozygous translocation has one pair of chromosomes that is normal and one pair that is reciprocally interchanged (part B). Crossing two translocation heterozygotes can produce an individual that is homozygous for the translocation (Figure 6-14C), in which both pairs of chromosomes are interchanged. The translocation in Figure 6-14 is properly called a **reciprocal translocation.**

Individuals that are heterozygous for a translocation produce only about half as many offspring as normal; this is called **semisterility.** The reason for the semisterility is shown in Figure 6-15. When meiosis occurs in a translocation heterozygote, the normal and translocated chromosomes must undergo synapsis as shown in Figure 6-15A. Segregation from this configuration can occur in the three ways indicated in Figure 6-15B. The frequencies of the three types of segregation depend on the positions of the breakpoints and on the particular organism. Some of the gametes contain both parts of the reciprocal translocation or both normal homologous chromosomes (bottom row in Figure 6-15B), and these gametes result in viable zygotes. The rest of the gametes contain one part of the translocation and one normal chromosome, which give lethal zygotes because they contain duplications or deficiencies.

A special type of *non*reciprocal translocation is a **Robertsonian translocation,** in which the centromeric regions of two nonhomologous acrocentric chromosomes become fused to form a single centromere (Figure 6-16). Robertsonian translocations are important in human genetics because, when one of the acrocentrics is chromosome 21, it leads to a high-risk familial type

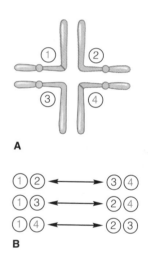

A

B

FIGURE 6-15   Segregation in a translocation heterozygote produces aneuploid gametes. A. Cross-shaped pairing configuration of the parts of the reciprocal translocation (1 and 4) and the normal homologous chromosomes (2 and 3). Only one of the two chromatids of each chromosome is shown. B. The three ways in which chromosomes can separate at anaphase I. The only euploid gametes result from the type of segregation shown in the bottom row in which 1 + 4 separate from 2 + 3; all the other possible gametes contain large deletions and deficiencies.

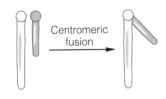

Normal chromosomes → Robertsonian translocation

Centromeric fusion

FIGURE 6-16 Formation of a Robertsonian translocation by fusion in the centromeric region of two acrocentric chromosomes. The centromere of the translocated chromosome is shown in white but actually includes sequences from both centromeres.

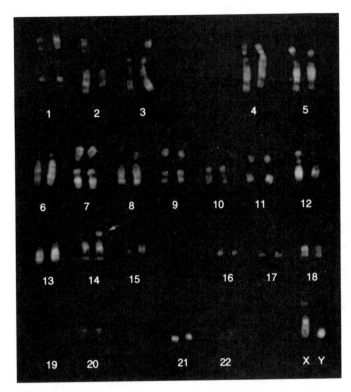

FIGURE 6-17 A karyotype of a child with Down syndrome, carrying a Robertsonian translocation of chromosomes 14 and 21 (arrow). Chromosomes 19 and 22 are faint in this photograph; this has no significance. (Courtesy of Irene Uchida.)

of Down syndrome. A Robertsonian translocation that joins chromosome 21 with chromosome 14 is shown in Figure 6-17 (arrow). The heterozygous carrier is phenotypically normal, but a high risk of Down syndrome results from the occurrence of aberrant segregation in meiosis. Of the several possible types of gametes that can arise, one contains a normal chromosome 21 along with the 14/21 Robertsonian translocation. If this aberrant gamete is used in fertilization, the fetus will contain two copies of the normal chromosome 21 plus the 14/21 translocation. In effect, the fetus contains three copies of chromosome 21 genes and hence has Down syndrome.

6-7 Chromosomal Abnormalities and Cancer

Cancer refers to an unrestrained proliferation and migration of cells. In all known cases, cancer cells derive from repeated division of an individual mutant cell whose growth has become unregulated, and so cancer cells initially constitute a clone. With continual growth of the clone, many cells within such clones often develop chromosomal abnormalities—extra chromosomes, missing chromosomes, deletions, duplications, or translocations. The chromosomal abnormalities found in cancer cells are diverse, and they may differ among cancer cells in the same individual or among individuals having the same type of cancer. The accumulation of chromosomal abnormalities is evidently one accompaniment of unregulated growth.

Amid the large number of apparently random chromosomal abnormalities found in cancer cells, a small number of aberrations occur consistently in certain types of cancer, particularly in blood diseases such as the leukemias. For example, chronic myelogenous leukemia is frequently associated with an apparent deletion of part of the long arm of chromosome 22. The abnormal chromosome 22 in this disease is called the **Philadelphia chromosome**; this chromosome is actually one part of a reciprocal translocation in which the missing segment of chromosome 22 is attached to either chromosome 8 or chromosome 9. Similarly, a deletion of part of the short arm of chromosome 11 is frequently associated with a kidney tumor called Wilms tumor, usually found in children.

Many of these characteristic chromosomal abnormalities have a breakpoint near the chromosomal location of a **cellular oncogene.** An **oncogene** is a gene associated with cancer. Cellular oncogenes, also called *proto-oncogenes,* are the cellular homologs of **viral oncogenes** contained in certain cancer-causing viruses. The distinction is one of location: cellular oncogenes are part of the normal genome, viral oncogenes are derived from cellular oncogenes through some rare mechanism in which the cellular oncogenes become incorporated into virus particles. More than fifty different cellular oncogenes are known. They are apparently normal developmental genes that predispose to cancer when mutated or abnormally regulated. Many of the genes function in normal cells as growth factors that promote and regulate cell division. When a chromosome rearrangement occurs near a cellular oncogene (or when the gene is incorporated into a virus), the gene may become expressed abnormally and result in unrestrained proliferation of the cell containing it. However, abnormal expression, by itself, is usually not sufficient to produce cancerous growth. One or more additional mutations must also occur in the same cell, but the nature of these additional mutations is as yet poorly understood.

A sample of characteristic chromosomal abnormalities found in certain cancers are illustrated in Figure 6-18, along with the locations of known cellular oncogenes on the same chromosomes. The symbols are those conventionally used in human genetics; they are:

1. The long arm of a chromosome is symbolized as q, the short arm as p. For example, 11p refers to the short arm of chromosome 11.

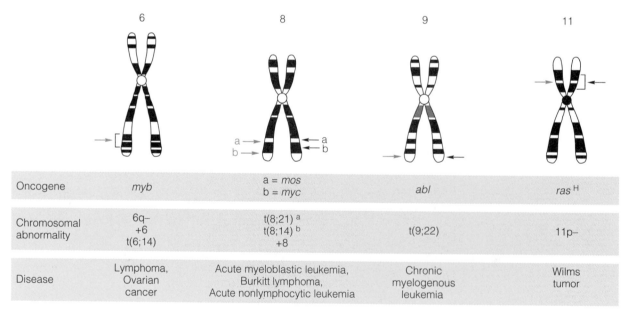

	6	8	9	11
Oncogene	*myb*	a = *mos* b = *myc*	*abl*	*ras* H
Chromosomal abnormality	6q– +6 t(6;14)	t(8;21) a t(8;14) b +8	t(9;22)	11p–
Disease	Lymphoma, Ovarian cancer	Acute myeloblastic leukemia, Burkitt lymphoma, Acute nonlymphocytic leukemia	Chronic myelogenous leukemia	Wilms tumor

FIGURE 6-18 Correlation between oncogene positions (red arrows) and chromosome breaks (black arrows) in aberrant human chromosomes frequently found in cancer cells.

2. A + (or −) sign *preceding* a symbol denotes an extra copy (or a missing copy) of the entire designated chromosome or arm. For example, +6 means the presence of an extra chromosome 6.

3. A + (or −) sign *following* a symbol denotes extra material (or missing material) corresponding to *part* of the designated chromosome or arm. For example, 11p− refers to a deletion of part of the short arm of chromosome 11.

4. The symbol "t" refers to a reciprocal translocation. Thus, t(9;22) refers to a reciprocal translocation between chromosome 9 and chromosome 22.

In Figure 6-18, the location of the cellular oncogene is indicated by a red arrow on the left, and the chromosomal breakpoint is indicated by a black arrow on the right. In most cases for which sufficient information is available, the correspondence between the breakpoint and the cellular-oncogene location is very close and, in some cases, the breakpoint is within the cellular oncogene itself. The significance of this correspondence is that the chromosomal rearrangement disturbs normal cellular-oncogene regulation and ultimately leads to the onset of cancer.

A second class of genes associated with inherited cancers are known as **tumor-suppressor genes,** or *antioncogenes*. These are genes whose presence is necessary to suppress tumor formation. Absence of both normal alleles, either through mutational inactivation or deletion, results in tumor formation. An example of a tumor-suppressor gene is the human gene *Rb-1*, located in chromosome 13 in band 13q14. When the normal *Rb-1* gene product is absent, malignant tumors form in the retinas and surgical removal of the eyes becomes necessary. The disease is known as **retinoblastoma.**

Retinoblastoma is unusual in that the predisposition to retinal tumors is dominant in pedigrees, but the *Rb-1* mutation is recessive at the cellular level; that is, persons who inherit one copy of the *Rb-1* mutation through the germ line are heterozygous, and the penetrance of retinoblastoma in these people is 100 percent. However, the retinal cells that become malignant have the genotype *Rb-1/Rb-1*. The explanation for this apparent paradox is illustrated in Figure 6-19. The diagram at the left shows the genotype of an *Rb-1* heterozygote, along

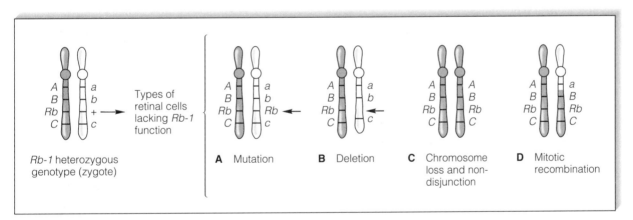

FIGURE 6-19 Mechanisms by which a single copy of *Rb-1* inherited through the germ line can become homozygous in cells of the retina: (A) new mutation; (B) deletion; (C) loss of the normal homologous chromosome and replacement by nondisjunction of the *Rb-1*-bearing chromosome; (D) mitotic recombination. Each of these events is rare, but there are so many cells in the retina that an average of three such events occurs per eye.

with a few other genes in the same chromosome. Four possible ways in which a second genetic event can result in *Rb-1/Rb-1* cells in the retina are shown at the right. The simplest (Figure 6-19A) is mutation of the wildtype gene in the other chromosome. The wildtype allele could also be deleted (part B). Part C illustrates a situation in which the normal homolog of the *Rb-1* chromosome is lost and replaced by nondisjunction of the *Rb-1*-bearing chromosome. A fourth possibility for making *Rb-1* homozygous is mitotic recombination (part D). Although each of these events has a low rate of occurrence per cell division, there are so many cells in the retina (approximately 10^8) that homozygosity is expected in at least one cell. (In fact, the average number of tumors per retina is three.)

CHAPTER SUMMARY

A typical chromosome contains a single centromere, the position of which determines the shape of the chromosome as it is pulled to the poles of the cell during anaphase. Rare chromosomes with no centromere, or those with two or more centromeres, are usually lost within a few cell generations because of aberrant separation during anaphase.

Polyploid organisms contain more than two complete sets of chromosomes. Polyploidy is widespread among higher plants and uncommon otherwise. An autopolyploid organism contains multiple sets of chromosomes from a single ancestral species; allopolyploid organisms contain complete sets of chromosomes from two or more ancestral species. Organisms occasionally arise in which an individual chromosome is either missing or present in excess; in either case, the number of copies of genes in such a chromosome is incorrect. Departures from normal gene dosage (aneuploidy) often result in reduced viability of the zygote in animals or of the gametophyte in plants.

The normal human chromosome complement consists of 22 pairs of autosomes, which are assigned numbers 1 through 22 from longest to shortest, and one pair of sex chromosomes (XX in females and XY in males). Fetuses containing an abnormal number of autosomes usually fail to complete normal embryonic development or die shortly after birth, though some people with Down syndrome (trisomy 21) survive for several decades. Persons with excess sex chromosomes survive, because the Y chromosome contains relatively few genes and because only one X chromosome is genetically active in cells of females (dosage compensation).

Most structural abnormalities in chromosomes are of one of four types—duplications, deletions, inversions, and translocations. A duplication is two copies of a chromosomal segment containing one or more genes or a part of a gene. In a deletion, one or more genes is missing. An organism can often tolerate an imbalance of gene dosage resulting from small duplications or deletions, but large duplications or deletions are almost always harmful. A chromosome containing an inversion has a group of adjacent genes in reverse of the normal order. Expression of the genes is usually unaltered, and so inversions rarely affect viability. However, crossing-over between an inverted chromosome and its noninverted homolog in meiosis yields abnormal chromatids. Two nonhomologous chromosomes that have undergone an exchange of parts constitute a reciprocal translocation. Organisms that contain a reciprocal translocation, as well as the normal homologous chromosomes of the translocation, produce fewer offspring (this is called semisterility) because of abnormal segregation of the chromosomes in meiosis. A translocation may also be nonreciprocal. A Robertsonian translocation is a type of nonreciprocal translocation in which the long arms of two acrocentric chromosomes are attached to a common centromere.

The terms monosomy or trisomy followed by a number indicate the chromosome that is lacking or extra. For example, monosomy 3 would mean that the cell is lacking one copy of chromosome 3. Trisomy 13 means that cells contain three copies of chromosome 13. An alternative notation is also employed in human genetics to describe monosomic and trisomic persons: the number of chromosomes present is stated, followed by the number of copies of the particular chromosome of interest. Thus, for a human being, who normally has 23 pairs or 46 chromosomes, the notation 45,X indicates that the cells have only 45 chromosomes and that the missing chromosome is an X. Similarly, a 47,XXX person has 47 instead of 46 chromosomes, and the extra chromosome is an X, because XXX shows that three X chromosomes are present. A female trisomic for chromosome 21 could also be designated 47,XX+21, and a male monosomic for chromosome 3 could be denoted 45,XY−3. Parts of chromosomes are sometimes absent. The p,q notation describes this anomaly. The short arm of a chromosome is denoted p and the long arm q. Thus, the short arm of chromosome 8 would be written as 8p, the long arm as 8q.

Malignant cells in many types of cancer contain specific types of chromosome abnormalities. Frequently, the breakpoints coincide with the chromosomal location of one of a group of cellular oncogenes coding for cell-growth factors. Abnormal oncogene expression is implicated in cancer. Viral oncogenes, found in certain cancer-causing viruses, are derived from cellular oncogenes. Cells also contain tumor-suppressor genes (antioncogenes), the absence of which predisposes to cancer. The gene for retinoblastoma codes for a tumor suppressor in the retina of the eyes. The gene is dominant in predisposing to retinal malignancy but recessive at the cellular level. People who inherit one copy of the gene through the germ line develop retinal tumors when the gene becomes homozygous in cells in the retina. Homozygosity in the retina can result from any number of genetic events, including new mutation, deletion, chromosome loss and nondisjunction, and mitotic recombination.

KEY TERMS

acentric chromosome
acrocentric chromosome
allopolyploid
amniocentesis
aneuploid
autopolyploidy
Barr body
cellular oncogene
colchicine
deletion
dicentric chromosome
dosage compensation
double-Y syndrome
Down syndrome
duplication
embryoid

euploid
hexaploid
inversion
karyotype
Klinefelter syndrome
metacentric chromosome
monoploid
monosomic
mosaic
octoploid
oncogene
Philadelphia chromosome
polyploidy
polysomy
reciprocal translocation
retinoblastoma

Robertsonian translocation
semisterility
submetacentric chromosome
tandem duplication
testis-determining factor (TDF)
tetraploid
translocation
triplication
triploid
trisomic
trisomy-X syndrome
tumor-suppressor gene
Turner syndrome
unequal crossing-over
viral oncogene

EXAMPLES OF WORKED PROBLEMS

Problem 1: The first artificial allotetraploid was created by the Russian agronomist G. D. Karpechenko in the 1930s by crossing the radish *Raphanus sativus* with the cabbage *Brassica oleracea*. Both species have a diploid chromosome number of 18. The initial F_1 hybrid was virtually sterile, but among the offspring was a rare, fully fertile allotetraploid that he called *Raphanobrassica*. What is the chromosome number of the F_1 of the cross *R. sativus* × *B. oleracea*? What is the chromosome number of *Raphanobrassica*?

Answer: Both *R. sativus* and *B. oleracea* produce gametes having 9 chromosomes, and so the F_1 has 18 chromosomes. The allotetraploid results from a doubling of the F_1 chromosome complement, and so *Raphanobrassica* has 36 chromosomes.

Problem 2: Genes *a*, *b*, *c*, *d*, *e*, and *f* are closely linked in a chromosome, but their order is unknown. Three deletions in the region are found to uncover recessive alleles of the genes as follows:

Deletion 1: uncovers *a*, *b*, and *d*
Deletion 2: uncovers *a*, *d*, *c*, and *e*
Deletion 3: uncovers *e* and *f*

What is the order of the genes? In this problem, we will see that there is enough information to order most, but not all, of the genes. Suggest what you might do to complete the ordering.

Answer: Problems of this sort are worked by noting that genes uncovered by a single deletion must be contiguous. The gene order is deduced from the overlaps between the deletions. The overlaps are *a* and *d* between the first and second deletions and *e* between the second and third. The gene order—as far as can be determined from these data—is diagrammed in part A of the illustration below. The deletions are shown in red. Gene *b* is at the far left, then *a* and *d* (in unknown order), then *c*, *e*, and *f*. (Gene *c* must be to the left of *e*, otherwise *c* would be uncovered by Deletion 3.) Part B is a completely equivalent map with gene *b* at the right. The ordering can be completed with a three-point cross between *b*, *a*, and *d* or *a*, *d*, and *c* or by examining additional deletions. Any deletion that uncovers either *a* or *d* (but not both), plus at least one other marker on either side would provide the information to complete the ordering.

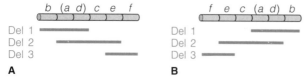

A   B

Problem 3: What are the consequences of a single crossover within the inverted region of a pair of homologous chromosomes with the gene order *A B C D* in one and *a c b d* in the other, in each of the following cases: **(a)** if the centromere is not included within the inversion? **(b)** if the centromere is included within the inversion?

Answer: These kinds of problems are almost impossible to solve without drawing a diagram. The illustration at the right shows how pairing within a heterozygous inversion results in a looped configuration. Crossing-over occurs at the four-strand stage of meiosis, but for simplicity only the chromatids participating in the crossover are diagrammed. Part A illustrates the situation when the centromere is not included within the inversion. The crossover chromatids consist of a dicentric (two centromeres) and an acentric (no centromere), and the products are duplicated for the terminal region containing A and deficient for the terminal region containing D, or the other way around. The noncrossover chromatids are the parental *A B C D* and *a c b d* monocentric chromosomes. Part B illustrates the

situation when the centromere is included in the inversion. The duplications and deficiencies are the same as in part A, but in this case both products are monocentrics. As before, the noncrossover chromatids are the parental *A B C D* and *a c b d* configurations.

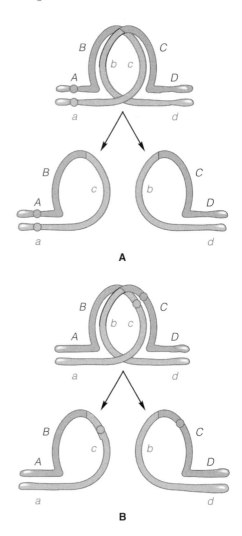

A

B

PROBLEMS

6-1. What type of abnormal chromosome forms a bridge at anaphase? What kind of chromosome cannot move during anaphase?

6-2. Which abnormality in chromosome structure can change a metacentric chromosome into a submetacentric chromosome? Which can fuse two acrocentric chromosomes into one metacentric chromosome?

6-3. How can a gamete be diploid? What gametes could be produced by a tetraploid plant with the genotype *AAaa BBbb*?

6-4. An autopolyploid series, like *Chrysanthemum*, contains five species, and the basic haploid chromosome number is 5. What chromosome numbers would be expected among the species?

6-5. A plant species S coexists with two related species A and B. All species are fully fertile. In meiosis, S forms 26 pairs of chromosomes, and A and B form 14 and 12 pairs of chromosomes, respectively. Hybrids between S and A form 14 pairs of chromosomes leaving 12 unpaired chromosomes, and hybrids between S and B form 12 pairs of chromosomes leaving 14 unpaired chromosomes. Suggest a likely evolutionary origin of species S.

6-6. A spontaneously aborted human fetus was found to have the karyotype 92,XXYY. What might have happened to the chromosomes in the zygote to result in this karyotype?

6-7. A spontaneously aborted human fetus is found to have 45 chromosomes. What is the most-probable karyotype? Had the fetus survived, what genetic disorder would it have?

6-8. Color blindness in human beings is an X-linked trait. A man who is color-blind has a 45,X (Turner syndrome) daughter who also is color-blind. Did the nondisjunction that led to the 45,X child occur in the mother or the father? How can you tell?

6-9. A phenotypically normal woman has a child with Down syndrome. The woman is found to have 45 chromosomes. What kind of chromosomal abnormality can account for these observations? How many chromosomes does the affected child have? How does this differ from the usual chromosome number and karyotype of a child with Down syndrome?

6-10. A chromosome has the gene sequence ABCDEFG. What sequence does the chromosome have after a C-through-E inversion? After a C-through-E deletion? Two chromosomes with the sequences ABCDEFG and MNOPQRSTUV undergo a reciprocal translocation after breaks in E-F and S-T. What are the possible products? Which products are genetically stable?

6-11. What does it mean to say that a deletion "uncovers" a recessive mutation?

6-12. Recessive genes a, b, c, d, e, and f are closely linked in a chromosome, but their order is unknown. Three deletions in the region are examined. One deletion uncovers a, d, and e; another uncovers c, d, and f; and the third uncovers b and c. What is the order of the genes?

6-13. Six bands in a salivary-gland chromosome of *Drosophila* are shown in the figure below, along with the extent of five deletions (Del 1–Del 5).

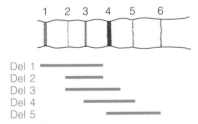

Recessive alleles a, b, c, d, e, and f are known to be in the region, but their order is unknown. When the deletions are heterozygous with each allele, the following results are obtained:

	a	b	c	d	e	f
Del 1	−	−	−	+	+	+
Del 2	−	+	−	+	+	+
Del 3	−	+	−	+	−	+
Del 4	+	+	−	−	−	+
Del 5	+	+	+	−	−	−

In this table, the − means that the deletion is missing the corresponding wildtype allele (the deletion uncovers the recessive) and + means that the corresponding wildtype allele is still present. Use these data to infer the position of each gene relative to the salivary chromosome bands.

6-14. A strain of corn that has been maintained by self-fertilization for many generations, when crossed with a normal strain, produces an F₁ in which many meiotic cells have dicentric chromatids and acentrics, and the amount of recombination in chromosome 6 is greatly reduced. However, in the original strain, no dicentrics or acentrics are found. What kind of abnormality in chromosome structure can explain these results?

6-15. Four strains of *Drosophila melanogaster* are isolated from different localities. The banding patterns of a particular region of salivary chromosome 2 have the configurations shown below (each letter denotes a band):

(a) a b f e d c g h i j
(b) a b c d e f g h i j
(c) a b f e h g i d c j
(d) a b f e h g c d i j

Assuming that part c is the ancestral sequence, deduce the evolutionary ancestry of the other chromosomes.

6-16. Two species of Australian grasshoppers coexist side by side. In meiosis, each has eight pairs of chromosomes. When the species are crossed, the chromosomes in the hybrid form six pairs of chromosomes and one group of four. What alteration in chromosome structure could account for these results?

6-17. Why are translocation heterozygotes semisterile? Why are translocation homozygotes fully fertile? If a translocation homozygote is crossed with an individual having normal chromosomes, what fraction of the F₁ is expected to be semisterile?

6-18. Curly wings (Cy) is a dominant mutation in the second chromosome of *Drosophila*. A Cy / + male was irradiated with x rays and crossed with + / + females, and the Cy / + sons were mated individually with + / + females. From one cross, the progeny were

Curly males	146
wildtype males	0
Curly females	0
wildtype females	163

What abnormality in chromosome structure is the most likely explanation for these results? (Note: Crossing-over does not occur in male *Drosophila*.)

6-19. Yellow body (*y*) is a recessive mutation near the tip of the X chromosome of *Drosophila*. A wildtype male was irradiated with x rays and crossed with *y*/*y* females, and one *y*⁺ son was observed. This male was mated with *y*/*y* females, and the offspring were

yellow females	256
yellow males	0
wildtype females	0
wildtype males	231

The yellow females were found to be chromosomally normal, and the *y*⁺ males were found to breed in the same manner as their father. What type of chromosome abnormality could account for these results?

6-20. In the homologous chromosomes shown here, the colored region represents an inverted segment of chromosome.

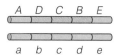

An individual of this genotype was crossed with an *a b c d e* homozygote. Most of the offspring were either *A B C D E* or *a b c d e*, but a few rare offspring were obtained that were *A B c D E*. What events occurring in meiosis in the inversion heterozygote can explain these rare progeny? Is the gene sequence in the rare progeny normal or inverted?

6-21. A wildtype strain of yeast, thought to be a normal haploid, was crossed with a different haploid strain carrying the mutation *his7*. This mutation is located in chromosome 2 and is an allele of a gene normally required for synthesis of the amino acid histidine. Among 15 tetrads analyzed from this cross, the following types of segregation were observed:

4 wildtype : 0 *his7*	4 tetrads
3 wildtype : 1 *his7*	4 tetrads
2 wildtype : 2 *his7*	1 tetrad

However, when the same wildtype strain was crossed with haploid strains with recessive markers on other chromosomes, segregation in the tetrads was always 2 : 2. What type of chromosome abnormality in the wildtype strain might account for the unusual segregation when the strain is mated with *his7*?

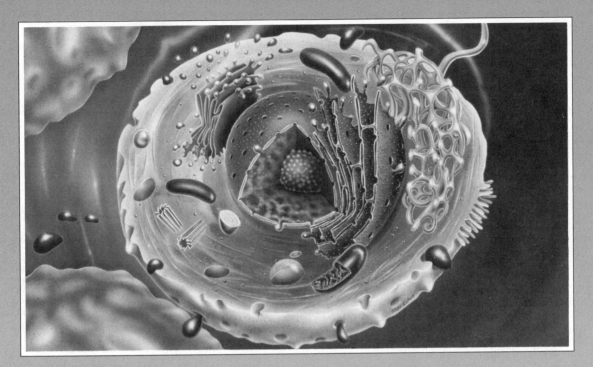

Extranuclear Inheritance

Most traits in eukaryotes show Mendelian inheritance because they are determined by genes in the nucleus that segregate in meiosis. Less common are traits determined by factors located outside the nucleus. These include traits determined by the DNA in mitochondria or chloroplasts—the self-replicating cytoplasmic **organelles** specialized for respiration or photosynthesis, respectively. Each type of organelle contains its own DNA that codes for certain proteins and RNA molecules. The DNA is replicated within the organelles and transmitted to daughter organelles as they form. The organelle genetic system is separate from that of the nucleus, and traits determined by organelle genes show patterns of inheritance quite different from the familiar Mendelian ratios.

Many organisms also contain intracellular parasites or symbionts, including cytoplasmic bacteria, viruses, or other elements. In some cases, inherited traits of the infected organism are determined by such cytoplasmic entities. Organelle heredity and other examples of the diverse phenomena that make up **extranuclear,** or **cytoplasmic, inheritance** are presented in this chapter.

Above: An artist's rendition of the inside of a eukaryotic cell. (Vincent Perez Studio— Shay Cohen, artist.)

7-1 Recognition of Extranuclear Inheritance

Other than the occurrence of non-Mendelian patterns of inheritance, no criterion is universally applicable to distinguish extranuclear from nuclear inheritance. In higher organisms, transmission of a trait through only one parent, **uniparental inheritance**, usually indicates extranuclear inheritance. Genetic transmission of cytoplasmic factors, such as mitochondria or chloroplasts, is determined by maternal and paternal contributions at the time of fertilization, by mechanisms of elimination from the zygote, and by the irregular sorting of the elements in cell division. Uniparental transmission through the mother constitutes **maternal inheritance**; uniparental transmission through the father is **paternal inheritance**.

Organelle inheritance in higher plants is often uniparental, but the particular type of uniparental inheritance depends on the organelle and the organism. For example, among angiosperms, mitochondria typically show maternal inheritance, but chloroplasts may be inherited maternally (the predominant mode), paternally, or from both parents, depending on the species. In conifers, chloroplast DNA is usually inherited paternally, but the chloroplast DNA in a few offspring results from maternal transmission. The redwood *Sequoia semipervirens* shows paternal transmission of both mitochondrial and chloroplast DNA. Whatever the pattern of cytoplasmic transmission, most species have a few exceptional progeny, indicating that the predominant mode of transmission is usually not absolute.

In higher animals, mitochondria are typically inherited through the mother because the egg is the major contributor of cytoplasm to the zygote. One genetic consequence of maternal inheritance is that reciprocal crosses produce different phenotypes of progeny; that is, the progeny from a mutant mother and normal father are mutant, whereas progeny from a normal mother and a mutant father are normal.

A maternal pattern of inheritance usually indicates extranuclear inheritance, but not always. The difficulty is in distinguishing between maternal inheritance and maternal effects. A **maternal effect** occurs when the genotype of the mother influences the phenotype of the progeny, either through substances present in the egg that affect early development or through effects of nurture—for example, the intrauterine environment of female mammals or their ability to produce milk. The distinction is highlighted in the following:

1. *Maternal inheritance* occurs when the hereditary determinants of a trait are extranuclear and genetic transmission is only through the maternal cytoplasm.
2. *Maternal effect* occurs when the nuclear genotype of the mother determines the phenotype of the progeny. The hereditary determinants are nuclear genes transmitted by both sexes, and in suitable crosses the trait undergoes Mendelian segregation.

7-2 Organelle Heredity

Maternal inheritance often results from genes contained in mitochondria (respiratory organelles) or chloroplasts (photosynthetic organelles). The DNA of these organelles is usually in the form of supercoiled circles of double-stranded molecules (Figure 7-1). The chloroplast DNA in most plants ranges in size from 120 to 160 kb. Mitochondrial genomes are usually smaller and have a greater size range. For example, the mitochondrial genome in mammals is about 16.5

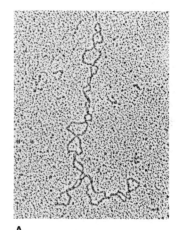

A

B

FIGURE 7-1 DNA from rat liver mitochondria: (A) native supercoiled configuration; (B) relaxed configuration produced by a nick in a DNA strand in the process of isolation. The size of each molecule is 16,298 nucleotide pairs. (Courtesy of David Wolstenholme.)

kb, that in *D. melanogaster* is about 18.5 kb, and that in some higher plants is more than 100 kb. The mitochondrial genomes of higher plants are exceptional in consisting of two or more circular DNA molecules of different sizes. Several copies of the genome are present in each organelle. There are typically from 2 to 10 copies in mitochondria and from 20 to 100 copies in chloroplasts. In addition, most cells contain multiple copies of each organelle.

Organelle genes code for DNA polymerases that replicate the organelle DNA. They also code for other components essential to their function or replication, but many organelle components are determined by nuclear genes. These components are synthesized in the cytoplasm and transported into the organelle. Mitochondrial DNA contains a relatively small number of genes. For example, both the human mitochondrial genome and that of yeast contain approximately 40 genes. Chloroplast genomes are generally larger than those of mitochondria and contain more genes. Many genes in mitochondria and chroroplasts code for the ribosomal RNA and transfer RNA components used in protein synthesis (Chapter 11). Figure 7-2 shows the organization of genes in the 120-kb chloroplast genome of the liverwort *Marchantia polymorpha*.

Because the structure and function of mitochondria and chloroplasts depend on both nuclear and organelle genes, it is sometimes difficult to distinguish these contributions. In addition, numerous mitochondria or chloroplasts are present in a zygote, but only one of them may contain a mutant allele. The result is that traits determined by organelle genes exhibit a pattern of determination and transmission very different from simple Mendelian traits determined entirely by nuclear genes.

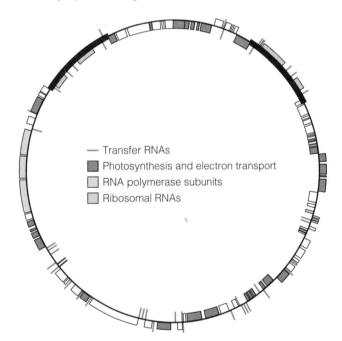

FIGURE 7-2   Organization of genes in the chloroplast genome of the liverwort *Marchantia polymorpha*. The large inverted repeat segments (black) that include the genes coding for ribosomal RNA (gray) are a feature of many other chloroplast genomes. The red lines show the locations of the genes for transfer RNA. Red bars are the genes for components of the photosynthetic and electron-transport systems; pink bars are the genes for subunits of RNA polymerase. Genes shown inside the circle are transcribed in a clockwise direction, those outside are transcribed counterclockwise. The length of the entire molecule is 121,024 nucleotide pairs. (Data from K. Ohyama, H. Fukuzawa, T. Kohchi, H. Shirai, T. Sano, S. Sano, K. Umesono, Y. Shiki, M. Takeuchi, Z. Chang, S. Aota, H. Inokuchi, and H. Ozeki. 1986. *Nature* 322:752–574.)

In this section, we briefly examine examples of traits determined by chloroplast and mitochondrial genes.

Leaf Variegation in Four-o'clock Plants

Chloroplast transmission accounts for the unusual pattern of inheritance observed with leaf variegation in a strain of the four-o'clock plant *Mirabilis jalapa*. Variegation refers to the appearance of white regions in the leaves and stems of plants resulting from the lack of green chlorophyll. The plants have some branches that are completely green, some completely white, and others variegated (Figure 7-3). All branches produce flowers, and these flowers can be used in crosses.

Nine distinct crosses can be made according to the white, green, or variegated pattern of the branch used for the pollen source or the egg (Table 7-1). From each cross, the seeds are collected and planted, and the phenotypes of the progeny are examined. The results of the crosses are summarized in Table 7-1. Two significant observations are:

1. Reciprocal crosses yield different results. For example, the cross green ♀ × white ♂ yields green plants, whereas the cross white ♀ × green ♂ yields white plants (which usually die shortly after germination because they lack chlorophyll and hence photosynthetic activity).
2. The phenotype of the female plants in each case determines the phenotype of the progeny (compare columns 1 and 3 in the table). The male makes no contribution to the phenotype of the progeny.

Furthermore, when flowers present on either the green or the variegated progeny plants are used in subsequent crosses, the patterns of transmission are identical with those of the original crosses. These observations suggest direct transmission of the trait through the mother, or maternal inheritance.

The explanation for the variegation is as follows:

1. Green color depends on the presence of chloroplasts, and pollen contains no chloroplasts.

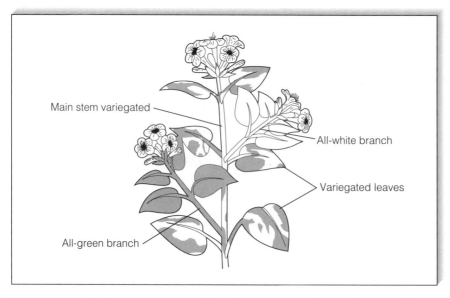

FIGURE 7-3   Leaf variegation in *Mirabilis jalapa*. Branches that are all green, all white, or variegated form on the same plant. Flowers form on all three kinds of branches.

TABLE 7-1 Results of crosses of variegated four-o'clock plants

Phenotype of branch bearing egg parent	Phenotype of branch bearing pollen parent	Phenotype of progeny
white	white	white
white	green	white
white	variegated	white
green	white	green
green	green	green
green	variegated	green
variegated	white	variegated, green, or white
variegated	green	variegated, green, or white
variegated	variegated	variegated, green, or white

2. Segregation of chloroplasts into daughter cells is determined by cytoplasmic division and is somewhat irregular.
3. The cells of white tissue contain mutant chloroplasts that lack chlorophyll.

These points allow one to draw the model shown in Figure 7-4. The flowers on green branches produce eggs with normal chloroplasts, and so all progeny are green. Flowers on white branches produce eggs containing only mutant

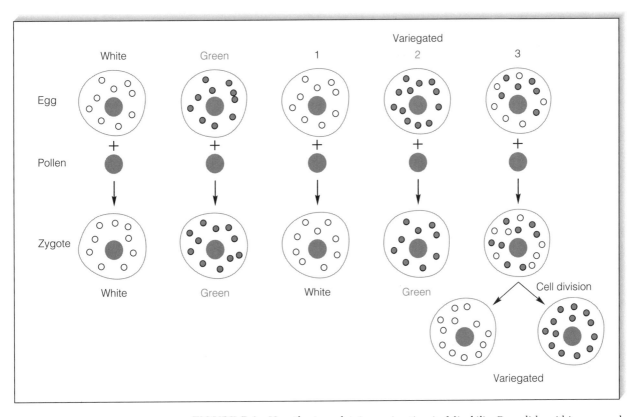

FIGURE 7-4 Hypothesis explaining variegation in *Mirabilis*. Gray disks within eggs and zygotes represent nuclei. Small red circles are normal chloroplasts; small open circles are mutant chloroplasts lacking chlorophyll. (Pollen contains no chloroplasts.) Numbers 1, 2, and 3 represent the three types of ovules produced on variegated branches.

chloroplasts. The eggs formed by flowers on variegated branches are of three types, those containing chlorophyll, those lacking chlorophyll, and those containing both normal and chlorophyll-free chloroplasts. The third class yields variegated plants because, in the development of the plant, cytoplasmic segregation sorts out normal and mutant chloroplasts.

Drug Resistance in Chlamydomonas

Cytoplasmic inheritance has been studied extensively in the unicellular green alga *Chlamydomonas*. Cells of this organism have a single large chloroplast containing from 75 to 80 copies of a double-stranded DNA molecule approximately 195 kb in size. The life cycle of *Chlamydomonas* is shown in Figure 7-5. There are two mating types—mt^+ and mt^-. These cells are of equal size and appear to contribute the same amount of cytoplasm to the zygote. Cells of opposite mating type fuse to form a diploid zygote, which undergoes meiosis to form a group of four haploid cells. The cells can be grown on solid growth medium to form visible clusters of cell **colonies** that can indicate the genotype. Reciprocal crosses between certain antibiotic-resistant strains (mutant) and antibiotic-sensitive strains (wildtype) yield different results. For example, for streptomycin-resistance (*str-r*)

$$str\text{-}r\,mt^+ \times str\text{-}s\,mt^- \text{ yields only } str\text{-}r \text{ progeny}$$
$$str\text{-}s\,mt^+ \times str\text{-}r\,mt^- \text{ yields only } str\text{-}s \text{ progeny}$$

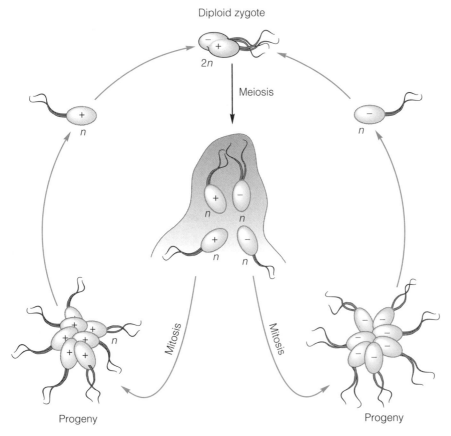

FIGURE 7-5   Life cycle of *Chlamydomonas reinhardtii*.

This is a clear case of uniparental inheritance of streptomycin resistance or sensitivity. In contrast, the *mt* alleles behave in strictly Mendelian fashion, yielding 1:1 progeny ratios, as expected of nuclear genes.

A large number of antibiotic-resistance markers with uniparental inheritance have been examined in *Chlamydomonas*. In each case, the antibiotic resistance has been traced to the chloroplast. Haploid *Chlamydomonas* contains only one chloroplast. If the chloroplast could be derived from the mt^+ or mt^- cell with equal probability, uniparental inheritance would not be observed. However, studies using physical markers to distinguish the chloroplast DNA of the two mating types have shown that the chloroplast of the mt^- parent is preferentially lost after mating, which accounts for the fact that the antibiotic-resistance phenotype of the mt^+ parent is the one transmitted to the progeny.

About 5 percent of the progeny from *Chlamydomonas* crosses do not exhibit cytoplasmic segregation. These progeny retain both parental chloroplasts, although they ultimately segregate in later cell divisions. These cells have proved to be quite valuable for genetic analysis, because recombination can occur between the DNA in the two chloroplasts. Analysis of the recombination frequencies for the various phenotypes has allowed the construction of fairly detailed maps of the chloroplast genome.

Respiration-defective Mitochondrial Mutants

Very small colonies of the yeast *Saccharomyces cerevisiae* are occasionally observed when the cells are grown on solid medium. These are called **petite** mutants. Microscopic examination indicates that the cells are of normal size. Physiological studies show that the cells grow very slowly and that the slow growth results from a defect in oxygen-requiring respiration that normally occurs in the metabolism of carbon compounds. Petites instead grow by glucose fermentation, which does not require oxygen but is extremely inefficient compared with aerobic metabolism, hence the slow growth. There are several types of petite mutants, two of which are discussed here.

Genetic analysis of crosses between petites and wildtype distinguishes the two types of petites illustrated in Figure 7-6. One type, called **segregational petite**, exhibits normal Mendelian segregation; that is, in a cross with wildtype,

FIGURE 7-6 Behavior of petite mutations in genetic crosses. The pink circles represent petite cells. The red nucleus in part A contains an allele resulting in petite colonies. The gray cells have normal mitochondria.

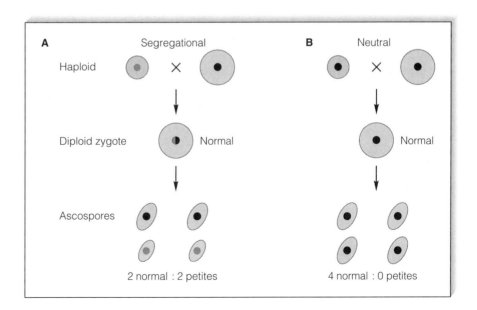

the diploid progeny are normal, and if the diploids are allowed to undergo meiosis half of the spores in an ascus produce petite colonies and half form wildtype colonies. The 1:1 ratio (actually 2:2) indicates that these petites are the result of nuclear mutation and that the determining allele is recessive.

The second major class, **neutral petite,** is quite different: in a cross with wildtype, all spores produce wildtype colonies (4:0 segregation). The same pattern is found if the progeny from such a cross are backcrossed with neutral petites. Thus, the phenotype that is inherited is that of the normal parent, and so the inheritance is uniparental. The explanation for these results is that the majority of neutral petites lack most or all mitochondrial DNA, in which many of the genes determining oxidative respiration are encoded. When a neutral-petite cell mates with a wildtype cell, the cytoplasm of the latter is the source of the normal mitochondrial DNA in the resulting progeny spores.

Petites arise at a frequency of roughly 10^{-5} per generation. The petite phenotype is the result of large deletions in mitochondrial DNA. For unknown reasons, yeast mitochondria frequently fuse and fragment, which may cause occasional deletions of DNA. Production of petites is apparently a result of segregation of aberrant mitochondrial DNA from normal DNA and the further sorting out of mutant mitochondria during cell division.

Cytoplasmic Male Sterility in Plants

An example of extranuclear inheritance important in agriculture is **cytoplasmic male sterility** in plants, a condition in which the plant does not produce active pollen, but the female reproductive process and fertility are normal. This type of sterility, used extensively in the production of hybrid corn seed, is not controlled by nuclear genes but is transmitted through the egg cytoplasm from generation to generation. The pattern of inheritance of cytoplasmic male sterility was first observed by Marcus Rhoades in an experiment summarized in Figure 7-7. A cross was made between a plant of a male-sterile variety and a normal male-fertile plant; the male-sterile progeny were then backcrossed for several generations with normal male-fertile plants. The resulting plants, in which all or nearly all nuclear genes from the male-sterile variety had been exchanged for those of the male-fertile plants, remained male sterile. The production of a small amount of functional pollen by the male-sterile variety made it possible to carry out the reciprocal male-fertile (female) × male-sterile (male) cross; it was observed that the progeny were fully male-fertile. Thus, male sterility or fertility in this case is maternally inherited.

Although cytoplasmic male sterility in maize is maternally inherited, certain nuclear genes called **fertility restorers** can suppress the male-sterilizing effect of the cytoplasm. Different types of cytoplasmic male sterilty can be classified according to response to restorer genes. For example, in one case, restoration of fertility requires the presence of dominant alleles of two different nuclear genes, and plants with male-sterile cytoplasm and both restorer alleles produce normal pollen. In another case, suppression requires the dominant allele of a different restorer gene, but this acts only in the gametophyte, and so plants with male-sterile cytoplasm that are heterozygous for the restorer allele produce a 1:1 ratio of normal:aborted pollen grains.

Cytoplasmic male sterility in maize results from alterations in mitochondrial DNA. Corresponding to the several classes of cytoplasmic male sterility differing in response to restorer genes, several different types of male-sterile mitochondrial DNA rearrangements have been identified. In at least one case, the sterility results from a DNA rearrangement that fuses two mitochondrial genes, creating a novel protein that causes male sterility.

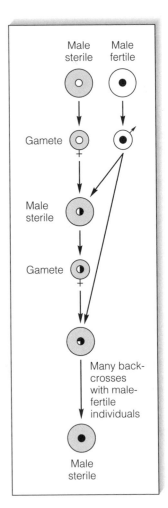

FIGURE 7-7 Diagram of an experiment demonstrating maternal inheritance of male sterility (pink cytoplasm) in maize.

7-3 Evolutionary Origin of Organelles

A widely accepted theory holds that organelles originally evolved from prokaryotic cells that lived inside primitive eukaryotic cells as symbionts. This is the **endosymbiont theory.** Mitochondria and chloroplasts share numerous features with prokaryotes that distinguish them from eukaryotic cells. Among the common features are:

1. The genomes are composed of circular DNA not extensively complexed with histonelike proteins.
2. The genomes are organized with functionally related genes close together and often expressed as a single unit.
3. The ribosome particles on which protein synthesis takes place have major subunits of similar size in organelles and in prokaryotes, which differ from those in eukaryotes.
4. The nucleotide sequences of the key RNA constituents of ribosomes are similar in chloroplasts, cyanobacteria such as *Anacystis nidulans*, and even bacteria such as *E. coli.*

The genomes of today's organelles are small compared with those of bacteria and cyanobacteria (formerly called bluegreen algae). For example, chloroplast genomes are only from 3 to 5 percent as large as the genomes of cyanobacteria. Organelle evolution was probably accompanied by major genome rearrangements, with some genes transferred into the nuclear genome of the host and others eliminated. Transfer of organelle genes into the nucleus differed from one lineage to the next. For example, the gene for one subunit of the mitochondrial ATPase is located in the mitochondrial genome in yeast but in the nuclear genome in *Neurospora.*

7-4 Cytoplasmic Transmission of Symbionts

In eukaryotes, a variety of cytoplasmically transmitted traits result from the presence of bacteria and viruses living in the cytoplasm of certain cells. One of the best known is the killer phenomenon found in certain strains of the protozoan *Paramecium aurelia*. Killer strains of *Paramecium* release to the surrounding medium a substance that is lethal to many other strains of the protozoan. The killer phenotype requires the presence of cytoplasmic bacteria referred to as **kappa particles,** whose maintenance is dependent on a dominant nuclear gene, *K.*

Paramecium is a diploid protozoan that undergoes sexual exchange through conjugation. The conjugation cycle is an unusual one (Figure 7-8). Initially, the cell has two diploid micronuclei. Many protozoans contain micronuclei (small nuclei) and a macronucleus (a large nucleus with specialized functions), but only the micronuclei are relevant to the genetic processes described here. Two cells come into contact and, before any conjugation, the two micronuclei in each cell undergo meiosis, forming eight micronuclei in each cell. Seven of the micronuclei and the macronucleus disintegrate, leaving each cell with one micronucleus, which then undergoes one mitotic division. The cell membrane between the two cells breaks down slightly and the cells exchange a single micronucleus, after which the nuclei fuse to form a diploid nucleus. After nuclear exchange, the two new cells (the exconjugants) are genotypically identical. In this sequence of events, only the micronuclei are exchanged, without any mixing of cytoplasm.

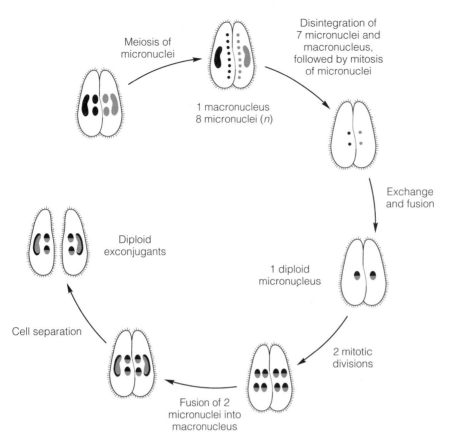

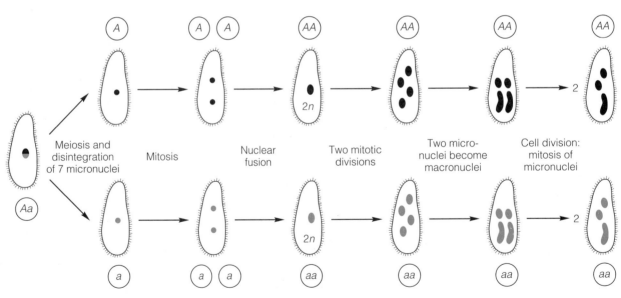

FIGURE 7-8   Conjugation in *Paramecium*.

FIGURE 7-9   Autogamy in *Paramecium*. The alleles in the heterozygous *Aa* cell undergo segregation in meiosis, with the result that the progeny cell becomes homozygous for either *A* or *a*. The initial arrows represent multiple steps in which the micronuclei undergo meiosis, after which seven of the products and the macronucleus disintegrate.

Single cells of *Paramecium* also occasionally undergo an unusual nuclear phenomenon called **autogamy** (Figure 7-9). Meiosis occurs and again all of the micronuclei disintegrate except one. This surviving nucleus then undergoes mitosis and nuclear fusion to recreate the diploid state. The critical point about autogamy is that, even if the initial cell was heterozygous, the newly formed diploid nucleus becomes homozygous because it is derived from a single haploid meiotic product. In a population of cells undergoing autogamy, the surviving micronucleus is selected randomly and so, for any pair of alleles, half of the new cells will contain one allele and half will contain the other.

Conjugation usually consists *only* of an exchange of micronuclei at the surface of contact between the two cells. Occasionally, it also includes cytoplasmic mixing followed eventually by autogamy. A comparison of the phenotypes following conjugation of killer (*KK*) cells and sensitive (*kk*) cells with and without cytoplasmic mixing and autogamy yields the evidence for cytoplasmic inheritance of the killer phenotype (Figure 7-10). Without cytoplasmic mixing (Figure 7-10A), the expected 1:1 ratio of killer to sensitive cells results; when autogamy occurs, the production of homozygosity converts

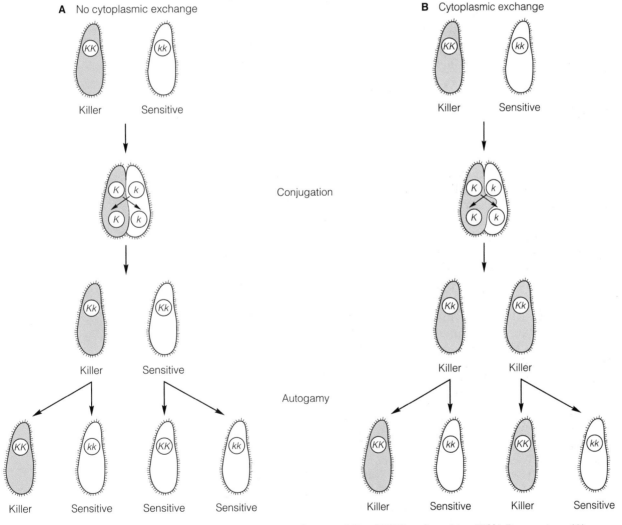

FIGURE 7-10   Crosses between killer (ITKK) and sensitive (ITkk) *Paramecium:* (A) result when no cytoplasmic exchange occurs in conjugation; (B) result when cytoplasmic exchange occurs. Cells containing kappa are shown in pink. The genotype and phenotype of each cell is indicated.

the *Kk* killer cells into one *KK* killer cell and one *kk* cell that cannot maintain the kappa particle and so becomes sensitive. The situation is quite different when cytoplasmic mixing occurs (Figure 7-10B). In this case, conjugation yields two killer cells because each cell has kappa particles derived from the cytoplasm of the killer parent. The results of autogamy in each of the exconjugants are one killer cell and one sensitive cell, indicating that the exconjugants were *Kk* heterozygotes.

Microbiological studies of killer strains have shown that the symbiotic bacterium identified with kappa particles is *Caedobacter taeniospiralis*. These bacteria produce the killer substance. Why the killer strains are not killed by the substance has not yet been determined.

Another example of a symbiont-related cytoplasmic effect is a condition in *Drosophila* called **maternal sex ratio**, which is characterized by the production of almost no male progeny. The daughters of sex-ratio females pass on the trait, whereas the occasional sons that are produced do not. Cytoplasm taken from the eggs of sex-ratio females transmits the condition when injected into females from unaffected cultures of the same or other *Drosophila* species. A bacteria has been isolated from the cytoplasm of sex-ratio females; when this bacteria is allowed to infect other *Drosophila* females, they acquire the sex-ratio trait. The causative agent of the sex-ratio condition is not the bacteria themselves, but a virus that multiplies in them. This virus, when released by the bacterial cells, kills most male *Drosophila* embryos. Why female embryos are not killed by the virus is still a mystery.

7-5   Maternal Effect in Snail-Shell Coiling

Maternal transmission usually, but not always, indicates cytoplasmic inheritance. An example is the determination of the direction of coiling of the shell of the snail *Limnaea peregra*. The direction of coiling, as viewed by looking into the opening of the shell, may be either to the right (dextral coiling) or to the left (sinistral coiling). Reciprocal crosses between homozygous strains give the following results:

rightward ♀ × leftward ♂ → all F_1 coil rightward
leftward ♀ × rightward ♂ → all F_1 coil leftward

In these crosses, the F_1 individuals have the same genotype, but the reciprocal crosses give different results in that the coiling of the progeny is the same as that of the mother. These results are typical of traits with maternal inheritance. However, in this case, all F_2 progeny from both crosses coil rightward, a result that is inconsistent with cytoplasmic inheritance. The F_3 generation provides the explanation.

The F_3 generation, obtained by self-fertilization of the F_2 snails (the snail is hermaphroditic), indicates that the inheritance of coiling direction depends on nuclear genes, rather than extranuclear factors. Three-fourths of the F_3 progeny have rightward coiled shells and one-fourth have leftward coiled shells (Figure 7-11). This is a typical 3:1 ratio, which indicates Mendelian segregation in the F_2. Rightward coiling ($+/+$ and $s/+$) is dominant to leftward coiling (s/s). The reason that reciprocal crosses give different results, and that the occurrence of the 3:1 Mendelian ratio is delayed for a generation, is as follows:

The coiling phenotype of an individual is determined by the genotype of its mother.

FIGURE 7-11 Inheritance of the direction of shell coiling in the snail *Limnea*. Leftward coiling (pink) is determined by the recessive allele *s*, and rightward coiling by the wildtype + allele. However, the direction of coiling is determined by the nuclear genotype of the mother, not the genotype of the snail itself or extranuclear inheritance. The F_2 and F_3 generations can be obtained by self-fertilization because the snail is hermaphroditic and can undergo either self-fertilization or cross-fertilization.

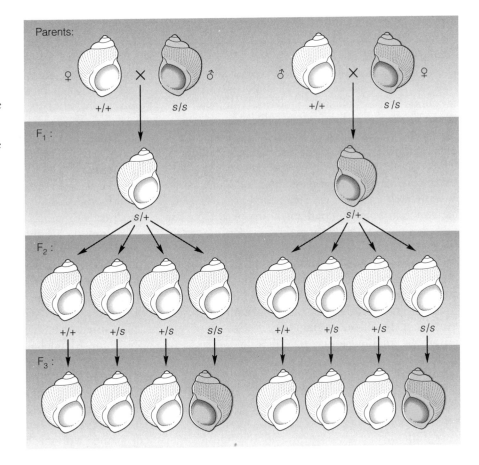

This means that the direction of shell coiling is not a case of maternal inheritance but rather a case of maternal effect (Section 7-1). Cytological analysis of developing eggs has provided the explanation: the genotype of the mother determines the orientation of the spindle in the initial mitotic division following fertilization, and this in turn controls the direction of shell coiling of the offspring (Figure 7-12). This classical example of maternal effect indicates that more than a single generation of crosses is needed to provide conclusive evidence of extranuclear inheritance of a trait.

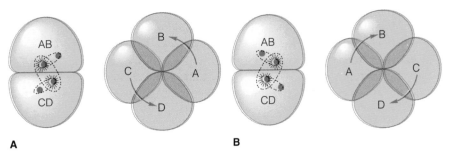

FIGURE 7-12 The direction of shell coiling in *Limnea* is determined by the orientation of the mitotic spindles during the second cleavage division in the zygote. The orientation is predetermined in the egg cytoplasm as a result of the mother's genotype. The spiral patterns established by the cleavage result in (A) leftward (sinistral) or (B) rightward (dextral) coiling.

Eukaryotic cells contain complex particulate structures in their cytoplasm. The most prevalent are mitochondria, which are found in both plant and animal cells, and chloroplasts, which are found in plant cells. Mitochondria are responsible for synthesis of the enzymes required for respiratory metabolism. Chloroplasts are responsible for photosynthesis. Both mitochondria and chloroplasts contain DNA molecules, which encode a variety of proteins. The genes contained in mitochondrial and chloroplast DNA are not present in chromosomal DNA. Hence, certain traits determined by these genes are not inherited according to Mendelian principles but show extranuclear (cytoplasmic) inheritance. Extranuclear inheritance is also observed with traits that are determined by bacteria and viruses that are transmitted in the cytoplasm from one generation to the next. In eukaryotic microorganisms in which there is no clear male-female distinction, cytoplasmic mixing usually occurs during zygote formation, and both parents contribute cytoplasmic particles to the zygote. However, in multicellular eukaryotes, the egg contains a large amount of cytoplasm and the sperm contains very little cytoplasm; so cytoplasmic transmission results in maternal inheritance.

Progeny resembling the mother are characteristic of both maternal inheritance and maternal effects. Maternal inheritance results from the transmission of extranuclear factors by the mother. Maternal effects result when the nuclear genotype of the mother determines the phenotype of the offspring: for example, through the organization of the egg or by nurturing ability. The direction of shell coiling in snails results from a maternal effect on the orientation of the spindle in the first mitotic division in the zygote.

Variegation in the four-o'clock is caused by defects in the chloroplasts; because of mutations in the chloroplast DNA, some chloroplasts lack chlorophyll. This deficiency leads to the formation of stems and leaves that either are completely white, because of lack of chlorophyll, or contain green and white patches (variegated), owing to development of tissue from some normal cells and some that contain chlorophyll-lacking chloroplasts. Because pollen contains no chloroplasts, the variegated genotype depends on the chloroplast content of the egg. The inheritance of resistance to certain antibiotics in *Chlamydomonas* is also cytoplasmic and depends on mutations in the chloroplast.

Mitochondrial defects account for cytoplasmic inheritance of some petite mutations in yeast. Petites produce very small colonies because the defective mitochondria are unable to provide the enzymes for aerobic metabolism. Petite cells can only use anaerobic metabolism, which is so inefficient that the cells grow very slowly, thereby producing small colonies. There are two major types of petites. Segregational petites result from a mutation in a nuclear gene and show strictly Mendelian inheritance. Neutral petites show non-Mendelian inheritance (4:0 segregation in ascospores) and result from large deletions in mitochondrial DNA. Male sterility in maize also results from defective mitochondria containing certain types of deletions in the DNA. Male sterility is very useful in plant breeding because it enables cross-pollination of plants without the need to remove immature anthers from the recipient plants.

The killer phenotype in *Paramecium* exemplifies cytoplasmic inheritance based on the presence of a symbiotic bacterium that produces a product that is toxic to many other paramecium. Cells containing the bacterium are resistant to the killer substance. *Paramecium* mates by two cells coming in contact and exchanging haploid nuclei. After the nuclei exchange, the cells separate. Normally the exchange occurs without any cytoplasmic mixing, in which case all segregating traits are inherited according to Mendelian principles. Sometimes cytoplasmic mixing occurs, and in this case one sees cytoplasmic inheritance. Maternal sex ratio in *Drosophila*, a deficiency of males among progeny, is another phenomenon determined by the presence of a symbiotic bacteria. In this case, the bacteria contain a virus that inhibits development of male embryos, whereas female embryos are unaffected.

KEY TERMS

autogamy
conjugation
cytoplasmic inheritance

cytoplasmic male sterility
endosymbiont theory
extranuclear inheritance

fertility restorer
kappa particle
maternal effect

maternal inheritance
maternal sex ratio
neutral petite

organelle
paternal inheritance
petite mutant

segregational petite
uniparental inheritance

Problem 1: Two strains of maize exhibit male sterility. One is cytoplasmically inherited, the other is Mendelian. What crosses could you make to identify which is which?

Answer: Cytoplasmic male sterility is transmitted through the female, and so outcrosses (crosses to unrelated strains) using the cytoplasmic male-sterile strain as the female parent will yield male-sterile progeny. Repeated outcrosses using the strain with Mendelian male sterility will result in either no male sterility (if the sterility gene is recessive) or 50 percent sterility (if it is dominant).

Problem 2: A snail in which the shell coils to the left undergoes self-fertilization and all of the progeny coil to the right. What is the genotype of the individual? If the progeny are individually self-fertilized, what coiling phenotypes are expected?

Answer: Shell coiling is determined by the genotype of the mother. Because the coiling of the individual in question is leftward, its mother had the genotype s/s, and so the individual must carry at least one s allele. Because the individual's own progeny coil to the right, the individual's genotype must be $+/s$. The progeny genotypes are 1/4 $+/+$, 1/2 $+/s$, and 1/4 s/s. When self-fertilized, all the $+/+$ and $+/s$ individuals yield right-coiling progeny and the s/s individuals yield left-coiling progeny.

Problem 3: Consider an inherited trait in *Drosophila* with the following characteristics: (1) it is transmitted from father

to son to grandson, (2) it is not expressed in females, (3) it is not transmitted through females. (a) Is the trait likely to result from extranuclear inheritance? (b) Can male-limited expression explain the data? (c) What hypothesis of Mendelian inheritance can explain the data? (d) What observations could be made to confirm the Mendelian hypthesis? (Hint: Think about the consequences of nondisjunction.)

Answer: (a) The trait is very unlikely to result from extranuclear inheritance. First, it would be very unusual for extranuclear inheritance in animals to go through the male; and, second, if it were transmitted through sperm, both sexes should be affected. (b) The trait is also unlikely to have male-limited expression. Although the observation that only males are affected is consistent with this hypothesis, a trait with male-limited expression could be transmitted through females. (c) The father–son–grandson pattern of inheritance parallels that of the Y chromosome. Therefore, the data fit a Y-linked model of inheritance. (d) Nondisjunction in an XY male results in XXY zygotes (which in *Drosophila* are female) and XO zygotes (which in *Drosophila* are male). Therefore, if the trait is Y-linked, then XXY females resulting from nondisjunction should be affected and XO males resulting from nondisjunction should be unaffected. Furthermore, the affected XXY females should give rise to affected sons (XY) and some affected daughters (XXY).

7-1. What are the normal functions of mitochondria and chloroplasts? What eukaryotic cells can survive without mitochondria? Can plant cells survive without chloroplasts?

7-2. What is meant by variegation?

7-3. Distinguish between maternal inheritance and maternal effect.

7-4. What does the occurrence of segregational petites in yeast imply about the relation between the nucleus and mitochondria?

7-5. A mutant plant is found with yellow instead of green leaves. Microscopic and biochemical analyses show that the cells contain very few chloroplasts and that the plant manages to grow by making use of other

pigments. The plant does not grow very well, but it does breed true. The inheritance of yellowness is studied in the following crosses:

green ♂ × yellow ♀ → all yellow
yellow ♂ × green ♀ → all green

What do these results suggest about the type of inheritance?

7-6. For the phenotype in Problem 7-5, suppose that variegated progeny are found in the crosses at a frequency of about 1 per 1000 progeny. How might this be explained?

7-7. What kinds of progeny result from the following crosses with *Mirabilis*?

(a) green ♀ × white ♂
(b) white ♀ × green ♂
(c) variegated ♀ × green ♂
(d) green ♀ × variegated ♂

7-8. In the maternally inherited leaf variegation in *Mirabilis*, how does the F_2 of the cross green ♀ × white ♂ differ from that expected if white leaves were due to a conventional X-linked recessive gene in animals?

7-9. Assuming that chloroplasts duplicate and segregate randomly in cell division, what is the probability that both replicas of a single mutant chloroplast will segregate to the same daughter cell in mitosis? After two divisions, what is the probability that at least one daughter cell will lack the mutant chloroplast?

7-10. If *a* and *b* are chloroplast markers in *Chlamydomonas*, what genotypes of progeny are expected from the cross $a^+ b^- m^+ \times a^- b^+ m^-$?

7-11. What is the phenotype of a diploid yeast produced by crossing a segregational petite with a neutral petite?

7-12. An antibiotic-resistant haploid stain of yeast is isolated. Mating with a wildtype (antibiotic-sensitive) strain produces a diploid, which when grown for some generations and induced to undergo sporulation produces tetrads containing either four antibiotic-resistant spores or four antibiotic-sensitive spores. What can you conclude about the inheritance of the antibiotic resistance?

7-13. What are the possible genotypes of the progeny of an *Aa Paramecium* that undergoes autogamy?

7-14. Equal numbers of *AA* and *aa Paramecium* undergo conjugation in random pairs. What are the expected matings? What genotypes and frequencies are expected after the conjugating pairs separate? What genotypes and frequencies are expected after the cells undergo autogamy?

7-15. Corn plants of genotype *AA bb* with cytoplasmic male sterility and normal *aa BB* plants are planted in alternate rows and pollination is allowed to occur at random. What progeny genotypes would be expected from the two types of plants?

7-16. Females of a certain strain of *Drosophila* are mated with wildtype males. All progeny are female. Three possible explanations are: (1) the females are homozygous for an autosomal allele that produces lethal male embryos (that is, a maternal effect); (2) the females are homozygous for an X-linked allele that is lethal in males; (3) the females carry a cytoplasmically inherited factor that kills male embryos. A cross is carried out between F_1 females and wildtype males, and only female progeny are produced. What mode of inheritance does this result imply?

7-17. A rightward coiled snail has all leftward coiled progeny when self-fertilized. What is the genotype of the rightward coiled parent? What was the genotype of the parent's mother?

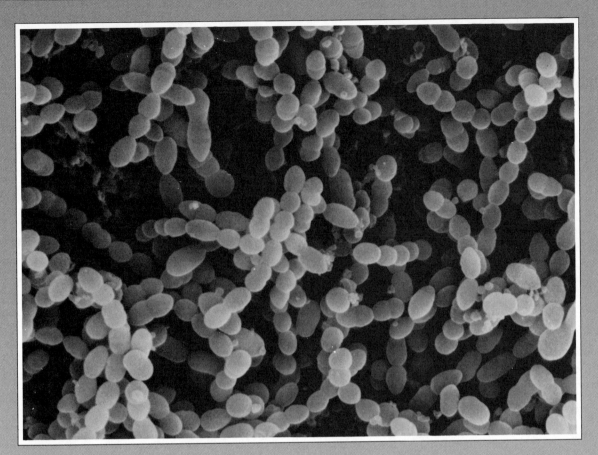

Population Genetics
and Evolution

In the genetic analyses examined so far, matings have been deliberately designed by the geneticist. However, in most organisms, matings are not under the control of an investigator, and often the familial relationships between individuals are not known. This situation is typical of studies of organisms in their natural habitat. Most organisms do not live in discrete family groups but exist as part of populations of individuals of unknown genetic relation. At first, it might seem that classical genetics with its simple Mendelian ratios could have

Above: Scanning electron micrograph of a population of bacteria. (From G. Shih and R. Kessel. 1983. *Living Images*. Jones and Bartlett.)

little to say about such complex situations, but this is not the case. The principles of Mendelian genetics can be used both to interpret data collected in natural populations and to make predictions about the genetic composition of populations. Application of genetic principles to entire populations of organisms constitutes the subject of **population genetics.**

8-1 Allele Frequencies and Genotype Frequencies

The term **population** refers to a group of organisms of the same species living within a prescribed geographical area. Sometimes the area is large, as in reference to the population of sparrows in North America; it may even include the entire earth, as in reference to the "human population." More commonly, the area is considered to be of a size within which individual members of the population are likely to find mates. Such a group of interbreeding organisms living within a defined geographical area is often called a **local population.** The complete set of genetic information contained within the members in a population is called the **gene pool.** This pool includes all alleles present in the population. This section begins with an analysis of a local population with respect to a phenotype determined by two codominant alleles (Section 1-5).

Calculation of Allele Frequency

The genetic properties of the human MN blood group system are exceptionally simple. In this system, there are three possible phenotypes—M, MN, and N—corresponding to the combinations of M and N antigens that can be present on the surface of red blood cells. These antigens are unrelated to ABO and other red-cell antigens. The M, MN, and N phenotypes correspond to three genotypes of one gene—*MM*, *MN*, and *NN*, respectively. In one study of the phenotypes of 1000 British people, 298 were M, 489 were MN, and 213 were N. From the simple genotype-phenotype correspondence in this system, the genotypes can be directly inferred to be

298 MM 489 MN 213 NN

These numbers contain a surprising amount of information about the population—for example, whether there may be mating between relatives—and one of the goals of population genetics is to be able to interpret this information. First, note that the sample contains two types of data—the number of occurrences of the three genotypes, and the number of occurrences of the individual M and N alleles. Furthermore, the 1000 people represent 2000 alleles of the gene because each person is diploid. These alleles break down in the following way:

298 MM persons = 596 M alleles
489 MN persons = 489 M alleles + 489 N alleles
213 NN persons = <u>   426 N alleles</u>
Totals = 1085 M alleles + 915 N alleles

Usually, it is more convenient to analyze the data in terms of relative frequency rather than with the actual numbers. For genotypes, the **genotype frequency** in a population is the proportion of individuals having the particular genotype. For individual alleles, the **allele frequency** of a specified allele is the proportion of all alleles of the specified type. For a sample of the type being discussed here, the genotype frequencies are obtained by dividing the observed numbers by the total sample size, in this case 1000. Therefore, the genotype frequencies are

$$0.298 \ MM \qquad 0.489 \ MN \qquad 0.213 \ NN$$

Similarly, the allele frequencies are obtained by dividing the observed numbers by the total (in this case 2000), and so

Allele frequency of M = 1085/2000 = 0.5425
Allele frequency of N = 915/2000 = 0.4575

Note that the genotype frequencies add up to 1.0, as do the allele frequencies; this is a consequence of their definition in terms of proportions, which must add up to 1.0 when all of the possibilities are taken into account. *Allele and genotype frequencies must always be between 0 and 1.* A population having an allele frequency of 1.0 for some allele is said to be **fixed** for that allele.

Allele frequencies can be used to make inferences about matings in a population and to predict the genetic composition of future generations. Allele frequencies are often more useful than genotype frequencies because individual alleles, not genotypes, form the bridge between generations. Alleles rarely undergo mutation in a single generation, and so they are relatively stable in their transmission from one generation to the next. Genotypes are not permanent; genotypes are totally broken up in the processes of segregation and recombination that occur in each reproductive cycle. Moreover, we know from simple Mendelian considerations what types of gametes must be produced from the MM, MN, and NN genotypes:

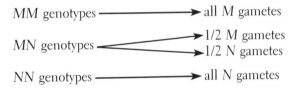

Consequently, the M-bearing gametes produced in the population will represent all the gametes from MM people and half the gametes from MN people. Likewise, the N-bearing gametes will represent all the gametes from NN people and half the gametes from MN people. Therefore, for gametes, the population produces the following frequencies:

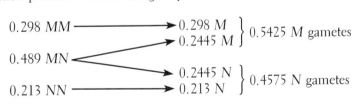

Note that the allele frequencies among *gametes* equal the allele frequencies among *adults* calculated earlier, which must be true whenever each adult in the population produces the same number of functional gametes.

Electrophoretic Variation and Allele Frequencies

An important method for studying genes in natural populations is *electrophoresis*, or the separation of charged molecules in an electric field (Section 4-9).

This method is commonly applied to study genetic variation in proteins coded by alternative alleles or to study the occurrence of particular restriction sites in homologous DNA fragments (Section 4-10).

In protein electrophoresis, protein-containing tissue samples from individuals are placed near the edge of a gel and a voltage is applied for several hours. All charged molecules in the sample move in response to the voltage, and a variety of techniques can be used to locate particular substances. For example, an enzyme can be located by staining the gel with a reagent that is converted into a colored product by the enzyme; wherever the enzyme is located, a colored band appears.

Figure 8-1 is a photograph of a gel stained to reveal the enzyme phosphoglucose isomerase in tissue samples from sixteen mice, illustrating typical raw data from an electrophoretic study. The pattern of bands varies from individual to individual, but only three patterns are observed. Samples from individuals 1, 2, 4, 8, 10, 11, and 15 have two bands, one that moves fast (upper band) and one that moves slowly (lower band). A second pattern is seen with the samples from individuals 3, 5, 9, 12, and 16, in which only the fast band appears. The samples from individuals 6, 7, 13, and 14 show a third pattern, in which only the slow band appears.

Patterns of enzyme mobility are *phenotypes*, not genotypes, and so the type of phenotypic variation illustrated in Figure 8-1 does not indicate the genetic basis of the variation. However, analysis of phenotypic variation of many enzymes in a wide variety of organisms has shown that such electrophoretic variation almost always has a simple genetic basis: each electrophoretic form of the enzyme contains one or more polypeptides with a genetically determined *amino acid replacement* that changes the electrophoretic mobility of the enzyme. Alternative forms of an enzyme coded by alleles of a single gene are known as **allozymes.** Alleles coding for allozymes are usually codominant (Section 1-5), which means that heterozygotes express the allozyme corresponding to each allele. Therefore, in Figure 8-1, individuals with only the fast allozyme are *FF* homozygotes, those with only the slow allozyme are *SS* homozygotes, and those with both fast and slow allozymes are *FS* heterozygotes. The 16 genotypes in Figure 8-1 are consequently

$$5\ FF \qquad 7\ FS \qquad 4\ SS$$

and the allele frequency of *F* in this small sample is

$$\text{Allele frequency of } F = [(2 \cdot 5) + 7]/(2 \cdot 16) = 0.53$$

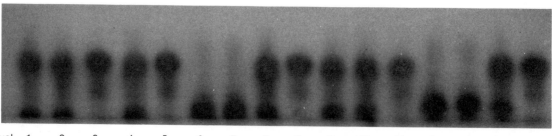

Individual	1	2	3	4	5	6	7	8	9	10	11	12	13	14	15	16
Genotype	F/S	F/S	F/F	F/S	F/F	S/S	S/S	F/S	F/F	F/S	F/S	F/F	S/S	S/S	F/S	F/F

Direction of movement

FIGURE 8-1 Electrophoretic mobility of glucose phosphate isomerase in a sample of sixteen mice. (Courtesy of S. E. Lewis and F. M. Johnson.)

Because there are only two alleles represented in Figure 8-1, the allele frequency of S must be $1 - 0.53 = 0.47$, which can equivalently be calculated directly as

$$\text{Allele frequency of } S = [7 + (2 \cdot 4)]/(2 \cdot 16) = 0.47$$

Many genes in natural populations are **polymorphic,** which means that they have two or more relatively frequent alleles. A typical quantitative definition is that a gene is polymorphic if the frequency of the *most-common* allele is less than 0.95. Using this definition, the gene in Figure 8-1 is polymorphic because the most-common allele (in this example, F) has a frequency of less than 0.95.

The proportion of genes that are polymorphic in a population is one widely used index of the amount of genetic variation present in the population. Among plants and vertebrate animals, approximately 20 percent of enzyme genes are polymorphic. The implication of this finding is that genetic variation among individuals is very common in natural populations of most organisms. Furthermore, among plants and vertebrate animals, the average proportion of genes that are heterozygous within individuals is about 0.05, which again emphasizes the prevalence of alternative alleles in populations. Invertebrate animals such as *Drosophila* have even more genetic variation.

Restriction Fragment Length Polymorphisms

Polymorphisms can also be detected directly in DNA molecules. This method uses a *probe* DNA usually obtained from bacteria or phage cells into which DNA fragments from the organism of interest have been cloned (Section 4-10). The probe DNA is complementary in sequence to a segment of DNA in the organism of interest, but its presence in clones allows the probe to be obtained in large quantities and pure form. In the *Southern blot* procedure, in which genomic DNA is digested with a restriction enzyme and fragments of different size separated by electrophoresis, radioactive probe DNA hybridizes only with DNA fragments containing complementary sequences and so identifies these restriction fragments (Section 4-10).

When this procedure is used to study natural populations, it is common to find restriction fragments complementary to a probe that differ in size among individuals. These differences result from differences in the locations of restriction sites along the DNA. Each region of the genome that hybridizes with the probe generates a restriction fragment. The restriction fragments that derive from corresponding positions in homologous chromosomes are analogous to alleles in that they segregate in meiosis: the offspring of a heterozygote may inherit a chromosome with either fragment size, but not both. Moreover, allelic restriction fragments of different sizes are codominant because both fragments are detected in heterozygotes, and this is the best kind of situation for genetic studies.

Two types of restriction fragment length variations are illustrated in Figure 8-2. Part A shows the simplest type, in which a single nucleotide difference in one chromosome knocks out a restriction site, and so the restriction fragment complementary to the probe stretches farther downstream to the next restriction site. As with any other type of allelic variation, individuals may be homozygous for allele 1, homozygous for allele 2, or heterozygous. Figure 8-2B shows a slightly more complex situation in which the restriction fragments differ in length because they contain variable numbers of tandemly repeated units within

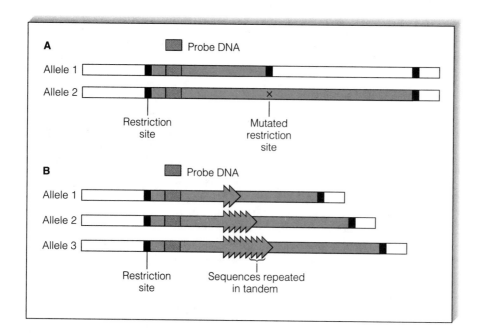

FIGURE 8-2 Two types of restriction fragment size variation in DNA. A. Nucleotide substitution in DNA molecule (marked with ×) eliminates a restriction site. The result is that, in the region detected by the probe DNA, alleles 1 and 2 differ in the size of a restriction fragment (red). B. Tandemly repeated subsequences in a nonessential region vary in number in different alleles. Again, probe DNA detects restriction fragments of different sizes.

a nonessential region. This type of variation can generate many alleles differing in the number of repeating units present.

When the sizes of restriction fragments differ among the alleles of a gene, this constitutes a **restriction fragment length polymorphism, or RFLP.** RFLPs are extremely common in the human genome and in the genomes of most other organisms; hence, they provide an extensive set of highly polymorphic, codominant genetic markers scattered throughout the genome, which can be studied with identical Southern blotting procedures except for varying the restriction enzyme and the type of probe DNA. Figure 8-3 illustrates the segregation of alleles of a restriction fragment polymorphism within a human pedigree.

Because of the convenience and power of the RFLP method, there has been a tremendous effort to identify large numbers of RFLPs in the human genome that are sufficiently scattered to provide a complete linkage map. This effort has been successful in that more than 1000 RFLP polymorphisms have been identified, and a subset of these covers the genome sufficiently well that any unmapped gene of interest is likely to lie within 5-to-10 percent recombination (5–10 map units) from one of the RFLP genetic markers. Other RFLPs have been identified that are very closely linked to disease genes—for example, Huntington disease and cystic fibrosis. These have been important in efforts to isolate and identify the disease genes and of practical benefit in allowing prenatal detection of the affected genotypes.

Some RFLP probes hybridize with many regions scattered throughout the genome, and most of the hybridization sites are themselves highly polymorphic with many alleles. These probes produce patterns of hundreds of bands. These patterns are unique to every individual except for identical twins. These highly informative probes are the basis of **DNA fingerprinting,** or specifying an individual's unique pattern of bands, which has become increasingly important in criminology because DNA fingerprints can be generated from minute quantities of dried blood, semen, hair roots, and other body materials containing DNA. DNA fingerprinting can establish *positively* whether such evidence came from a particular suspect.

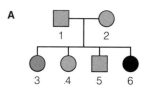

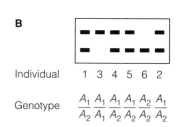

FIGURE 8-3 Segregation of a restriction fragment length polymorphism like that in Figure 8-2A. A. Both parents 1 and 2 are heterozygous A_1 / A_2. Expected progeny resulting from segregation are A_1 / A_1, A_1 / A_2, and A_2 / A_2 in the proportions $1 : 2 : 1$. B. Expected patterns of bands in Southern blots. Homozygous individuals 3 (A_1 / A_1) and 6 (A_2 / A_2) each yield a single restriction fragment that hybridizes with the probe DNA. The heterozygous individuals show both bands.

8-2 Random Mating

Random mating occurs when individuals pair up independently with respect to genotype; that is, genotypes are paired as mates exactly as would be expected by chance. Random mating is by far the most-important mating system in most animals and plants (except for plants that regularly reproduce through self-fertilization), and in these organisms it often determines the distribution of allozyme or blood-group genotypes.

With random mating, the relation between allele frequency and genotype frequency is particularly simple, because *random mating of individuals is equivalent to random union of gametes.* Conceptually, we may imagine all the gametes of a population to be present in a large container. To form zygote genotypes, pairs of gametes are withdrawn from the container at random. To be specific, consider the alleles M and N in the MN blood groups, whose allele frequencies are p and q, respectively (remember that $p + q = 1$). The genotype frequencies expected with random mating can be deduced from the following diagram:

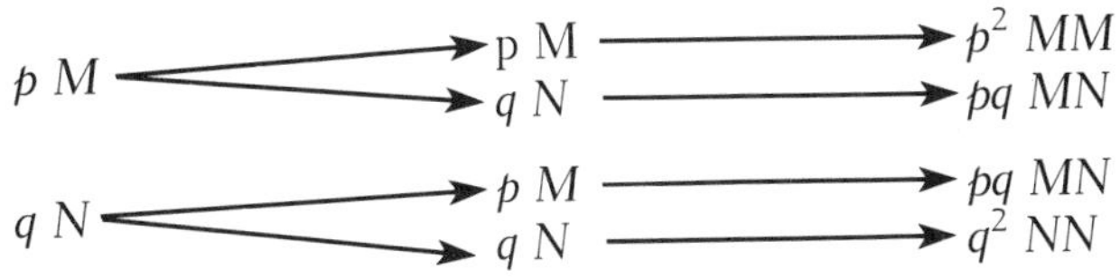

$$p\,M \longrightarrow \begin{array}{c} p\,M \longrightarrow p^2\,MM \\ q\,N \longrightarrow pq\,MN \end{array}$$

$$q\,N \longrightarrow \begin{array}{c} p\,M \longrightarrow pq\,MN \\ q\,N \longrightarrow q^2\,NN \end{array}$$

In this diagram, the gametes at the left represent the sperm and those in the middle the eggs. The genotypes that can be formed with two alleles are shown at the right, and with random mating the frequency of each genotype is calculated by multiplying the allele frequencies of the corresponding gametes. However, the genotype MN can be formed in two ways—the M allele could have come from the father (top part of diagram), or from the mother (bottom part of diagram). In each case, the frequency of the MN genotype is pq; considering both possibilities, the frequency of MN is $pq + pq = 2pq$. Consequently, the overall genotype frequencies expected with random mating are

$$MM:\ p^2 \qquad MN:\ 2pq \qquad NN:\ q^2 \qquad (8\text{-}1)$$

The frequencies p^2, $2pq$, and q^2 that result from random mating for a gene with two alleles constitute what is called the **Hardy-Weinberg principle.** Sometimes the Hardy-Weinberg principle is demonstrated by a Punnett square, as illustrated in Figure 8-4. Such a square is completely equivalent to the tree diagram used earlier. Although the Hardy-Weinberg principle is exceedingly simple, it has a number of important implications that are not obvious. They are described in the next sections.

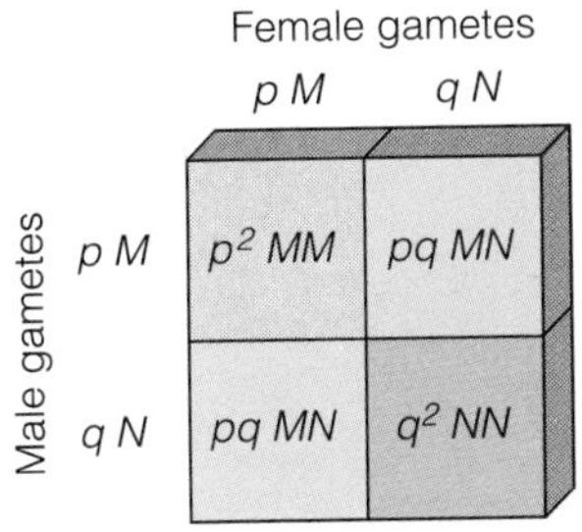

Female gametes

	$p\,M$	$q\,N$
$p\,M$	$p^2\,MM$	$pq\,MN$
$q\,N$	$pq\,MN$	$q^2\,NN$

Male gametes

FIGURE 8-4 A cross-multiplication square (Punnett square) showing the result of random union of male and female gametes, which is equivalent to random mating of individuals.

8-3 Implications of the Hardy-Weinberg Principle

An important implication of the Hardy-Weinberg principle is that *the allele frequencies remain constant from generation to generation.* Consider a gene with two alleles, A and a, having frequencies p and q, respectively ($p + q = 1$). With random mating, the frequencies of genotypes AA, Aa, and aa among zygotes are p^2, $2pq$, and q^2, respectively. Assuming equal *viability* (ability to survive) among the genotypes, these frequencies equal those among adults. If all of the genotypes are equally fertile, then the frequency p' of allele A among gametes of the next generation will be

$$p' = 2pq/2 = p(p + q) = p \quad \text{(because } p + q = 1\text{)}$$

This argument shows that the frequency of allele A remains constant at the value of p through the passage of one (or any number of) complete generations. This principle depends on certain assumptions, of which the most important are:

1. Mating is random.
2. Allele frequencies are the same in males and females.
3. All genotypes are equal in viability and fertility (that is, *selection* does not occur).
4. Mutation does not occur.
5. Migration into the population does not occur.
6. The population is sufficiently large that the frequencies of alleles do not change from generation to generation because of chance.

Whether random mating occurs can often be deduced from genotype and allele frequencies by the use of the Hardy-Weinberg principle. To illustrate the method, consider again the MN blood groups among British people, discussed in Section 8-1. The frequencies of alleles M and N among 1000 adults were 0.5425 and 0.4575, respectively; assuming random mating, we can calculate the *expected* genotype frequencies from Equation 8-1 as

$$MM: (0.5425)^2 = 0.2943$$
$$MN: 2(0.5425)(0.4575) = 0.4964$$
$$NN: (0.4575)^2 = 0.2093$$

and so the expected number of individuals in the population with each genotype would be 294.3 *MM*, 496.4 *MN*, and 209.3 *NN*. The *observed* numbers are 298 *MM*, 489 *MN*, and 213 *NN*. Goodness of fit between observed and expected would normally be determined by means of the χ^2 test described in Chapter 2. We will not do so here because the agreement between expected and observed values is quite good. (The χ^2 test yields a probability of about 0.67, and so the hypothesis of random mating can account for the data.) Note that, although the British population may be undergoing random mating with respect to MN blood groups, the same population may be undergoing nonrandom mating with regard to other genes.

Another important implication of the Hardy-Weinberg principle is that, *for a rare allele, the frequency of heterozygotes far exceeds the frequency of the rare homozygote.* For example, when the frequency of the rare allele is $q = 0.1$, the ratio of heterozygotes to homozygotes equals $2pq/q^2 = 2(0.9)/(0.1)$, or approximately 20; when $q = 0.01$, this ratio is about 200; and when $q = 0.001$, it is about 2000. Clearly:

> When an allele is rare, there are many more heterozygotes than there are homozygotes for the rare allele.

One practical implication of this principle is seen in the example of **cystic fibrosis,** an inherited secretory disorder of the pancreas and lungs, which is one of the most-common recessively inherited severe disorders among Caucasians. Cystic fibrosis affects about 1 in 1700 newborns. In this case, the heterozygotes cannot readily be identified by phenotype; so a method of calculating allele frequencies that is different from the gene-counting method used earlier must be used. The new method is straightforward because with random mating the frequency of recessive homozygotes must correspond to q^2. Thus, for cystic fibrosis,

$$q^2 = 1/1700 = 0.00059 \quad \text{or} \quad q = (0.00059)^{1/2} = 0.024$$

and, consequently,

$$p = 1 - q = 1 - 0.024 = 0.976$$

The frequency of heterozygotes that carry the allele for cystic fibrosis is calculated as

$$2pq = 2(0.976)(0.024) = 0.047 = 1/21$$

Therefore, for cystic fibrosis, although only 1 person in 1700 is affected with the disease (homozygous), about 1 person in 21 is a carrier (heterozygous). Considerations like these are important in predicting the outcome of population screening for the detection of carriers of harmful recessive alleles, which is essential in evaluating the potential benefits of such programs.

8-4 Extensions of the Hardy-Weinberg Principle

So far, the Hardy-Weinberg principle has been applied to two alleles of a single autosomal gene. However, the principle is easily extended to more-complex cases. Two of these cases are multiple alleles of an autosomal gene and X-linked genes.

Multiple Alleles

Extension of the Hardy-Weinberg principle to multiple alleles is illustrated by the three-allele case. With three alleles, the Punnett square for random mating is shown in Figure 8-5. The alleles are designated A_1, A_2, and A_3, with the uppercase letter representing the gene and the subscript designating the particular allele. The allele frequencies are p_1, p_2, and p_3, respectively. With three alleles (as with any number of alleles), the allele frequencies of all alleles must sum to 1—that is, $p_1 + p_2 + p_3 = 1.0$. As in Figure 8-4, each square is

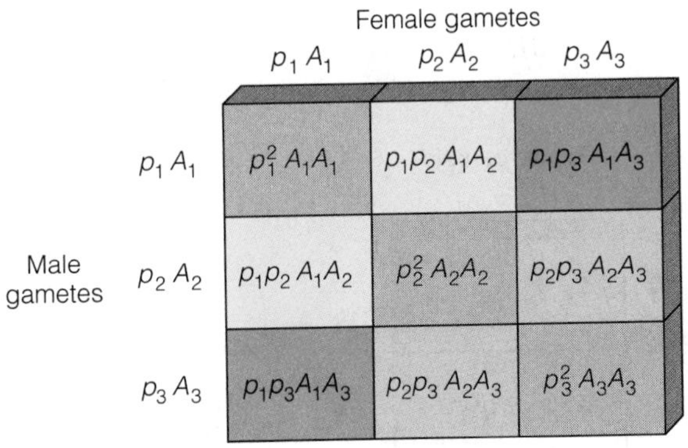

FIGURE 8-5 Punnett square showing the results of random mating with three alleles.

obtained by multiplying the frequencies of the alleles at the corresponding margins; any homozygote (such as A_1A_1) has a random-mating frequency equal to the square of the corresponding allele frequency (in this case, p_1^2). The various degrees of colored shading represent the heterozygotes. Any heterozygote (such as A_1A_2) has a random-mating frequency equal to twice the product of the corresponding allele frequencies (in this case, $2p_1p_2$). The extension to any number of alleles is straightforward:

Frequency of any homozygote = square of allele frequency
Frequency of any heterozygote = 2 × product of allele frequencies

Let us use the gene controlling the human ABO blood groups (Chapter 1) to illustrate these considerations. This gene has three principal alleles, designated I^A, I^B, and I^O. In one study of 3977 Swiss people, the allele frequencies were found to be 0.27 I^A, 0.06 I^B, and 0.67 I^O. Applying the rules for multiple alleles, we can expect the genotype frequencies resulting from random mating to be

$$
\left.
\begin{aligned}
I^A I^A &= (0.27)^2 = 0.0729 \\
I^A I^O &= 2(0.27)(0.67) = 0.3618
\end{aligned}
\right\} \text{Type A} = 0.4347
$$

$$
\left.
\begin{aligned}
I^B I^B &= (0.06)^2 = 0.0036 \\
I^B I^O &= 2(0.06)(0.67) = 0.0804
\end{aligned}
\right\} \text{Type B} = 0.0840
$$

$$
I^O I^O = (0.67)^2 = 0.4489 \quad \text{Type O} = 0.4489
$$

$$
I^A I^B = 2(0.27)(0.06) = 0.0324 \quad \text{Type AB} = 0.0324
$$

Because both I^A and I^B are dominant to I^O, the expected frequency of blood-group *phenotypes* is that shown on the right. Note that the majority of A and B phenotypes are heterozygous for the I^O allele.

X-linked Genes

Random mating for two X-linked alleles (*H* and *h*) is illustrated in Figure 8-6. The principles are the same as those considered earlier, but male gametes carrying the X chromosome (Figure 8-6A) must be distinguished from those carrying the Y chromosome (Figure 8-6B). When the male gamete carries an X chromosome, the Punnett square is exactly the same as for the two-allele autosomal gene in Figure 8-4. However, because the male gamete carries an X chromosome, all the offspring in question are female. Consequently, among females, the genotype frequencies are

$$HH: p^2 \qquad Hh: 2pq \qquad hh: q^2$$

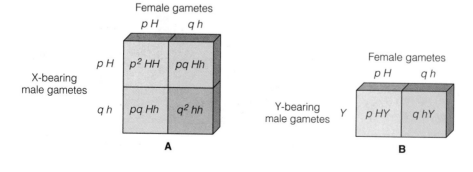

FIGURE 8-6 The results of random mating for an X-linked gene: (A) genotype frequencies in females; (B) genotype frequencies in males.

When the male gamete carries a Y chromosome, the outcome is quite different (Figure 8-6B). All the offspring are male and receive only one X chromosome, which comes from their mothers. Therefore, each male receives only one copy of each X-linked gene, and the genotype frequencies among males are the same as the allele frequencies—namely, $H{:}p$ and $h{:}q$

An important implication of Figure 8-6 is that, if h is a rare recessive allele, then there will be many more males exhibiting the trait than females; that is, the frequency of affected females (q^2) will be much smaller than the frequency of affected males (q). For example, the common form of X-linked color blindness in human beings affects about one in 20 males, and so $q = 1/20 = 0.05$. The frequency of color-blind females is calculated as $q^2 = (0.05)^2 = 0.0025$, or about one in 400.

8-5 Inbreeding

Inbreeding means mating between relatives, such as first cousins. The principal consequence of inbreeding is that the frequency of heterozygous offspring is smaller than with random mating. This effect is most dramatic in repeated self-fertilization, such as occurs naturally in certain plants. Consider a hypothetical population consisting exclusively of Aa heterozygotes. With self-fertilization, each plant would produce offspring in the proportions 1/4 AA, 1/2 Aa, and 1/4 aa. Thus, one generation of self-fertilization reduces the proportion of heterozygotes from 1 to 1/2. In the second generation, only the heterozygous plants can again produce heterozygous offspring, and only half of their offspring will again be heterozygous. Heterozygosity is therefore reduced to 1/4 of what it was originally. Three generations of self-fertilization reduce the heterozygosity to $1/4 \times 1/2 = 1/8$, and so forth. The remainder of this section demonstrates how the reduction in heterozygosity owing to inbreeding may be expressed quantitatively and measured.

Repeated self-fertilization is a particularly intense form of inbreeding, but weaker forms of inbreeding are qualitatively similar in that they also lead to a reduction in heterozygosity. A convenient measure of the effect of inbreeding is based on the reduction in heterozygosity. Suppose that H_I is the frequency of heterozygous genotypes in a population of inbred individuals. The most widely used measure of inbreeding is called the **inbreeding coefficient**, symbolized F, which is defined as the proportionate reduction in H_I compared with the value of $2pq$ that would be expected with random mating. That is,

$$F = (2pq - H_I)/2pq \tag{8-2}$$

Equation 8-2 can be rearranged as

$$H_I = 2pq(1 - F)$$

The homozygous genotypes increase in frequency as the heterozygotes decrease in frequency, and overall the genotype frequencies in the inbred population are

$$
\begin{aligned}
&AA{:}\ p^2(1 - F) + pF \\
&Aa{:}\ 2pq(1 - F) \\
&aa{:}\ q^2(1 - F) + qF
\end{aligned}
\tag{8-3}
$$

These frequencies amount to a modification of the Hardy-Weinberg principle in order to take inbreeding into account. When $F = 0$ (no inbreeding), the genotype frequencies are the same as those given in the Hardy-Weinberg principle in Equation 8-1—namely, p^2, $2pq$, and q^2. At the other extreme, when $F = 1$ (complete inbreeding), the inbred population consists entirely of AA and aa individuals occurring in the frequencies p and q, respectively.

As an illustration, consider a population of plants that frequently undergoes inbreeding through self-pollination. A sample of 140 plants is examined and the observed numbers of genotypes are 60 AA, 24 Aa, and 56 aa. The allele frequencies are then $(24 + 60 + 60)/280 = 0.51$ A $= p$ and $1 - 0.51$ $= 0.49$ $a = q$. The observed frequency of heterozygotes in the population is $24/140 = 0.17 = 2pq(1 - F)$, and so $0.17 = 2(0.51)(0.49)(1 - F)$, from which $F = 0.66$. This high value of F indicates that a significant fraction of the population undergoes self-fertilization in each generation.

Effects of Inbreeding

The effects of inbreeding differ according to the normal mating system of an organism. At one extreme, in regularly self-fertilizing plants, inbreeding is already so intense and the organisms are so highly homozygous that additional inbreeding has virtually no effect. However:

> In most species, inbreeding is harmful, and much of the effect is due to rare recessive alleles that would not otherwise become homozygous.

Among human beings, inbreeding is usually rare because of social conventions, though in small isolated populations (aboriginal groups, religious communities) it does occur, mainly through matings between remote relatives. The most-common type of close inbreeding is between first cousins. The effect is always an increase in the frequency of genotypes that are homozygous for rare, usually harmful recessives. For example, among American whites, the frequency of albinism among offspring of matings between nonrelatives is approximately 1 in 20,000 but, among offspring of first-cousin matings, the frequency is approximately 1 in 2000. The reason for the increased risk can be understood by comparing the genotype frequencies of homozygous recessives in Equation 8-3 (for inbreeding) and Equation 8-1 (for random mating). In the most-common form of inbreeding among human beings (that is, mating between first cousins), $F = 0.062 (= 1/16)$ among the offspring. Therefore, the frequency of homozygous recessives produced by first-cousin mating will be

$$q^2(1 - 0.062) + q(0.062)$$

whereas, among nonrelatives, the frequency of homozygous recessives is simply q^2. For albinism, $q = 0.007$ (approximately), and the calculated frequencies are 5×10^{-4} for the offspring of first cousins and 5×10^{-5} for the offspring of nonrelatives. The increased frequency with first-cousin mating is a principal consequence of inbreeding.

8-6   Genetics and Evolution

Evolution refers to the progressive adaptation of populations to their environment. Evolution occurs because genetic variation exists in populations and because there is a natural selection of organisms that are best adapted to the environment. Genetic variation and natural selection are population phenomena and so are conveniently discussed in terms of allele frequencies.

Four processes account for most of the changes in allele frequency in populations. They form the basis of cumulative change in the genetic characteristics of populations, leading to the descent with modification that characterizes the process of evolution. Although the point has yet to be proved, most evolutionary biologists believe that these same processes, when carried out continuously through geological time, can also account for the formation of new species and higher taxonomic categories. These processes are:

1. **Mutation,** the origin of new genetic capabilities in populations by means of spontaneous heritable changes in genes.
2. **Migration,** the movement of individual organisms among subpopulations within a larger population.
3. **Natural selection,** resulting from the different abilities of individual organisms to survive and reproduce in their environment.
4. **Random genetic drift,** the random, undirected changes in allele frequency that occur by chance in all populations, but particularly in small ones.

Aspects of these processes pertaining to population genetics are considered in the following sections.

8-7 Mutation and Migration

Mutation is the ultimate source of genetic variation. It is an essential process in evolution, but it is a relatively weak force for changing allele frequencies, primarily because typical mutation rates are too low. Moreover, most newly arising mutations are harmful to the organism. Although some mutations may be **selectively neutral,** meaning that they do not affect the ability of the organism to survive and reproduce, only a very few mutations are favorable for the organism and contribute to evolution. The low mutation rates that are observed are thought to have evolved as a compromise: the mutation rate is high enough to generate the favorable mutations that a species requires to continue evolving, but it is not so high that the species suffers too much genetic damage from the preponderance of harmful mutations.

Migration is similar to mutation in that new alleles can be introduced into a local population, although the alleles derive from another population rather than from new mutations. Ultimately, all alleles must originate from mutation but, if the migration of individual organisms among populations is sufficiently rare, local populations can undergo considerable genetic differentiation. Genetic differentiation among local populations occurs if there are differing frequencies of common alleles or if each population possesses one or more rare alleles not found in other populations. Genetic differences among local populations can obviously be swamped if the populations exchange individuals (undergo migration) at a sufficiently high rate. However, only a relatively small amount of migration among local populations—on the order of just a few migrant individuals in each local population in each generation—is usually an impediment to the evolution of high levels of genetic differentiation. On the other hand, genetic differentiation can occur in spite of migration if other evolutionary forces, such as natural selection for adaptation to the local environments, are sufficiently strong.

The driving force of adaptive evolution is natural selection, which is a consequence of hereditary differences in the ability of individual organisms to survive and reproduce in the prevailing environment. Since first proposed by Charles Darwin, the concept of natural selection has been revised and extended, most notably by the incorporation of genetic concepts. In its modern formulation, the occurrence of natural selection rests on three premises:

1. In all organisms, more offspring are produced than survive and reproduce.
2. Organisms differ in their ability to survive and reproduce, and some of these differences are due to genotype.
3. In every generation, genotypes that promote survival in the prevailing environment (favored genotypes) are present in excess among individuals of reproductive age and hence they contribute disproportionately to the offspring of the next generation. In this way, the alleles that enhance survival and reproduction increase in frequency from generation to generation, and the population becomes progressively better for survival and reproduction in the environment. This progressive improvement in populations constitutes the process of evolutionary **adaptation.**

Selection in Escherichia coli: An Example

Selection is easily studied in bacterial populations because of the short generation time (about 30 minutes). Figure 8-7 shows the result of competition between two bacterial genotypes, A and B. Genotype A is the superior competitor under the particular conditions, and so the frequency of the A strain increases. In the experiment, the competition was allowed to continue for 290 generations, during which time the proportion of A genotypes (p) increased from 0.60 to 0.9995 and that of B genotypes decreased from 0.40 to 0.0005.

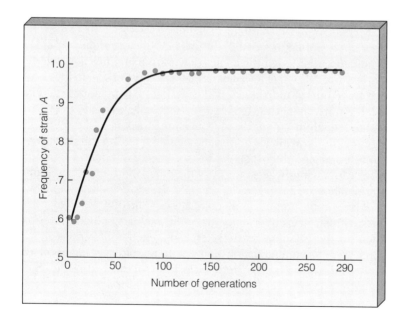

FIGURE 8-7   Increase in frequency of a favored strain of *E. coli* resulting from selection in a continuously growing population. The y-axis is the number of A cells at any time divided by the total number of cells (A + B). Note that the changes in frequency are the greatest when the frequency of the favored strain, A, is at intermediate values.

The data points give a satisfactory fit to an equation of the form

$$p_n/q_n = (p_0/q_0)\,(1/w)^n \qquad (8\text{-}4)$$

in which p_0 and q_0 are the initial frequencies of A and B (in this case 0.6 and 0.4, respectively), p_n and q_n are the frequencies after n generations of competition, n is the numer of generations of competition, and w is a measure of the competitive ability of B when competing against A under the conditions of the experiment. The theoretical derivation of Equation 8-4 is beyond the scope of this book, but the black curve in Figure 8-7 shows the goodness of fit with the experimental data when $w = 0.958$.

The value of $w = 0.958$ is called the **relative fitness** of the B genotype relative to the A genotype under these particular conditions. Relative fitness measures the comparative contribution of each parental genotype to the pool of offspring genotypes produced in each generation; that is, for each offspring cell produced by an A genotype, a B genotype produces an average of 0.958 offspring cells.

In population genetics, relative fitnesses are usually calculated with the most-favored genotype (A in this case) taken as the standard with a fitness of 1.0. However, the selective disadvantage of a genotype is often of greater interest than its relative fitness. The selective disadvantage of a disfavored genotype is called the **selection coefficient** associated with the genotype, and it is calculated as the difference between the fitness of the standard (taken as 1.0) and the relative fitness of the genotype in question. In the case at hand, the selection coefficient against B, denoted s, is

$$s = 1.000 - 0.958 = 0.042 \qquad (8\text{-}5)$$

The meaning of s is that the selective disadvantage of strain B is 4.2 percent per generation. When the fitnesses are known, Equation 8-5 also permits the prediction of the allele frequencies in any future generation, given the original frequencies. Alternatively, it can be used to calculate the number of generations required for selection to change the allele frequencies from any specified initial values to any later ones. For example, from the relative fitnesses of A and B just estimated, one can calculate the number of generations required to change the frequency of A from 0.1 to 0.8. In this example, $p_0/q_0 = 0.1/0.9$, $p_n/q_n = 0.8/0.2$, and $w = 0.958$. A little manipulation of Equation 8-4 gives

$$n = [\log (0.1/0.9) - \log (0.8/0.2)]/\log (0.958) = 83.5 \text{ generations}$$

Selection in Diploids

Selection in diploids is analogous to that in haploids, but there is an additional complication that comes from the occurrence of heterozygous genotypes. Without going through the analysis, let us consider the essential conclusion. Figure 8-8 shows the change in allele frequencies for both a favored dominant and a favored recessive. The striking feature of the figure is that the frequency of the favored dominant allele changes very slowly when the allele is common, and the frequency of the favored recessive allele changes very slowly when the allele is rare. The reason is that rare alleles occur much more frequently in heterozygotes than in homozygotes. With a favored dominant at high frequency, most of the recessive alleles occur in heterozygotes, and the heterozygotes are not exposed to selection and hence do not contribute to change in allele frequency. Conversely, with a favored recessive at low

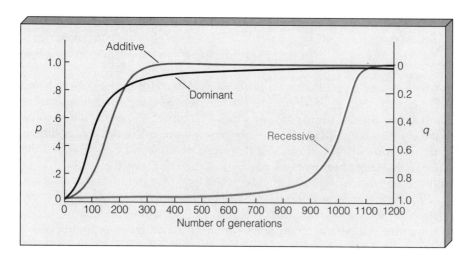

FIGURE 8-8 Theoretically expected change in frequency of an allele favored by selection in a diploid organism undergoing random mating when the favored allele is dominant or recessive or when the heterozygote is intermediate in fitness (additive). In each case, the selection coefficient against the least-fit homozygous genotype is 5 percent. The data are plotted in terms of p, the allele frequency of the beneficial allele. These curves demonstrate that selection for or against a recessive allele is very inefficient when the recessive allele is rare.

frequency, most of the favored alleles are in heterozygotes, and again the heterozygotes are not exposed to selection and do not contribute to change in allele frequency. The principle is quite general:

> Selection for or against recessive alleles is very inefficient when the recessive allele is rare.

One simple example of this principle is selection against a recessive lethal. In this case, the number of generations required to reduce the frequency of the recessive allele from q to $q/2$ equals $1/q$ generations. For example, if $q = 0.01$, successive halvings of the recessive allele frequency require 100, 200, 400, 800, 1600, ... generations, and so the recessive lethal allele is eliminated very slowly.

The inefficiency of selection against rare recessive alleles has an important practical implication. There is a widely held belief that medical treatment to save the lives of persons with rare recessive disorders will cause a deterioration of the human gene pool because of the reproduction of persons who carry the harmful genes. This belief is unfounded. With rare alleles, the proportion of homozygotes is so small that reproduction of the homozygotes contributes a negligible amount to change in allele frequency. Considering their low frequency, it matters little whether homozygotes reproduce or not. Similar reasoning applies to eugenic proposals to "cleanse" the human gene pool of harmful recessives by preventing the reproduction of affected persons. People with severe genetic disorders rarely reproduce anyway, and, even when they do, they have essentially no effect on allele frequency. *The largest reservoir of harmful recessive alleles is in the genomes of phenotypically normal carrier heterozygotes.*

Selection-Mutation Balance

It is apparent from Figure 8-8 that selection tends to eliminate harmful alleles from a population. However, harmful alleles can never be eliminated totally, because recurrent mutation of the normal allele continually creates new harmful alleles. These new mutations tend to replenish the harmful alleles eliminated by selection, and eventually the population will attain a state of equilibrium in which the new mutations exactly balance the selective eliminations. Two important cases—complete recessive and partial dominant—must be considered. In both cases, equilibrium results from the balance of new mutations against those alleles eliminated by selection.

The allele frequency of a harmful allele maintained at equilibrium depends strongly on whether the allele is completely recessive. Because selection against a complete recessive is so inefficient when the allele is rare, even a small amount of selection against heterozygous carriers results in a dramatic reduction in the equilibrium allele frequency. For example, consider a homozygous lethal allele that has an equilibrium frequency of 0.01 when completely recessive; if the heterozygous carriers had a relative fitness of 0.99 instead of 1.0, the equilibrium frequency would decrease to 0.001. This decrease occurs because there are many more heterozygotes than homozygotes for the rare allele, and so a small amount of selection against heterozygotes affects such a relatively large number of individuals that its overall effect is very great.

Heterozygote Superiority

So far, we have considered only cases in which the fitness of the heterozygote is intermediate between those of the homozygotes (or possibly equal in fitness to one homozygote). In these cases, the allele associated with the most-fit homozygote eventually becomes fixed, unless the selection is opposed by mutation. In this section, we consider the possibility of **heterozygote superiority,** which occurs when the fitness of the heterozygote is greater than that of both homozygotes.

When there is heterozygote superiority, neither allele can be eliminated by selection. In each generation, the heterozygotes produce more offspring than do the homozygotes, and the selection for heterozygotes keeps both alleles in the population. Selection eventually produces an equilibrium in which the allele frequencies no longer change.

Heterozygote superiority does not appear to be a particularly common form of selection in natural populations. However, there are several well-established cases, the best known of which involves the sickle-cell hemoglobin mutation (Hb^S) and its relation to a type of malaria caused by the parasitic protozoan *Plasmodium falciparum*. In the absence of modern medical care, the Hb^S allele is virtually lethal when homozygous, yet in certain parts of Africa and the Middle East the allele frequency reaches 10 percent or even higher. Such frequencies are much too high to be explained by recurrent mutation. The explanation is that heterozygous persons carrying the Hb^S allele are less susceptible to malaria than are homozygous normal persons, and therefore the heterozygous genotypes have the highest fitness.

8-9 Random Genetic Drift

Random genetic drift is one of the most subtle processes in population genetics and, indeed, in all of genetics. It comes about because populations are not infinitely large, as we have been assuming all along, but finite (limited in size). The breeding individuals in any one generation produce a potentially infinite pool of gametes. Barring fertility differences, the allele frequencies among gametes would equal the allele frequencies among adults. However, because of the finite size of the population, only relatively few of the gametes participate in fertilization to form the zygotes of the next generation. In other words, there is a process of *sampling* that occurs in going from one generation to the next; because there is chance variation among samples, the allele frequencies among gametes and those among zygotes may differ.

An extreme example illustrates the essential features of random genetic drift. Consider several populations of annual plants maintained at a size of two plants in each generation by the random selection of two seeds from the plants of the preceding generation. Suppose that, initially, all populations were established with two heterozygous genotypes. In that case, the initial allele frequencies are 0.5, but they will not remain constant, because of the two-seed mode of propagation. The possible genotypes of each pair of offspring in the second generation and the probabilities of each are shown in Table 8-1. Among 16 populations an expected 6 (pairs 3 and 4) would retain the allele frequencies of 0.5; 8 (pairs 5 and 6) would have frequencies of 0.75 and 0.25; and 2 (pairs 1 and 2) would have lost one or the other of the alleles. In subsequent matings, allele frequencies would continue to change in 14 of the populations (entries 3 through 6) because of the two-seed propagation, but in 2 (entries 1 and 2) one allele is fixed—either A or a. Fixation and random changes in allele frequencies would not have occurred if each mating, including the first, had resulted in an infinite number of offspring, instead of only two.

A less-extreme example is illustrated in Table 8-2. Here 12 subpopulations (a through l), each initially consisting of 8 diploid individuals and containing 8 copies of allele A and 8 copies of a, have been allowed to mate at random. A computer, programmed to produce random matings and to take into account the probability of different offspring genotypes in each mating that it selects, has been used to calculate the number of A alleles in each subpopulation. The vertical columns show the number of A alleles in each population as time passes, and the dispersion of allele frequencies resulting from random genetic drift are apparent. These changes in allele frequency would be less pronounced and would require more time in larger populations than in the very small populations illustrated here, but the overall effect would be the same. That is, the dispersion of allele frequency resulting from random genetic drift depends on population size; the smaller the population, the greater the dispersion and the more rapidly it occurs.

In Table 8-2, the principal effect of random genetic drift is evident in the first generation—the allele frequencies have begun to spread out. By generation 7, the spreading is extreme, and the number of A alleles ranges from

---

TABLE 8-1   Possible pairs of offspring produced by two heterozygous *(Aa)* parents

Genotypes of pair		Probability	Allele frequency	
			A	a
1*	AA, AA	(1/4) (1/4) = 1/16	1.00	0.00
2†	aa, aa	(1/4) (1/4) = 1/16	0.00	1.00
3	AA, aa	2(1/4) (1/4) = 2/16	0.50	0.50
4	Aa, Aa	(2/4) (2/4) = 4/16	0.50	0.50
5	AA, Aa	2(1/4) (2/4) = 4/16	0.75	0.25
6	aa, Aa	2(1/4) (2/4) = 4/16	0.25	0.75

*Fixed for A
†Fixed for a

TABLE 8-2   Effects of random genetic drift

Generation	a	b	c	d	e	f	g	h	i	j	k	l	$\bar{p}$	H_s
				Population designation									Averages	
0	8	8	8	8	8	8	8	8	8	8	8	8	0.500	0.500
1	11	7	9	8	8	6	8	7	8	6	11	10	0.516	0.478
2	10	9	10	8	8	8	6	7	6	7	13	11	0.536	0.465
3	7	11	6	5	12	5	8	5	8	7	14	9	0.505	0.438
4	8	11	7	4	12	5	8	8	7	4	15	6	0.495	0.421
5	8	8	5	3	13	2	8	12	9	5	15	6	0.490	0.387
6	11	5	3	1	13	3	10	13	6	7	15	3	0.469	0.337
7	11	8	4	3	11	1	8	14	3	7	15	2	0.453	0.334
8	14	7	4	3	10	1	9	14	3	9	16	1	0.474	0.300
9	15	5	3	5	7	0	12	14	2	11	16	0	0.469	0.251
10	16	6	5	9	8	0	9	13	3	10	16	0	0.495	0.288
11	16	1	5	11	6	0	10	13	3	10	16	0	0.474	0.249
12	16	0	5	12	6	0	9	13	2	9	16	0	0.458	0.232
13	16	0	3	12	7	0	9	13	1	11	16	0	0.458	0.210
14	16	0	5	15	7	0	9	11	1	12	16	0	0.479	0.204
15	16	0	3	14	7	0	8	12	3	13	16	0	0.479	0.208
16	16	0	2	14	9	0	11	14	2	14	16	0	0.510	0.168
17	16	0	1	15	6	0	12	14	2	13	16	0	0.495	0.152
18	16	0	1	15	4	0	13	15	5	13	16	0	0.510	0.147
19	16	0	1	16	2	0	14	16	6	14	16	0	0.526	0.104
20	16	0	1	16	2	0	15	16	9	16	16	0	0.557	0.079
21	16	0	2	16	3	0	15	16	10	16	16	0	0.573	0.092

1 to 15. In general, *random genetic drift causes differences in allele frequency among subpopulations and therefore causes genetic divergence among subpopulations.*

Although allele frequencies among individual subpopulations spread out because of random genetic drift, the *average* allele frequency among subpopulations remains approximately constant. This point is illustrated by the column headed $\bar{p}$ in Table 8-2. The average allele frequency stays close to 0.5, its initial value. Indeed, if an infinite number of subpopulations were being considered instead of the 12 in Table 8-2, the average allele frequency would be exactly 0.5 in every generation. That is, in spite of the random drift of allele frequency in individual subpopulations, the average allele frequency among a large number of subpopulations remains constant and equal to the average allele frequency among the original subpopulations.

After a sufficient number of generations of random genetic drift, some of the subpopulations become fixed for A (red) or fixed for *a* (gray). Because we are excluding the occurrence of mutation, a population that becomes fixed for an allele remains fixed thereafter. After 21 generations in Table 8-2, only four of the populations are still segregating; eventually, these too will become fixed. Because the average allele frequency of A remains constant, it follows that a fraction p_0 of the populations (p_0 represents the allele frequency of A in the initial generation) will ultimately become fixed for A and a fraction $1 - p_0$ will become fixed for a. That is, *the probability of ultimate fixation of a particular allele is equal to the frequency of that allele in the original population*. In Table 8-2, five of the fixed populations are fixed for A and three for a, which is not very different from the equal numbers expected theoretically with an infinite number of subpopulations.

If random genetic drift were the only force at work, all alleles would become either fixed or lost and there would be no polymorphism. On the other hand, many factors can act to retard or prevent the effects of random genetic drift, of which the following are the most important: (1) large population size; (2) mutation and migration, which impede fixation because alleles lost by random genetic drift can be reintroduced by either process; (3) natural selection, particularly those modes of selection that tend to maintain genetic diversity, such as heterozygote superiority.

CHAPTER SUMMARY

Population genetics is the application of Mendel's laws and other principles of genetics to populations of organisms. The population unit is a group of organisms of the same species living within a geographical region of such size that most matings occur between members of the group. In most natural populations, many genes are polymorphic in that they have two or more common alleles. One of the goals of population genetics is to determine the nature and extent of genetic variation in natural populations.

The relation between the relative proportions of particular alleles (allele frequencies) and genotypes (genotype frequencies) is determined in part by the frequencies with which particular genotypes form mating pairs. In random mating, mating is independent of genotype. When a population undergoes random mating for an autosomal gene with two alleles, the frequencies of the genotypes are given by the Hardy-Weinberg principle. If the alleles of the gene are A and a, and their allele frequencies are p and q, respectively, then the Hardy-Weinberg principle states that the genotype frequencies with random mating are: AA p^2; Aa $2pq$; and aa q^2. Hardy-Weinberg proportions are strictly obeyed only when there is no migration, differential survival or reproduction, or changes in allele frequency owing to chance. The Hardy-Weinberg genotype frequencies provide a convenient approximation to genotype frequencies that are actually found in many natural populations, and goodness of fit with Hardy-Weinberg frequencies can be evaluated with a χ^2 test. An important implication of the Hardy-Weinberg principle is that rare alleles occur much more frequently in heterozygotes than in homozygotes ($2pq$ versus q^2).

Inbreeding means mating between relatives, and the extent of inbreeding is measured by the inbreeding coefficient. The main consequence of inbreeding is that a rare harmful allele present in a common ancestor may be transmitted to both parents of an inbred individual in a later generation and become homozygous in the inbred offspring. Among inbred individuals, the frequency of heterozygous genotypes is smaller, and that of homozygous genotypes greater, than would occur with random mating.

Evolution is the progressive increase in the degree to which a species becomes adapted to its environment. A principal mechanism of evolution is natural selection, in which individuals superior in survival or reproductive ability in the prevailing environment contribute a disproportionate share of genes to future generations, thereby gradually increasing the frequency of the favorable alleles in the whole population. However, at least three other processes can also change allele frequency—mutation (heritable change in a gene), migration (movement of individuals among local populations), and random genetic drift (resulting from restricted population size). Spontaneous mutation rates are generally so low that the effect of mutation on changing allele frequency is minor, except for rare alleles. Migration can have significant effects on allele frequency because migration rates may be very large. The main effect of migration is the tendency to equalize allele frequencies among the populations that exchange migrants.

Selection occurs through differences in viability (the probability of survival of a genotype) and in fertility (the probability of successful reproduction).

Populations maintain harmful alleles at low frequencies as a result of a balance between selection, which tends to eliminate the alleles, and mutation, which tends to increase their frequencies. The equilibrium allele frequency that occurs with selection-mutation balance is usually significantly greater for alleles that are completely recessive than for alleles that are partially dominant. This difference arises because selection is quite ineffective in affecting the frequency of a completely recessive allele when the allele is rare, owing to the almost exclusive occurrence of the allele in heterozygotes.

A few examples are known in which the heterozygous genotype has a greater fitness than either of the homozygous genotypes (heterozygote superiority). Heterozygote superiority results in an equilibrium in which both alleles are maintained in the population. An example is sickle-cell anemia in regions of the world where falciparum malaria is endemic. Heterozygous persons have an increased resistance to malaria and only a mild anemia, which results in greater fitness.

Random genetic drift is a statistical process of change in allele frequency in small populations, resulting from the inability of every individual to contribute equally to the offspring of successive generations. In a subdivided population, random genetic drift is a principal cause of divergence in allele frequency. In an isolated population, barring mutation, an allele will ultimately become fixed or lost as a result of random genetic drift.

KEY TERMS

adaptation	heterozygote superiority	random genetic drift
allele frequency	inbreeding	random mating
allozyme	inbreeding coefficient	relative fitness
cystic fibrosis	local population	restriction fragment length
DNA fingerprinting	migration	polymorphism
evolution	mutation	RFLP
fixed allele	natural selection	selection coefficient
gene pool	polymorphic gene	selection-mutation balance
genotype frequency	population	selectively neutral mutation
Hardy-Weinberg principle	population genetics	

EXAMPLES OF WORKED PROBLEMS

Problem 1: A sample of 300 plants from a population are examined for the electrophoretic mobility of an enzyme that varies according to the genotype determined by two alleles, F and S, of a single gene. The results are 7 individual plants with genotype FF, 106 of genotype FS, and 187 of genotype SS. What are the allele frequencies of F and S? What are the expected numbers of the three genotypes, assuming random mating?

Answer The allele frequencies are determined by counting the genes. The 7 FF individual plants represent 14 F alleles, the 106 FS individuals represent 106 F and 106 S alleles, and the 187 SS individuals represent 374 S alleles, for a total of $300 \times 2 = 600$ alleles altogether. The allele frequency p of F is $(14 + 106)/600 = 0.2$, and the allele frequency q of S is $(106 + 374)/600 = 0.8$. As a check on the calculations, note that the allele frequencies sum to unity, as they should. For the second part of the question,

the expected genotype frequencies with random mating are p^2 FF, $2pq$ FS, and q^2 SS, and so the expected numbers are as follows:

$$
\begin{array}{lll}
FF\text{:} & (0.2)^2 \times 300 = 12 \\
FS\text{:} & 2(0.2)(0.8) \times 300 = 96 \\
SS\text{:} & (0.8)^2 \times 300 = 192
\end{array}
$$

As a check on the calculations, note that $12 + 96 + 192 = 300$.

Problem 2: Excessive secretion of male sex hormones results in premature sexual maturation in males and masculinization of the sex characters in females. This disorder is called the adrenogenital syndrome, and in Switzerland there is an autosomal recessive form of the disease that affects about one in 5000 newborns. **(a)** Assuming random mating, what is the allele frequency of the recessive? **(b)** What is the frequency of heterozygous carriers? **(c)** What is the expected frequency of the condition among the offspring of first cousins?

Answer: **(a)** Set $q^2 = 1/5000$, which implies that $q = (1/5000)^{1/2} = 0.014$. **(b)** The frequency of heterozygotes is $2pq$ in which $p = 1 - 0.014 = 0.986$, and so $2(0.014)(0.096) = 0.028 = 1/36$; that is, almost 3 percent of the population are carriers, even though only about 0.02 percent of the population are affected. **(c)** With first-cousin mating, the inbreeding coefficient $F = 1/16 = 0.062$, and the expected proportion of affected people is $q^2(1 - F) + qF = (0.014)^2(0.938) + (0.014)(0.062) = 0.001$; that is, there is about a fivefold greater risk of homozygosity for this allele among the offspring of first cousins.

Problem 3: Warfarin is a rat killer that acts by hindering blood coagulation. Continued use of the poison has resulted in the evolution of resistance in many rat populations due to selection favoring a resistance mutation R. The normal, sensitive allele may be denoted S. In the absence of warfarin, the relative fitness of SS, RS, and SS are in the ratio $1.00 : 0.77 : 0.46$, and in the presence of warfarin the ratio of fitnesses is $0.68 : 1.00 : 0.37$. **(a)** With regard to persistence or loss of the R and S alleles, what is the expected result of the continued use of the chemical? **(b)** What would happen if warfarin were no longer used? **(c)** Under what circumstances might the R allele become fixed?

Answer: **(a)** In the presence of warfarin, there is heterozygote superiority, and so continued use would result in a stable equilibrium in which both R and S would persist in the population. **(b)** In the absence of warfarin, the genotype SS is favored and RS is intermediate in fitness, and so allele S would become fixed and R would become lost. **(c)** The allele R could become fixed in a small population where, because of chance fluctuations in allele frequency, the S allele might become lost. However, small populations would also increase the chance that the R allele might be lost.

PROBLEMS

8-1. Name four evolutionary processes that can change allele frequencies in natural populations? How do allele frequencies change in the absence of these processes?

8-2. In terms of allele frequency, what does it mean to say that an allele is fixed? That an allele is lost?

8-3. What is random genetic drift and why does it occur? In the absence of any counteracting forces, what is the ultimate effect of random genetic drift on allele frequency? Are random changes in allele frequency from one generation to the next greater in small populations or in large ones, and why?

8-4. Why is natural selection very inefficient with rare recessive alleles?

8-5. What is the fitness of an organism that dies before the age of reproduction? What is the fitness of an organism that is sterile?

8-6. If the genotype AA is an embryonic lethal and the genotype aa is fully viable but sterile, what genotype frequencies would be found in adults in an equilibrium population containing the A and a alleles? Is it necessary to assume random mating?

8-7. For an X-linked recessive allele maintained by mutation-selection balance, would you expect the equilibrium frequency of the allele to be greater or smaller than that of an autosomal recessive allele, assuming that the relative fitnesses are the same in both cases? Why?

8-8. How many A and a alleles are present in a sample of organisms consisting of 10 AA, 15 Aa, and 4 aa individuals? What are the allele frequencies in this sample?

8-9. Allozyme phenotypes of alcohol dehydrogenase in the flowering plant *Phlox drummondii* are determined by codominant alleles of a single gene. In one sample of 35 plants, the following data were obtained:

Genotype AA AB BB BC CC AC
Number   2 5 12 10 5 1

What are the frequencies of the alleles A, B, and C in this sample?

8-10. DNA from 100 unrelated people was digested with the restriction enzyme HindIII and the resulting fragments separated and probed with a sequence for a particular gene. Four fragment lengths that hybridized with the probe were observed—namely, 5.7, 6.0, 6.2, and 6.5 kb—each fragment defining a different restriction-fragment allele. The adjoining illustration shows the gel patterns observed, with the number of individuals with each combination shown across the top. Estimate the allele frequencies of the four restriction-fragment alleles.

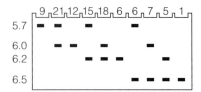

8-11. Which of the following genotype frequencies of *AA*, *Aa*, and *aa*, respectively, satisfy the Hardy-Weinberg principle?
(a) 0.25, 0.50, 0.25
(b) 0.36, 0.55, 0.09
(c) 0.49, 0.42, 0.09
(d) 0.64, 0.27, 0.09
(e) 0.29, 0.42, 0.29

8-12. If the frequency of a homozygous dominant genotype in a randomly mating population is 0.09, what is the frequency of the dominant allele? What is the combined frequency of all the other alleles of this gene?

8-13. Hartnup disease is an autosomal-recessive disorder of intestinal and renal transport of amino acids. The frequency of affected newborn infants is about 1 in 14,000. Assuming random mating, what is the frequency of heterozygotes?

8-14. A randomly mating population of dairy cattle contains an autosomal-recessive allele causing dwarfism. If the frequency of dwarf calves is 10 percent, what is the frequency of heterozygous carriers of the allele in the entire herd? What is the frequency of heterozygotes among nondwarf individuals?

8-15. In certain grasses, the ability to grow in soil contaminated with the toxic metal nickel is determined by a dominant allele.
(a) If 60 percent of the seeds in a randomly mating population are able to germinate in contaminated soil, what is the frequency of the resistance allele?
(b) Among plants that germinate, what proportion are homozygous?

8-16. In Caucasians, the M-shaped hairline that recedes with age (sometimes to nearly complete baldness) is due to an allele that is dominant in males but recessive in females. The frequency of the baldness allele is approximately 0.3. Assuming random mating, what frequencies of the bald and nonbald phenotypes are expected in males and females?

8-17. In a Pygmy group in Central Africa, the frequencies of alleles determining the ABO blood groups were estimated as 0.74 for I^O, 0.16 for I^A, and 0.10 for I^B. Assuming random mating, what are the expected frequencies of ABO genotypes and phenotypes?

8-18. Among 35 individuals of the flowering plant *Phlox roemariana*, the following genotypes were observed for a gene determining electrophoretic forms of the enzyme phosphoglucose isomerase: 2 *AA*, 13 *AB*, 20 *BB*.
(a) What are the frequencies of the alleles *A* and *B*?
(b) Assuming random mating, what are the expected numbers of the genotypes?

8-19. If an X-linked recessive trait is present in 2 percent of the males in a population with random mating, what is the frequency of the trait in females? What is the frequency of carrier females?

8-20. In a population of *Drosophila*, an X-linked recessive allele causing yellow body color is present in genotypes at frequencies typical of random mating; the frequency of the recessive allele is 0.20. Among 1000 females and 1000 males, what are the expected numbers of the yellow and wildtype phenotypes in each sex?

8-21. How does the frequency of heterozygotes in an inbred population compare with that in a randomly mating population with the same allele frequencies?

8-22. Which of the following genotype frequencies of *AA*, *Aa*, and *aa*, respectively, are suggestive of inbreeding?
(a) 0.25, 0.50, 0.25
(b) 0.36, 0.55, 0.09
(c) 0.49, 0.42, 0.09
(d) 0.64, 0.27, 0.09
(e) 0.29, 0.42, 0.29

8-23. Galactosemia is an autosomal-recessive condition associated with liver enlargement, cataracts, and mental retardation. Among the offspring of unrelated individuals the frequency of galactosemia is 8.5×10^{-6}. What is the expected frequency among the offspring of first cousins ($F = 1/16$) and among the offspring of second cousins ($F = 1/64$)?

8-24. Self-fertilization in the annual plant *Phlox cuspidata* results in an inbreeding coefficient of $F = 0.66$.
(a) What frequencies of the genotypes for the enzyme phosphoglucose isomerase would be expected in a population with alleles *A* and *B* at respective frequencies 0.43 and 0.57?
(b) What frequencies of the genotypes would be expected with random mating?

8-25. Two strains of *Escherichia coli*, A and B, are inoculated into a chemostat in equal frequencies and undergo competition. After 40 generations, the frequency of the B strain is 35 percent. What is the fitness of strain B relative to strain A under the particular experimental conditions, and what is the selection coefficient against strain B?

8-26. What is the ultimate fate of the A allele with random mating when the relative fitnesses of *AA*, *Aa*, and *aa* are as follows: (1) 1.00, 0.98, 0.96; (2) 1.00, 1.02, 1.02; (3) 1.00, 1.02, 0.98; (4) 1.00, 1.01, 1.02

8-27. With an autosomal gene with two alleles in a randomly mating population, what is the probability that a heterozygous female will have a heterozygous offspring?

8-28. In a randomly mating population containing *n* equally frequent alleles, what is the expected frequency of heterozygotes?

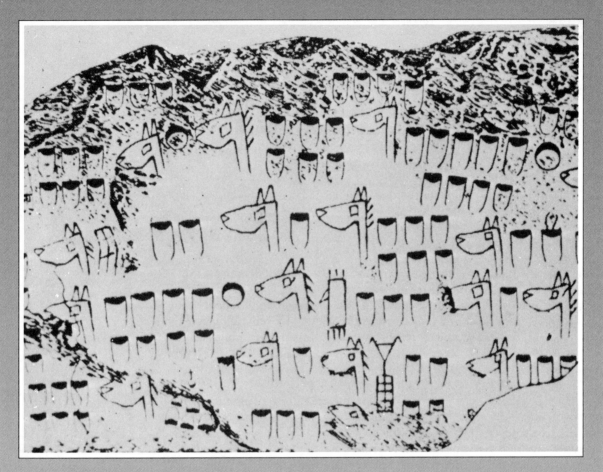

Quantitative Genetics

Earlier chapters have emphasized traits in which differences in phenotype result from alternative genotypes of a single gene. Examples include green versus yellow peas, red eyes versus white eyes in *Drosophila*, normal versus sickle-cell hemoglobin, and the ABO blood groups. These traits are particularly suited for genetic analysis through the study of pedigrees because of the small number of genotypes and phenotypes and because of the simple correspondence between genotype and phenotype. However, many traits of importance in plant breeding, animal breeding, and medical genetics are influenced by *multiple*

Above: A carving, about 4000 years old, showing the pedigree of horses raised by a breeder. The carved stone, found in Asia, shows that breeders kept records of desirable traits in their horses, as do modern breeders. (Courtesy of Dorsey Stuart.)

genes and by the effects of environment. With these traits, a single genotype can have many possible phenotypes (depending on the environment), and a single phenotype can include many possible genotypes. Genetic analysis of such complex traits requires special concepts and methods, which are introduced in this chapter.

9-1 Quantitative Inheritance

Many traits are influenced not only by the alleles of two or more genes but also by the effects of environment. Such traits are called **quantitative traits,** and with quantitative traits the phenotype of an individual is potentially influenced by

1. **Genetic factors** in the form of alternative genotypes of one or more genes, and
2. **Environmental factors**—for example, the effect of nutrition on the growth rate of animals or those of fertilizer, rainfall, and planting density on yield in crop plants.

With some quantitative traits, differences in phenotype result largely from differences in genotype, and the environment plays a minor role. With others, differences in phenotype result largely from the effects of environment, and genetic factors play a minor role. However, most quantitative traits fall between these extremes, and both genotype and environment must be taken into account in their analysis. Quantitative traits are often referred to as **multifactorial traits** to emphasize the many genetic and environmental factors in their determination.

In a genetically heterogeneous population, many genotypes are formed by the processes of segregation and recombination. Variation in genotype can be eliminated by studying inbred lines, which are homozygous for most genes, or the F_1 progeny from a cross of inbred lines, which are uniformly heterozygous for all genes at which the parental inbreds differ. In contrast, complete elimination of environmental variation is impossible, no matter how hard the experimenter may try to render the environment identical for all members of a population. With plants, for example, small variations in soil quality or exposure to the sun will produce slightly different environments, sometimes even for adjacent plants. Similarly, highly inbred *Drosophila* still show variation in phenotype (for example, in body size) brought about by environmental differences among animals within the same culture bottle. Therefore, traits that are susceptible to small environmental effects will never be uniform, even in inbred lines.

Most traits of importance in plant and animal breeding are quantitative traits. In agricultural production, one economically important quantitative trait is yield—for example, the harvest of corn, tomatoes, soybeans, or grapes. In domestic animals, important quantitative traits include milk production, egg-laying, fleece weight, litter size, and carcass quality. Important quantitative traits in human genetics include infant growth rate, adult weight, blood pressure, serum cholesterol, and length of life. In evolutionary studies, fitness is the preeminent quantitative trait.

Most quantitative traits cannot be studied by means of the usual pedigree methods because the effects of segregation of alleles of one gene may be concealed by effects of other genes, and environmental effects may cause identical genotypes to have different phenotypes. Therefore, individual pedigrees of quantitative traits do not fit any simple pattern of dominance, recessiveness, or X linkage. Nevertheless, genetic effects on quantitative traits can be assessed by comparing the phenotypes of relatives who, because of their familial relationship, must have a certain proportion of their genes in common. Such studies utilize many of the concepts of population genetics discussed in Chapter 8.

Three categories of traits are frequently found to have quantitative inheritance. They are described in the following subsection.

Continuous, Meristic, and Threshold Traits

Most phenotypic variation in populations is not manifested in a few easily distinguished categories. Instead, the traits vary continuously from one phenotypic extreme to the other with no clear-cut breaks in between. Such traits are called **continuous traits**; some examples are milk production in cattle, growth rate in poultry, yield in corn, and blood pressure in human beings. For a trait like milk production, there is a continuous range in phenotype from minimum to maximum with no clear separation between one phenotype and the next. The distinguishing characteristic of continuous traits is that the phenotype of an individual can fall anywhere on a continuous scale of measurement; so the number of possible phenotypes is virtually unlimited.

Meristic traits are traits in which the phenotype is determined by counting. Some examples are number of skin ridges forming fingerprints, number of kernels on an ear of corn, number of eggs laid by a hen, number of bristles on the abdomen of a fly, and number of puppies in a litter.

Threshold traits are traits having only two, or a few, phenotypic classes, but their inheritance is determined by the effects of multiple genes together with the environment. Examples of threshold traits are twinning in cattle and parthenogenesis (development of unfertilized eggs) in turkeys. In many threshold-trait disorders, the phenotypic classes are "affected" versus "not affected." Examples of human threshold-trait disorders include adult diabetes, schizophrenia, and many congenital abnormalities, such as spina bifida. Threshold traits can be interpreted as continuous traits by imagining that each individual has an underlying risk or *liability* toward manifestation of the condition. A liability above a certain cutoff, or *threshold*, results in expression of the condition; a liability below the threshold results in normality. The liability of an individual toward a threshold trait cannot be observed directly, but inferences about liability can be drawn from the incidence of the condition among individuals and their relatives.

Distributions

The **distribution** of a trait in a population is a description of the population in terms of the proportion of individuals that have each possible phenotype. Characterizing the distribution of some traits is straightforward because the number of phenotypic classes is small. For example, the distribution of progeny in a certain pea cross may consist of 3/4 green seeds and 1/4 yellow seeds, and the distribution of ABO blood groups among Greeks consists of 42 percent O, 39 percent A, 14 percent B, and 5 percent AB. However, with continuous traits, the large number of possible phenotypes makes such summaries impractical. Often, it is convenient to reduce the number of phenotypic classes by grouping similar phenotypes together. Data for an example pertaining to the distribution of height among 4995 British women are given in Table 9-1 and in Figure 9-1. You can imagine the bar graph in Figure 9-1 being built step by step as the women are measured by placing a small square along the x-axis at the location

TABLE 9-1 Distribution of height among British women

Interval number (i)	Height interval (in inches)	Midpoint (x_i)	Number of women (f_i)
1	53–55	54	5
2	55–57	56	33
3	57–59	58	254
4	59–61	60	813
5	61–63	62	1340
6	63–65	64	1454
7	65–67	66	750
8	67–69	68	275
9	69–71	70	56
10	71–73	72	11
11	73–75	74	4
		Total N	4995

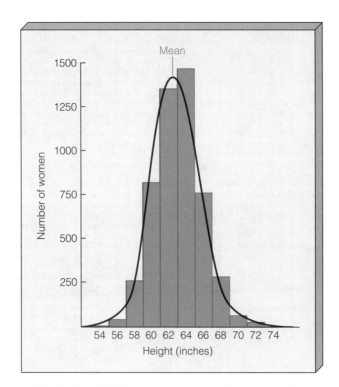

FIGURE 9-1 Distribution of height among 4995 British women and the smooth normal distribution that approximates the data.

corresponding to the height of each woman. As sampling proceeds, the squares begin to pile up in certain places, leading ultimately to the bar graph shown.

Displaying a distribution completely, as in Table 9-1 or Figure 9-1, is always adequate but often unnecessary; frequently, a description of the distribution in terms of two major features is sufficient. The height intervals in Table 9-1 are numbered from 1 (53–55 inches) to 11 (73–75 inches). The symbol x_i designates the midpoint of the height interval numbered i; for example, $x_1 = 54$ inches, $x_2 = 56$ inches, and so forth. The number of women in height interval i is designated f_i; for example, $f_1 = 5$ women, $f_2 = 33$ women, and so forth. The total size of the sample, in this case 4995, is denoted N. Two important quantities serve to characterize the distribution of height among these women (and the distribution of many other quantitative traits):

1. The **mean,** or average, is the peak of the distribution. The mean of a population is estimated from a sample of individuals from the population, as follows:

$$\bar{x} = \Sigma \, f_i \, x_i / \Sigma \, f_i \qquad (9\text{-}1)$$

in which $\bar{x}$ is the estimate of the mean and Σ symbolizes summation over all classes of data (in this example, summation over all 11 height intervals). In Table 9-1, the mean height in the sample of women is 63.1 inches.

2. The **variance** is a measure of the spread of the distribution and is estimated in terms of the squared *deviation* (difference) of each observation from the mean. The variance is estimated from a sample of individuals as follows:

$$s^2 = \Sigma \, f_i (x_i - \bar{x})^2 / (N - 1) \qquad (9\text{-}2)$$

in which s^2 is the estimated variance and x_i, f_i, and N are as in Table 9-1. Note that $(x_i - \bar{x})$ is the difference from the mean, and the denominator is the total number of individuals minus 1. The variance describes the extent to which the phenotypes are clustered around the mean, as shown in Figure 9-2. A large value implies that the distribution is spread out, and a small value implies that it is clustered near the mean. From the data in Table 9-1, the variance of the population of British women is estimated as $s^2 = 7.24 \text{ in}^2$.

A quantity closely related to the variance—the **standard deviation** of the distribution—is defined as the square root of the variance. For the data in Table

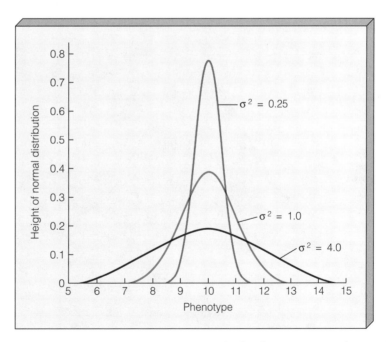

FIGURE 9-2 Graphs showing that the variance of a distribution measures the spread of the distribution around the mean. The area under each curve covering any range of phenotypes equals the proportion of individuals having phenotypes within the range.

9-1, the estimated standard deviation s is obtained from Equation 9-2 as $s = (s^2)^{1/2} = (7.24 \text{ in}^2)^{1/2} = 2.29$ inches. Note that the mean and the standard deviation have the same units, in this case, inches.

When the data are symmetrical, or approximately symmetrical, the distribution of a trait can often be approximated by a smooth arching curve of the type shown on Figure 9-1. The arch-shaped curve is called the **normal distribution.** Because the normal curve is symmetrical, half of its area is determined by points that have values greater than the mean and half by points with values less than the mean, and thus the proportion of phenotypes that exceed the mean is 1/2. One important characteristic of the normal distribution is that it is completely determined by the value of the mean and the variance.

The mean and standard deviation (square root of the variance) of a normal distribution provide a great deal of information about the distribution of phenotypes in a population, as is illustrated in Figure 9-3. Specifically, for a normal distribution,

1. Approximately 68 percent of the population have a phenotype within *one* standard deviation of the mean (in the symbols of Figure 9-3, between $\mu - \sigma$ and $\mu + \sigma$).
2. Approximately 95 percent lie within *two* standard deviations of the mean (between $\mu - 2\sigma$ and $\mu + 2\sigma$).
3. Approximately 99.7 percent lie within *three* standard deviations of the mean (between $\mu - 3\sigma$ and $\mu + 3\sigma$).

Applying these rules to the data in Figure 9-1, in which the mean and standard deviation are 63.1 and 2.69 inches, approximately 68 percent of the women are expected in the range $63.1 - 2.69$ inches to $63.1 + 2.69$ inches (that is, $60.4 - 65.8$) and approximately 95 percent in the range $63.1 - 2(2.69)$ to $63.1 + 2(2.69)$ inches (that is, $57.7 - 68.5$).

Real data frequently conform to the normal distribution. Normal distributions are usually the rule when the phenotype is determined by the cumulative effect of many individually small independent factors. This is the case for many multifactorial traits.

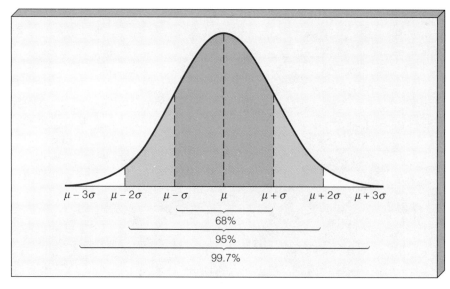

FIGURE 9-3 Features of a normal distribution. The proportion of individuals found within one, two, or three standard deviations from the mean is approximately 68 percent, 95 percent, and 99.7 percent, respectively. In this normal distribution the mean is symbolized μ and the standard deviation σ.

In considering the genetics of multifactorial traits, an important objective is to assess the relative importance of genotype versus environment. In some cases in experimental organisms, it is possible to separate genotype and environment with respect to their effects on the mean. For example, a plant breeder may study the yield of a series of inbred lines grown in a group of environments differing in planting density or amount of fertilizer. It would then be possible (1) to compare yields of the same genotype grown in different environments and thereby rank the *environments* relative to their effects on yield, and (2) to compare yields of different genotypes grown in the same environment and thereby rank the *genotypes* relative to their effects on yield.

Such a fine discrimination between genetic and environmental effects is not usually possible, particularly in human quantitative genetics. For example, with regard to the height of the women in Figure 9-1, environment could be considered favorable or unfavorable for tall stature only in comparison with the mean height of a genetically identical population reared in a different environment. This population does not exist. Likewise, the genetic composition of the population could be judged as favorable or unfavorable for tall stature only in comparison with the mean of a genetically different population reared in an identical environment. This population does not exist either.

Without such standards of comparison, it is impossible to determine the genetic versus environmental effects on the mean. However, it is still possible to assess genetic versus environmental contributions to the *variance*, because, instead of comparing the means of two or more populations, the phenotypes of individuals within the *same* population can be compared. Some of the differences in phenotype result from differences in genotype and others from differences in environment, and it is often possible to separate these effects.

In any distribution of phenotypes, such as the one in Figure 9-1, four sources contribute to phenotypic variation:

1. Genotypic variation.
2. Environmental variation.
3. Variation due to genotype–environment interaction.
4. Variation due to genotype–environment association.

Each of these sources of variation is discussed in the following sections.

Genotypic Variance

The variation in phenotype caused by differences in genotype among individuals is termed **genotypic variance.** Figure 9-4 illustrates the genetic variation expected among the F_2 generation from a cross of two inbred lines differing in genotype at three unlinked genes. The alleles of the three genes are represented as A/a, B/b, and C/c, and the genetic variation in the F_2 caused by segregation and recombination is evident in the shading. Relative to a meristic trait, if we assume that each uppercase allele is favorable for the expression of the trait and adds one unit to the phenotype and each lowercase allele is without effect, then the *aa bb cc* genotype has a phenotype of 0 and the *AA BB CC* genotype has a phenotype of 6. Thus there are seven possible phenotypes (0–7) in the F_2 generation. The distribution of phenotypes in the F_2 generation is shown in the bar graph having diagonal lines in Figure 9-5. The normal distribution approximating the data has a mean of 3 and a variance

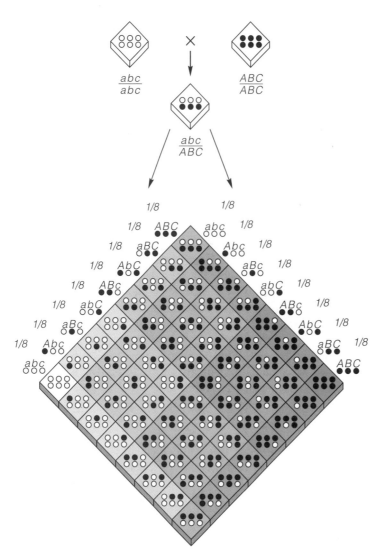

FIGURE 9-4   Segregation of three independent genes affecting a quantitative trait. Each uppercase allele in a genotype contributes one unit to the phenotype.

of 1.5. In this case, we are assuming that *all* of the variation in phenotype in the population results from differences in genotype among the individuals.

Figure 9-5 also includes a colored bar graph representing the theoretical distribution when the trait is determined by thirty unlinked genes segregating in a randomly mating population, grouped into the same number of phenotypic classes as the three-gene case. We assume that fifteen of the genes are nearly fixed for the favorable allele and fifteen nearly fixed for the unfavorable allele. The contribution of each favorable allele to the phenotype has been chosen to make the mean of the distribution equal to 3 and the variance equal to 1.5. Moreover, the distribution with thirty genes is virtually identical with that with three genes, and both are approximated by the same normal curve. If such distributions were obtained in research, an experimenter would not be able to distinguish between them. That is:

> Even in the absence of environmental variation, the distribution of phenotypes, by itself, provides no information about the number of genes influencing a trait and no information about the dominance relations of the alleles.

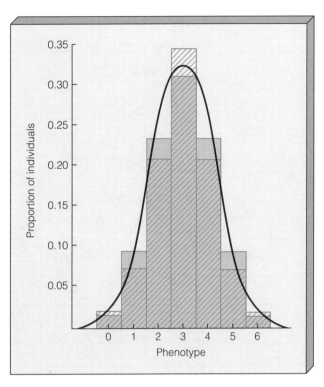

FIGURE 9-5   The bar graph with the diagonal lines is the distribution of phenotypes determined by the segregation of three genes illustrated in Figure 9-4. The shaded bar graph is the theoretical distribution expected from the segregation of thirty independent genes. Both distributions are approximated by the same normal distribution (black curve).

However, the number of genes influencing a quantitative trait is important in determining the potential for genetic improvement of a population. For example, in the three-gene case in Figure 9-5, the best-possible genotype would have a phenotype of 6, but in the thirty-gene case the best-possible genotype (homozygous for the favorable allele of all thirty genes) would have a phenotype of 33.

Later in this chapter, some methods for estimating the number of genes affecting a quantitative trait will be presented. All the methods depend on comparing the phenotypic distributions in the F_1 and F_2 generations of crosses between nearly or completely homozygous lines.

Environmental Variance

The variation in phenotype among individuals caused by differences in environment is termed **environmental variance.** Figure 9-6 is an example showing the distribution of seed weight in edible beans. The mean of the distribution is 500 mg and the standard deviation is 95 mg. However, all of the beans in the population are genetically identical and homozygous because they are highly inbred. Therefore, *all* of the phenotypic variation in seed weight in this population results from environmental variance. Comparison of Figures 9-5 and 9-6 demonstrates the following principle:

> The distribution of a trait in a population provides no information about the relative importance of genotype and environment. Variation in the trait can be entirely genetic, entirely environmental, or a combination of both influences.

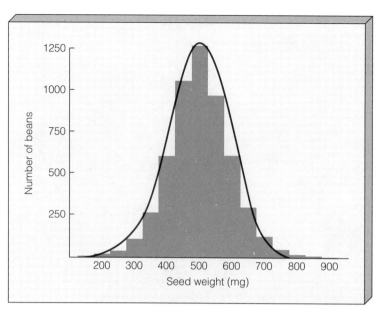

FIGURE 9-6   Distribution of seed weight in a homozygous line of edible beans. All variation in phenotype results from environmental differences among individuals.

Genotypic and environmental variance are seldom separated as clearly as in Figures 9-5 and 9-6, because usually they occur together. Their combined effects are illustrated for a simple hypothetical case in Figure 9-7. At the upper left is the distribution of phenotypes for three genotypes assumed to be uninfluenced by environment. As depicted, the trait is a discrete trait with three phenotypes determined by the effects of two additive alleles. The genotypes are in random-mating proportions for an allele frequency of 1/2, and the distribution of phenotypes has mean 5 and variance 2. Because it results solely from differences in genotype, this variance is *genotypic variance*, which is symbolized σ_g^2. At the upper right is the distribution of phenotypes that would be obtained in the presence of environmental variation, but only the heterozygote is illustrated. This distribution corresponds to the one in Figure 9-6, and its variance is 1. Because this variance results solely from differences in environment, it is *environmental variance*, which is symbolized σ_e^2. When the effects of genotype and environment are combined in the same population, then all three genotypes occur, each genotype is affected by environmental variation, and the distribution shown in the lower part of the figure results. The variance of this distribution is the **total variance** in phenotype, which is symbolized σ_t^2. Because we are assuming that genotype and environment have separate independent effects on phenotype, we expect σ_t^2 to be greater than either σ_g^2 or σ_e^2 alone. In fact,

$$\sigma_t^2 = \sigma_g^2 + \sigma_e^2 \tag{9-3}$$

In words, Equation 9-3 states that

> When genetic and environmental effects contribute independently to phenotype, then the total variance equals the sum of the genotypic and environmental variance.

Equation 9-3 is one of the most-important relations in quantitative genetics, and how it can be used to analyze data will be explained shortly.

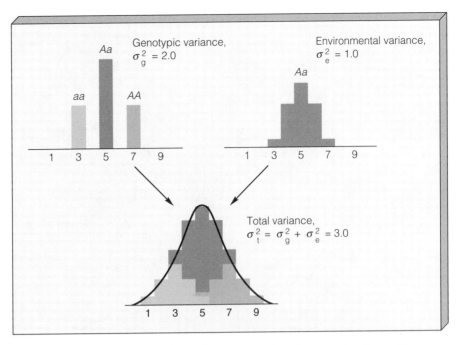

FIGURE 9-7   The combined effects of genotypic and environmental variance. Upper left: population affected only by genotypic variance, σ_g^2. Upper right: population of *Aa* genotypes affected only by environmental variance, σ_e^2. Bottom: population affected by both genotypic and environmental variance, in which the total phenotypic variance, σ_t^2, equals the sum of σ_g^2 and σ_e^2.

Although the equation serves as an excellent approximation in very many cases, it is valid in an exact sense only when genotype and environment are independent in their effects on phenotype. The two most-important departures from independence are discussed in the next subsection.

Genotype-Environment Interaction and Genotype-Environment Association

In the simplest cases, environmental effects on phenotype are additive, and each environment adds or detracts the same amount from the phenotype, independent of the genotype. When this is not true, then environmental effects on phenotype differ according to genotype, and a **genotype-environment interaction (G-E interaction)** is said to occur. In some cases, G-E interaction can even change the relative rank of genotypes, and genotypes that are superior in certain environments may become inferior in others. An example of extreme genotype-environment interaction in maize is illustrated in Figure 9-8. The two strains of corn are hybrids formed by crossing different pairs of inbred lines, and their overall means are approximately the same. However, the strain designated A clearly outperforms B in the negative stressful environments (environmental quality is judged on the basis of soil fertility, moisture, and other factors), whereas the performance is reversed when the environment is of high quality. In some organisms, particularly plants, experiments like those illustrated in Figure 9-8 can be carried out to determine the contribution of G-E interaction to total phenotypic variance. In other organisms, particularly human beings, the effect cannot be evaluated separately.

Interaction of genotype and environment is common and very important in both plants and animals. Because of interaction, no one plant variety will

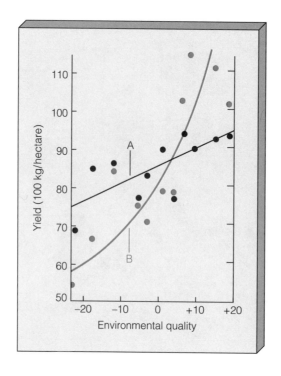

FIGURE 9-8 Genotype-environment interaction in maize. Strain A is superior when environmental quality is low (negative numbers), but strain B is superior when environmental quality is high. (Data from W. A. Russell. 1974. *Annual Corn & Sorghum Research Conference* 29: 81.)

outperform all others in all types of soil and climate, and therefore plant breeders must develop special varieties that are suited to each growing area.

When the different genotypes are not distributed at random in all the possible environments, there is **genotype-environment association (G-E association)**. In these circumstances, certain genotypes are preferentially associated with certain environments, which may either increase or decrease the average phenotype of these genotypes compared with what would result in the absence of G-E association. An example of deliberate genotype-environment association is found in dairy husbandry, in which some farmers feed their cattle according to milk yield. Because of this practice, cows with superior genotypes with respect to milk production also receive a superior environment in the form of more feed. In plant or animal breeding, genotype-environment association can often be eliminated or minimized by appropriate randomization of genotypes within the experimental plots. In other cases, human genetics again being a prime example, the possibility of G-E association cannot usually be controlled.

9-3 Analysis of Quantitative Traits

Equation 9-3 can be used to separate the effects of genotype and environment on the total phenotypic variance. Two types of data are required: (1) the phenotypic variance of a genetically uniform population, which provides an estimate of σ_e^2 because a genetically uniform population has a value of $\sigma_g^2 = 0$, and (2) the phenotypic variance of a genetically heterogeneous population, which provides an estimate of $\sigma_g^2 + \sigma_e^2$. An example of a genetically uniform population is the F_1 generation from a cross between two highly homozygous strains, such as inbred lines. An example of a genetically heterogeneous population is the F_2 generation from the same cross. If the environments of both populations are the same, and if there is no G-E interaction, then the estimates may be combined to extract a value for σ_g^2.

As a numerical illustration, let us consider the eye size of a cave-dwelling fish, *Astyanax*. The variances in eye diameter in the F_1 and F_2 generations from a cross of two highly homozygous strains were estimated as 0.057 and 0.563, respectively. Written in terms of the components of variance, these are

$$F_2: \sigma_t^2 = \sigma_g^2 + \sigma_e^2 = 0.563$$
$$F_1: \sigma_e^2 = 0.057$$

The estimate of genotypic variance, σ_g^2, is obtained by subtracting the second equation from the first; that is,

$$0.563 - 0.057 = 0.506$$

or

$$(\sigma_g^2 + \sigma_e^2) - \sigma_e^2 = \sigma_g^2$$

Hence, the estimate of σ_g^2 is 0.506, whereas that of σ_e^2 is 0.057. In this example, the genotypic variance is much greater than the environmental variance, but this is not always the case.

The next subsection shows what information can be obtained from these numbers.

The Number of Genes Affecting a Quantitative Trait

When the number of genes influencing a quantitative trait is not too large, knowledge of the genotypic variance can be used to estimate the number of genes. All that is needed are the means and variances of two phenotypically divergent strains and their F_1, F_2, and backcrosses. In ideal cases, the data appear as in Figure 9-9, in which P_1 and P_2 represent the divergent strains—for example, inbred lines. The points lie on a triangle, with increasing variance according to the increasing genetic heterogeneity (genotypic variance) of the populations. The F_1 and backcross means lie exactly between their parental means, which implies that the alleles affecting the trait are *additive*; that is, for each gene, the phenotype of the heterozygote is the average of the phenotypes

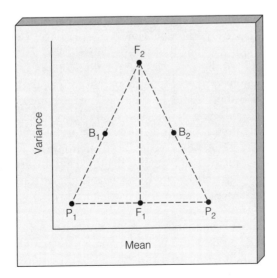

FIGURE 9-9 Means and variances of parents (P), backcross (B), and hybrid (F) progeny of inbred lines for an ideal quantitative trait affected by unlinked and completely additive genes. (After R. Lande. 1981. *Genetics* 99: 541.)

of the corresponding homozygotes. In such a simple situation, it may be shown
that the number n of genes contributing to the trait is

$$n = D^2/8\sigma_g^2 \tag{9-4}$$

in which D represents the difference between the means of the original parental
strains, P_1 and P_2. This equation may be verified by applying it to the ideal case
in Figure 9-7. The parental means are 7 (AA) and 3 (aa), giving $D = 7 - 3 =$
4, and in this example $\sigma_g^2 = 2$. Consequently, $n = 16/(8 \times 2) = 1$, which is
correct.

Applied to actual data, Equation 9-4 requires several assumptions that are
not necessarily valid. The theory assumes that (1) the alleles of each gene are
additive, (2) the genes contribute equally to the trait, (3) the genes are unlinked,
and (4) the original parental strains are homozygous for alternative alleles of
each gene. However, when the assumptions are invalid, the outcome is that n
is smaller than the actual number of genes affecting the trait. Thus, n is the
minimum number of genes that can account for the data. For the cave-dwelling
Astyanax fish discussed in the preceding section, the parental strains had
average phenotypes of 7.05 and 2.10, giving $D = 4.95$. The estimated value of
$\sigma_g^2 = 0.506$, and so the minimum number of genes affecting eye diameter is
$n = (4.95)^2/(8 \times 0.506) = 6.0$. Therefore, at least six different genes affect the
diameter of the eye of the fish.

The number of genes that affect a quantitative trait is important because it
influences the amount by which a population can be genetically improved by
selective breeding. With traits determined by a small number of genes, the
potential for change in a trait is small, and a population consisting of the best-
possible genotypes may have a mean value that is only two or three standard
deviations above the mean of the original population. However, traits
determined by a large number of genes have a large potential for improvement.
For example, after a population of *Tribolium* was bred for increased pupa
weight, the mean value for pupa weight was found to be 17 standard deviations
above the mean of the original population. Thus, determination of traits by a
large number of genes implies that selective breeding can create an improved
population in which the value of *every* individual greatly exceeds that of the *best*
individuals that existed in the original population.

RFLP Methods in Quantitative Genetics

Although quantitative traits are influenced by multiple genes, calculations like
those in the preceding subsection indicate that the number of genes is not
necessarily large. Also, some genes may be more important than others in
having a greater effect on the phenotype. These considerations imply that
important genes affecting quantitative traits could be identified in pedigrees,
manipulated in breeding programs, and even cloned and studied like other
genes. However, they cannot be identified directly because their individual
effects are obscured by the segregation of other genes and environmental
variation. Even so, genes affecting quantitative traits can be identified if they are
genetically linked with simple Mendelian genetic markers, because the effects
of the genotype affecting the quantitative trait are then correlated with the
genotype of the genetic marker.

In most organisms, the problem has been that there are not enough genetic
markers. This deficiency can be overcome by the use of restriction fragment
length polymorphisms, or RLFPs, which result from variation in the positions

of restriction sites in homologous chromosomes that are identified in Southern blots using appropriate complementary probe DNA (Section 8-1). RFLPs are abundant, distributed throughout the genome, and often have multiple codominant alleles, making them ideally suited for linkage studies of quantitative traits. In RFLP studies, as many widely scattered RFLPs as possible are monitored, along with the quantitative trait, in successive generations of a genetically heterogeneous population. Statistical studies are then carried out to identify which RFLP alleles are the best predictors of phenotype of the quantitative trait—for example, alleles whose presence is consistently accompanied by superior performance with respect to the quantitative trait. These RFLPs identify regions of the genome containing one or more genes having important effects on the quantitative trait, and the RFLPs can be used to trace the segregation of the important regions in breeding programs and even as entry points for cloning genes with particularly large effects.

Broad-Sense Heritability

Estimates of the number of genes that determine quantitative traits are frequently unavailable because the necessary experiments are impractical or have not been carried out. Another attribute of quantitative traits, which requires less data to evaluate, makes use of the ratio of the genotypic variance to the total phenotypic variance. This ratio of σ_g^2 to σ_t^2 is called **broad-sense heritability,** symbolized as H^2, and it measures the importance of genetic variation, relative to environmental variation, in causing variation in the phenotype of a trait of interest. Broad-sense heritability is a ratio of variances, specifically

$$H^2 = \sigma_g^2/\sigma_t^2 = \sigma_g^2/(\sigma_g^2 + \sigma_e^2) \tag{9-5}$$

Using the data for eye diameter in *Astyanax*, in which $\sigma_g^2 = 0.506$ and $\sigma_g^2 + \sigma_e^2 = 0.563$, Equation 9-5 yields $H^2 = 0.506/0.563 = 0.90$ for the estimate of broad-sense heritability. This value implies that 90 percent of the variation in eye diameter in the population results from differences in genotype among individuals.

Knowledge of heritability is useful in the context of plant and animal breeding because heritability can be used to predict the magnitude and speed of population improvement. The broad-sense heritability defined in Equation 9-5 is used in predicting the outcome of selection practiced among clones, inbred lines, or varieties. Analogous predictions for random-bred populations utilize another type of heritability, different from H^2, and this will be discussed shortly. Broad-sense heritability measures how much of the total variance in phenotype results from differences in genotype. For this reason, H^2 is often of interest in regard to human quantitative traits.

Twin Studies

In human beings, twins would seem to be ideal subjects for separating genotypic and environmental variance because **identical twins,** which arise from the splitting of a single fertilized zygote, are genetically identical and are often strikingly similar in such traits as facial features and body build. **Fraternal twins,** which arise from two fertilized eggs, have the same genetic relationship as ordinary siblings and thus only half of the genes in either twin are identical with those in the other. Theoretically, the variance between members of an identical-twin pair would be equivalent to σ_e^2, because the twins are genetically identical, whereas the variance between members of a fraternal-twin pair would include not only σ_e^2, but also part of the genotypic variance (approximately $\sigma_g^2/2$, because of the identity of half of the genes in fraternal twins). Consequently, both σ_g^2 and σ_e^2 could be estimated from twin data and combined as in Equation 9-5 to estimate H^2. Table 9-2 summarizes estimates of H^2 based on twin studies of several traits.

Unfortunately, twin studies are subject to several important sources of error, most of which increase the similarity of identical twins, and so the numbers in Table 9-2 should be considered very approximate and probably too high. Four of the sources of error are the following:

1. Genotype-environment interaction, which increases the variance in fraternal twins but not in identical twins.
2. Frequent sharing of embryonic membranes between identical twins, resulting in a more-similar intrauterine environment.
3. Greater similarity in the treatment of identical twins by parents, teachers, and peers, resulting in a decreased environmental variance in identical twins.
4. Different sexes in half of the pairs of fraternal twins, in contrast with the same sex of identical twins.

These pitfalls and others imply that data from human twin studies should be interpreted with caution and reservation.

9-4 Artificial Selection

The practice of breeders in choosing a select group of individual organisms from a population to become the parents of the next generation is termed **artificial selection.** When artificial selection is carried out by choosing among clones, inbred lines, or varieties, then the broad-sense heritability permits an assessment of how rapidly progress can be achieved. Broad-sense heritability is important in this context because with clones, inbred lines, or varieties, superior genotypes—however they are defined—can be perpetuated.

In sexually reproducing populations that are genetically heterogeneous, broad-sense heritability is not relevant in predicting progress resulting from artificial selection, because superior genotypes must necessarily be broken up by the processes of segregation and recombination. To the extent that high genetic

TABLE 9-2 Broad-sense heritability based on twin studies

Trait	Heritability (H^2)	Trait	Heritability (H^2)
Longevity	29	Verbal ability	63
Height	85	Numerical ability	76
Weight	63	Memory	47
Amino acid excretion	72	Sociability index	66
Serum lipid levels	44	Masculinity index	12
Maximum blood lactate	34	Temperament index	58
Maximum heart rate	84		

Note: Most of these estimates are based on small samples and should be considered to be very approximate and probably too high.

merit may depend on particular combinations of alleles, each generation of artificial selection results in a slight setback in that the offspring of superior parents are generally not quite as good as the parents themselves. Progress under selection can still be predicted, but the prediction makes use of another type of heritability—narrow-sense heritability—discussed in the following subsection.

Narrow-Sense Heritability

Figure 9-10 illustrates a typical form of artificial selection and its result; the trait is the length of corolla tube in the flower of *Nicotiana longiflora* (tobacco). Data in the upper graph is from the parental generation, and that in the lower graph is from the offspring generation. The parental generation is the population from which the parents were chosen for breeding. The type of selection is called **individual selection** because each member of the population is evaluated according to its own individual phenotype. The selection is practiced by choosing some arbitrary level of phenotype—called the **truncation point**—that determines which individuals will be saved for breeding purposes. All

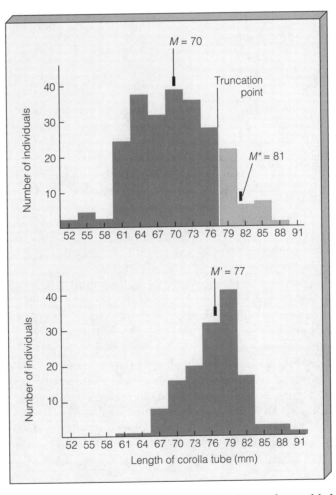

FIGURE 9-10   Selection for increased length of corolla tube in tobacco. M, M^*, and M' are the means of the parental generation, of selected parents (individuals with phenotype measurements that exceed the truncation point), and of the offspring of selected parents, respectively.

individuals with a phenotype above the threshold (colored area in the figure) are randomly mated among themselves to produce the next generation.

In evaluating progress through individual selection, three distinct phenotypic means are important. In Figure 9-10, these means are symbolized as M, M^*, and M' and are defined as follows:

1. M is the mean phenotype of the entire population in the parental generation, including both the selected and the nonselected individuals.
2. M^* is the mean phenotype among those individuals selected as parents (phenotypes above the truncation point).
3. M' is the mean phenotype among the progeny of selected parents.

The relation between these three means is given by

$$M' = M + h^2(M^* - M) \qquad (9\text{-}6)$$

in which the symbol h^2 is the **narrow-sense heritability** of the trait in question.

Later in this chapter, how narrow-sense heritability can be estimated from the similarity in phenotype among relatives will be explained. In Figure 9-10, h^2 is the only unknown quantity, and so it can be estimated from the data themselves. Rearranging Equation 9-6 and substituting the values for the means from Figure 9-10 leads to

$$h^2 = (M' - M)/(M^* - M) = (77 - 70)/(81 - 70) = 0.64$$

Analogous to the way in which total phenotypic variance can be split into the sum of the genotypic variance and the environmental variance (Equation 9-3), the genotypic variance can be split into parts in accord with the additive effects of alleles, dominance effects, and effects of interaction between alleles of different genes. The difference between the broad-sense heritability, H^2, and the narrow-sense heritability, h^2, is that H^2 includes all of these genetic contributions to variation, whereas h^2 includes only the additive effects of alleles. From the standpoint of animal improvement, h^2 is the heritability of interest because *the narrow-sense heritability, h^2, is the proportion of the variance in phenotype that can be used to predict changes in population mean with individual selection according to Equation 9-6.*

In general, the narrow-sense heritability of a trait is smaller than the broad-sense heritability. For example, in the parental generation in Figure 9-10, the broad-sense heritability of corolla-tube length was $H^2 = 0.82$. The two types of heritability are equal only when the alleles affecting the trait are additive in their effects.

Equation 9-6 is of fundamental importance in quantitative genetics because of its predictive value. This can be seen in the following example. The selection in Figure 9-10 was carried out for several generations. After two generations, the mean of the population was 83, and parents having a mean of 90 were selected. By use of the estimate $h^2 = 0.64$, the mean in the next generation could be predicted. The information provided is that $M = 83$ and $M^* = 90$. Therefore, Equation 9-6 implies that the predicted mean is

$$M' = 83 + (0.64)(90 - 83) = 87.5$$

This value is in good agreement with the observed value of 87.9.

Long-Term Artificial Selection

Artificial selection is analogous to natural selection in that both types of selection cause an increase in the frequency of alleles that improve the selected trait (or traits). Thus, the principles of natural selection discussed in Chapter 8 also apply to artificial selection. For example, artificial selection is most effective in changing the frequency of alleles that are in an intermediate range of frequency $(0.2 < p < 0.8)$. Alleles with frequencies outside this range respond more slowly to selection, and rare recessive alleles respond the slowest of all. With quantitative traits, including fitness, the total selection is shared among all the genes that affect the trait, and the selection coefficient for each allele is determined by (1) the magnitude of the effect of the allele, (2) the frequency of the allele, (3) the total number of genes affecting the trait, (4) the narrow-sense heritability of the trait, and (5) the proportion of the population that is selected for breeding.

The value of heritability is determined by both the magnitude of effects and the frequency of alleles. If all favorable alleles were fixed $(p = 1)$ or lost $(p = 0)$, the heritability of the trait would be 0. Therefore, the heritability of a quantitative trait is expected to decrease over many generations of artificial selection as a result of favorable alleles becoming nearly fixed. For example, ten generations of selection for less fat in a population of Duroc pigs decreased the heritability of fatness from 73 to 30 percent.

Population improvement by means of artificial selection cannot continue indefinitely. A population may respond to selection until its mean is many standard deviations different from the mean of the original population, but eventually the population reaches a **selection limit** in which successive generations show no further improvement. Progress may stop because all alleles affecting the trait are either fixed or lost, and so the narrow-sense heritability of the trait becomes 0. However, a more-common reason for a selection limit is that natural selection counteracts artificial selection. Many genes that respond to artificial selection as a result of their favorable effect on a selected trait also have indirect harmful effects on fitness. For example, selection for increased size of eggs in poultry results in a decrease in the number of eggs, and selection for extreme body size (large or small) in most animals results in a decrease in fertility. When one trait (for example, number of eggs) changes in the course of selection for a different trait (for example, size of eggs), the unselected trait is said to have undergone a **correlated response** to selection. Correlated response of fitness is typical in long-term artificial selection. Each increment of progress in the selected trait is partially offset by a decrease in fitness because of correlated response; eventually artificial selection for the trait of interest is exactly balanced by natural selection against the trait, and so a selection limit is reached and no further progress is possible without changing the strategy of selection.

Inbreeding Depression and Heterosis

Inbreeding can have harmful effects on economically important traits such as yield of grain or egg production. This decline in performance is called **inbreeding depression,** and it results principally from rare harmful recessive alleles becoming homozygous because of inbreeding (Chapter 8). Figure 9-11 shows inbreeding depression in yield of corn.

Most highly inbred strains suffer from many genetic defects, as might be expected from the uncovering of deleterious recessive alleles. One would also

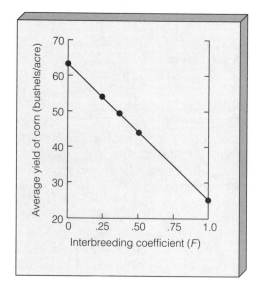

FIGURE 9-11 Inbreeding depression for yield in corn. (Data from N. Neal. 1935. *J. Amer. Soc. Agron.* 27: 666–670.)

expect that, if two different inbred strains were crossed, the F_1 would show improved features, because a harmful recessive allele inherited from one parent would be likely to be covered up by a normal dominant allele from the other parent. This is indeed the case, and the phenomenon is called **heterosis,** or **hybrid vigor.** The phenomenon, which is widely used in the production of corn and other agricultural products, yields genetically identical hybrid plants with traits that are sometimes more favorable than those of the ancestral plants from which the inbreds were derived. The most-common features of these hybrid plants are their rapid growth, larger size, and greater yield than the inbred parents. Furthermore, the F_1 plants have a fairly uniform phenotype (because $\sigma_g^2 = 0$). Genetically heterogeneous crops with high yields or certain other desirable features can also be produced by traditional plant-breeding programs, but growers often prefer hybrid plants because of their uniformity. For example, uniform height and time of maturity facilitates machine harvesting, and plants bearing fruit at the same time accommodates picking and shipping schedules.

Hybrid varieties of corn are used almost exclusively in the United States for commercial crops. A farmer cannot plant the seeds from his crop, because the F_2 generation consists of a variety of forms, most of which do not show hybrid vigor. Thus, the production of hybrid seeds is a major industry in corn-growing sections of the United States.

9-5 Correlation Between Relatives

Quantitative genetics relies extensively on similarity among relatives to assess the importance of genetic factors. Particularly in the study of such traits as human behavior, interpretation of familial resemblance is not always straight-forward because of the possibility of nongenetic, but nevertheless familial, sources of resemblance. However, in plant and animal breeding, the situation is usually less complex because genotypes and environments are under experimental control.

Covariance and Correlation

Genetic data about families are frequently pairs of numbers—pertaining to pairs of parents, pairs of twins, or pairs consisting of a single parent and offspring. An important issue in quantitative genetics is the degree to which the numbers in each pair are associated. The usual way to measure the association is to calculate a statistical quantity called the **correlation coefficient** between the variables.

The correlation coefficient among relatives is based on the covariance in phenotype among them. Much as the variance describes the tendency of a set of measurements to vary (Equation 9-2), the **covariance** describes the tendency of pairs of numbers to vary together (co-vary). Calculation of the covariance is similar to that of the variance in Equation 9-2 except that the squared deviation term $(x_i - \bar{x})^2$ is replaced with the product of the deviations of the pairs of measurements from their respective means, that is $(x_i - \bar{x})(y_i - \bar{y})$. For example, $x_i - \bar{x}$ could be the deviation of a father's height from the overall father mean, and $y_i - \bar{y}$ the deviation of his son's height from the overall son mean. In symbols, let f_i be the number of pairs of relatives with phenotypic measurements x_i and y_i. Then the estimated **covariance (Cov)** of the trait among the relatives is

$$Cov = \Sigma\, f_i(x_i - \bar{x})(y_i - \bar{y})/(N - 1) \tag{9-7}$$

in which N is the total number of pairs of relatives studied.

The **correlation coefficient (r)** of the trait between the relatives is calculated from the covariance as follows:

$$r = Cov/(s_x\, s_y) \tag{9-8}$$

in which s_x and s_y are the standard deviations of the measurements in the relatives estimated from Equation 9-2. The correlation coefficient can range from -1.0 to $+1.0$. A value of $+1.0$ means perfect association. When $r = 0$, x and y are not associated.

Estimation of Narrow-Sense Heritability

Covariance and correlation are important in quantitative genetics because the correlation coefficient of a trait between individuals with various degrees of genetic relationship is related fairly simply to the narrow-sense or broad-sense heritability. Theoretical values of the correlation coefficient for various pairs of relatives are given in Table 9-3, in which h^2 represents the narrow-sense

---

TABLE 9-3   Theoretical correlation coefficient in phenotype between relatives

Degree of relationship	Correlation coefficient*
Offspring and one parent	$h^2/2$
Offspring and average of parents	$h^2/2$
Half siblings	$h^2/4$
First cousins	$h^2/8$
Monozygotic twins	H^2
Full siblings	$\sim H^2/2$
Double first cousins†	$\sim H^2/2$

*Contributions from interactions among alleles of different genes have been ignored. For this and other reasons H^2 correlations are very approximate.
†Double first cousins are the offspring of matings between siblings from two sibships. They are cousins through both of their parents.

heritability and H^2 the broad-sense heritability. Considering parent-offspring, half-sibling, or first-cousin pairs, narrow-sense heritability can be estimated directly by multiplication. Specifically, h^2 can be estimated as twice the parent-offspring correlation, four times the half-sibling correlation, or eight times the first-cousin correlation.

With full siblings, identical twins, and double first cousins, the correlation coefficient is related to broad-sense heritability, H^2, because phenotypic resemblance depends not only on additive effects, but also on dominance. In these relatives, dominance contributes to resemblance because the relatives can share *both* of their alleles because of their common ancestry, whereas parents and offspring, half siblings, and first cousins can share at most a single allele of any gene because of common ancestry. Therefore, to the extent that phenotype depends on dominance effects, full siblings can resemble each other more than they resemble their parents.

The potentially greater resemblance between siblings than between parents and offspring may be understood by considering an autosomal recessive trait caused by a rare recessive allele. In a randomly mating population, the probability that an offspring will be affected, if the mother is affected, is q (the allele frequency), which corresponds to the random-mating probability that the sperm giving rise to the offspring carries the recessive allele. In contrast, the probability that both of two siblings will be affected, if one of them is affected, is 1/4, because most of the matings that produce affected individuals will be between heterozygous parents. Clearly, when the trait is rare, the parent-offspring resemblance will be very small, whereas the sibling-sibling resemblance will be substantial. This discrepancy is entirely a result of dominance, and it arises only because full siblings can share both of their alleles.

9-6 Genetics of Behavior

One can hardly imagine a subject more controversial than the inheritance of human behavioral differences, particularly of socially undesirable behavior. Since Mendel, commentators have been divided, some claiming the simple and direct inheritance of virtually all forms of socially unacceptable behavior— "feeble-mindedness," habitual drunkenness, criminality, prostitution, and so forth—and others arguing for environmental causation of such behavior.

The extreme hereditarian views were supported by studies of certain families with a high incidence of undesirable traits, the most well known being the Jukes and Kallikak families (both pseudonyms). Figure 9-12 is part of the Jukes pedigree published in 1902 purporting to show the inheritance of "shiftlessness" and "partial shiftlessness." The solid and shaded symbols are as in the original, and they give the false impression of objectivity. However, the diagnosis of "shiftlessness" or "partial shiftlessness" is entirely subjective, as indicated by the descriptions of some of the individuals in the pedigree. Individual I-1 is described as "a lazy mulatto," I-2 as "a nonindustrious harlot, but temperate," II-10 as "lazy, in poorhouse," and III-18 as a "bad boy." Today, this sort of analysis is rejected as subjective and prejudicial, but at the time it was taken quite seriously as an element in the "inheritance" of poverty.

Even accepting the pedigree at face value, all it says is that no family exists outside of its environment, and that social environments, like genes, tend to be perpetuated from generation to generation. Familial association of certain characteristics (for example, poverty) and certain types of behavior is not evidence that the traits are genetically, rather than socially, transmitted. An obvious example is English grammar and pronunciation. Children tend to speak with the same quirks and localisms as their parents and other relatives, yet

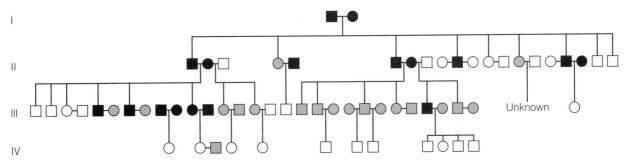

FIGURE 9-12 A part of the Jukes pedigree illustrating hereditarian prejudices of some early investigators. Blackened symbols purported to represent "shiftlessness" and shaded symbols "partial shiftlessness." (From R. L. Dugdale. 1902. *The Jukes: A Study in Crime, Pauperism, Disease and Heredity.* Putnam; and C. B. Davenport. 1911. *Heredity in Relation to Eugenics.* Holt.)

nobody would presume that English usage is genetically transmitted. Nevertheless, some early behaviorists were extreme hereditarians.

The opposite view was held by the extreme environmentalists, whose thinking is typified by the following example:

> So let us hasten to admit—yes, there are heritable differences in form, in structure. . . . These differences are in the germ plasm and are handed down from parent to child. . . . But do not let these undoubted facts of inheritance lead us astray as they have some of the biologists. The mere presence of these structures tells us not one thing about function. . . . Our hereditary structure lies ready to be shaped in a thousand different ways—the same structure—depending on the way in which the child is brought up. . . . We have no real evidence of the inheritance of behavioral traits. (J. B. Watson, 1925)

Such extreme hereditarian and environmentalist views are now in the minority, and most geneticists and psychologists are willing to concede the importance of both heredity *and* environment in human behavioral variation. The important questions relate to the relative importance of "nature" and "nurture," and how best to assess the situation experimentally.

Genetic and Cultural Effects on IQ Scores

Modern "intelligence" tests such as the Stanford-Binet and the Wechsler Adult tests derive from attempts in France in the early 1900s to develop simple procedures to identify children with mild learning disabilities. The tests actually assess a variety of skills, such as vocabulary, short-term memory, deductive reasoning, and the ability to perceive patterns in geometrical designs. Although no single test can adequately assess all aspects of what is commonly understood to be intelligence, the tests can be useful in identifying children who may need special attention in school. Although score on an IQ test is a statistical abstraction rather than a definitive measure of intelligence, IQ scores are relatively stable. The correlation between IQ scores of the same person tested as a child and again as an adult is approximately 0.8. Consequently, an IQ score can be treated as a phenotype like any other quantitative trait, quite apart from possible misgivings concerning the true relation between IQ and intelligence.

Genetic analysis of IQ data requires that special attention be given to cultural transmission of nongenetic factors that potentially affect IQ. These familial factors increase the resemblance between relatives, and, unless properly taken into account, they inflate the apparent genetic heritability. Traditional types of studies, which ignore cultural transmission, typically lead to heritability values of 60 to 80 percent.

Modern approaches yield values for genetic and cultural heritability of IQ in American whites as:

$$h^2 = 0.297 \pm 0.023 \text{ (genetic)}$$
$$b^2 = 0.289 \pm 0.016 \text{ (cultural)}$$

in which h^2 is the narrow-sense heritability and b^2 is a corresponding term for the familial transmission of strictly cultural effects (cultural inheritance). Genetic and transmissible cultural influences are approximately equal in accounting for variation in IQ among whites, both being about 30 percent. However, the largest single contributor to variation in IQ among whites is that resulting from *non*transmissible environmental influences; these nontransmissible effects account for 32 percent of the variation in IQ. The remaining 8 percent results from correlations between genetic and transmissible cultural factors—an example of genotype-environment association or possibly genotype-environment interaction.

These estimated values of genetic and cultural factors in IQ transmissibility imply that, if the environments of all American whites were equalized, leaving the genetic variation as it is, the degree of variation in IQ within this population would be reduced by 70 percent.

Race and IQ

The results of the IQ studies are frequently misinterpreted. The most-common mistake is to apply the genetic and cultural heritabilities to differences between populations. For essentially the same reasons that identical twins reared in different environments may have different phenotypes, two genetically identical populations that differ in their environments may differ in average phenotype. Both genetic and cultural heritablity are relevant to interpreting the variation in phenotypes *within* a specified population. Use of these quantities for comparison *between* populations is unjustified and may often lead to incorrect conclusions.

Heritability has mistakenly been applied to racial differences in average IQ. Such tests have been standardized so that the average score of American whites is 100. American blacks average about 85 on the same tests, and in some Japanese studies the average is about 110. The question that has been debated is whether these averages reveal genetic differences between the races. *No methods are currently known that can answer such a question.* Because the emotional overtones of racial differences in IQ are so great, the central issue may become clearer by means of analogy. Instead of considering human IQ, one might just as well ask why chicken-farmer Brown's hens lay an average of 230 eggs per year but farmer Smith's average only 210. Lacking any knowledge of the genetic relationship between the flocks, the reason for the difference cannot be specified. This is true despite the fact that one might know with certainty that the genetic heritability within each flock is 30 percent, because heritability within flocks is irrelevant to the comparison. On the one hand, farmer Brown's hens could be genetically superior in egg laying. On the other hand, farmer Smith's chicken feed or husbandry could be inferior. To determine the reason for the difference, some of Brown's chickens would have to be reared on Smith's farm, and vice versa. If the egg-laying of the transplanted hens did not change, then the difference between the two groups of hens would be entirely genetic; if the egg-laying of the transplanted hens did change, then the difference between the groups would be environmental. Experiments of this type cannot be carried out with human beings.

Many environmental factors can affect IQ-test performance between races, among which the following three are particularly important:

1. *Differences in socioeconomic status.* IQ scores are correlated with socioeconomic status, and so comparisons of populations differing in socioeconomic status will inevitably lead to differences in average IQ. For example, the difference between American blacks and whites is much reduced when comparable socioeconomic levels are considered.

2. *Differences in language skills.* IQ tests are largely verbal, and so differences in language skills will be reflected in test performance. IQ tests are written in standard American English, which may create difficulty for some black Americans.

3. *Differences in motivation.* Test performance is affected by the environmental surroundings in which the test is taken, the attitude and skill of the examiner, the mood and motivation of the subject, and other factors. Different populations, particularly those with different cultural backgrounds, may react differently with respect to one or more of these factors and so perform differently on the tests.

CHAPTER SUMMARY

Many traits that are important in agriculture and human genetics are determined by the effects of multiple genes and by the environment. Such traits are multifactorial, and their analysis is known as quantitative genetics. There are three types of multifactorial traits—quantitative, meristic, and threshold traits. Quantitative traits are expressed according to a continuous scale of measurement, like height. Meristic traits are traits that are expressed in whole numbers, like the number of grains on an ear of corn. Threshold traits have an underlying risk and are either expressed or not expressed in each individual; an example is diabetes. The genes affecting quantitative traits are no different from those affecting simple Mendelian traits, and the genes can have multiple alleles, partial dominance, and so forth. When several genes affect a trait, the pattern of genetic transmission need not fit a simple Mendelian pattern because the effects of one gene can be obscured by other genes or the environment. However, the number of genes can be estimated and many of them can be mapped by the use of linkage to restriction fragment length polymorphisms.

Many quantitative and meristic traits have a distribution that approximates the bell-shaped curve of a normal distribution. A normal distribution can be completely described by two quantities—the mean and the variance. The standard deviation of a distribution is the square root of the variance. In a normal distribution, approximately 68 percent of the individuals have a phenotype within one standard deviation from the mean, and approximately 95 percent of the individuals have a phenotype within two standard deviations from the mean.

Variation in phenotype of multifactorial traits among individuals in a population derives from four principal sources: (1) variation in genotype, which is measured by the genotypic variance; (2) variation in environment, which is measured by the environmental variance; (3) variation resulting from the interaction between genotype and environment (G-E interaction); and (4) variation resulting from nonrandom association of genotypes and environments (G-E association). The ratio of genotypic variance to the total phenotypic variance of a trait is called the broad-sense heritability: this quantity is useful in predicting the outcome of artificial selection practiced among clones, inbred lines, or varieties. When artificial selection is practiced in a randomly mating population, then a second type of heritability, the narrow-sense heritability, is used for prediction. The value of the narrow-sense heritability can be determined from the correlation in phenotype among groups of relatives.

One common type of artificial selection is called truncation selection, in which only those individuals that have a phenotype above a certain value (the truncation point) are saved for breeding the next generation. Artificial selection usually results in improvement of the selected population. However, progress often slows or ceases when selection is carried out for many generations because (1) some of the favorable genes become nearly fixed in the population, which decreases the narrow-sense heritability, and (2) natural selection may counteract the artificial selection.

K E Y   T E R M S

artificial selection
broad-sense heritability
continuous trait
correlated response
correlation coefficient
covariance
distribution
environmental variance
fraternal twins
genotype-environment association

genotype-environment interaction
genotypic variance
heterosis
hybrid vigor
identical twins
inbreeding depression
individual selection
mean
meristic trait
multifactorial trait

narrow-sense heritability
normal distribution
quantitative trait
selection limit
standard deviation
threshold trait
total variance
truncation point
variance

E X A M P L E S   O F   W O R K E D   P R O B L E M S

Problem 1: The following data are the values of a quantitative trait measured in a sample of 100 individuals taken at random from a population. (a) Estimate the mean, variance, and standard deviation in the population. (b) Assuming a normal distribution, what proportion of individuals in the entire population would be expected to have a phenotype in the range 93–107? Above 114?

Phenotype	Number of individuals
86–90	9
91–95	14
96–100	26
101–105	29
106–110	12
111–115	10

Answer: The data are tabulated in ranges, but for purposes of calculation it is better to retabulate the data as the midpoints of the ranges, as is done in the table at the right. **(a)** The mean is estimated from Equation 9-1 as (88 × 10 + 93 × 16 + 98 × 26 + 103 × 29 + 108 × 10 + 113 × 9)/100 = 10,000/100 = 100. The variance is estimated from Equation 9-2 as $[(88 - 100)^2 \times 10 + (93 - 100)^2 \times 16 + (98 - 100)^2 \times 26 + (103 - 100)^2 \times 29 + (108 - 100)^2 \times 10 + (113 - 100)^2 \times 9]/99 = 4750/99 = 48$. The standard deviation is then estimated as $(48)^{1/2} = 7$. **(b)** The

range 93–107 happens to be the mean plus or minus one standard deviation (100 ± 7), and approximately 68 percent of the population is expected to be within this range. To estimate the proportion with a phenotype above 114, note that this is two standard deviations above the mean. Because approximately 95 percent of the population have phenotypes within the range 100 ± 14 (that is, two standard deviations), the remaining 5 percent will have phenotypes greater than 114 or less than 86; because the normal distribution is symmetric, half of the 5 percent—or 2.5 percent—will be expected to have phenotypes greater than 114.

Phenotype	Number of individuals
88	10
93	16
98	26
103	29
108	10
113	9

Problem 2: A genetically heterogeneous population of wheat has a variance in maturation time (days) of 40, whereas two inbred populations derived from it have a variance in maturation time of 10. **(a)** What is the genotypic variance, σ_g^2, the environmental variance, σ_e^2, and the broad-sense heritability, H^2, of maturation time in this

population? **(b)** If the inbred lines were crossed, what would the predicted variance in maturation time of the F_1 generation be?

Answer: (a) The total variance, σ_t^2, in the genetically heterogeneous population is the sum $\sigma_g^2 + \sigma_e^2 = 40$. The variance in the inbred lines equals the environmental variance (because $\sigma_g^2 = 0$ in genetically homogeneous populations); hence $\sigma_e^2 = 10$. Therefore, σ_g^2 in the heterogeneous population equals $40 - 10 = 30$. The broad-sense heritability H^2 equals $\sigma_g^2/(\sigma_g^2 + \sigma_e^2) = 30/40 = 75$ percent. **(b)** If the inbred lines were crossed, the F_1 would be genetically uniform, and consequently the predicted variance would be $\sigma_e^2 = 10$.

Problem 3: A breeder aims to increase the height of maize plants by artificial selection. In a population with an average plant height of 150 cm, plants averaging 180 cm are selected and mated at random to give the next generation. If the narrow-sense heritability of plant height in this population is 40 percent, what is the expected average plant height among the progeny generation?

Answer: Use Equation 9-6 with $M = 150$, $M^* = 180$, and $h^2 = 0.40$. Then the predicted mean is calculated as $M' = 150 + 0.40 \times (180 - 150) = 162$ cm.

PROBLEMS

9-1. What is a quantitative trait? What is the difference between a continuous trait, a meristic trait, and a threshold trait?

9-2. What is the genotypic variance of a quantitative trait? What is the environmental variance?

9-3. Distinguish between the variance due to genotype-enviroment interaction and the variance due to genotype-environment association? Which type of variance is most easily controlled by the experimentalist?

9-4. Two varieties of corn, A and B, are field tested in Indiana and North Carolina. Strain A is more productive in Indiana, but strain B is more productive in North Carolina. What phenomenon in quantitative genetics does this example illustrate?

9-5. How are the variance and the standard deviation related? When does the variance of a distribution of phenotypes equal zero?

9-6. The following questions pertain to a normal distribution:
   (a) What term applies to the value along the x-axis corresponding to the peak of the distribution?
   (b) If two normal distributions have the same mean but different variances, which is the broader?
   (c) What proportion of the population is expected to lie within one standard deviation of the mean? Within two standard deviations?

9-7. Distinguish between the broad-sense heritability of a quantitative trait and the narrow-sense heritability. If a population is fixed for all genes that affect a particular quantitative trait, what is the value of the narrow-sense and broad-sense heritabilities?

9-8. In comparing a quantitative trait in the F_1 and F_2 generations obtained by crossing two highly inbred strains, which set of progeny provides an estimate of

the environmental variance? What determines the variance of the other set of progeny?

9-9. Ten female mice had the following number of live-born offspring in their first litters: 11, 9, 13, 10, 9, 8, 10, 11, 10, 13. Considering these females to be representative of the total population from which they came, estimate the mean, variance, and standard deviation of size of the first litter in the entire population.

9-10. Values of IQ score are distributed approximately according to a normal distribution with mean 100 and standard deviation 15. What proportion of the population has a value above 130? Below 85? Above 85?

9-11. The data in the adjoining table pertain to milk production over an eight-month lactation among 304 two-year-old Jersey cows.

Pounds of milk produced	Number of cows
1000–1500	1
1500–2000	0
2000–2500	5
2500–3000	23
3000–3500	60
3500–4000	58
4000–4500	67
4500–5000	54
5000–5500	23
5500–6000	11
6000–6500	2

Data have been grouped by intervals, but for purposes of computation each cow may be treated as if the milk production were equal to the midpoint of the interval (for example, 1250 for cows in the 1000–1500 interval).

(a) Estimate the mean, variance, and standard deviation in milk yield.

(b) What range of yield would be expected to include 68 percent of the cows?

(c) Round the limits of this range to the nearest 500 pounds and compare the observed with the expected number of animals.

(d) Do the same calculations as in parts b and c for the range expected to include 95 percent of the animals.

9-12. In the F_2 generation of a cross between two cultivated varieties of tobacco, the number of leaves per plant was distributed according to a normal distribution with mean 18 and standard deviation 3. What proportion of the population is expected to have the following phenotypes: (1) between 15 and 21 leaves; (2) between 12 and 24 leaves; (3) fewer than 15 leaves; (4) more than 24 leaves; (5) between 21 and 24 leaves?

9-13. Two highly homozygous inbred strains of mice are crossed and the six-week weight of the F_1 progeny determined.

(a) What is the magnitude of the genotypic variance in the F_1?

(b) If all alleles affecting six-week weight were additive, what is the expected mean phenotype of the F_1 compared with the average six-week weight of the original inbred strains?

9-14. In a cross between two cultivated inbred varieties of tobacco, the variance in leaf number per plant in the F_1 generation is 1.46 and in the F_2 generation it is 5.97. What are the genotypic and environmental variances? What is the broad-sense heritability in leaf number?

9-15. Two inbred lines of *Drosophila* are crossed, and the F_1 population has a mean number of abdominal bristles of 20 and a standard deviation of 2. The F_2 generation has a mean of 20 and a standard deviation of 3. What is the environmental variance, the genetic variance, and the broad-sense heritability of bristle number in this population?

9-16. In an experiment with weight gain between ages 3 and 6 weeks in mice, the difference in mean phenotype between two strains was 17.6 grams and the genotypic variance was estimated as 0.88. Estimate the minimum number of genes affecting this trait.

9-17. A flock of broiler chickens has a mean weight gain of 700 g between ages 5 and 9 weeks, and the narrow-sense heritability of weight gain in this flock is 0.80. Selection for increased weight gain is carried out for five consecutive generations, and in each generation the average of the parents is 50 g greater than the average of the population from which the parents were derived. Assuming that the heritability of the trait remains constant at 80 percent, what is the expected mean weight gain after the five generations?

9-18. To estimate the heritability of maze-learning ability in rats, a selection experiment was carried out. From a population in which the average number of trials necessary to learn the maze was 10.8, with a variance of 4.0, animals were selected that managed to learn the maze in an average of 5.8 trials. Their offspring required an average of 8.8 trials to learn the maze. What is the estimated narrow-sense heritability of maze-learning ability in this population?

9-19. A replicate of the population in Problem 9-18 was reared in another laboratory under rather different conditions of handling and other stimulation. The mean number of trials required to learn the maze was still 10.8, but the variance was increased to 9.0. Animals with a mean learning time of 5.8 trials were again selected, and the mean learning time of the offspring was 9.9.

(a) What is the heritability of the trait under these conditions?

(b) Is this result consistent with that in Problem 9-18, and how can the difference be explained?

9-20. In terms of the narrow-sense heritability, what is the theoretical correlation coefficient in phenotype between first cousins who are the offspring of monozygotic twins?

9-21. If the correlation coefficient of a trait between first cousins is 0.09, what is the estimated narrow-sense heritability of the trait?

9-22. A representative sample of lamb weights at the time of weaning in a large flock is shown below. If the narrow-sense heritability of weaning weight is 20 percent, and the half of the flock consisting of the heaviest lambs is saved for breeding for the next generation, what is the best estimate of the average weaning weight of the progeny? (Note: If a normal distribution has mean μ and standard deviation σ, then the mean of the upper half of the distribution is given by $\mu + 0.8\sigma$.)

81	81	83	101	86
65	68	77	66	92
94	85	105	60	90
94	90	81	63	58

9-23. Consider a trait determined by n unlinked, additive genes with two alleles of each, in which each favorable allele contributes one unit to the phenotypic value and each unfavorable allele contributes nothing. What is the value of the genetic variance?

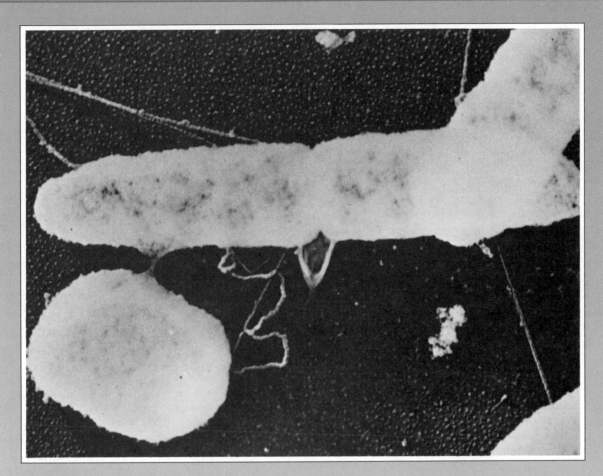

Genetics of Bacteria and Viruses

In earlier chapters, we examined the genetic properties of several eukaryotic organisms. These organisms are diploid and multichromosomal and, in the course of sexual reproduction, genotypic variation among the progeny is achieved both by random assortment of the chromosomes and by crossing-over, processes taking place mainly in meiosis. Two important features of crossing-over in eukaryotes are that it results in a reciprocal exchange of material between two homologous chromosomes and that both products of a single

Above: Electron micrograph of *E. coli* bacteria in the act of mating. The left-hand member of the pair of dividing cells is the F^+ donor and the small, nearly spherical cell at the lower left is the F^- recipient. The conjugation bridge appears as a thin fiber between the mating cells. (Courtesy of T. F. Anderson.)

exchange can often be recovered. The situation is quite different in prokaryotes, as will be seen in this chapter.

A bacterial cell contains a major DNA molecule that almost never encounters another complete molecule. Instead, genetic exchange is usually between a chromosomal fragment and an intact chromosome. Furthermore, a clear donor-recipient relation exists—that is, a donor cell is the source of the DNA fragment, which is transferred to the recipient cell by one of several mechanisms, and exchange of genetic material takes place in the recipient. Incorporation of a part of the transferred donor DNA into the chromosome requires at least two exchange events; because the recipient molecule is circular, only an even number of exchanges results in a viable product. The usual outcome of these events is the recovery of *only one* of the crossover products. However, in some situations the transferred DNA is also circular, and a single exchange results in total incorporation of this DNA into the chromosome of the recipient.

Three major types of genetic transfer will be described in the first part of this chapter: **transformation,** in which DNA is taken up from the environment and incorporated into the genome; **transduction,** in which DNA is transferred from one bacterial cell to another by a bacterial virus; and **conjugation,** in which donor DNA is transferred from one bacterial cell to another by direct contact. Recombination frequencies that result from these processes are used to produce genetic maps of bacteria. Although the maps are exceedingly useful, they differ in major respects from the types of maps generated by crosses in eukaryotes.

Bacteriophages are bacterial viruses able to multiply only within bacterial cells. They possess several systems for exchanging genetic material and undergoing genetic recombination. In the life cycle of a bacteriophage (the term is usually shortened to phage), a phage particle attaches to a host bacterium and releases its nucleic acid into the cell; the nucleic acid replicates many times; finally, newly synthesized nucleic acid molecules are packaged into protein

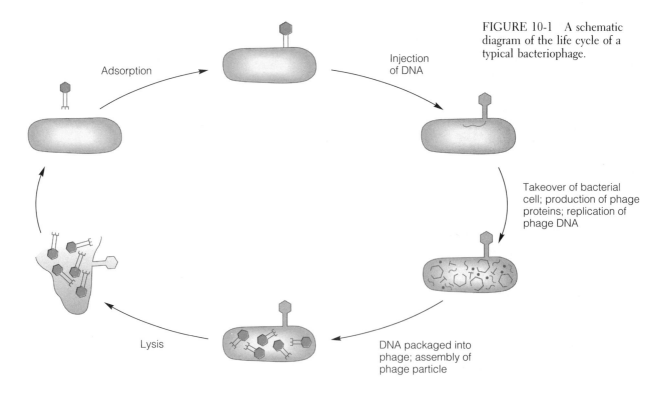

Adsorption

Injection of DNA

Takeover of bacterial cell; production of phage proteins; replication of phage DNA

DNA packaged into phage; assembly of phage particle

Lysis

FIGURE 10-1 A schematic diagram of the life cycle of a typical bacteriophage.

shells—forming progeny phage—and then the particles are released from the cell (Figure 10-1). Phage progeny from a bacterium infected by one phage have the parental genotype, except for very rare mutations occurring in the phage DNA. However, if two phage particles having *different* genotypes infect a single bacterial cell, new genotypes can arise by genetic recombination. This process differs significantly from genetic recombination in eukaryotes in two ways: (1) the number of participating DNA molecules varies from one cell to the next, and (2) reciprocal recombinants are not always recovered in equal frequencies from a single infected cell. Some phages also possess systems that enable phage DNA to recombine with bacterial DNA. Phage-phage and phage-bacterium recombination will be the topic of the second part of this chapter. The best-understood bacterial and phage systems are those of *E. coli*, and we will concentrate on these systems.

In Chapter 5, it was pointed out that the repetitive sequences of eukaryotic DNA include transposable elements—DNA sequences capable of relocation within a genome. Transposable elements are also present in bacteria and in some phages and are responsible for a variety of genetic alterations, such as mutation, deletion, gene inversion, and fusion of circular DNA molecules. These elements have been extensively studied in bacteria and are the final topic of this chapter.

10-1 Mutants of Bacteria

Bacteria can be grown both in liquid medium and on the surface of a semisolid growth medium hardened with agar. Bacteria used in genetic analysis are usually grown on the latter. A single bacterial cell placed on a solid medium will grow and divide many times, forming a visible cluster of cells called a *colony* (Figure 10-2). The number of bacterial cells in a suspension can be determined by spreading a known volume of the suspension on a solid medium and counting the number of colonies that form. Typical *E. coli* cultures contain up to 10^9 cells/ml. The appearance of colonies, or the ability or inability to form colonies on particular media can, in some cases, be used to identify the genotype of the cell that produced the colony.

As we have seen in earlier chapters, genetic analysis requires mutants; with bacteria, three types are particularly useful:

1. **Antibiotic-resistant mutants.** These mutants are able to grow in the presence of an antibiotic, such as streptomycin (Str) or tetracycline (Tet). For example, streptomycin-sensitive (Str-s) cells have the wildtype phenotype and fail to form colonies on medium containing streptomycin, but streptomycin-resistant (Str-r) mutants can form colonies.

2. **Nutritional mutants.** Wildtype bacteria can synthesize most of the complex nutrients that they need from simple molecules present in the growth medium. The wildtype cells are said to be **prototrophs.** The ability to grow in simple medium can be lost by mutations that disable the enzymes used in synthesizing the complex nutrients. Mutant cells are unable to synthesize an essential nutrient and thus cannot grow unless the required nutrient is supplied in the medium. Such a mutant bacterium is said to be an **auxotroph** for the particular nutrient. For example, a methionine auxotroph cannot grow on a **minimal medium** containing only inorganic salts and a source of energy and carbon atoms (such as glucose), but the methionine auxotroph can grow if the minimal medium is supplemented with methionine.

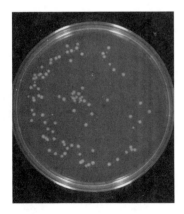

FIGURE 10-2 A petri dish with bacterial colonies that have formed on a solid medium. (Courtesy of Gordon Edlin.)

3. **Carbon-source mutants** cannot utilize particular substances as sources of energy or carbon atoms. For example, Lac⁻ mutants cannot utilize the sugar lactose for growth and are unable to form colonies on minimal medium containing lactose as the only carbon source.

A medium on which all wildtype cells form colonies is called a **nonselective medium.** Mutants and wildtype may or may not be distinguishable by growth on a nonselective medium. If the medium allows growth of only one type of cell (either wildtype or mutant), it is said to be **selective.** For example, a medium containing streptomycin is selective for the Str-r phenotype and selects against the Str-s phenotype, and minimal medium containing lactose as the sole carbon source is selective for Lac⁺ cells and against Lac⁻ cells.

In bacterial genetics, phenotype and genotype are designated in the following way. A phenotype is designated by three letters, the first of which is capitalized, with a superscript + or − to denote presence or absence of the designated character, and with s and r for sensitivity or resistance, respectively. A genotype is designated by lowercase italicized letters. Thus, a cell unable to grow without a supplement of leucine (a leucine auxotroph) has a Leu⁻ phenotype, and this would usually result from a *leu⁻* mutation. Often the − superscript is omitted, but its use prevents ambiguity.

$10\text{-}2$ Bacterial Transformation

Bacterial transformation is a process in which recipient cells acquire genes from free DNA molecules in the surrounding medium. In the laboratory, donor DNA is usually isolated from donor cells and then added to a suspension of recipient cells but, in nature, DNA can become available by spontaneous breakage (lysis) of donor cells. In Chapter 4, a classical transformation experiment was described in which the rough-colony phenotype of *Streptococcus pneumoniae* was changed into the smooth-colony phenotype by exposure of the cells to DNA from a smooth-colony strain. This experiment demonstrated that DNA is the genetic material.

Transformation begins with uptake of a DNA fragment from the surrounding medium by a recipient cell and terminates with *one strand* of donor DNA replacing the homologous segment in the recipient DNA. Most bacterial species are probably capable of the recombination step, but the ability of most bacteria to take up DNA efficiently is limited. Even in a species capable of transformation, DNA is able to penetrate only some of the cells in a growing population. However, appropriate treatment of cells of these species yields a population of cells that are competent to take up DNA.

Transformation is a convenient technique for gene mapping in some species. The technique is related to that described in Chapter 4. When DNA is isolated from a donor bacterium, it is invariably broken into small fragments. In most species, with suitable recipient cells and excess DNA, transformation occurs at a frequency of about one transformed cell per 10^3 cells. If two genes, *a* and *b*, are so widely separated in the donor chromosome that they are always contained in two different DNA fragments, then the probability of simultaneous transformation (**cotransformation**) of an $a^-\ b^-$ recipient into wildtype is the product of the probabilities of transformation of each marker, or roughly $10^{-3} \times 10^{-3}$, which equals one a^+b^+ transformant per 10^6 recipient cells. However, if the two genes are so near one another that they are often present in a single donor fragment, the frequency of cotransformation is nearly the same as the

FIGURE 10-3
Cotransformation of linked
markers. Markers *a* and *b* are
near enough to each other that
they are often present on the
same donor fragment, as are
markers *b* and *c*. Markers *a* and
c are not near enough to
undergo cotransduction. The
gene order must therefore be
a b c.

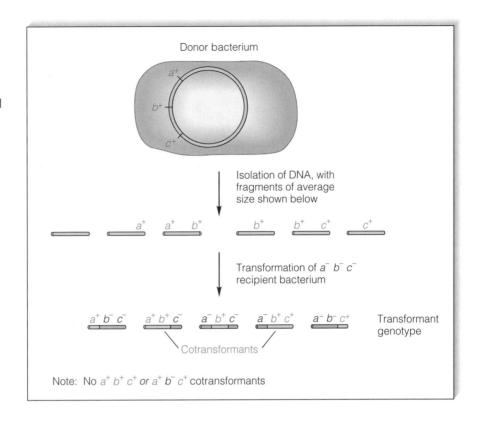

frequency of single-gene transformation, or one wildtype transformant per 10^3 recipients. The general principle is as follows:

> Cotransformation of two genes at a frequency substantially greater than the product of the single-gene transformations implies that the two genes are close together in the bacterial chromosome.

Studies of the ability of various pairs of genes to be cotransformed also yields gene order. For example, if genes *a* and *b* can be cotransformed, and genes *b* and *c* can be cotransformed, but genes *a* and *c* cannot, the gene order must be *a b c* (Figure 10-3). Note that cotransformation frequencies are not equivalent to recombination frequencies used in mapping eukaryotes because they are determined by the size distribution of donor fragments and the likelihood of recombination between baterial DNA molecules rather than the occurrence of chiasmata in synapsed homologous chromosomes (Section 3-2).

10-3 Conjugation

Conjugation is a process in which DNA is transferred from a donor cell to a recipient cell by cell-to-cell contact. It has been observed in many bacterial species, and is best understood in *E. coli*, in which it was discovered by Joshua Lederberg in 1951.

When bacteria conjugate, DNA is transferred to a recipient cell from a donor cell under the control of a set of genes that give the donor cell its transfer capability. These genes are usually present either in a nonchromosomal circular DNA molecule called a **plasmid** or as a block of genes in the chromosome. In the latter case, the plasmid is said to have been **integrated** into the chromosome.

Conjugation begins with physical contact between a donor cell and a recipient cell. Then, a passageway forms between the cells, and a copy of the donor DNA moves from the donor to the recipient. In the final stage, which requires recombination if the donor contains an integrated plasmid, a segment of the transferred donor DNA becomes part of the genetic complement of the recipient. If the donor contains a free plasmid, only the plasmid DNA is transferred and takes up residence in the recipient.

Let us begin with a description of the genetic properties of plasmids and their transfer and then examine plasmid-mediated chromosomal transfer.

Plasmids

Plasmids are circular DNA molecules, capable of replicating independently of the chromosome and ranging in size from a few kilobases to a few hundred kilobases (Figure 10-4). They have been observed in many bacterial species and are usually nonessential for growth of the cells. For the study of plasmids in the laboratory, a culture of cells derived from a single plasmid-containing cell is used; because plasmids replicate and are inherited, all cells of the culture contain the plasmid of interest. Plasmids contain a diversity of genes for their own maintenance, which are not present in the bacterial chromosome, but plasmids can also acquire chromosomal genes by several mechanisms. The presence of certain plasmids in bacterial cells is made evident by phenotypic characteristics of the host cell conferred by genes in the plasmid. For example, a plasmid containing the *tet-r* gene will make the bacterial cell resistant to tetracycline.

Plasmids rely on the DNA-replication enzymes of the host cell for their reproduction, but *initiation* of replication is controlled by plasmid genes. As a result, the number of copies of a particular plasmid in a cell varies from one plasmid to the next, depending on its particular mode of regulation of initiation. High-copy-number plasmids are found in as many as fifty copies per host cell, whereas low-copy-number plasmids are present to the extent of one or two per

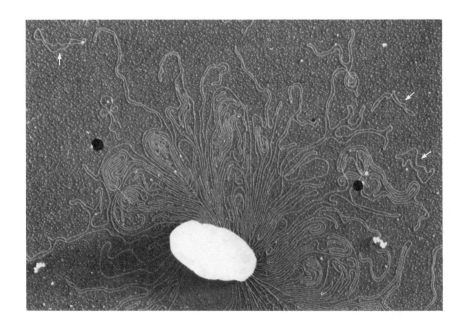

FIGURE 10-4 Electron micrograph of a ruptured *E. coli* cell showing released chromosomal DNA and several plasmid molecules (small circular molecules, indicated by arrows). (Courtesy of David Dressler and Huntington Potter.)

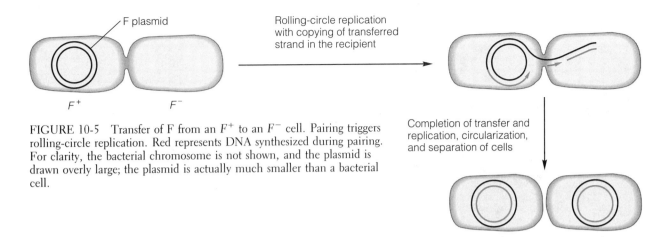

FIGURE 10-5 Transfer of F from an F^+ to an F^- cell. Pairing triggers rolling-circle replication. Red represents DNA synthesized during pairing. For clarity, the bacterial chromosome is not shown, and the plasmid is drawn overly large; the plasmid is actually much smaller than a bacterial cell.

cell. Plasmid DNA can be taken up by cells competent for transformation and can become permanently established in the bacteria. This ability has made plasmids important in genetic engineering (Chapter 16).

From a genetic point of view, the **F plasmid** (F for Fertility) is of greatest interest, for this plasmid is one of those that contains transfer genes that mediate conjugation in *E. coli*. Cells that contain F are donors and are designated F^+; those lacking F are recipients and are designated F^-. The F plasmid is a low-copy-number plasmid. A typical F^+ cell contains one or two copies of F. These replicate once per cell cycle and segregate to both daughter cells in cell division.

The F plasmid contains a set of genes for establishing conjugation between cells and for transferring DNA from donor to recipient. A copy of the F plasmid can be transferred during conjugation from an F^+ cell to an F^- cell. Transfer is always accompanied by replication of the plasmid. Contact between an F^+ and an F^- cell initiates rolling-circle replication of F (Chapter 4), which results in the transfer of a single-stranded linear branch of the rolling circle to the recipient cell. During transfer, DNA synthesis occurs in both donor and recipient (Figure 10-5). Synthesis in the donor replaces the transferred single strand, and synthesis in the recipient converts the transferred single strand into double-stranded DNA. When transfer is complete, the linear F strand becomes circular again in the recipient cell. Note that, because one replica remains in the donor whereas the other strand is transferred to the recipient, after transfer *both cells contain F and can function as donors.* The transfer of F requires only a few minutes. In laboratory cultures, if a small number of donor cells is mixed with an excess of recipient cells, in a few hours F spreads throughout the population and ultimately all cells become F^+. Transfer is not so efficient under natural conditions, and only a few percent of naturally occurring *E. coli* cells contain the F factor.

Hfr Cells

The F plasmid occasionally becomes integrated into the *E. coli* chromosome by an exchange between a sequence in F and a sequence in the chromosome (Figure 10-6). The bacterial chromosome remains circular, though enlarged by the F DNA. Integration of F is an infrequent event, but single cells containing integrated F can be isolated and cultured. The cells in such a strain are called **Hfr cells.** Hfr stands for high frequency of recombination, which refers to the relatively high frequency with which donor genes are transferred to the

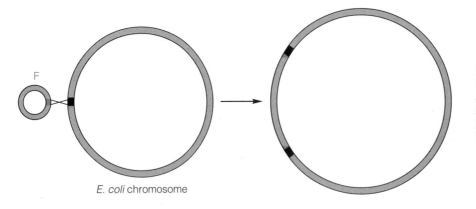

FIGURE 10-6 Integration of F (red circle) by a reciprocal exchange between a base sequence in F and a homologous sequence in the bacterial chromosome.

E. coli chromosome

recipient. Integrated F mediates the transfer of DNA from the bacterial chromosome in an Hfr cell; thus a replica of part of the bacterial chromosome, as well as part of the plasmid, is transferred to the F^- cell.

Hfr × F^- conjugation is illustrated in Figure 10-7. The stages of transfer are much like those by which F is transferred to F^- cells—namely, pairing of donor and recipient, rolling-circle replication in the donor, and conversion of the transferred single-stranded DNA into double-stranded DNA by replication in the recipient—but the transferred DNA does not become circular and is not capable of further replication in the recipient. Furthermore, the replication and associated transfer of the Hfr DNA is controlled by F and is initiated in the Hfr

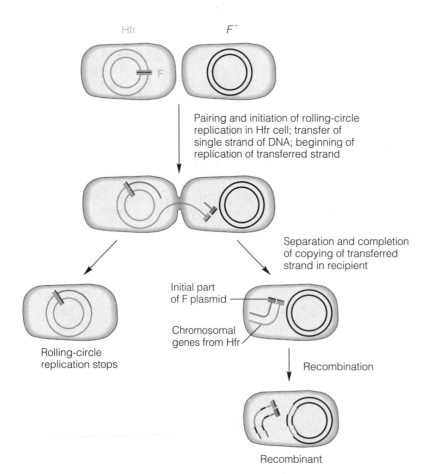

FIGURE 10-7 Stages in transfer and production of recombinants in an Hfr × F^- mating. Pairing initiates rolling-circle replication within the F sequence in the Hfr cell, resulting in transfer of a single strand of DNA. The single strand is converted into double-stranded DNA in the recipient. The mating cells usually break apart before the entire chromosome is transferred. Recombination occurs between the Hfr fragment and the F^- chromosome and leads to recombinants containing genes from the Hfr chromosome. Note that only a part of F is transferred. This part of F is not incorporated into the recipient chromosome. The recipient remains F^-.

Hfr F^-

F

Pairing and initiation of rolling-circle replication in Hfr cell; transfer of single strand of DNA; beginning of replication of transferred strand

Separation and completion of copying of transferred strand in recipient

Rolling-circle replication stops

Initial part of F plasmid

Chromosomal genes from Hfr

Recombination

Recombinant

chromosome at the same point in F at which replication and transfer begin within an unintegrated F plasmid. Thus, a part of F is the first DNA transferred, chromosomal genes are transferred next, and the remaining part of F is the last DNA to enter the recipient. However, the conjugating cells usually break apart long before the entire bacterial chromosome is transferred.

Several differences between F transfer and Hfr transfer are notable.

1. It takes 100 minutes under the usual conditions for an entire bacterial chromosome to be transferred, in contrast with about 2 minutes for the transfer of F. The difference in time is a result of the relative sizes of F and the chromosome.
2. In the transfer of Hfr DNA into a recipient cell, the mating pair usually breaks apart before the entire chromosome is transferred. Under usual conditions, several hundred genes are transferred before the cells separate.
3. In a mating between Hfr and F^- cells, the F^- recipient remains F^- because cell separation usually occurs before the final segment of F is transferred.
4. In Hfr transfer, some regions in the transferred DNA fragment become incorporated into the recipient chromosome. The incorporated regions replace homologous regions in the recipient chromosome. The result is the occurrence of recombinants in the F^- cell that contain one or more genes from the Hfr donor cell. For example, in a mating between Hfr leu^+ and F^- leu^-, some F^- leu^+ cells arise. *The genotype of the donor Hfr cell remains unchanged.*

Genetic analysis requires that recombinant recipients be identified. Because the recombinants derive from recipient cells, a method is needed to eliminate the donor cells. The most-common procedure is to use an F^- recipient containing an allele that can be selected. The selected allele should be located at such a place in the chromosome that most mating cells will have broken apart before transfer of the selected allele occurrs, and the selected allele must not be present in the Hfr cell. The selective agent can then be used to select the F^- cells and eliminate the Hfr donors. Genes conferring antibiotic resistance are especially useful for this purpose. For instance, after a mating between Hfr leu^+ str-s and F^- leu^- str-r cells, the Hfr Str-s cells can be selectively killed by plating the mating mixture on medium containing streptomycin. A selective medium that lacks leucine can then be used to distinguish between the nonrecombinant and recombinant recipients. The F^- leu^- parent cannot grow in medium lacking leucine, but recombinant F^- leu^+ cells can grow because they possess a leu^+ gene. Thus, only recombinant recipients—that is, cells having the genotype leu^+ str-r—form colonies on a selective medium containing streptomycin and lacking leucine.

When a mating is done in this way, the transferred marker that is selected by the growth conditions (leu^+ in this case) is called a **selected marker,** and the marker used to prevent growth of the donor (str-s in this case) is called the **counterselected marker.** Selection and counterselection are necessary in bacterial matings because recombinants constitute only a small proportion of the entire population of cells (in spite of the name high frequency of recombination).

Time-of-Entry Mapping

Genes can be mapped by Hfr × F^- matings. However, the genetic map is quite different from all maps that we have seen so far in that it is not a linkage map but a transfer-order map. It is obtained by deliberate interruption of DNA transfer during mating—for example, by violent agitation of the suspension of mating cells in a kitchen blender. The time at which a particular gene is transferred can be determined by breaking the mating cells apart at various times and noting the earliest time at which breakage no longer prevents recombinants from appearing. This procedure is called the **interrupted-mating technique.** When this is done with Hfr $\times$ F^- matings, it is observed that the number of recombinants of any particular allele increases with the time during which the cells are in contact. This phenomenon is illustrated in Table 10-1. The reason for the increase is that different Hfr $\times$ F^- pairs initiate conjugation and chromosome transfer at slightly different times.

A greater understanding of the transfer process can be obtained by observing the results of a mating with several genetic markers. For example, consider the mating

$$\text{Hfr } a^+ b^+ c^+ d^+ e^+ \text{ str-s} \times F^- a^- b^- c^- d^- e^- \text{ str-r}$$

in which a^- cells require nutrient A, b^- cells require nutrient B, and so forth. At various times after mixing of the cells, samples are agitated violently and then plated on a series of media containing streptomycin and different combinations of the five substances A through E (in each medium one of the five is left out). Colonies that form on the medium lacking A are a^+ str-r, those growing without B are b^+ str-r, and so forth. All of these data can be plotted on a single graph to give a set of curves, as shown in Figure 10-8A. Four features of this set of curves are notable:

1. The number of recombinants in each curve increases with length of time of mating.
2. For each marker, there is a time (the **time of entry**) before which no recombinants are detected.
3. Each curve has a linear region that can be extrapolated back to the time axis, defining the time of entry of each gene a^+, b^+, . . . , e^+.

TABLE 10-1 Data showing the production of Leu$^+$ Str-r recombinants in a cross between Hfr *leu$^+$ str-s* and F^- *leu$^-$ str-r* cells when mating is interrupted at various times

Minutes after mating	Number of Leu$^+$ Str-r recombinants per 100 Hfr cells
0	0
3	0
6	6
9	15
12	24
15	33
18	42
21	43
24	43
27	43

Note: Minutes after mating means minutes after the Hfr and F^- cell suspensions are mixed. Extrapolation of the recombination data to a value of zero recombinants indicates that the earliest time of entry of the *leu$^+$* marker is 4 min.

FIGURE 10-8 Time-of-entry mapping: (A) time-of-entry curves for one Hfr strain; (B) the linear map derived from the data in part A; (C) a linear map obtained with the same Hfr but with a different F^- strain containing the alleles $b^-\ e^-\ f^-\ g^-\ h^-$; (D) a composite map formed from the maps in parts B and C; (E) a linear map from another Hfr strain; (F) the circular map (red) obtained by combining the two (black) maps of parts D and E.

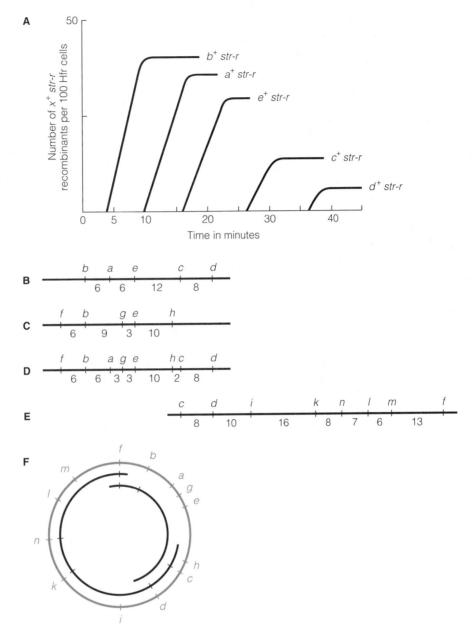

4. The number of recombinants of each type reaches a maximum, the value of which decreases with successive times of entry.

The explanation for the time-of-entry phenomenon is the following. All donor cells do not start transferring DNA at the same time, and so the number of recombinants increases with time. Transfer begins at a particular point in the Hfr chromosome (the replication origin of F). Genes are transferred in linear order to the recipient, and the time of entry of a gene is the time at which that gene first enters a recipient in the population. Separation of a mating pair prevents further transfer and limits the number of recombinants seen at a particular time.

The times of entry of the genes used in the mating just described can be placed on a map, as shown in Figure 10-8B. The numbers on this map and the other are genetic distances between the markers, *measured as minutes between their times of entry*. Mating with another F^- with genotype

$b^- \, e^- \, f^- \, g^- \, h^- \, str\text{-}r$ could be used to locate the three genes f, g, and h. Data for the second recipient would yield a map such as that in Figure 10-8C. Because genes b and e are common to both maps, the two maps can be combined to form a more-complete map, as shown in Figure 10-8D.

Studies with different Hfr strains (Figure 10-8E) also are informative. It is usually found that different Hfr strains are distinguishable by their origins and directions of transfer, indicating that F can integrate at numerous sites in the chromosome and in two different orientations. Combining the maps obtained with different Hfr strains yields a composite map that is *circular*, as illustrated in Figure 10-8F. The circularity of the map is a result of the circularity of the *E. coli* chromosome in F^- cells and the multiple points of integration of the F plasmid; if F could integrate at only one site and in one orientation, the map would be linear.

Data from a great many mapping experiments have been combined to produce an accurate map of nearly 1000 genes throughout the *E. coli* chromosome.

F′ Plasmids

Occasionally, F is excised from Hfr DNA by an exchange between the same sequences used in the integration event. However, in some cases the excision process is not a precise reversal of integration. Instead, breakage and reunion is between the nonhomologous regions at a boundary of F and nearby chromosomal DNA (Figure 10-9). When such aberrant excision occurs, a plasmid containing chromosomal DNA—an **F′ plasmid**—is formed. By the use of Hfr strains having different origins of transfer, F′ plasmids having chromosomal segments from many regions of the chromosome have been isolated. These elements are extremely useful because they render any recipient diploid for the region of the chromosome carried by the plasmid. These diploid regions allow dominance tests and gene-dosage tests (studies of the effects of increasing the number of copies of a gene on gene expression). Because only a part of the genome is diploid, cells containing an F′ plasmid are **partial diploids.** Examples of genetic analysis using F′ plasmids are given in Chapter 12 in a discussion of the *E. coli lac* genes.

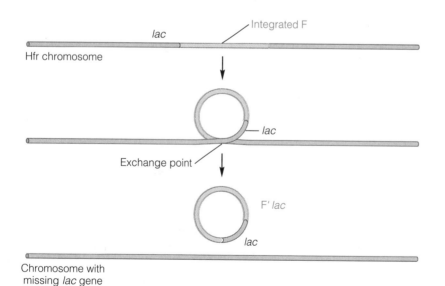

FIGURE 10-9 Formation of an F′*lac* plasmid by aberrant excision of F from an Hfr chromosome. Breakage and reunion is between nonhomologous regions.

In the process of **transduction**, bacterial DNA is transferred from one bacterial cell to another *by a phage particle containing the* DNA. Such a particle is called a **transducing phage.** Two types of transducing phages are known—generalized and specialized. A **generalized transducing phage** produces some particles that contain only DNA obtained from the host bacterium, rather than phage DNA; the bacterial DNA fragment can be derived from *any* part of the bacterial chromosome. A **specialized transducing phage** produces particles containing both phage and bacterial genes linked in a single DNA molecule, and the bacterial genes are obtained from a *particular* region of the bacterial chromosome. In this section, we consider *E. coli* phage P1, a well-studied generalized transducing phage. Specialized transducing particles will be discussed in Section 10-6.

During infection by P1, the phage makes a nuclease that cuts the bacterial DNA into fragments. Single fragments of bacterial DNA comparable in size to P1 DNA are occasionally packaged into phage particles in place of P1 DNA. The positions of the nuclease cuts in the host chromosome are random, and so a transducing particle may contain a fragment derived from any region of the host DNA. A large population of P1 phages will contain particles carrying any bacterial gene. On the average, any particular gene is present in roughly one transducing particle per 10^6 viable phages. When a transducing particle adsorbs to a bacterium, the bacterial DNA contained in the phage head is injected into the cell and becomes available for recombination with the homologous region of the host chromosome.

Let us now examine the events that follow infection of a bacterium by a generalized transducing particle obtained, for example, by growth of P1 on wildtype *E. coli* containing a *leu*⁺ gene (Figure 10-10). If such a particle adsorbs to a bacterial cell of *leu*⁻ genotype and injects the DNA that it contains into the cell, the cell survives because the phage head contained only bacterial genes and no phage genes. A recombination event exchanging the *leu*⁺ allele carried by the phage for the *leu*⁻ allele contained in the host converts the genotype of

FIGURE 10-10 Transduction. Phage P1 infects a *leu*⁺ donor, yielding predominately normal P1 phage with an occasional one carrying bacterial DNA instead of phage DNA. If the phage population infects a bacterial culture, the normal phages produce progeny phage, whereas the transducing particle yields a transductant. Notice that the recombination step requires two crossovers. For clarity, double-stranded DNA is drawn as a single line. The size of the DNA fragment in the transducing particle is not drawn to scale.

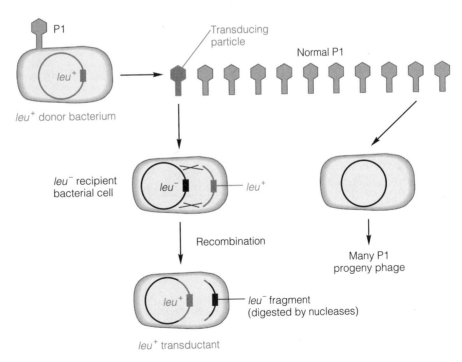

the host cell from *leu⁻* into *leu⁺*. In such an experiment, typically about one *leu⁻* cell in 10⁶ becomes *leu⁺*. Such frequencies are easily detected on selective growth medium. For example, if the infected cell is placed on solid medium lacking leucine, it is able to multiply and a *leu⁺* colony forms. A colony does not form unless recombination inserted the *leu⁺* allele.

The small fragment of bacterial DNA contained in a transducing particle includes about fifty genes, and so transduction is a valuable tool for genetic linkage studies of short regions of the bacterial genome. Consider a population of P1 prepared from a bacterium having a *leu⁺ gal⁺ bio⁺* genotype. This sample contains particles able to transfer any of these alleles to another cell; that is, a *leu⁺* particle can transduce a *leu⁻* cell to *leu⁺*, or a *gal⁺* particle can transduce a *gal⁻* cell to *gal⁺*. Furthermore, if a *leu⁻ gal⁻* culture is infected with phage, both *leu⁺ gal⁻* and *leu⁻ gal⁺* bacteria are produced. However, *leu⁺ gal⁺* colonies do not arise because the *leu* and *gal* genes are too far apart to be included in the same DNA fragment (Figure 10-11A).

The situation is quite different with a recipient cell with genotype *gal⁻ bio⁻*, because the *gal* and *bio* genes are so closely linked that both genes are sometimes present in a single DNA fragment carried in a transducing particle— namely, a *gal-bio* particle (Figure 10-11B). However, all *gal⁺* transducing particles do not also include *bio⁺*, and all *bio⁺* particles do not include *gal⁺*. The probability of both markers being in a single particle, and hence the probability of simultaneous transduction of both markers (**cotransduction**), depends on how close to each other the genes are. The closer they are, the greater the frequency of cotransduction. Cotransduction of the *gal-bio* pair can be detected by plating infected cells on the appropriate growth

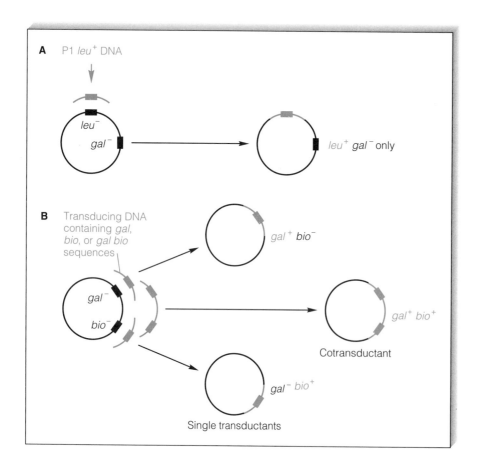

FIGURE 10-11 Demonstration of linkage of the *gal* and *bio* genes by cotransduction. A. A transducing particle carrying the *leu⁺* allele can convert a *leu⁻ gal⁻* cell into the *leu⁺ gal⁻* genotype (but could not produce the *leu⁺ gal⁺* genotype). B. The transductants that could be formed by three possible types of transducing particles—one carrying *gal⁺*, one carrying *bio⁺*, and one carrying the linked alleles *gal⁺ bio⁺*. The third type results in cotransduction.

medium. If *bio*$^+$ transductants are selected by spreading the infected cells on a glucose-containing medium lacking biotin, both *gal*$^+$ *bio*$^+$ and *gal*$^-$ *bio*$^+$ colonies will grow. If these colonies are tested for the *gal* marker, 50 percent are found to be *gal*$^+$ *bio*$^+$ and the rest *gal*$^-$ *bio*$^+$; similarly, if *gal*$^+$ transductants are selected, 50 percent are found to be *gal*$^+$ *bio*$^+$. In other words, the **frequency of cotransduction** of *gal* and *bio* is 50 percent, which means that 50 percent of all transducing particles that contain one gene also include the other gene.

Studies of cotransduction can be used to map closely linked genetic markers by means of three-factor crosses analogous to those described in Chapter 3. That is, P1 is grown on bacteria with three markers and used to transduce cells with different alleles of these markers. Cotransductants containing various combinations of pairs of alleles are examined. The central marker can be identified because it will almost always be cotransduced with the flanking markers. (See Figure 10-11B.)

10-5 Bacteriophage Genetics

The life cycles of phages fit into two distinct categories—the lytic and the lysogenic cycles. In the **lytic cycle**, phage nucleic acid enters a cell and replicates repeatedly, the bacterium is killed, and hundreds of phage progeny result (see Figure 10-1). All phage species can undergo a lytic cycle; a phage capable *only* of lytic growth is called **virulent**. In the alternative **lysogenic cycle**, no progeny particles are produced, the bacterium survives, and a phage DNA molecule is transmitted to each bacterial daughter cell. In most cases, transmission of this kind is accomplished by integration of the phage chromosome into the bacterial chromosome. A phage capable of such a life cycle is called **temperate**.

Several phage particles can infect a single bacterium and the DNA of each can replicate. In a multiply infected cell in which a lytic cycle is occurring, genetic exchanges occur between phage DNA molecules. If the infecting particles carry different mutations as genetic markers, recombinant phage result. There are several modes of recombination—general and site specific—in the lytic cycle. Only the former will be considered in this chapter. Recombination also occurs in the lysogenic cycle; however, the end result is not the production of recombinant phage chromosomes but the joining of phage and bacterial DNA molecules. This process, which for most temperate phages is site specific, is described in Section 10-6.

Plaque Formation and Phage Mutants

Phages are easily detected because in a lytic cycle an infected cell breaks open (**lyses**) and releases phage particles to the growth medium. The test is performed in the following way (Figure 10-12A).

A large number of bacteria (about 10^8) are placed on a solid medium. After a period of growth, a continuous turbid layer of bacteria results. If a phage is present at the time the bacteria are placed on the medium, it adsorbs to a cell and, shortly afterward, the infected cell lyses and releases many phages; each of these progeny adsorbs to a nearby bacterium, and after another lytic cycle these bacteria in turn release phages that can infect still other bacteria in the vicinity. These multiple cycles of infection continue, and after several hours the phages destroy all of the bacteria in a localized area, giving rise to a clear transparent region in the otherwise turbid layer of confluent bacterial growth—a **plaque**.

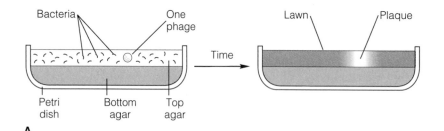
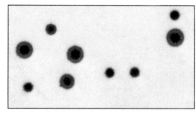

A

B

FIGURE 10-12   A. Plaque formation. Bacteria grow and form a translucent lawn. Because the bacteria have lysed, there are no bacteria in the vicinity of the site of the initial phage; this empty area, which remains transparent, is called a plaque. B. Plaques of *E. coli* phage T4. Two types of plaques are present. The smaller plaques are made by wildtype phage; the larger plaques are made by an *rII* mutant phage. Note the halo around the large plaques—it is a result of large amounts of the lysis enzyme diffusing outward and lysing uninfected cells.

Phages can multiply only in growing bacterial cells, and so exhaustion of nutrients in the growth medium limits phage multiplication and the size of the plaque. Because a plaque is a result of an initial infection by one phage particle, the number of individual phages originally present on the medium can be counted.

The genotypes of phage mutants can be determined by studying the plaques. In some cases the appearance of the plaque is sufficient. For example, phage mutations that decrease the number of phage progeny from infected cells often yield smaller plaques. Large plaques can be produced by mutants that cause premature lysis of infected cells, so that each round of infection proceeds more quickly (Figure 10-12B). Another type of phage mutation can be identified by the ability or inability of the phage to form plaques on a particular bacterial strain.

Genetic Recombination in the Lytic Cycle

If two phage particles with different genotypes infect a single bacterium, some phage progeny are genetically recombinant. Figure 10-13 shows the progeny of

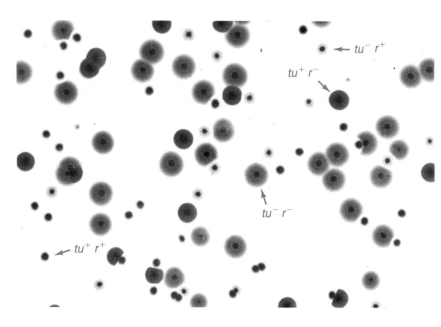

FIGURE 10-13   Progeny of a cross between *E. coli* T4 phage of genotype *tu⁻ r⁻* (a plaque is identified near the center) and wildtype (a plaque is identified at the lower left). The two expected types of recombinant plaques are found, and representative plaques are identified at the upper right. (Courtesy of A. H. Doermann.)

a mixed infection with *E. coli* phage T4 mutants. The r^- allele results in large plaques and the tu^- allele results in plaques with a light turbid halo. The cross is

$$tu^- \, r^- \text{ (turbid, large)} \times tu^+ \, r^+ \text{ (clear, small)}$$

Four plaque types can be seen in Figure 10-13—the two parental phages and the $tu^- \, r^+$ (turbid, small) and $tu^+ \, r^-$ (clear, large) recombinants. When many bacteria are infected, equal numbers of complementary recombinant types are usually found. The recombination frequency, expressed as a percentaage, is defined as

$$(\text{Number of recombinant genotypes/Total number of genotypes}) \times 100$$

Recombination experiments with many pairs of markers can yield a genetic map of the phage genome, with the use of techniques analogous to those presented in Chapter 3.

Arrangement of Genes in Phage Chromosomes

Genetic mapping has provided valuable information about the organization of the genome in several phages that have been intensively studied. Many of the gene products and their functions have been identified. A striking feature of the arrangement of phage genes is that they are often clustered according to function. An example can be seen in the genetic map of *E. coli* phage (Figure 10-14). Note that the right half of the map consists entirely of genes whose products—head and tail proteins—are required for assembly of the phage structure, and the head genes and the tail genes form subclusters. The left half of the map also shows several gene clusters—for example, for DNA replication, recombination, and lysis. These genes are also clustered according to the time

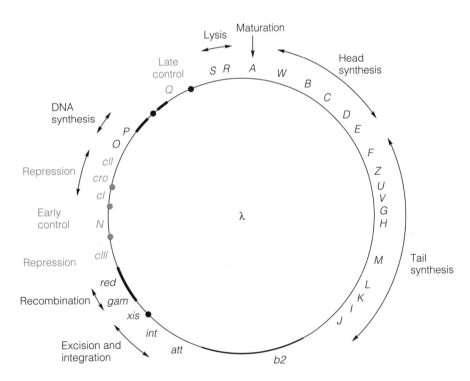

FIGURE 10-14 Genetic map of *E. coli* phage λ. The map is often drawn as a circle to relate it to the intracellular DNA, which is circular. Regulatory genes and functions are printed in red. All genes are not shown. Major regulatory sites are indicated by solid circles. These sites are frequently adjacent to the gene whose product acts on them. Repression refers to functions necessary for the lysogenic cycle or related to the choice between the lytic and lysogenic cycles. Early and late control refers to genes necessary for the expression of other genes early and late in the life cycle.

at which their products are synthesized. For example, the N gene acts early, genes O and P are active later, and genes Q, S, and R, and the head–tail cluster are expressed last. Some mammalian viruses also have a clustering of genes that are expressed at similar times.

Fine Structure of the rII Gene in Bacteriophage T4

In Chapter 3, experiments with *Drosophila* were described in which recombination within a gene was first discovered. These experiments gave the first indication that intragenic recombination could occur and that genes have fine structure. Other genes were studied in the years that followed this experiment, but none could equal the fine-structure mapping of the *rII* gene in bacteriophage T4 carried out by Seymour Benzer. Using 2400 independent mutations and novel mapping techniques that reduced the number of required crosses from more than a half million to several thousand, he was able to show the following:

1. Genetic exchange can occur within a gene and probably between any pair of adjacent nucleotides.
2. Mutations are not produced at equal frequencies at all sites within a gene. For example, the 2400 mutations were located at only 304 sites. One of these sites was represented 474 times, whereas mutations in most other sites were recovered only once or a few times.

The *rII* analysis was important because three distinct meanings of the word *gene* were distinguished experimentally for the first time. Most commonly, the word *gene* refers to a unit of function. Physically, this corresponds to a protein-coding segment of DNA. Benzer assigned the term **cistron** to this unit of function, and the term is still occasionally used. Two other distinct meanings of the term *gene* are: (1) the unit of genetic transmission that participates in recombination, and (2) the unit of genetic change or mutation. Physically, both the recombinational and the mutational units correspond to the individual nucleotides in a gene. Despite potential ambiguity, the term *gene* is still the most-important word in genetics, and in most cases the shade of meaning intended is clear from the context.

10-6 Lysogeny and *E. coli* Phage λ

A temperate phage has two alternate life cycles—a lysogenic cycle and a lytic cycle. The lytic cycle is that depicted in Figure 10-1. In the lysogenic cycle, phage are not produced and the host survives, and a replica of the infecting phage DNA becomes inserted (**integrated**) into the bacterial chromosome (Figure 10-15). The inserted DNA is called a **prophage,** and the surviving bacterial cell is called a **lysogen.** Many bacterial generations later, the prophage can be activated and excised from the chromosome, and a lytic cycle can begin. When activation occurs, the host cell is killed and progeny phage are released, exactly as in a normal lytic cycle. Two features of the lysogenic cycle will be considered here—how the DNA is integrated and how it is excised, and the discussion is confined to *E coli* phage λ.

Time-of-entry experiments and transduction analysis of *E. coli* cells lysogenic for phage λ have indicated that, first, the *gal* and *bio* genes and the prophage are linked, with the gene order *gal* λ *bio*. Second, the presence of the prophage increases the physical distance between *gal* and *bio*, because the *gal* and *bio* genes can be cotransduced by P1 phage grown on a nonlysogen, but not

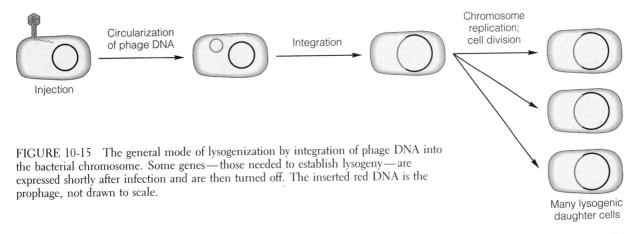

Injection   Circularization of phage DNA   Integration   Chromosome replication; cell division

Many lysogenic daughter cells

FIGURE 10-15 The general mode of lysogenization by integration of phage DNA into the bacterial chromosome. Some genes—those needed to establish lysogeny—are expressed shortly after infection and are then turned off. The inserted red DNA is the prophage, not drawn to scale.

by P1 grown on a lysogen. Third, mapping of prophage genes with respect to the *gal* and *bio* genes by P1 transduction yields a prophage genetic map that is a permutation of the genetic map of the phage obtained from standard phage crosses. That is, the genetic map from phage crosses is *A b O R*, and the prophage map is *gal O R A b bio* (Figure 10-16). These observations result from circularization of the λ chromosome; that is,

1. In a lysogen, the λ prophage is *linearly* inserted between the *gal* and *bio* genes.
2. Integration is the result of a breaking-and-rejoining event that occurs between a particular site in λ DNA and a site in the bacterial chromosome between the *gal* and *bio* genes.

These two features have been proved by studies of the structure of λ DNA and an analysis of the nature of the sites of exchange.

The DNA of λ is a linear molecule having **cohesive ends** twelve bases long. A cohesive end (*cos*) is a single-stranded region at each end of the linear molecule; the two ends are complementary in nucleotide sequence. These termini can base-pair, forming a circular molecule, as shown in Figure 10-17. Circularization, which occurs early in both the lytic and the lysogenic cycles, is a necessary event in prophage integration.

The sites of breakage and rejoining in the bacterial and phage DNA are called the **bacterial** and **phage attachment sites,** respectively. Each attachment site consists of three segments. The central segment has the same nucleotide sequence in both attachment sites and is the region in which the breakage and rejoining occurs. The phage attachment site is denoted *POP'*(P for phage), and the bacterial attachment site is denoted *BOB'*(B for bacteria). Comparison of the genetic maps of the phage and the prophage indicate that *POP'* is located near the middle of the linear form of the phage DNA molecule. A phage protein, **integrase,** catalyzes a site-specific recombination event—that is, it recognizes the phage and bacterial attachment sites and causes the physical

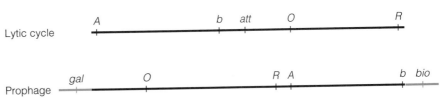

Lytic cycle   *A*   *b*   *att*   *O*   *R*

Prophage   *gal*   *O*   *R A*   *b bio*

FIGURE 10-16 The map order of genes in phage λ as determined by phage recombination (lytic cycle) and in the prophage (prophage order). The genes have been selected arbitrarily to provide reference points. Bacterial DNA is shown in red.

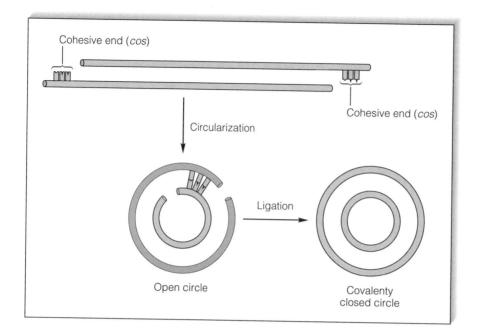

FIGURE 10-17 A diagram of a linear λ DNA molecule showing the cohesive ends (complementary single-stranded ends); circularization by means of base-pairing between the cohesive ends forming an open, or nicked, circle; and formation of a covalently closed (uninterrupted) circle by sealing (ligation) of the single-strand breaks.

exchange that results in integration of the λ DNA molecule into the bacterial DNA. The geometry of the exchange is shown in Figure 10-18. Note that the permutation of the phage and prophage maps is a consequence of the circularization of the phage DNA and the central location of *POP'*.

A lysogenic cell can replicate nearly indefinitely without release of phage. However, activation of the phage by means of prophage excision sometimes occurs followed by a lytic cycle with production of the usual number of phage progeny. This phenomenon is called **prophage induction,** and it is initiated by damage to the bacterial DNA. The damage sometimes occurs spontaneously but is more often caused by an environmental agent, such as chemicals or

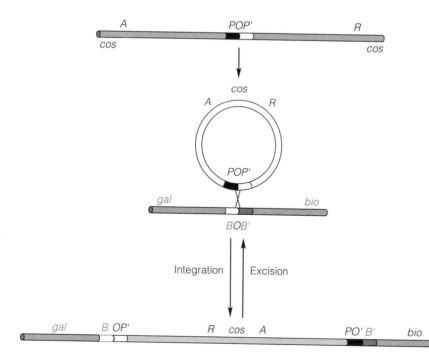

FIGURE 10-18 The geometry of prophage integration and excision of phage λ. The phage attachment site is *POP'*. The bacterial attachment site is *BOB'*. The prophage is flanked by two hybrid attachment sites denoted *BOP'* and *POB'*.

radiation. The ability to be induced is advantageous for the phage because the phage DNA can escape from a damaged cell. The biochemical mechanism of induction is complex and will not be discussed, but the actual excision of the phage is straightforward.

Excision is another site-specific recombination event that reverses the integration process. Excision requires integrase plus an additional phage protein called **excisionase.** Genetic evidence and studies of physical binding of purified excisionase, integrase, and λ DNA indicate that excisionase binds to integrase and thereby enables the latter to recognize the prophage attachment sites *BOP'* and *POB'*; once bound to these sites, integrase makes cuts in the *O* sequence and reforms the *BOB'* and *POP'* sites. This reverses the integration reaction, causing excision of the prophage (see Figure 10-18).

When a cell is lysogenized, a block of phage genes becomes part of the bacterial chromosome, and so it might be expected that the phenotype of the bacterium would change. Most phage genes in a prophage are kept in an inactive state by the product of one expressed phage gene called a **repressor** gene. This protein is coded in a gene that is active initially in the infecting phage and then continually in the prophage; the repressor gene is frequently the only prophage gene that is expressed in lysogens. If a lysogen is infected with a phage of the same type as the prophage—for example, λ infecting a λ lysogen—the repressor present within the cell from the prophage genes prevents expression of the genes of the infecting phage. This resistance to infection by a phage identical with the prophage, which is called **immunity,** is the usual criterion for determining whether a bacterial cell contains a particular phage. For example, λ will not form plaques on bacteria containing a λ prophage.

Specialized Transducing Phage

When a bacterium lysogenic for phage λ is subjected to DNA damage that leads to induction, the prophage is usually excised from the chromosome precisely. However, in about one cell per 10^6–10^7 cells, an excision error is made (Figure 10-19), and a chance breakage in two nonhomologous sequences occurs—one break within the prophage and the other in the bacterial DNA. The free ends of the excised DNA are then joined to create a DNA circle capable of

FIGURE 10-19 Aberrant excision leading to the production of specialized transducing phages λ *gal* (top) and λ *bio* (bottom).

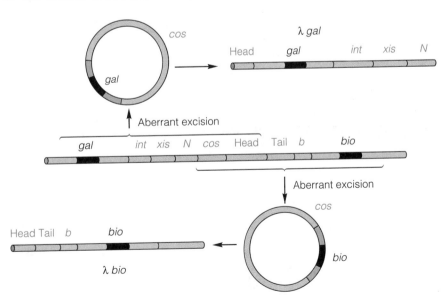

replication. The sites of breakage may not always be situated such that a length of DNA that can fit in a λ phage head is produced—the DNA may be too large or too small—but sometimes a molecule forms that can replicate and be packaged. In λ lysogens, the phrophage is located between the *gal* and *bio* genes and, because the aberrant cut in the host DNA can be either to the right or to the left of the prophage, particles can arise that carry either the *bio* genes (cut at the right) or the *gal* genes (cut at the left). These particles are called λ *bio* and λ *gal* transducing particles. They are **specialized transducing phages** because they can transduce only certain bacterial genes, in contrast with the P1-type generalized transducing particles, which can transduce any gene.

10-7 Transposable Elements

DNA sequences that are present in a genome in multiple copies and that have the capability of occasional movement or *transposition* to new locations in the genome were described in Chapter 5. These *transposable elements* are widespread in eukaryotes and bacteria.

Many transposable elements in bacteria have been extensively studied. The bacterial elements first discovered—called **insertion sequences** or **IS elements**—are small and do not contain any known host genes. They do possess inverted-repeat sequences at their termini, as do many transposable elements in eukaryotes, and most code for a **transposase** protein required for transposition and one or more additional proteins that regulate the rate of transposition. The DNA organization of the insertion sequence IS50 is diagrammed in Figure 10-20A. Insertion sequences are instrumental in the origin of Hfr bacteria from F$^+$ cells (Section 10-3), because integration of the F plasmid is normally through genetic exchange between insertion sequences present in F and homologous copies present at various sites in the bacterial chromosome.

Other transposable elements in bacteria contain one or more bacterial genes that can be shuttled between different bacterial hosts by transposing into bacterial plasmids (such as the F plasmid), which are capable of conjugational transfer. These gene-containing elements are called **transposons**; their length is typically several kilobases of DNA, but a few are much longer. Some

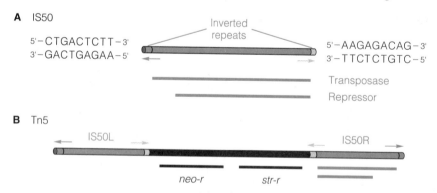

FIGURE 10-20 Transposable elements in bacteria. A. Insertion sequence IS50. The element is terminated by short inverted-repeat sequences, the terminal nine of which are indicated. IS50 contains a region coding for the transposase and for a repressor of transposition. The coding regions are identical in the region of overlap, but the repressor is somewhat shorter because it begins at a different place. B. Composite transposon Tn5. The central sequence contains genes for resistance to neomycin, *neo-r*, and streptomycin, *str-r*, and is flanked by two copies of IS50 in inverted orientation. The left-hand element (IS50L) contains mutations and is nonfunctional, and so the transposase and repressor are made by the right-hand element (IS50R).

transposons have composite structures with the bacterial genes sandwiched between insertion sequences, as is the case with the Tn5 element illustrated in Figure 10-20B, which terminates in two IS50 elements in inverted orientation. Transposons are usually designated by the abbreviation Tn followed by a number (for example, Tn5). When it is necessary to refer to genes carried in such an element, the usual designations for the genes are used. For example, Tn5(*neo-r str-r*) contains genes for neomycin and streptomycin resistance. Such genes provide markers, making it easy to detect transposition of the composite element, as is shown in Figure 10-21. An F'*lac*⁺ plasmid is transferred by conjugation to a bacterial cell containing a transposable element with the *neo-r* (neomycin-resistance) gene. The bacterial cell is allowed to grow and, in the course of multiplication, transposition of the transposon into the F' plasmid occasionally occurs in a progeny cell. Transposition yields an F' plasmid containing both the *lac*⁺ and *neo-r* genes. In a subsequent mating to an Neo-s Lac⁻ cell, the *lac*⁺ and *neo-r* markers are transferred together and so are genetically linked.

In nature, sequential transposition of transposons containing *different* antibiotic-resistance genes into the same plasmid results in the evolution of plasmids that confer resistance to multiple antibiotics. These multiple-resistance plasmids are called **R plasmids.** The evolution of R plasmids is promoted by the use (and regretable overuse) of antibiotics, which selects for resistant cells because, in the presence of antibiotics, resistant cells have a growth advantage over sensitive cells. Presence of multiple antibiotics in the environment selects for multiple-drug resistance. Serious clinical complications result when multiple-drug-resistance plasmids are transferred to bacterial pathogens. Infections caused by pathogens containing R factors are difficult to treat because the cells are resistant to the commonly used antibiotics.

When transposition occurs, the transposable element can be inserted in any one of a large number of positions. The existence of multiple insertion sites can be shown when a wildtype lysogenic *E. coli* culture is infected with a temperate phage that is identical with the prophage (immunity prevents the phage from killing the cell) but carries a transposable element with an antibiotic-resistance marker. Transposition events can be detected by the production of antibiotic-resistant bacteria containing new mutations resulting from insertion of the transposon into a chromosomal gene. For example, a

FIGURE 10-21 An experiment demonstrating transposition of transposon Tn5 containing a neomycin-resistance gene from the chromosome to an F' plasmid containing the bacterial gene for lactose utilization (F'*lac*). The bacterial chromosome is *lac*⁻ and the cells unable to grow on lactose unless the F'*lac* is present. After the transposition, the F'*lac* also contains Tn5, indicated by the linkage of the *neo-r* gene to the F' factor. Note that transposition to the F' does not eliminate the copy of the transposon in the chromosome.

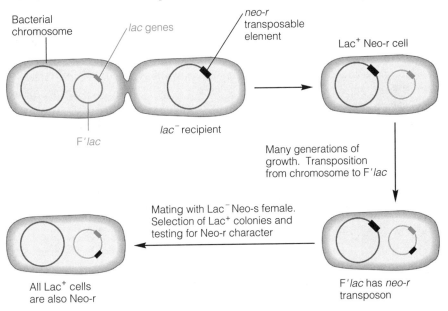

lac+ leu+ neo-s culture infected with a *neo-r* transposon can yield both *lac− neo-r* and *leu− neo-r* mutants. If many hundreds of Neo-r bacterial colonies are examined and tested for a variety of nutritional requirements and the ability to utilize different sugars as a carbon source, colonies can usually be found bearing a mutation in almost any gene that is examined. This observation indicates that potential insertion sites for transposons are scattered throughout the chromosome of *E. coli* and other bacterial species.

The end result of the transposition process is the insertion of a transposable element between two base pairs in a recipient DNA molecule. In inserting, most transposition events also create a duplication of a short (2–12 nucleotide pairs) sequence of host DNA, which after transposition flanks the insertion site (Section 5-8). The insertion of a transposable element does not require DNA sequence homology or the use of most of the enzymes of homologous recombination, because transposition occurs in cells in which the major system for homologous recombination has been eliminated. Direct nucleotide-sequence analysis of many transposable elements and their insertion sites confirms the absence of DNA sequence homology in the recipient with any sequence in the transposable element.

Transposons in Genetic Analysis

Transposons can be employed in a variety of ways in bacterial genetic analysis. Three features make them especially useful for this purpose. First, they can insert at a large number of potential target sites that are essentially random in their distribution around the genome. Second, many transposons code for their own transposase and require only a small number of host genes for mobility. Third, transposons contain one or more genes for antibiotic resistance serving as genetic markers that can be selected. Genetic analysis using transposons is particularly important in bacterial species that do not have readily exploitable systems for genetic manipulation or large numbers of identified and mapped genes. Many of these species are bacterial **pathogens,** meaning agents of disease, and transposons can be used to identify and manipulate the disease factors.

Use of a *neo-r* transposable element to identify a particular disease gene in a pathogenic bacterial species is diagrammed in Figure 10-22. In this example, the transposon is introduced into the pathogenic bacterium by the use of a mutant bacteriophage that cannot replicate in the pathogenic host. A variety of other methods of introduction are also possible. After introduction of the transposon and selection on medium containing the antibiotic, the only resistant cells are those in which the transposon inserted into the bacterial chromosome, because the DNA in which it was introduced is incapable of replication. Any of a number of screening methods are then used to identify cells in which a particular disease gene became inactivated (nonfunctional) as a result of transposon insertion into the gene. This method is known as **transposon tagging** because the gene with the insertion is tagged (marked) with the antibiotic-resistance phenotype of the transposon.

Once the disease gene has been tagged by transposon insertion, it can be used in many ways (for example, in genetic mapping) because the phenotype resulting from the presence of the tagged gene is antibiotic resistance, which is easily identified and selected. The lower part of Figure 10-22 shows how the transposon tag is used to transfer the disease gene into *E. coli*. In the first step, DNA from the pathogenic strain containing the tagged gene is purified, cut into fragments of suitable size, and inserted into a small plasmid capable of replication in *E. coli*. (Details of these genetic-engineering methods are

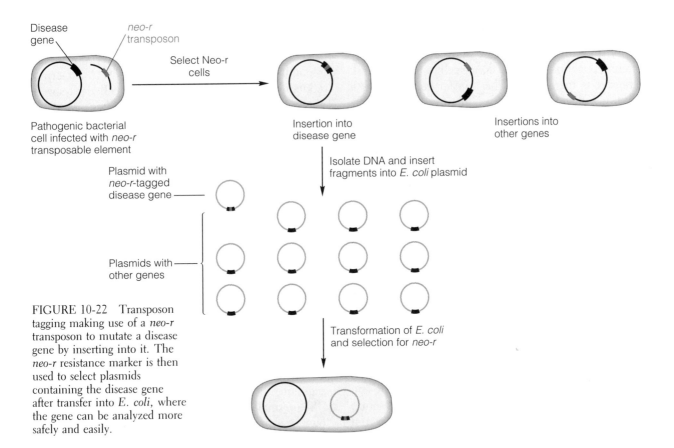

Disease gene

neo-r transposon

Select Neo-r cells

Pathogenic bacterial cell infected with *neo-r* transposable element

Insertion into disease gene

Insertions into other genes

Isolate DNA and insert fragments into *E. coli* plasmid

Plasmid with *neo-r*-tagged disease gene

Plasmids with other genes

Transformation of *E. coli* and selection for *neo-r*

FIGURE 10-22 Transposon tagging making use of a *neo-r* transposon to mutate a disease gene by inserting into it. The *neo-r* resistance marker is then used to select plasmids containing the disease gene after transfer into *E. coli*, where the gene can be analyzed more safely and easily.

discussed in Chapter 16.) The result is a heterogeneous collection of plasmids, most containing DNA fragments other than the one of interest. However, plasmids containing the fragment of interest also contain the transposon tag and so are able to confer antibiotic resistance on the host. The heterogeneous collection of plasmids is introduced into *E. coli* cells by transformation, and the cells are grown on solid medium containing the antibiotic. The only cells that survive contain the desired plasmid, which contains the transposon along with DNA sequences from the desired gene. The transposon-tagged gene is said to be **cloned** in *E. coli*.

Cloning of genes in *E. coli* greatly facilitates further genetic and molecular studies, such as DNA sequencing. The disease gene usually presents no hazard in *E. coli* for several reasons: (1) more than one gene in the pathogen is usually required to cause disease; (2) the cloned transposon-tagged gene is inactivated by the insertion; (3) the disease gene cloned in *E. coli* is usually incomplete or, if complete, often inactive in this host. In cases in which some danger might result from cloning, potential problems can be minimized by using special kinds of *E. coli* strains and procedures that ensure containment in the laboratory.

CHAPTER SUMMARY

DNA can be transferred between bacteria in three ways: transformation, transduction, and conjugation. In transformation, free DNA molecules, obtained from donor cells, are taken up by recipient cells; by a recombinational mechanism, a single-stranded segment becomes integrated into the recipient chromosome, replacing a homologous segment.

In conjugation, donor and recipient cells pair, and a single strand of DNA is transferred by rolling-circle replication from the donor cell to the recipient. Transfer is mediated by the transfer genes of the F plasmid. When F is a free plasmid, it becomes established in the recipient as an autonomously replicating plasmid; if F is integrated into the donor chromosome—that is, if the donor is an Hfr cell—only part of the donor chromosome is usually transferred, and it can be maintained in the recipient only after an exchange event. About 100 minutes is required to transfer the entire *E. coli* chromosome, but the mating cells normally break apart before transfer is complete. DNA transfer occurs from a particular point in the Hfr chromosome (the site of integration of F) and proceeds linearly. The times at which donor markers first enter recipient cells—the times of entry—can be arranged in order, yielding a map of the bacterial genome. This map is circular because of the multiple sites at which an F plasmid integrates into the bacterial chromosome. Occasionally, F is excised from the chromosome in an Hfr cell; aberrant excision, in which one cut is made at the end of F and the other cut is made in the chromosome, gives rise to F′ plasmids, which can transfer bacterial genes.

In transduction, a generalized transducing phage infects a donor cell, fragments the host DNA, and packages fragments of host DNA into phage particles. These transducing particles, which contain no phage DNA, can inject donor DNA into a recipient bacterium; by genetic exchange, the transferred DNA can replace homologous DNA of the recipient, producing a recombinant bacterium called a transductant.

When bacteria are infected with several phages, genetic exchange can occur between phage DNA molecules, creating recombinant phage progeny. Measurement of recombination frequency yields a genetic map of the phage. A common feature of these maps is clustering of phage genes with related function.

The temperate phages possess mechanisms for recombining phage and bacterial DNA. A bacterium containing integrated phage DNA is called a lysogen, the integrated phage DNA is a prophage, and the overall phenomenon is lysogeny. The integrative exchange is between particular sequences in the bacterial and phage DNA called attachment sites. The bacterial and phage attachment sites are not identical, but have a short common sequence in which the exchange occurs. Temperate phages circularize their DNA after infection. Because the phage has a single attachment site, which is not terminally located, the order of the prophage genes is a permutation of the order of the genes in the phage particle. The integrated prophage DNA is stable but, if the bacterial DNA is damaged, prophage induction is initiated, the phage DNA is excised, and the lytic cycle of the phage occurs.

Most bacteria possess transposable elements, which are capable of moving from one part of the DNA to another. Transposition does not utilize sequence homology. Bacterial insertion sequences (IS) are small transposable elements that code for their own transposase and that have inverted-repeat sequences at the ends. Some larger transposons have a composite structure consisting of a central region, often containing one or more antibiotic-resistance genes, flanked by two IS sequences in inverted orientation. Accumulation of such transposons in plasmids gives rise to multiple-drug-resistance R plasmids, which confer resistance to several chemically unrelated antibiotics.

Transposable elements are useful in genetic analysis because they can create mutations by integrating within a host gene and interrupting its continuity. Such mutations allow transposon tagging for genetic analysis of bacterial pathogens and other species.

K E Y   T E R M S

antibiotic-resistant mutant
attachment site
auxotroph
carbon-source mutant
cistron
cloned gene
cohesive end

conjugation
cotransduction
cotransformation
counterselected marker
excisionase
F plasmid
F′ plasmid

generalized transduction
Hfr strain
immunity
insertion sequence
integrase
interrupted-mating technique
IS element

lysis

lysogen

lysogenic cycle

lytic cycle

minimal medium

nonselective medium

nutritional mutant

partial diploid

pathogen

phage attachment site

phage repressor

plaque

plasmid

prophage

prophage induction

prototroph

R plasmid

selected marker

selective medium

specialized transduction

temperate phage

time of entry

transducing phage

transduction

transformation

transposase

transposon

transposon tagging

virulent phage

EXAMPLES OF WORKED PROBLEMS

Problem 1: In an Hfr × F⁻ cross in *E. coli*, the Hfr genotype is $a^+ b^+$ *str-s* and the F^- genotype is $a^- b^-$ *str-r*. Recombinants of genotype a^+ *str-r* are selected, and almost all of them prove to be b^+. How can this result be explained?

Answer: Because most colonies that are a^+ are also b^+, it is clear that *a* and *b* are close together in the bacterial chromosome.

Problem 2: Consider the following Hfr × F⁻ cross:

Hfr genotype: $a^+ b^+ c^+$ *str-s*

F^- genotype: $a^- b^- c^-$ *str-r*

The order of gene transfer is *a b c*, with *a* transferred at 9 minutes, *b* at 11 minutes, *c* at 30 minutes, and *str-s* at 40 minutes. Recombinants are selected by plating on a medium lacking particular nutrients and containing streptomycin. Which of the following statements are true, and why? (Note: Genetic markers introduced before a selected marker, and not closely linked to it, are incorporated into the recipient genome about 50 percent of the time.)
(a) a^+ *str-r* colonies ≈ b^+ *str-r* colonies.
(b) a^+ *str-r* colonies > c^+ *str-r* colonies.
(c) b^+ *str-r* colonies < c^+ *str-r* colonies.
(d) Among colonies selected for b^+ *str-r*,
$a^+ b^+$ *str-r* ≈ $a^- b^+$ *str-r*.
(e) Among colonies selected for c^+ *str-r*,
$a^+ c^+$ *str-r* ≈ $a^- c^+$ *str-r*.
(f) Among colonies selected for $a^+ c^+$ *str-r*,
$a^+ b^+ c^+$ *str-r* < $a^+ b^- c^+$ *str-r*.

Answer: (a) True, because *a* and *b* are only two minutes apart and, if chromosome transfer has gone far enough to include *a*, it will usually also include *b*. (b) True, because *a* enters long before *c*. (c) False, because *b* enters long before *c*. (d) False, because *a* and *b* are only two minutes apart, and strains selected for b^+ will usually also contain a^+. (e) True, because *a* is transferred long before *c*, and so the probability of incorporation of a^+ is about 50 percent.

(f) False, because *a* and *b* are only two minutes apart, and strains selected for a^+ will usually also contain b^+.

Problem 3: A temperate bacteriophage has the gene order *a b c d e f g h*, whereas the order of genes in the prophage present in the bacterial chromosome is *g h a b c d e f*. What information does this give you about the location of the attachment site in the phage?

Answer: The gene order in the prophage is permuted with respect to the gene order in the phage, and the attachment site defines the site of breakage and rejoining with the bacterial chromosome. Because the adjacent genes *f* and *g* are separated by prophage integration, the attachment site must be located between *f* and *g*.

Problem 4: An Hfr *str-s* that transfers genes in alphabetical order

$$a^+ b^+ c^+ \ . \ . \ . \ x^+ y^+ z^+$$

with *a* transferred early and *z* a terminal marker, is mated with $F^- z^-$ *str-r* cells. The mating mixture is agitated violently 15 minutes after mixing to break apart conjugating cells and then plated on a medium lacking Z and containing streptomycin. The *z* gene is far from the *str* gene. The yield of z^+ *str-r* colonies is about one per 10^7 Hfr cells. What are two possible modes of origin and genotypes of such a colony? How could the two possibilities be distinguished?

Answer: The time of 15 minutes is too short to allow transfer of a terminal gene. Thus, one explanation for a z^+ *str-r* colony is that aberrant excision of the F factor occurred in an Hfr cell, yielding an F′ plasmid carrying terminal markers. When the F′ z^+ plasmid is transferred into an F^- recipient, the genotype of the resulting cell would be F′ z^+ / z^- *str-r*. Alternatively, a z^+ *str-r* colony can originate by reverse mutation of $z^- \rightarrow z^+$ in the F^- recipient. The possibilities can be distinguished because the F′ z^+ / z^- *str-r* cell is able to transfer the z^+ marker to other cells.

10-1. What is the difference between a selected marker and a counterselected marker? Why are both necessary in an Hfr × F⁻ mating?

10-2. Distinguish between generalized and specialized transduction.

10-3. In carrying out specialized transduction of a *gal⁻ bio⁻* strain of *E. coli* using bacteriophage λ from a *gal⁺ bio⁺* lysogen, what medium would you use to select for *gal⁺* transductants without selecting for *bio⁺*?

10-4. Given that bacteriophage λ has a genome size of 50 kilobase pairs, what is the approximate genetic length of λ prophage in minutes? (Hint: There are 100 minutes in the entire *E. coli* genetic map.)

10-5. How could you obtain a lysate of bacteriophage P1 in which some of the transducing particles contained bacteriophage λ?

10-6. A temperate bacteriophage has gene order *A B C att D E F*. What is the gene order in the prophage?

10-7. Why are λ specialized-transducing particles produced only by inducing a lysogen to produce phages rather than by lytic infection?

10-8. If Leu⁺ Str-r recombinants are desired from the cross Hfr *leu⁺ str-s* × F⁻ *leu⁻ str-r*, on what kind of medium should the mating pairs be plated? Which are the selected and counterselected markers?

10-9. How many plaques can be formed by a single bacteriophage particle? A bacteriophage adsorbs to a bacterium in a liquid growth medium. Before lysis, the infected cell is added to a suspension of cells that are plated on solid medium to form a lawn. How many plaques will result?

10-10. Phage T2 (a relative of T4) normally forms small, clear plaques on a lawn of *E. coli* strain B. Another strain of *E. coli*, called B/2, is unable to adsorb T2 phage particles, and no plaques are formed. T2*h* is a host-range mutant capable of adsorbing to *E. coli* B and to B/2, and it forms normal-looking plaques. If *E. coli* B and the mutant B/2 are mixed in equal proportions and used to produce a lawn, how can plaques made by T2 and T2*h* be distinguished by their appearance?

10-11. If 10⁶ phages are mixed with 10⁶ bacteria and all phages adsorb, what fraction of the bacteria remain uninfected?

10-12. An Hfr strain transfers genes in alphabetical order. When tetracycline sensitivity is used for counterselection, the number of *h⁺ tet-r* colonies is 1000-fold lower than the number of *h⁺ str-r* colonies found

when streptomycin sensitivity is used for counterselection. Suggest an explanation for the difference.

10-13. An Hfr strain transfers genes in the order *a b c*. In an Hfr *a⁺ b⁺ c⁺ str-s* × F⁻ *a⁻ b⁻ c⁻ str-r* mating, do all *b⁺ str-r* recombinants receive the *a⁺* allele? Are all *b⁺ str-r* recombinants also *a⁺*? Why or why not?

10-14. If the genes in a bacterial chromosome are in alphabetical order and an Hfr cell transfers genes in the order *g h i . . . d e f*, what types of F′ plasmids can be derived from the Hfr?

10-15. In an Hfr *lac⁺ met⁺ str-s* × *lac⁻ met⁻ str-r* mating, *met* is transferred much later than *lac*. If cells are plated on minimal medium containing glucose and streptomycin, what fraction of the cells are expected to be *lac⁺*? If cells are plated on minimal medium containing lactose, methionine, and streptomycin, what fraction of the cells are expected to be *met⁺*?

10-16. In the mating

Hfr *met⁻ his⁺ leu⁺ trp⁺* × F⁻ *met⁺ his⁻ leu⁻ trp⁻*

the *met* marker is known to be transferred very late. After a short time, the mating cells are interrupted and the cell suspensions plated on four different growth media. The amino acids in the growth media and the number of colonies observed on each are as follows:

histidine + tryptophan	250 colonies
histidine + leucine	50 colonies
leucine + tryptophan	500 colonies
histidine	10 colonies

What is the purpose of the *met⁻* mutation in the Hfr strain? What is the order of transfer of the genes? Why is the number of colonies so small for the medium containing only histidine?

10-17. Bacterial cells of genotype *pur⁻ pro⁺ his⁺* were transduced with P1 bacteriophage grown on bacteria of genotype *pur⁺ pro⁻ his⁻*. Transductants containing *pur⁺* were selected and tested for the unselected markers *pro* and *his*. The number of *pur⁺* colonies with each of four genotypes is as follows:

pro⁺ his⁺	102
pro⁻ his⁺	25
pro⁺ his⁻	160
pro⁻ his⁻	1

What is the gene order?

10-18. For generalized transduction, the theoretical relation between map distance in minutes and frequency of cotransduction is as follows:

Map distance =
$$2 - 2 \times \text{(Cotransduction frequency)}^{1/3}$$

(a) What map distance in minutes corresponds to 75 percent cotransduction? To 50 percent cotransduction? To 25 percent cotransduction?

(b) What is the cotransduction frequency between genetic markers separated by half a minute? One minute? Two minutes? Greater than two minutes?

(c) If genes a, b, and c are in alphabetical order and the cotransduction frequencies are 30 percent for $a-b$ and 10 percent for $b-c$, what is the expected cotransduction frequency between a and c?

10-19. An ampicillin-resistant (Amp-r) strain of *E. coli* isolated from nature is infected with λ and a suspension of phage progeny is obtained. This phage suspension is used to infect a culture of laboratory *E. coli* that is lysogenic for λ, sensitive to the antibiotic, and unable to undergo homologous recombination. Because the λ repressor is present in the lysogen, infecting λ molecules cannot replicate and are gradually diluted out of the culture by growth and continued division of the cells. However, rare Amp-r cells are found among the lysogens. Explain how they might have arisen.

10-20. On continued growth of a culture of a λ lysogen, rare cells become nonlysogenic by spontaneous loss of the prophage. Such loss is a result of prophage excision without subsequent development of the phage. The lysogen is said to have been *cured*. A λ phage that carries a gene for tetracycline resistance is used to lysogenize a Tet-s cell. At a later time, a cell is isolated that is Tet-r but no longer contains a λ prophage. Explain how the cured cell could remain Tet-r.

10-21. A time-of-entry experiment is carried out in the mating

$$\text{Hfr } a^+ b^+ c^+ d^+ str\text{-}s \times F^- a^- b^- c^- d^- str\text{-}r$$

in which the spacing between the genes is equal. The data in the adjoining table are obtained. What are the times of entry of each gene? Suggest one possible reason for the low frequency of d^+ str-r recombinants.

Time of mating in minutes	Number of recombinants of indicated genotype per 100 Hfr			
	a^+ str-r	b^+ str-r	c^+ str-r	d^+ str-r
0	0.01	0.006	0.008	0.0001
10	5	0.1	0.01	0.0004
15	50	3	0.1	0.001
20	100	35	2	0.001
25	105	80	20	0.1
30	110	82	43	0.2
40	105	80	40	0.3
50	105	80	40	0.4
60	105	81	42	0.4
70	103	80	41	0.4

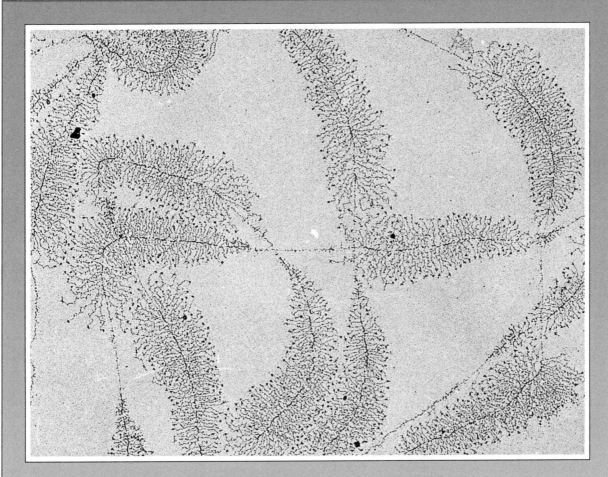

Gene Expression

Earlier chapters have been concerned with genetic analysis—genes as units of genetic information, their relation to chromosomes, and the chemical structure and replication of the genetic material. In this chapter, we shift our perspective and consider the process by which the information contained in genes is converted into molecules that determine the properties of cells and viruses—that is, the process of **gene expression**. It is accomplished by a sequence of

Above: Electron micrograph of a part of the DNA of *Triturus viridescens* containing tandem repeats of ribosomal RNA genes that are undergoing transcription. The thin strands forming the featherlike arrays are rRNA molecules. A gradient of lengths can be seen for each rRNA gene. Regions in the "feathers" that lack rRNA strands are nontranscribed spacers. (Courtesy of Oscar Miller.)

events in which the information contained in the base sequence of DNA is first copied into an RNA molecule and then used to determine the amino acid sequence of a protein molecule. The principle steps in gene expression can be summarized as follows:

1. RNA molecules are synthesized enzymatically by RNA *polymerase*, which uses the base sequence of a segment of a single strand of DNA as a template in a polymerization reaction similar to that used in replicating DNA. The overall process by which the segment corresponding to a particular gene is selected and an RNA molecule is made is called **transcription.**
2. In eukaryotes, the RNA usually undergoes chemical modification in the nucleus called **processing.**
3. Protein molecules are then synthesized by the use of the base sequence of a processed RNA molecule to direct the sequential joining of amino acids in a particular order, and so the amino acid sequence is a direct consequence of the base sequence. The production of an amino acid sequence from an RNA base sequence is called **translation,** and the protein made is called a **gene product.**

11-1   Proteins and Amino Acids

Proteins are the molecules responsible for catalyzing most intracellular chemical reactions (enzymes), for regulating gene expression (regulatory proteins), and for determining many features of the structures of cells, tissues, and viruses (structural proteins). A protein is composed of one or more chains of amino acids that are covalently joined. The chains of amino acids are called **polypeptides.** The 20 different amino acids commonly found in natural polypeptides can be in any number and any order. Because the number of amino acids in a polypeptide molecule usually ranges from 100 to 1000, the number of different protein molecules that is possible is enormous.

Each amino acid contains a carbon atom (the α carbon) to which is attached one carboxyl group ($—COOH$), one amino group ($—NH_2$), and a side chain commonly called an **R group** (Figure 11-1). The R groups are generally chains or rings of carbon atoms bearing various chemical groups. The simplest side chains are those of glycine ($—H$) and of alanine ($—CH_3$). For reference, the chemical structures of all of the 20 amino acids are shown in Figure 11-2. (For purposes of this book, it is not necessary to memorize them.)

Polypeptide chains are formed when the carboxyl group of one amino acid joins with the amino group of a second amino acid; the resulting chemical bond is an ordinary covalent bond called a **peptide bond** (Figure 11-3A). Thus, the basic unit of a protein is a polypeptide chain in which α-carbon atoms alternate with peptide units to form a backbone having an ordered array of side chains (Figure 11-3B).

The two ends of every polypeptide molecule are distinct. One end has a free $—NH_2$ group and is called the **amino terminus;** the other end has a free $—COOH$ group and is the **carboxyl terminus.** Polypeptides are synthesized by adding individual amino acids to the carboxyl end of the growing chain. Conventionally, the amino acids of a polypeptide chain are numbered starting with the amino acid at the amino end.

Most polypeptide chains are highly folded, and a variety of three-dimensional shapes have been observed. The manner of folding is determined primarily by the sequence of amino acids—in particular, by noncovalent interactions between the side chains—and so each polypeptide chain tends to

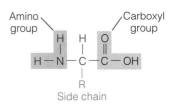

FIGURE 11-1 The general structure of an amino acid.

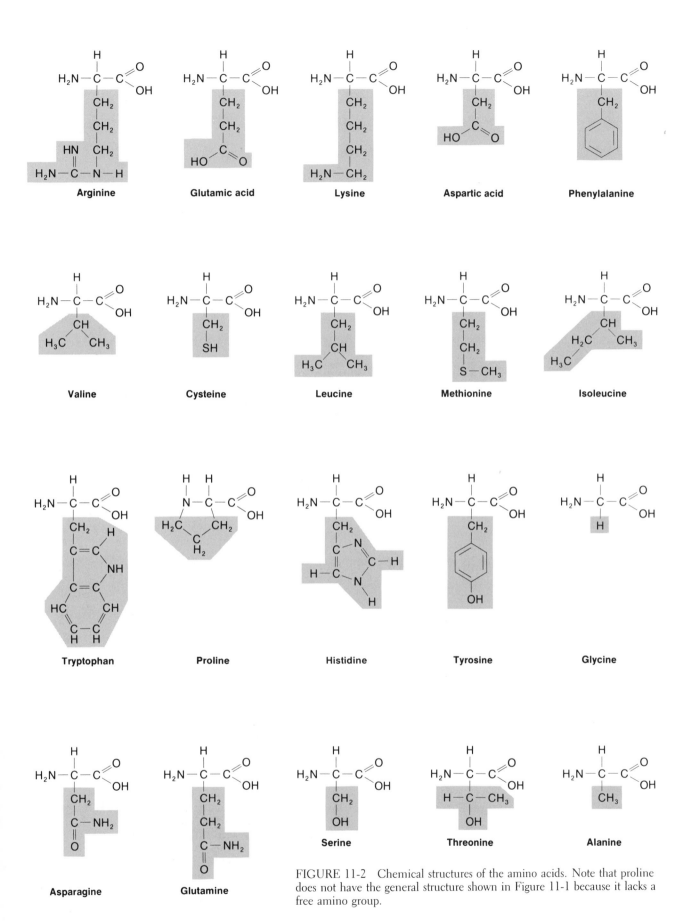

Arginine

Glutamic acid

Lysine

Aspartic acid

Phenylalanine

Valine

Cysteine

Leucine

Methionine

Isoleucine

Tryptophan

Proline

Histidine

Tyrosine

Glycine

Asparagine

Glutamine

Serine

Threonine

Alanine

FIGURE 11-2 Chemical structures of the amino acids. Note that proline does not have the general structure shown in Figure 11-1 because it lacks a free amino group.

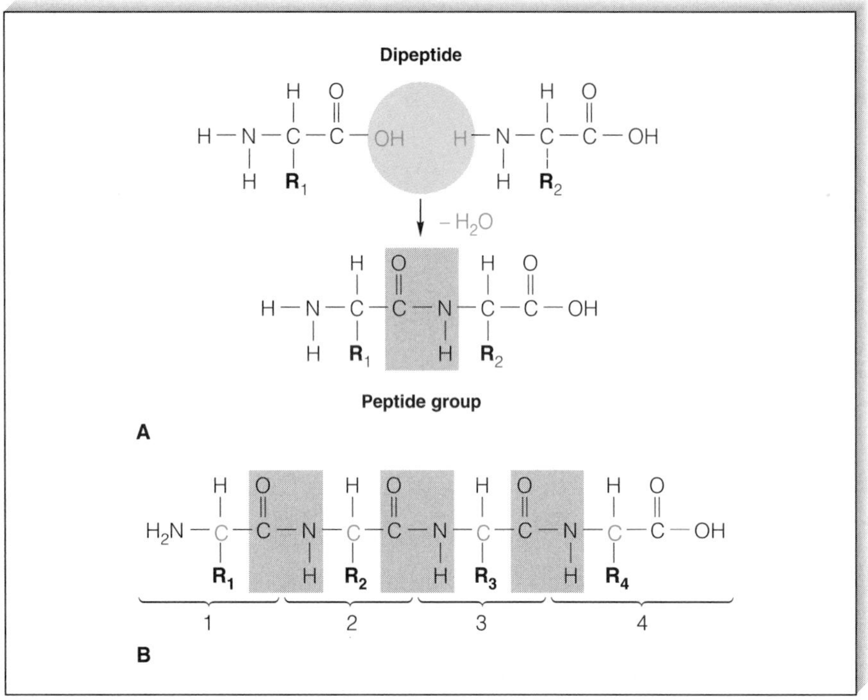

FIGURE 11-3 Properties of a polypeptide chain.
A. Formation of a dipeptide by reaction of the carboxyl group of one amino acid (left) with the amino group of a second amino acid (right). Water (pink circle) is eliminated to form a peptide group (gray rectangle).
B. A tetrapeptide showing the alternation of α-carbon atoms (red) and peptide groups (gray rectangles). The four amino acids are numbered below.

fold into a unique three-dimensional shape as it is being synthesized. In some cases, protein folding is assisted by interactions with other proteins in the cell. The rules of folding are complex and shape cannot usually be predicted from the amino acid sequence except for the simplest proteins. On the average, the molecules fold so that amino acids with charged side chains tend to be on the surface of the protein (in contact with water) and those with uncharged side chains tend to be internal. Specific folded configurations also result from hydrogen-bonding between peptide groups. Two fundamental polypeptide structures are the α helix and the β sheet, which are illustrated in Figure 11-4.

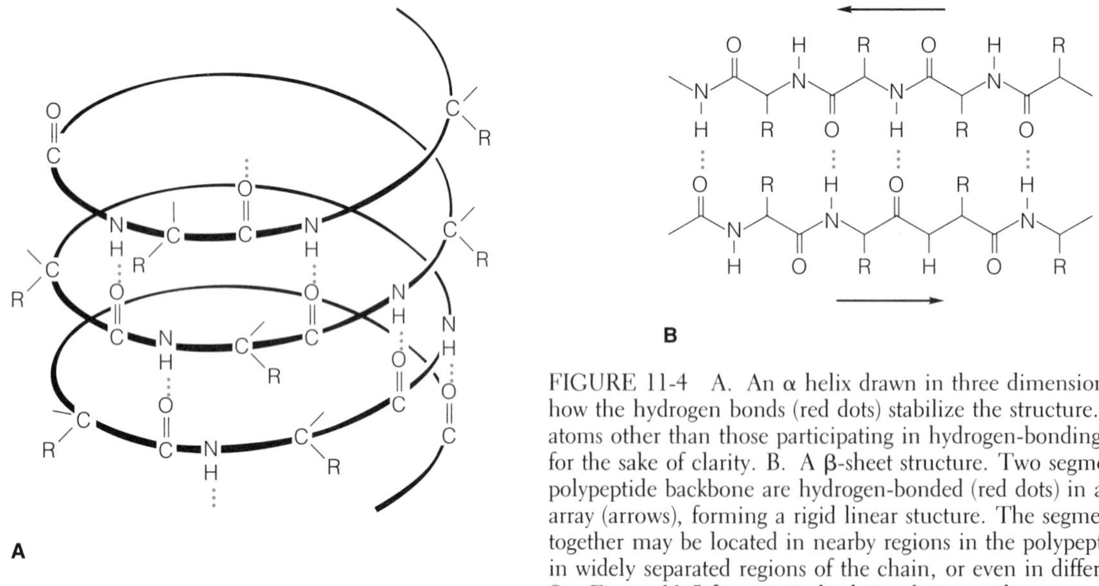

FIGURE 11-4 A. An α helix drawn in three dimensions, showing how the hydrogen bonds (red dots) stabilize the structure. Hydrogen atoms other than those participating in hydrogen-bonding are omitted for the sake of clarity. B. A β-sheet structure. Two segments of a polypeptide backbone are hydrogen-bonded (red dots) in an antiparallel array (arrows), forming a rigid linear stucture. The segments brought together may be located in nearby regions in the polypeptide chain, or in widely separated regions of the chain, or even in different chains. See Figure 11-5 for a typical relation between these segments.

FIGURE 11-5 A schematic diagram of the path of the backbone of a polypeptide, showing possible ways in which the polypeptide may be folded. Heavy black arrows represent β sheets; the red dots joining the arrows represent hydrogen bonds. Note that the interacting regions are not always located in nearby sequences within the polypeptide chain. Helical regions are drawn in red. Two heavy red bars represent disulfide bonds. The two shaded areas are hydrophobic clusters.

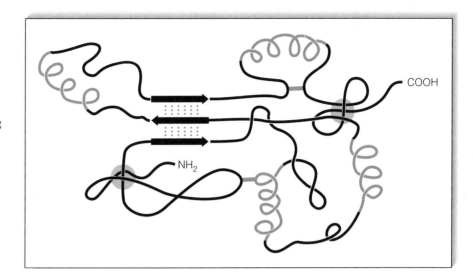

Covalent bonds may also form between the sulfur atoms of some pairs of cysteines. Figure 11-5 is a drawing of the three-dimensional structure of a hypothetical protein, showing several types of folding.

Many protein molecules consist of more than one polypeptide chain. When this is the case, the protein is said to contain **subunits.** The subunits may be identical or different. For example, hemoglobin, the oxygen carrier of blood, consists of four subunits, two each of two different types.

11-2 Relations Between Genes and Polypeptides

Most genes contain the information for the synthesis of only one polypeptide chain. Furthermore, the *sequence* of nucleotides in a gene determines the *sequence* of amino acids in a polypeptide. This point was first proved by studies of the tryptophan synthetase gene *trpA* in *E. coli*, a gene in which many mutations had been obtained and accurately mapped. The effects of numerous mutations on the amino acid sequence of the enzyme were determined by directly analyzing the amino acid sequences of the wildtype and mutant enzymes. Each mutation was found to result in a single amino acid substituting for the wildtype amino acid in the enzyme; more importantly, *the order of the mutations in the genetic map was the same as the order of the affected amino acids in the polypeptide chain* (Figure 11-6). This attribute of genes and

FIGURE 11-6 Correlation of the positions of mutations in the genetic map of the *E. coli* *trpA* gene with positions of amino acid substitutions in the TrpA protein.

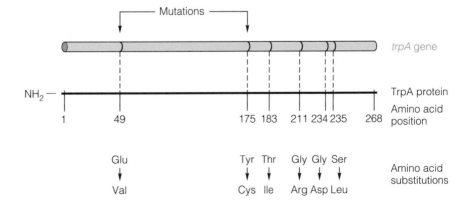

polypeptides is called **colinearity,** which means that the sequence of base pairs in DNA determines the sequence of amino acids in the polypeptide in a colinear or point-to-point manner. Colinearity is universally found in prokaryotes. However, we will see later that in eukaryotes noninformational DNA sequences interrupt the continuity of most genes; in these genes, the order but not the spacing between the mutations correlates with amino acid substitution.

11-3 Transcription

The first step in gene expression is the synthesis of an RNA molecule copied from the segment of DNA that constitutes the gene. The basic features of the production of RNA are described in this section.

General Features of RNA Synthesis

The essential chemical characteristics of the enzymatic synthesis of RNA resemble those of DNA synthesis (Chapter 4):

1. The precursors in the synthesis of RNA are the four ribonucleoside 5′-triphosphates—namely, adenosine triphosphate (ATP), guanosine triphosphate (GTP), cytidine triphosphate (CTP), and uridine triphosphate (UTP). They differ from the DNA precursors only in that the sugar is ribose rather than deoxyribose and the base uracil (U) replaces thymine (T) (Figure 11-7).

2. In the formation of RNA, a sugar-phosphate bond is formed between the 3′-hydroxyl group of one nucleotide and the 5′-triphosphate of a second nucleotide (Figure 11-8A). This is the same chemical reaction as that which occurs in the synthesis of DNA, but the enzyme is different.

3. The sequence of bases in an RNA molecule is determined by the base sequence of the DNA template. Each base added to the growing end of the RNA chain is chosen for its ability to base-pair with the DNA template strand; thus, the bases C, T, G, and A in a DNA strand cause G, A, C, and U, respectively, to be added to the growing end of an RNA molecule.

4. Nucleotides are added only to the 3′-OH end of the growing chain; as a result, the 5′ end of a growing RNA molecule bears a triphosphate group. (The 5′→3′ direction of chain growth is the same as that in DNA synthesis.)

A significant difference between DNA polymerase and RNA polymerase is that *RNA polymerase is able to initiate chain growth without a primer.*

An important feature of RNA synthesis is the following:

In any particular region of the DNA, only one strand serves as a template for RNA; that is, each RNA transcript derives from a single DNA strand.

The implications of this statement are shown in Figure 11-8B.

The synthesis of RNA can be described as consisting of four discrete stages:

1. **Promoter recognition.** RNA polymerase binds to DNA within a base sequence from 20 to 200 bases long called a **promoter.** Many promoter

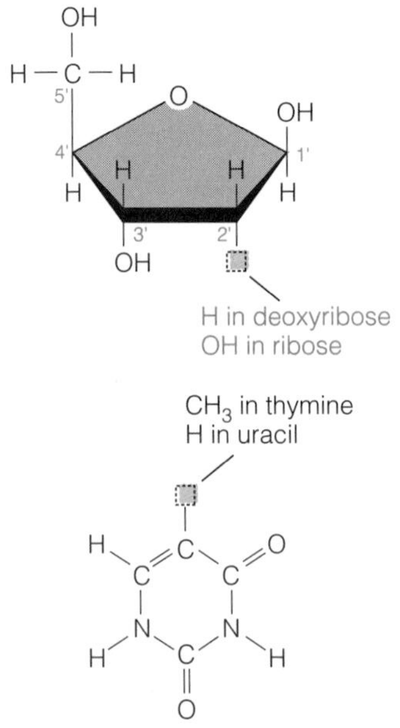

FIGURE 11-7 Differences in the structures of ribose and deoxyribose, and of uracil and thymine.

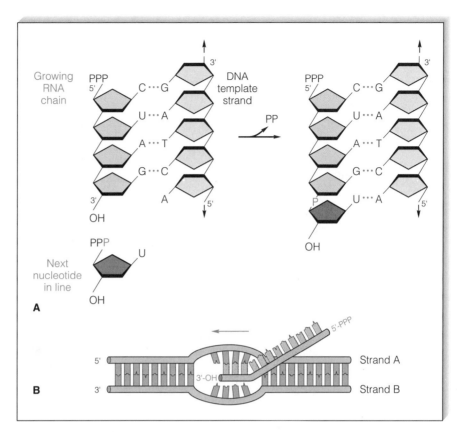

FIGURE 11-8   RNA synthesis. A. The polymerization step in RNA synthesis. The incoming nucleotide forms hydrogen bonds (red dots) with a DNA base. Reaction occurs between the OH group in the growing RNA chain and the red P in the next nucleotide in line. B. Geometry of RNA synthesis. Because RNA elongates in the $5' \rightarrow 3'$ direction, its synthesis moves along the DNA template in the $3' \rightarrow 5'$ direction; that is, the RNA molecule is antiparallel to the DNA strand being copied. The result is that RNA is copied only from one strand of a segment of a DNA molecule, in this example strand A. It is not copied from strand B in this region of the DNA. However, elsewhere—in a different gene, for example—strand B might be copied; in that case, strand A would not be copied in that region of the DNA.

sequences have been isolated and their base sequences determined. Although there is substantial sequence variation among promoter regions—in part corresponding to different strengths of the promoters in binding with the RNA polymerase—certain sequence motifs appear quite frequently. Two consensus sequences often found in promoter regions in *E. coli* are illustrated in Figure 11-9. A **consensus sequence** designates a pattern of bases determined from actual sequences by majority rule: each base in the consensus sequence is the base most often observed at that position in actual sequences. An individual sequence may resemble the consensus sequence quite closely or only very vaguely, depending on the sequence.

The consensus promoter regions in *E. coli* are TTGACA, centered approximately 35 base pairs upstream from the transcription start site ($+1$), and TATAAT, centered approximately 10 base pairs upstream from the $+1$ site. The -10 sequence, which is called the **TATA box,** is similar to sequences found at corresponding positions in

Gene	−35 Sequence	−10 Sequence	Transcription start
	TTGACA	TATAAT	+1
lac	TAGGCACCCCAGGCTTTACACTTTA	TGCTTCCGGCTCGTATGTTGTG	TGGAATTGTGAGC
lacI	GACACCATCGAATGGCGCAAAACTT	TTCGCGGTATGGCATGATAGCGCCCGGAAGAGAGT
trp	TCTGAAATGAGCTGTTGACAATTAA	TCATCGAACTAGTTAACTAGTACGCAAGTTCACGT
his	ATATAAAAAAGTTCTTGCTTTCTAA	CGTGAAAGTGGTTTAGGTTAAAAGACATCAGTTGAA
leu	GTTGACATCCGT	TTTTGTATCCAGTAACTCTAAAAGCATATCGCATT
gal	CTAATTTATTCCATGTCACACTTTTCGCATCTTTGTTATGCTATGGTTAT	TTCATACCATAAG
bio	GCCTTCTCCAAAACGTGTTTTTTGT	TGTTAATTCGGTGTAGACTTGT	AAACCTAAATCT
recA	TTTCTACAAAACACTTGATACTGTA	TGAGCATACAGTATAATTGC	TTCAACAGAACAT

FIGURE 11-9 Base sequences in promoter regions of several genes in *E. coli*. The consensus sequences located 10 and 35 nucleotides upstream from the transcription start site (+1) are indicated. Promoters vary tremendously in their ability to promote transcription. Much of the variation in promoter strength results from differences between the promoter elements and the consensus sequences at −10 and −35.

many eukaryotic promoters. The positions of the promoter sequences determine where the RNA polymerase begins synthesis, and an A or G is often the first nucleotide in the transcript.

The strength of the binding of RNA polymerase to different promoters varies greatly, which causes differences in the extent of expression from one gene to another. Most of the differences in promoter strength result from variations in the −35 and −10 promoter elements and of the spacing between them. Promoter strength among *E. coli* genes differs by a factor of 10^4, and most of the variation can be attributed to the promoter sequences themselves. In general, the more closely the promoter elements resemble the consensus sequence, the stronger the promoter. The situation is somewhat different in eukaryotes, where other types of DNA sequences (enhancers) interact with promoters to determine the level of transcription.

2. **Chain initiation.** After the initial binding step, RNA polymerase initiates RNA synthesis at a nearby transcription start site, denoted the +1 site in Figure 11-9. The first nucleoside triphosphate is placed at this site and synthesis begins in a 5′→ 3′ direction.

3. **Chain elongation.** RNA polymerase then moves along the DNA, adding nucleotides to the growing RNA chain. Only one DNA strand is transcribed, which is called the **template strand.**

4. **Chain termination.** RNA polymerase reaches a chain-termination sequence and both the newly synthesized RNA and the polymerase are released. Two kinds of termination events are known: those that are self-terminating and depend only on the DNA base sequence, and those that require the presence of a termination protein. Self-termination is the most-common case and occurs when the polymerase encounters a particular sequence of bases in the coding strand that causes the polymerase to stop. An example is shown in Figure 11-10.

Initiation of a second round of transcription need not await completion of the first, because the promoter becomes available after RNA polymerase has polymerized from 50 to 60 nucleotides. For a rapidly transcribed gene, such reinitiation occurs repeatedly, and a gene can be cloaked with numerous RNA molecules in various degrees of completion. The micrograph in Figure 11-11 shows a region of the DNA of the newt *Triturus* containing tandem repeats of a particular gene. Each gene is associated with growing RNA molecules. The

FIGURE 11-10 A. Base sequence of the transcription-termination region for the set of tryptophan-synthesizing genes in *E. coli*. The inverted repeat sequences (black arrows) are characteristic of termination sites. B. The 3′ terminus of the RNA transcript, folded to form a stem-and-loop structure. The sequence of U's found at the end of the transcript in this and many other prokaryotic genes is in red.

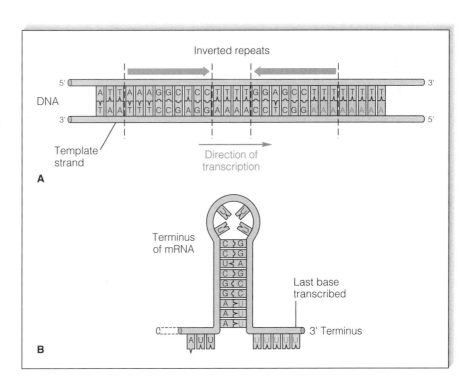

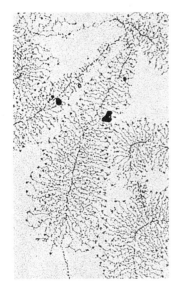

FIGURE 11-11 Electron micrograph of part of the DNA of the newt *Triturus viridescens* containing tandem repeats of ribosomal RNA genes being transcribed. The thin strands forming the featherlike arrays are RNA molecules. A gradient of lengths of these strands can be seen for each rRNA gene. Regions in the "feathers" that lack RNA strands are spacer DNA sequences, which are not transcribed. (Courtesy of Oscar Miller.)

shortest ones are at the promoter end of the gene, the longest are near the gene terminus.

The existence of promoters was first demonstrated in genetic experiments with *E. coli* by the isolation of particular Lac⁻ mutations, denoted p^-, that eliminate activity of the *lac* gene but *only when the mutations are adjacent to the gene in the same DNA molecule*. The need for the coupled genetic configuration can be seen by examining a cell with two copies of the gene *lacZ*—for example, a cell containing an F′*lacZ* plasmid, which contains *lacZ* both in the bacterial chromosome and in the F′ plasmid. Transcription of the *lacZ* gene enables the cell to synthesize the enzyme β-galactoidase. Table 11-1 shows that a wildtype *lacZ* gene (*lacZ⁺*) is inactive when it and a p^- mutation are present in the same DNA molecule (either in the chromosome or in an F′ plasmid); this can be seen by comparing entries 4 and 5. Analysis of the RNA shows that, in a cell with the genotype p^- *lacZ⁺*, the *lacZ⁺* gene is not

TABLE 11-1   Effect of promoter mutations on transcription of the *lacZ* gene

Genotype	Transcription of *lacZ⁺* gene
1. p^+ *lacZ⁺*	Yes
2. p^- *lacZ⁺*	No
3. p^+ *lacZ⁺* / p^+ *lacZ⁻*	Yes
4. p^- *lacZ⁺* / p^+ *lacZ⁻*	No
5. p^+ *lacZ⁺* / p^- *lacZ⁻*	Yes

Note: *lacZ⁺* is the wildtype gene, *lacZ⁻* is a mutant that produces a nonfunctional enzyme.

transcribed, though transcription of a mutant RNA does occur if the genotype is $p^+ lacZ^-$. The p^- mutations are called *promoter mutations*.

Mutations have also been instrumental in defining the transcription-termination region. For example, mutations have been isolated that create a new termination sequence ahead of the normal one. When such a mutation is present, an RNA molecule is made that is shorter than the wildtype RNA. Other mutations eliminate the terminator, resulting in a longer transcript.

The best-understood RNA polymerase is that of the bacterium *E. coli*. This enzyme, which consists of five protein subunits, is one of the largest enzymes known and can be easily seen by electron microscopy (Figure 11-12). In *E. coli*, all transcription is through the use of this enzyme. Eukaryotic cells have three distinct RNA polymerases, denoted I, II, and III, each of which makes a particular class of RNA molecule. RNA polymerases I and III catalyze the synthesis of RNA species needed in protein synthesis (ribosomal RNA and transfer RNA, respectively). RNA polymerase II is the enzyme responsible for synthesis of all RNA transcripts that contain information specifying amino acid sequences. These are called messenger RNA molecules, which are discussed in the next section.

Messenger RNA

Amino acids do not bind directly to DNA. Consequently, intermediate steps are needed for arranging the amino acids in a polypeptide chain in the order determined by the DNA base sequence. This process begins with transcription of the base sequence of the **coding strand** of DNA into the base sequence of an RNA molecule. In prokaryotes, this RNA molecule, which is called **messenger RNA,** or **mRNA,** is used directly in polypeptide synthesis. In eukaryotes, the RNA molecule is usually processed before it becomes mRNA. The amino acid sequence is then determined from the base sequence in mRNA by the protein-synthesizing machinery of the cell.

Not all base sequences in an mRNA molecule are translated into the amino acid sequences of polypeptides. For example, translation of an mRNA molecule rarely starts exactly at one end of the RNA molecule and proceeds to the other end; instead, initiation of polypeptide synthesis may begin many nucleotides distant the 5′ end of the RNA. The section of untranslated RNA before the region encoding the first polypeptide chain is called a **leader,** which

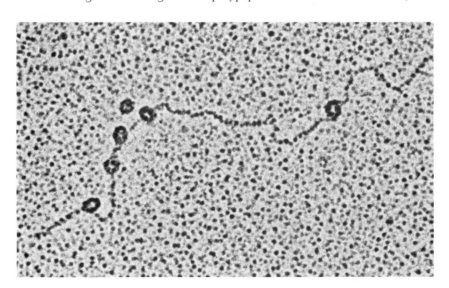

FIGURE 11-12   *E. coli* RNA polymerase molecules bound to DNA. (Courtesy of Robley Williams.)

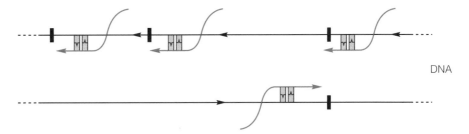

DNA

FIGURE 11-13 A typical arrangement of promoters (black arrowheads) and termination sites (black bars) in a segment of a DNA molecule. Promoters are present in both DNA strands. Termination sites are usually located so that transcribed regions do not overlap.

in some cases contains regulatory sequences that affect the rate of protein synthesis. The leader is followed by a **coding sequence,** which specifies the polypeptide chain. A typical coding sequence in an mRNA molecule is between 500 and 3000 bases long (depending on the number of amino acids in the protein). Following the coding sequence, the 3′ end of an mRNA molecule also is not translated.

The template strand of each gene corresponds to only one of the DNA strands present in the gene, but which DNA strand is the template can differ from gene to gene along a DNA molecule. That is, except for some small viruses, *not all mRNA molecules are transcribed from the same DNA strand.* Thus, in an extended segment of a DNA molecule, mRNA molecules would be seen growing in either of two directions (Figure 11-13), depending on which DNA strand functions as a template.

In prokaryotes, most mRNA molecules are degraded within a few minutes after synthesis. In eukaryotes, a typical lifetime is several hours, though some last only minutes, whereas others persist for days. In both kinds of organisms, the degradation enables cells to dispense with molecules that are no longer needed. The short lifetime of prokaryotic mRNA is an important factor in regulating gene activity (Chapter 12).

11-4 RNA Processing

The process of transcription in prokaryotes and eukaryotes is very similar, but there are major differences in the relation between the transcript and the mRNA used for polypeptide synthesis. In prokaryotes, the immediate product of transcription (the **primary transcript**) is mRNA; in contrast, *the primary transcript in eukaryotes must be converted into mRNA.* This conversion, which is called **RNA processing,** usually consists of two types of events—modification of the ends and excision of untranslated sequences embedded *within* coding sequences. These events are illustrated diagramatically in Figure 11-14.

Each end of a eukaryotic transcript is processed. The 5′ end is altered by the addition of a modified guanosine in an uncommon 5′-5′ (instead of 3′-5′) linkage; this terminal group is called a **cap.** The 3′ terminus of a eukaryotic mRNA molecule is usually modified by the addition of a polyadenosine sequence (the **poly-A tail**) as many as 200 nucleotides long. The 5′ cap is necessary for the ribosome to bind with the mRNA to begin protein synthesis, and the poly-A tail is thought to help determine mRNA stability.

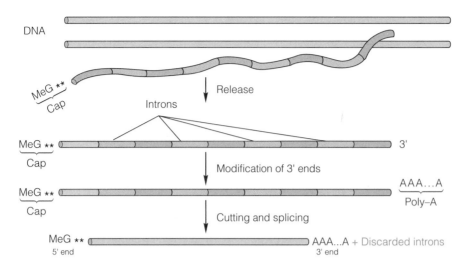

DNA

MeG **
Cap

Release

Introns

MeG ** Cap

3'

Modification of 3' ends

MeG ** Cap

AAA...A
Poly–A

Cutting and splicing

MeG ** ⬚⬚⬚⬚⬚⬚⬚⬚⬚⬚⬚⬚⬚⬚⬚⬚⬚⬚⬚ AAA...A + Discarded introns

5' end   3' end

FIGURE 11-14   A schematic drawing showing production of eukaryotic mRNA. The primary transcript is capped before it is released from the DNA. The 3' end is usually modified by the addition of consecutive adenines. Along the way, the introns are excised. MeG denotes 7-methylguanosine (a modified form of guanosine), and the two asterisks indicate two nucleotides whose riboses are methylated.

A second important feature peculiar to the primary transcript in eukaryotes, also shown in Figure 11-14, is the presence of segments of RNA, called **intervening sequences** or **introns,** which are excised from the primary transcript. Accompanying the excision is a rejoining of the coding fragments (**exons**) to form the mRNA molecule. The mechanism of excision and splicing is illustrated schematically in Figure 11-15. Part A shows the consensus sequence found at the **donor** (5') end and the **acceptor** (3') end of most introns. The symbols are: N, any nucleotide; R, any purine (A or G); Y, any pyrimidine (C or U); S, either A or C. After an initial cut in the donor site (part B), the G at the 5' end of the intron forms a loop by becoming attached to an A nucleotide located a short distance upstream from the group of consecutive pyrimidines (Y) near the acceptor site. The A–G linkage is unusual in being 2'-to-5' (instead of the usual 3'-to-5'). In the final step (part C), a cut is made in the acceptor site and the intron is freed. The remaining parts (the exons) are joined together. The excision of the introns and the joining of the exons to form the final mRNA molecule is called **RNA splicing.** The freed intron is said to be a **lariat** structure because it has a loop and a tail. The lariat is rapidly degraded into individual nucleotides by nucleases.

The existence and the positions of introns in a particular primary transcript are readily demonstrated by renaturing the transcribed DNA with the fully

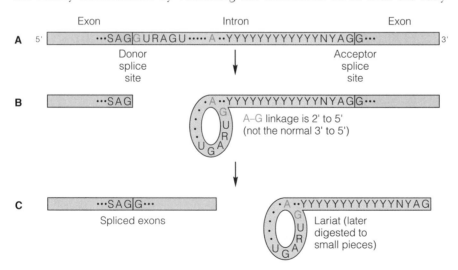

FIGURE 11-15   A schematic diagram showing removal of one intron from a primary transcript. In reality, one intron terminus interacts with an RNA-protein particle, which causes one end of the intron to be cut, forming an unusual looped structure. The cut end is later brought to the site of cleavage of the uncut end, a second cut is made, and the exon termini are joined and sealed. The loop is released as a lariat-shaped structure that is degraded. Because the loop includes most of the intron, the loop of the lariat is usually very much longer than the tail.

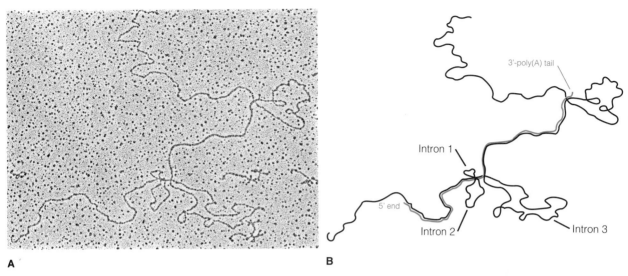

A **B**

FIGURE 11-16   A. An electron micrograph of a DNA-RNA hybrid obtained by annealing a single-stranded segment of adenovirus DNA with one of its mRNA molecules. The loops are single-stranded DNA. (Courtesy of Tom Broker and Louise Chow.) B. An interpretive drawing. RNA and DNA strands are shown in red and black, respectively. Four regions do not anneal—three single-stranded DNA segments corresponding to the introns, and the poly-A tail of the mRNA molecule.

processed mRNA molecule. The DNA-RNA hybrid can then be examined by electron microscopy. An example of adenovirus mRNA (fully processed) and the corresponding DNA are shown in Figure 11-16. The DNA copies of the introns appear as single-stranded loops in the hybrid molecule, because no corresponding RNA sequence is available for hybridization.

The number of introns per RNA molecule varies considerably from one gene to the next. For example, 2 introns are present in the primary transcript of α-globin and 52 introns are in collagen RNA. Furthermore, within a particular RNA molecule, the introns are widely distributed and have many different sizes (Figure 11-17). Most introns range in size from 100 to 10,000 base pairs and, in the processing of a typical primary transcript, the amount of discarded RNA ranges from about 50 percent to nearly 90 percent of the primary transcript.

Introns and splicing are recognized in nuclear particles known as **spliceosomes**. These abundant particles are composed of protein and several types of specialized small RNA molecules ranging in size from 100 to 200 bases. The specificity of splicing comes from the small RNAs, which contain sequences that are complementary to the splice junctions. Introns also are present in some genes in organelles. In one class of organelle introns, the intron contains a sequence coding for a protein that participates in removing the intron that codes for it. The situation is even more remarkable in the splicing of a

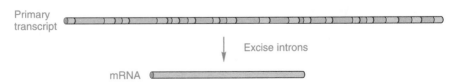

Primary transcript

Excise introns

mRNA

FIGURE 11-17   A diagram of the primary transcript and the processed mRNA of the conalbumin gene. The seventeen introns, which are excised from the primary transcript, are shown in black.

ribosomal RNA precursor in the ciliate *Tetrahymena*. In this case, the splicing reaction is intrinsic to the folding of the precursor; that is, the RNA precursor is *self-splicing* because the folded precursor RNA creates its own RNA-splicing activity. This is one of the few examples of RNA molecules having enzymelike activities of their own.

Most introns appear to have no function in themselves. When an artificial gene is created that lacks a particular intron, in most cases the gene functions normally. In those cases in which an intron seems to be required for function, it is usually not because the interruption of the gene is necessary, but because the intron happens to include certain nucleotide sequences that regulate the timing or tissue specificity of transcription. The implication is that many mutations in introns, including small deletions and insertions, should have essentially no effect on gene function, and this is indeed the case. Moreover, the nucleotide sequence in a particular intron is found to change extremely rapidly in the course of evolution (including small deletions and insertions), and this lack of sequence conservation is another indication that most of the nucleotide sequences present within introns are not important.

Mutations that affect any of the critical splicing signals do have important consequences because they interfere with the splicing reaction. Two possible outcomes are illustrated in Figure 11-18. In part A, the intron with the mutated splice site fails to be removed, and it remains in the processed mRNA. The result is the production of a mutant protein with a normal sequence of amino acids up to the splice site but an abnormal sequence afterward. Most introns are long enough that, by chance, they contain a stop sequence that terminates protein synthesis (Section 12-6), and once a stop is encountered the protein grows no further. A second kind of outcome is shown in Figure 11-18B. In this case, splicing does occur, but at an alternative splice site. (The example shows the alternative site downstream from the mutation, but alternative sites can also be upstream.) The alternative site is called a **cryptic splice site** because it is not normally used. The cryptic splice site is usually a poorer match with the consensus sequence and is ignored when the normal splice site is available. The result of using the alternative splice site is again an incorrectly processed mRNA and a mutant protein. In some splice-site mutations, both outcomes (Figure 11-18A and B) can occur: some transcripts leave the intron unspliced, whereas others are spliced at cryptic splice sites.

Although introns play no essential role in regulating gene expression, they may play a role in gene evolution. In some cases, the exons in a gene code for

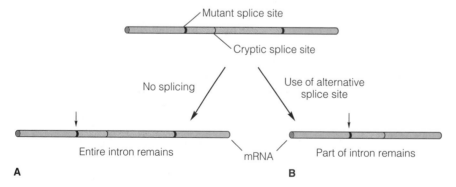

FIGURE 11-18   Possible consequences of mutation in the donor splice site of an intron: (A) no splicing occurs, and the entire intron remains in processed transcript; (B) splicing occurs at a downstream cryptic splice site, and only the upstream part of the original intron still remains in the processed transcript. Neither outcome results in a normal protein product.

segments of the completed protein that are relatively independent in their folding characteristics. For example, the central exon of the β-globin gene codes for the segment of the protein that folds around an iron-containing molecule of heme. Relatively autonomous folding units are known as folding **domains,** and the correlation between exons and domains found in some genes suggests that the genes were originally assembled from smaller pieces. In some cases, the ancestry of the exons can be traced. For example, the human gene for the low-density lipoprotein receptor participating in cholesterol regulation shares exons with certain blood-clotting factors and epidermal growth factor. The model of protein evolution through the combination of different exons is called the **exon-shuffle** model. The mechanism for combining exons from different genes is not known. Although some genes support the model, in other genes the boundaries of the folding domains do not coincide with exons.

Exon shuffling could have been important early in gene evolution because introns may be an ancient feature of gene structure. For example, some of the genes for fundamental metabolic processes have introns in the same places in plants and animals. This implies that the introns were in place before plants and animals became separate lineages. Probably introns are as old as genes themselves. The occurrence of self-splicing RNAs means that introns could have existed long before the evolution of the spliceosome mechanism. It has even been suggested that, rather than eukaryotes acquiring introns in the course of evolution, prokaryotes may have lost their introns in their evolution.

11-5 Translation

The synthesis of every protein molecule in a cell is directed by an mRNA originally copied from DNA. Protein production includes two kinds of processes: (1) information-transfer processes in which the RNA base sequence determines an amino acid sequence, and (2) chemical processes, in which the amino acids are linked together. The complete series of events is called **translation.**

The translation system consists of four major components:

1. **Ribosomes.** These components are particles on which protein synthesis is carried out; they contain the enzymes needed to form a peptide bond between amino acids, a site for binding one mRNA molecule, and two sites for bringing in and aligning the tRNAs bearing amino acids in preparation for assembly into the finished polypeptide chain. Ribosomes consist of two subunit particles. An electron micrograph and a plastic model are shown in Figure 11-19.

FIGURE 11-19 Ribosomes. A. An electron micrograph of 70S ribosomes from *E. coli.* Some of the ribosomes are oriented as in the model shown in part B. Individual ribosomal subunits are identified by the letters S and L, which stand for the small and large ribosomal subunits. B. A three-dimensional model of the *E. coli* 70S ribosome. The 30S particle is white and the 50S particle is red. (Courtesy of James Lake.)

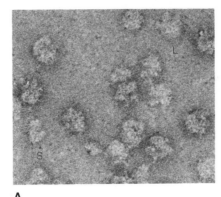

A

B

2. **Transfer RNA, or tRNA.** Amino acids do not bind directly to mRNA, but the order in which they are present in a polypeptide is determined by the base sequence in the mRNA molecule. This ordering is accomplished by a set of adaptor molecules, the tRNA molecules. A tRNA molecule reads the base sequence of mRNA according to a language called the *genetic code*, which consists of a set of relations between sequences of three adjacent bases in an mRNA molecule and particular amino acids. Each group of three adjacent bases in the coding region of the mRNA is a **codon**.

3. **Aminoacyl tRNA synthetases.** This set of enzymes catalyzes the attachment of each amino acid to its corresponding tRNA molecule. A tRNA attached to its amino acid is called a **charged** tRNA.

4. **Initiation, elongation,** and **termination factors.** Polypeptide synthesis can be divided into three stages—(1) initiation, (2) elongation, and (3) termination. Each stage requires specialized molecules. These molecules will not be discussed in this book.

In prokaryotes, all of these components are present throughout the cell; in eukaryotes, they are located in the cytoplasm.

The mechanism of protein synthesis can be depicted as in Figures 11-20 through 11-22. In overview, the process is that an mRNA molecule binds to the surface of a ribosome. Appropriate charged tRNAs bind sequentially, one by one, to the mRNA molecule that is attached to the ribosome. Peptide bonds are made between successively aligned amino acids, each time joining the amino group of the incoming amino acid to the carboxyl group of the amino acid at the growing end. Finally, the chemical bond between the last tRNA and its attached amino acid is broken and the completed protein is removed.

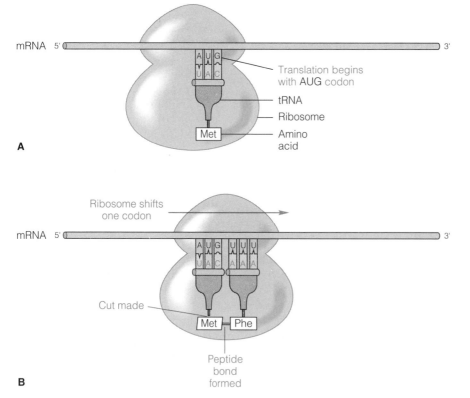

FIGURE 11-20 Initiation of protein synthesis. A. The ribosome and the mRNA first form a complex in which the tRNA-binding site is occupied by a charged tRNA^{Met}. Binding of the tRNA molecule is determined by base-pairing between three bases in the mRNA (the codon) and three bases in the tRNA (the anticodon). B. A second charged tRNA (in this example, tRNA^{Phe}) joins the complex from the right as the ribosome shifts one codon, the amino acid is cleaved from the tRNA^{Met} at the left, and a peptide bond is formed.

The main features of the initiation step in polypeptide synthesis are the binding of mRNA to the ribosome, selection of the initiation codon (**start codon, AUG**), and the binding of charged tRNA bearing the first amino acid (Figure 11-20A). Binding is accomplished by hydrogen-bonding between the codon in the mRNA and a three-base **anticodon** in the tRNA. In the elongation stage, there are two processes: joining together two amino acids by peptide-bond formation, and moving the mRNA along the ribosome so that the codons can be translated successively (Figure 11-20B). This is accomplished by successive binding of charged tRNA molecules to the ribosome. The amino acid brought in by each new tRNA is linked to the growing polypeptide by the formation of a peptide bond. The termination stage occurs when a stop codon is encountered.

The initiation reaction is of some importance in understanding many features of gene expression. The process is quite complex and will not be described in detail. However, in prokaryotes, a specific base sequence in each mRNA molecule is used to bind ribosomes. It is called the **ribosome-binding site**, and this sequence precedes the AUG initiation codon by a few bases. In prokaryotes, mRNA molecules commonly contain information for the amino acid sequences of several different polypeptide chains; in this case, such a molecule is called a **polycistronic mRNA**. (*Cistron* is a term often used to mean a base sequence encoding a single polypeptide chain.) In a polycistronic mRNA, each polypeptide coding region is preceded by its own ribosome-binding site and AUG initiation codon. After the synthesis of one polypeptide is finished, the next along the way is translated. The genes contained in a polycistronic mRNA molecule often encode the different proteins of a metabolic pathway. For example, in *E. coli*, the ten enzymes needed to synthesize histidine are coded by one polycistronic mRNA molecule. The use of polycistronic mRNA is an economical way for a cell to regulate the synthesis of related proteins in a coordinated manner. For example, in prokaryotes, the usual way to regulate the synthesis of a particular protein is to control the synthesis of the mRNA molecule that codes for it (Chapter 12). With a polycistronic mRNA molecule, the synthesis of several related proteins can be regulated by a single signal, so that appropriate quantities of each protein are made at the same time; this is termed **coordinate regulation**.

In eukaryotes, the 5′ terminus of an mRNA molecule binds to the ribosome, after which the mRNA molecule slides along the ribosome until the AUG codon nearest the 5′ terminus is in contact with the ribosome. Then protein synthesis begins. There is no mechanism for initiating polypeptide synthesis at any AUG other than the first one encountered. Eukaryotic mRNA is always monocistronic (Figure 11-21).

FIGURE 11-21 Different products are translated from a three-cistron mRNA molecule by the ribosomes of prokaryotes and eukaryotes. The prokaryotic ribosome translates all of the genes, but the eukaryotic ribosome translates only the gene nearest the 5′ terminus of the mRNA. Translated sequences are shown in black, stop codons are white, and the ribosome-binding sites and spacer sequences are gray.

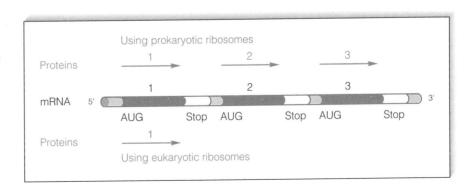

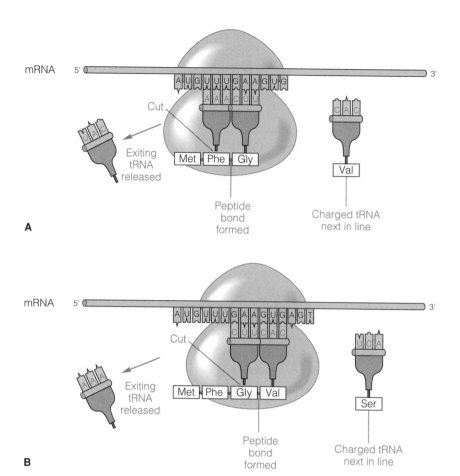

mRNA 5'   3'

Cut

Exiting
tRNA
released

Met | Phe | Gly

Val

Peptide
bond
formed

Charged tRNA
next in line

A

FIGURE 11-22   Elongation
cycle in protein synthesis.
A. In each cycle, the ribosome
shifts one codon along the
ribosome. The shifting frees a
tRNA at the left and opens up
the site for a new charged
tRNA at the right. The bond
connecting the polypeptide with
the preceding tRNA is broken
and a new peptide bond is
formed with the next amino
acid in line. B. Result of the
next cycle: the polypeptide Met-
Phe-Gly elongates to Met-Phe-
Gly-Val.

mRNA   5'   3'

Cut

Exiting
tRNA
released

Met | Phe | Gly | Val

Ser

Peptide
bond
formed

Charged tRNA
next in line

B

After polypeptide synthesis has been initiated and the elongation steps are underway, the events in Figure 11-22A take place at each step in elongation:

1. An uncharged tRNA for a preceding amino acid (in this case, methionyl-tRNA) is released from the ribosome.
2. The bond connecting the polypeptide to the tRNA for the most recently added amino acid (in this case, phenylalanine) is cut.
3. The polypeptide chain is attached to the next amino acid in line (in this case, glycine) by means of a peptide bond.
4. The ribosome shifts along the mRNA by one codon, which shifts the rightmost tRNA (carrying the polypeptide) to the position on the left.
5. A charged tRNA carrying the next amino acid in line (in this case, valine) comes into the right-hand site because of base-pairing between the anticodon in the tRNA and the codon in the mRNA.

Polypeptide elongation may be considered a cycle of events repeated again and again. For example, the next cycle in the synthesis of the polypeptide in Figure 11-22A would be as depicted in Figure 11-22B. Starting with the polypeptide Met-Phe-Gly, the result of the next cycle is the polypeptide Met-Phe-Gly-Val. In this cycle, the phenylalanyl-tRNA is released, the bond connecting the glycine molecule to the tRNA^{Gly} is broken, a new peptide bond is formed that attaches glycine to the valine molecule on the tRNA^{Val}, the ribosome shifts, and the tRNA^{Ser} comes onto the ribosome for the next cycle.

Elongation steps occur repeatedly until a stop codon for termination is reached. The **stop codons** are UAA, UAG, or UGA. No tRNA exists that can bind to a stop codon. Interruption of the successive binding of tRNA molecules starts a chain of events that results in dissociation of the completed polypeptide

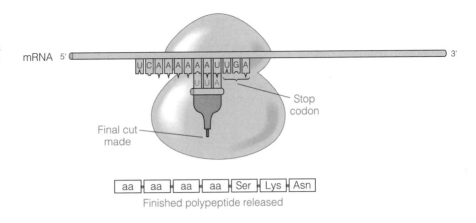

FIGURE 11-23 Termination of protein synthesis. When a stop codon is reached, no tRNA is available to bind to the site at the right, which causes release of the newly formed polypetide and the remaining bound tRNA.

mRNA 5'

Stop codon

Final cut made

| aa | aa | aa | aa | Ser | Lys | Asn |

Finished polypeptide released

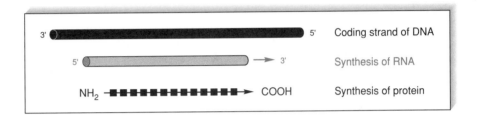

FIGURE 11-24 Direction of synthesis of RNA with respect to the coding strand of DNA, and of synthesis of protein with respect to mRNA.

3'   5'   Coding strand of DNA

5'   3'   Synthesis of RNA

NH_2 COOH   Synthesis of protein

from the ribosome (Figure 11-23), which is then available to begin another cycle of synthesis.

An important feature of translation is that it proceeds in a particular direction along the mRNA and the polypeptide:

The mRNA is translated from an initiation codon to a stop codon in the 5'-to-3' direction. The polypeptide is synthesized from the amino end toward the carboxyl end by the addition of amino acids, one by one, to the carboxyl end.

For example, a polypeptide with the sequence NH_2-Met-Pro- . . . -Gly-Ser-COOH would have started with methionine, and serine would have been the last amino acid added to the chain. The directions of synthesis are illustrated schematically in Figure 11-24.

In writing nucleotide sequences, it is conventional to place the 5' end at the left and, in writing amino acid sequences, to place the amino end at the left. Thus, polynucleotides are generally written so that both synthesis and translation proceed from left to right, and polypeptides are written so that synthesis proceeds from left to right. This convention is used in all of the following sections concerning the genetic code.

11-6 The Genetic Code

Only four bases in DNA serve to specify twenty amino acids in proteins; so some combination of the bases is needed for each amino acid. In fact, more than twenty distinct combinations are needed for polypeptide synthesis, because signals are required for starting and stopping the synthesis of each polypeptide chain. Each codon in mRNA is a sequence of three adjacent bases that specifies a particular amino acid (or chain termination). The **genetic code** is the set of all codons. Before the genetic code was elucidated, it was reasoned that, if all

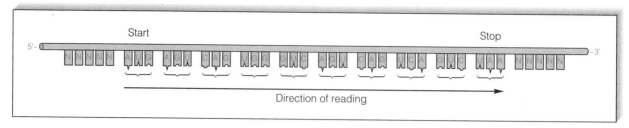

FIGURE 11-25   Bases in an RNA molecule are read sequentially in the 5' → 3' direction, in groups of three.

codons were assumed to have the same number of bases, then each codon would have to contain at least three bases. Codons consisting of pairs of bases would be insufficient because four bases can form only $4^2 = 16$ pairs; triplets of bases would suffice because these can form $4^3 = 64$ triplets. In fact, the genetic code is a **triplet code** and all 64 possible codons carry information of some sort. Most amino acids are coded by more than one codon. Furthermore, in translating mRNA molecules, the codons do not overlap but are used sequentially (Figure 11-25).

Genetic Evidence for a Triplet Code

So far, the argument that each codon must contain at least three letters has been given, but codons having more than three letters have not been ruled out. The first widely accepted proof for a triplet code came from a genetic experiment that examined the effect of inserting or deleting particular numbers of bases in a coding sequence.

When *E. coli* phage T4 is grown in a medium containing the substance proflavin, a molecule that interleaves between the DNA base pairs, errors are occasionally made in DNA replication, and daughter molecules are formed that either have an additional nucleotide pair or lack a nucleotide pair. Because bases are read sequentially, a mutation consisting of a base-pair addition or deletion upsets the phase of the units by which the code is read for a specific protein (the **reading frame**), and the sequence of amino acids downstream from the added or deleted base is different from the normal sequence (Figure 11-26). Such mutations are called **frameshift** mutations, and they are found among the progeny phages. Frameshift mutations arise at a frequency of about one mutation per gene per 10^5-to-10^6 phages. If a proflavin-induced mutant is

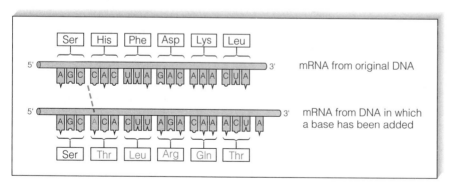

FIGURE 11-26   The change in the amino acid sequence of a protein caused by the addition of an extra base, which shifts the reading frame. A deleted base also shifts the reading frame.

grown again in a medium containing proflavin, some phages appear in which the wildtype phenotype is restored. An experiment was undertaken to determine the mutational events resulting in such restoration.

A possible mechanism for the reversion of frameshift mutations might be the deletion of the same base originally added, but such an event would seem unlikely. Mutations are produced randomly, and so removal of the *particular* added base that gave rise to the mutation should occur at a very low frequency. However, the observed frequencies of mutation and reverse mutation were comparable. Another possible explanation is that the second mutation deletes a base pair sufficiently close to the inserted base pair. This would restore the reading frame and in certain cases might result in the production of a biologically functional protein, even though one small segment of the original wildtype protein and the double mutant might have different amino acid sequences. This explanation proved to be right. Study of a collection of proflavin-induced mutants in the *rIIB* gene of *E. coli* phage T4 showed that such combinations of mutations can have this effect.

The *rIIB* protein contains a region in which numerous amino acid substitutions can be tolerated without total loss of function. However, even in this region, proflavin induces frameshift mutations that inactivate the protein. Normally, when two phage strains carrying different mutations in the same gene are crossed by mixed infection of a bacterium, some wildtype phage progeny arise by genetic recombination between the sites of the two mutations. However, when two randomly selected proflavin-induced *rII* mutants were crossed, wildtype phage progeny did not always arise. The recombination results indicated that the mutants were of two distinct classes, termed (+) and (−). Crosses between two mutants in different classes (that is, one (+) and one (−) mutant) often yielded recombinants with the wildtype phenotype, but no such recombinants were formed by crossing mutants in the same class [(+) × (+) or (−) × (−)]. These results were interpreted in the following way. The (+) mutants were considered to have one additional pair of bases and the (−) mutants to lack one pair, and a double mutant of the type (+)(+) or (−)(−) to have two additional bases or to lack two bases, respectively. [At the time, it was not known whether the (+) class was a base addition or deletion, but sequencing studies later showed it to be an addition.] Each double mutation would shift the reading frame by two bases. Such a shift does not yield a wildtype phenotype because a functional protein is not made. In a (+)(−) double mutant arising by recombination, although the shifted reading frame *following* the (+) mutation would be incorrect, the correct reading frame would be restored at the site of the (−) mutation. Between the two sites of mutation, the amino acid sequence would not be wildtype but, as long as both mutations were in a region that tolerated amino acid changes, the (+)(−) and (−)(+) phage recombinants would be functional. That is, *if the reading frame is restored before reaching a region intolerant to change, a functional protein can be produced.* By this reasoning, reversal of a (+) mutation must often result from the production of a (−) mutation at a second nearby site.

Because double mutants of the types (+)(+) or (−)(−) never have the wildtype phenotype, the genetic code cannot be a two-letter code. If it were a two-letter code, the reading frame would be restored in these combinations, and the protein would contain one additional amino acid (or one less) in the tolerant region. The triplet code was tested by the construction of triply mutant recombinants. The result was that the triple mutants (+)(+)(+) and (−)(−)(−) have the wildtype phenotype, whereas the mixed triples, (+)(+)(−) and (+)(−)(−), remain mutant. How the combination of three mutants of the same type can yield a wildtype phage is shown in Figure 11-27.

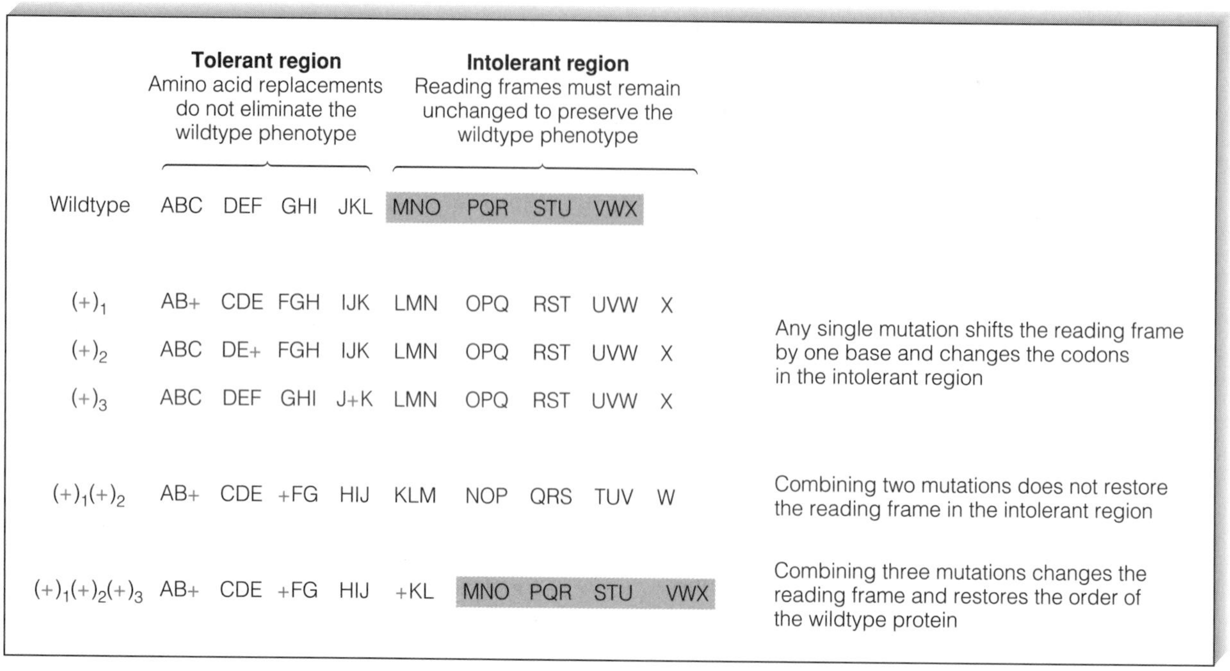

Tolerant region
Amino acid replacements do not eliminate the wildtype phenotype
Intolerant region
Reading frames must remain unchanged to preserve the wildtype phenotype
Wildtype ABC DEF GHI JKL MNO PQR STU VWX
(+)1 AB+ CDE FGH IJK LMN OPQ RST UVW X
(+)2 ABC DE+ FGH IJK LMN OPQ RST UVW X
(+)3 ABC DEF GHI J+K LMN OPQ RST UVW X
Any single mutation shifts the reading frame by one base and changes the codons in the intolerant region
(+)1(+)2 AB+ CDE +FG HIJ KLM NOP QRS TUV W
Combining two mutations does not restore the reading frame in the intolerant region
(+)1(+)2(+)3 AB+ CDE +FG HIJ +KL MNO PQR STU VWX
Combining three mutations changes the reading frame and restores the order of the wildtype protein

FIGURE 11-27 A diagram showing that, in a triplet code, it is possible to combine three single-base base additions [$(+)_1$, $(+)_2$, $(+)_3$, shown in red] to restore the reading frame in the intolerant region (pink). Two single-base insertions do not restore the reading frame. The $(+)_1(+)_2(+)_3$ triple mutants have an extra amino acid in the tolerant region, but this does not give a mutant phenotype.

Elucidation of the Base Sequences of the Codons

Polypeptide synthesis can be carried out in *E. coli* cell extracts obtained by breaking cells open. Various components can be isolated and a functioning protein-synthesizing system can be reconstituted by mixing ribosomes, tRNA molecules, mRNA molecules, and various protein factors. If radioactive amino acids are added to the extract, radioactive polypeptides are made. Synthesis continues for only a few minutes because mRNA is gradually degraded by various nucleases in the mixture. The elucidation of the genetic code began with the observation that when the degradation of mRNA was allowed to go to completion and the synthetic polynucleotide polyuridylic acid (poly-U) was added to the mixture as an mRNA molecule, a polypeptide consisting only of phenylalanine (Phe-Phe-Phe-...) was synthesized. From this simple result and knowledge that the code is a triplet code, it was concluded that UUU must be a codon for the amino acid phenylalanine. Variations on this basic experiment identified other codons. For example, when a long sequence of guanines was added at the terminus of the poly-U, the polyphenylalanine was terminated by a sequence of glycines, indicating that GGG is a glycine codon (Figure 11-28).

5'
3'
U U U U U U U U U U U U U U U U U U G G G G G G G
Phe Phe Phe Phe Phe Phe Gly Gly
Phe Phe Phe Phe Phe Leu Gly Gly
Phe Phe Phe Phe Phe Trp Gly

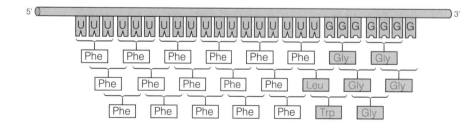

FIGURE 11-28 Polypeptide synthesis by the use of UUUU . . . UUGGGGGGG as an mRNA in three different reading frames, showing the origin of incorporation of glycine, leucine, and tryptophan.

A trace of leucine or tryptophan was also present in the glycine-terminated polyphenylalanine. Incorporation of these amino acids was directed by the codons UUG and UGG at the transition point between U and G. When a single guanine was added to the terminus of a poly-U chain, the polyphenylalanine was terminated by leucine. Thus, UUG is a leucine codon, and UGG must be a codon for tryptophan. Similar experiments were carried out with poly-A, which yielded polylysine, and poly-C, which produced polyproline.

Other experiments led to a complete elucidation of the code. Three codons

UAA UAG UGA

were found to be stop signals for translation, and one codon, AUG, which encodes methionine, was shown to be the initiation codon. AUG also codes for internal methionines.

A Summary of the Code

The *in vitro* translation experiments, which used components isolated from the bacterium *E. coli*, have been repeated with components obtained from many species of bacteria, yeast, plants, and animals. The same codon assignments can usually be made for all organisms that have been examined. Thus, the genetic code is considered to be nearly universal. The complete code is shown in Table 11-2. Note that four codons—the three stop codons and the start codon—are signals. Altogether 61 codons specify amino acids, and in many cases several

TABLE 11-2   The "universal" genetic code

First position (5' end)	Second position				Third position (3' end)
	U	C	A	G	
U	Phe	Ser	Tyr	Cys	U
	Phe	Ser	Tyr	Cys	C
	Leu	Ser	Stop	Stop	A
	Leu	Ser	Stop	Trp	G
C	Leu	Pro	His	Arg	U
	Leu	Pro	His	Arg	C
	Leu	Pro	Gln	Arg	A
	Leu	Pro	Gln	Arg	G
A	Ile	Thr	Asn	Ser	U
	Ile	Thr	Asn	Ser	C
	Ile	Thr	Lys	Arg	A
	Met	Thr	Lys	Arg	G
G	Val	Ala	Asp	Gly	U
	Val	Ala	Asp	Gly	C
	Val	Ala	Glu	Gly	A
	Val	Ala	Glu	Gly	G

Note: The codon AUG, which codes for methionine (boxed), is usually used for initiation. The codons are conventionally written with the 5' base on the left and the 3' base on the right.

codons direct the insertion of the same amino acid into a polypeptide chain. This feature means that the genetic code is redundant, or **degenerate.** All amino acids except tryptophan and methionine are specified by more than one codon. The redundancy is not random. For example, with the exception of serine, leucine, and arginine, all codons corresponding to the same amino acid are in the same box of Table 11-2; that is, *synonymous codons usually differ in only the third base.* For example, GGU, GGC, GGA, and GGG all code for glycine. Moreover, in all cases in which two codons code for the same amino acid, the third base is either A or G (both purines) or T or C (both pyrimidines).

The codon assignments shown in Table 11-2 are completely consistent with all chemical observations and with the amino acid sequences of wildtype and mutant proteins. In virtually every case in which a mutant protein differs by a single amino acid from the wildtype form, the amino acid substitution can be accounted for by a single base change between the codons corresponding to the two different amino acids. For example, substitution of serine for proline, which is a common mutational change, can be accounted for by the single base changes CCC → UCC, CCU → UCU, CCA → UCA, and CCG → UCG.

Transfer RNA and Aminoacyl Synthetase Enzymes

The decoding operation by which the base sequence within an mRNA molecule becomes translated into an amino acid sequence of a protein is accomplished by charged tRNA molecules, each of which is linked to the correct amino acid by an aminoacyl tRNA synthetase.

The tRNA molecules are small, single-stranded nucleic acids ranging in size from about 70 to 90 nucleotides. Like all RNA molecules, they have a 3'-OH terminus, but the opposite end terminates with a 5'-monophosphate rather than a 5'-triphosphate, because tRNA molecules are cut from a large primary transcript. Internal complementary base sequences form short double-stranded regions, causing the molecule to fold into a structure in which open loops are connected to one another by double-stranded stems (Figure 11-29). In two dimensions, a tRNA molecule is drawn as a planar cloverleaf. Its three-dimensional structure is more complex, as is shown in Figure 11-30 in which part A shows a skeletal model of a yeast tRNA molecule that carries phenylalanine and part B is an interpretive drawing.

Three regions of each tRNA molecule are used in the decoding operation. One of these regions is the anticodon sequence consisting of three bases that can form base pairs with a codon sequence in the mRNA. No normal tRNA molecule has an anticodon complementary to any of the stop codons UAG, UAA, or UGA, which is why these codons are stop signals. A second critical site is at the 3' terminus of the tRNA molecule, where the amino acid attaches. A specific aminoacyl tRNA synthetase matches the amino acid with the anticodon; to do so, the enzyme must be able to distinguish one tRNA molecule from another. The necessary distinction is made by recognition regions that encompass many parts of the tRNA molecule.

The different tRNA molecules and synthetases are designated by stating the name of the amino acid that is linked to a particular tRNA molecule by a specific synthetase; for example, leucyl-tRNA synthetase attaches leucine to tRNA[Leu]. When an amino acid has become attached to a tRNA molecule, the tRNA is said to be charged. An uncharged tRNA lacks an amino acid. At least one, and usually only one, aminoacyl synthetase exists for each amino acid.

FIGURE 11-29 A tRNA cloverleaf configuration with its bases numbered. A few bases present in almost all tRNA molecules are indicated. The names of regions found in all tRNA molecules are in red. DHU refers to a base, dihydrouracil, found in one loop; the Greek letter ψ is a symbol for the unusual base pseudouridine.

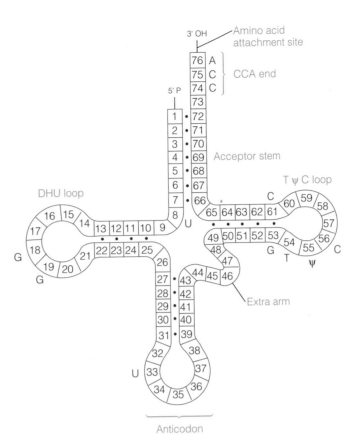

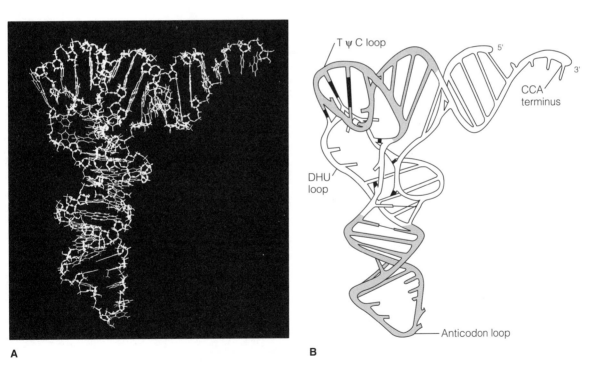

FIGURE 11-30 Yeast phenylalanine tRNA (called tRNA^Phe): (A) a skeletal model; (B) a schematic diagram of the three-dimensional structure. (Courtesy of Sung-Hou Kim.)

Redundancy and Wobble

Several features of the genetic code and of the decoding system suggest that something is missing in the explanation of codon-anticodon binding. First, the code is highly redundant. Second, the identity of the third base of a codon is often unimportant. In some cases, any nucleotide will do, in others any purine or any pyrimidine. Third, the number of distinct tRNA molecules that have been isolated from a single organism is less than the number of codons; because all codons are used, the anticodons of some tRNA molecules must be able to pair with more than one codon. Experiments with several purified tRNA molecules showed this to be the case.

Much of the redundancy in the genetic code exists because the requirement for base-pairing at the third position of the codon is less stringent than at the first two positions. This is called **wobble**; that is, the first two bases must form pairs of the usual type (A with U or G with C), but the third base pair can be of a different type (for example, G with U). This observation was derived from the discovery that the anticodon of yeast tRNA^{Ala} contains the base **inosine, I,** in the position that pairs with the third base of the codon. Later analyses of other tRNA molecules showed that inosine was common in this position, though not always present. Inosine can form hydrogen bonds with A, U, and C. In the wobble hypothesis, all pairs of bases that can form hydrogen bonds are considered to be possible in the third position of the codon, except purine–purine base pairs, which would cause excessive distortion in the region of the pairing. The possible base pairs are listed in Table 11-3. Wobble, which has been confirmed by direct sequencing of many tRNA molecules, explains the pattern of redundancy in the code in that certain anticodons (for example, those containing U, I, and G in the first position of the anticodon) can pair with several codons.

TABLE 11-3	Allowed pairings due to wobble
First base in anticodon (5' position)	Allowed bases in third codon position (3' position)
A	U
C	G
U	A or G
G	C or U
I	A or C or U

The Sequence Organization of a Typical Prokaryotic mRNA Molecule

Most prokaryotic mRNA molecules are polycistronic; that is, they contain sequences specifying the synthesis of several proteins. Thus, a polycistronic mRNA molecule must possess a series of start and stop codons for use in translation. If an mRNA molecule encodes three proteins, the minimal coding requirement would be the sequence

AUG (start)/protein 1/stop–AUG/protein 2/stop–AUG/protein 3/stop

The stop codons might be UAA, UAG, or UGA. Actually, such an mRNA molecule is probably never so simple in that the leader sequence preceding the first start signal may be several hundred bases long and spacer sequences containing ribosome-binding sites are usually present between one stop codon and the next start codon.

11-7 Overlapping Genes

The idea that several reading frames might exist within a single coding segment of DNA was not considered for many years. The reason is that a mutation in a gene that overlaps another gene would often produce defects in both gene products, but double mutations are very rare. Furthermore, the existence of overlapping reading frames was thought to place severe constraints on the amino acid sequences of two proteins translated from the same part of an mRNA molecule. However, because the code is highly redundant, the constraints are actually not so rigid.

If multiple reading frames were used, a single DNA segment would be utilized with maximal efficiency. However, a disadvantage is that evolution might be more difficult because random mutations would rarely improve the function of both proteins. Nevertheless, some cases of **overlapping genes** have been found. Most examples are in transposable elements (Figure 10-20) or in small viruses in which there is a premium on packing the largest amount of genetic information into a small DNA molecule. Some of the best examples of overlapping genes are in the *E. coli* phage φX174.

Phage φX174 contains a single strand of DNA consisting of 5386 nucleotides of known base sequence. If a single reading frame were used, at most 1795 amino acids could be encoded in the sequence and, with an average protein size of about 400 amino acids, only 4 or 5 proteins could be made. However, φX174 makes 11 proteins containing a total of more than 2300 amino acids. This paradox was resolved when it was shown that translation occurs in several reading frames from three mRNA molecules (Figure 11-31). For example, the sequence for protein B is contained totally in the sequence for

FIGURE 11-31 Physical map of *E. coli* phage φX174 showing the start and stop points for mRNA synthesis and the boundaries of the individual protein products (identified by uppercase letters). The solid regions are spacers.

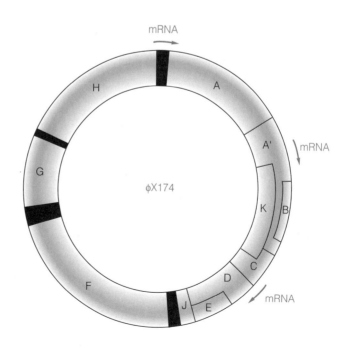

protein A but translated in a different reading frame. Similarly, the protein-E sequence is totally within the sequence for protein D. Protein K is initiated near the end of gene A, includes the base sequence of gene B, and terminates in gene C; synthesis is not in phase with either gene A or gene C. Of note is protein A', which is formed by initiating translation within the mRNA for protein A and using the same reading frame, so that it terminates at the same stop codon as that for protein A. Thus, the amino acid sequence of A' is identical with a segment of protein A. In total, five different proteins obtain some or all of their primary structure from shared base sequences in φX174.

11-8 Complex Translation Units

In most prokaryotes and eukaryotes, the unit of translation is almost never simply one ribosome traversing an mRNA molecule but is a more-complex structure, of which there are several forms. Two examples are given in this section.

After about 25 amino acids have been joined together in a polypeptide chain, an AUG initiation codon is completely free of the ribosome and a second initiation complex can form. The overall configuration is that of two ribosomes moving along the mRNA at the same speed. When the second ribosome has moved along a distance similar to that traversed by the first, a third ribosome can attach to the initiation site. The process of movement and reinitiation continues until the mRNA is covered with ribosomes at a density of about one ribosome per 80 nucleotides. This large translation unit is called a **polysome**, and this is the usual form of the translation unit. An electron micrograph of a polysome is shown in Figure 11-32.

An mRNA molecule being synthesized has a free 5′ terminus. Because translation is in the 5′→3′ direction (Figure 11-24), the mRNA is synthesized in a direction appropriate for immediate translation. That is, the ribosome-binding site (in prokaryotes) and the 5′ terminus (in eukaryotes) is transcribed first, followed in order by the initiating AUG codon, the region encoding the amino acid sequence, and finally the stop codon. Thus, in prokaryotes, in which no nuclear membrane separates the DNA and the ribosome, the initiation complex can form before the mRNA is released from the DNA. This allows the simultaneous occurrence, or **coupling**, of transcription and translation. Figure 11-33 shows an electron micrograph of a DNA molecule with a number of attached mRNA molecules, each associated with ribosomes (part A), and an interpretation (part B). Transcription of DNA begins in the upper-left part of the micrograph. The lengths of the polysomes increase with distance from the transcription initiation site because the mRNA is farther from that site and hence longer. *Coupled transcription and translation does not occur in eukaryotes*, because the mRNA is synthesized and processed in the nucleus and later transported through the nuclear envelope to the cytoplasm, where the ribosomes are located.

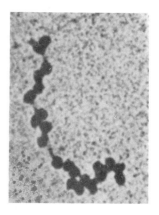

FIGURE 11-32 Electron micrograph of *E. coli* polysomes. (Courtesy of Barbara Hamkalo.)

11-9 The Overall Process of Gene Expression

In this chapter, various features of the process of gene expression have been described. Parts of it are quite complex, and indeed even more complex than described because much detail has been eliminated. Nonetheless, the basic mechanism is a simple one:

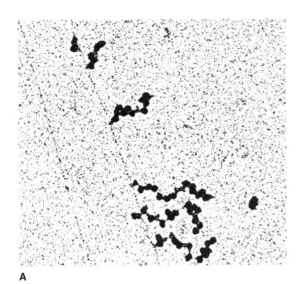

A

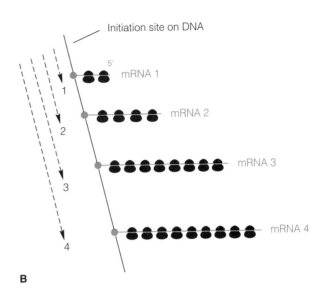

B

FIGURE 11-33 Visualization of transcription and translation. A. Transcription of a section of the DNA of *E. coli* and translation of the nascent mRNA. Only part of the chromosome is being transcribed. The dark spots are ribosomes, which coat the mRNA. (From O. L. Miller, B. A. Hamkalo, and C. A. Thomas. 1977. *Science*, 169, 392. Copyright 1977 by the American Association for the Advancement of Science.) B. An interpretation of the electron micrograph in part A. The mRNA is red and is coated with black ribosomes. The large red spots are the RNA polymerase molecules; they are too small to be seen in the photograph. The dashed arrows show the distances of each RNA polymerase from the transcription-initiation site. Arrows 1, 2, and 3 have the same length as mRNA 1, 2, and 3; mRNA 4 is shorter than arrow 4 because its 5′ end has been partially digested by nuclease.

A base sequence in a DNA molecule is converted into a complementary base sequence in an intermediate (mRNA), and then the base sequence in the mRNA is converted into an amino acid sequence of a polypeptide chain.

Both of these steps, which have a multitude of substeps, utilize the simplest of principles: (1) the rules of base-pairing provide the base sequence of the mRNA, and (2) a two-ended molecule (tRNA), with an amino acid attached at one end and the ability to base-pair with RNA bases at the other, converts each set of three bases into one amino acid. Various recognition regions are needed to ensure that the correct base sequence is read and that the correct amino acid is put in the appropriate position in the protein. As is always the case when the information in nucleic acid molecules is used, base sequences provide the information for the first process. That is, specific sequences in the DNA are recognized as the beginning (promoter) and end (transcription-termination site) of a gene, and these sequences are recognized by an enzyme (RNA polymerase) that makes the copy of the gene that is used by the protein-synthesizing machinery. To ensure that the correct amino acid sequence is assembled, one codon (AUG) is used to tell the system where to start reading, and a stop codon defines the end of the polypeptide chain. Particular recognition sites in tRNA enable the aminoacyl tRNA synthetase enzymes to connect amino acids to the tRNA molecules with the correct anticodons.

An essential feature of the entire process of gene expression is that both DNA and RNA are scanned by molecules that move in a single direction. That is, RNA polymerase moves along the DNA as it polymerizes nucleotides, and the ribosome and the mRNA move with respect to one another as different amino acids are brought in for covalent linking.

The flow of information from a gene to its product is from DNA to RNA to protein. The properties of the different protein products of genes are determined by the sequence of amino acids of the polypeptide chain and by the way in which the chain is folded. Each gene is usually responsible for the synthesis of a single polypeptide.

Gene expression begins by the enzymatic synthesis of an RNA molecule that is a copy of one strand of the DNA segment corresponding to the gene. This process is called transcription and is carried out by the enzyme RNA polymerase. This enzyme joins ribonucleoside triphosphates by the same chemical reaction used in DNA synthesis. RNA polymerase differs from DNA polymerase in that a primer is not needed to initiate synthesis. Transcription is initiated when RNA polymerase binds to a promoter sequence. Each promoter consists of several subregions, of which two are the polymerase-binding site and the polymerization start site. Polymerization continues until a termination site is reached. The product of transcription is an RNA molecule. In prokaryotes, this molecule is used directly as messenger RNA (mRNA) in polypeptide synthesis. In eukaryotes, the RNA is processed: noncoding sequences called introns are removed, the exons are spliced together, and the termini are modified by the formation of a 5′ cap and the addition of a poly-A tail at the 3′ end.

After mRNA is formed, polypeptide chains are synthesized by translation of the mRNA molecule. Translation is the successive reading of the base sequence of an mRNA molecule in groups of three bases called codons. There are 64 codons; 61 correspond to the 20 amino acids, of which one (AUG) is a start codon. The remaining three codons (UAA, UAG, UGA) are stop codons. The code is highly redundant, in that many amino acids correspond to several codons. The codons in mRNA are recognized by tRNA molecules, which contain a three-base sequence complementary to a codon and called an anticodon. When used in polypeptide synthesis, each tRNA molecule possesses a terminally bound amino acid (charged tRNA). The correct amino acid is attached to each tRNA species by specific enzymes called aminoacyl tRNA synthetases. Polypeptides are synthesized on particles called ribosomes. Synthesis begins by binding of a ribosome and an mRNA molecule. Then, charged tRNA molecules are successively hydrogen-bonded to the mRNA by a codon-anticodon interaction. The amino acids are linked together. Successive binding of charged tRNA molecules and joining of the bound amino acid to the growing polypeptide chain results in synthesis of a polypeptide chain. This process continues until a stop codon in the mRNA is reached; at this point, the polypeptide is released. Several ribosomes can translate an mRNA molecule simultaneously, forming a polysome. In prokaryotes, translation often begins before synthesis of mRNA is completed; in eukaryotes, this does not occur because mRNA is made in the nucleus, whereas the ribosomes are located in the cytoplasm. Prokaryotic mRNA molecules are often polycistronic, encoding several different polypeptides. Translation proceeds sequentially along the mRNA molecule from the start codon nearest the ribosome-binding site, terminating at stop codons and reinitiating at the next start codon. This is not possible with eukaryotes, because only the AUG site nearest the 5′ terminus of the mRNA can be used to initiate polypeptide synthesis; thus, eukaryotic mRNA is monocistronic.

KEY TERMS

amino terminus
aminoacyl tRNA synthetase
anticodon
AUG
cap
carboxyl terminus
chain elongation
chain initiation
chain termination
charged tRNA

coding sequence
coding strand
codon
colinearity
consensus sequence
coordinate regulation
coupled transcription-translation
cryptic splice site
degenerate code
exon

exon shuffle
folding domain
frameshift mutation
gene expression
gene product
genetic code
inosine (I)
intervening sequence
intron
lariat structure

leader
messenger RNA
mRNA
overlapping genes
peptide bond
poly-A tail
polycistronic mRNA
polypeptide chain
polysome
primary transcript
promoter

protein subunit
R group
reading frame
ribosome
ribosome-binding site
RNA processing
RNA splicing
splice acceptor
splice donor
spliceosome
start codon

stop codon
TATA box
template strand
transcription
transfer RNA
translation
triplet code
tRNA
uncharged tRNA
wobble

EXAMPLES OF WORKED PROBLEMS

Problem 1: The following is the nucleotide sequence of a strand of DNA:

TACGTCTCCAGCGGAGATCTTT-
TCCGGTCGCAACTGAGGTTGATC

The strand is transcribed from left to right and codes for a small peptide.
(a) Which end is the 3′ and which the 5′ end?
(b) What is the sequence of the complementary DNA strand?
(c) What is the sequence of the transcript?
(d) What is the amino acid sequence of the peptide?
(e) In which direction along the transcript does translation occur?
(f) Which is the amino (–NH₂) end of the peptide? The carboxyl (–COOH) end?

Answer: (a) Because the DNA strand is transcribed from left to right, and the first nucleotides incorporated into an RNA transcript form its 5′ end, the 3′ end of the DNA strand must be at the left. The original strand (**a**), the complementary strand (**b**), the transcript (**c**), and the peptide (**d**) line up as follows:
(a) 3′-TACGTCTCCAGCGGAGACCTTTTCCGGTC-
GCAACTGAGGTTGATC-5′
(b) 5′-ATGCAGAGGTCGCCTCTGGAAAAGGCCAG-
CGTTGACTCCAACTAG-3′
(c) 5′-AUGCAGAGGUCGCCUCUGGAAAAGGCCA-
GCGUUGACUCCAACUAG-3′
(d) Met-Gln-Arg-Ser-Pro-Leu-Glu-Lys-Ala-Ser-Val-Asp-Ser-Asn
(e) Translation is from 5′ → 3′ along the mRNA, and so, in this example, translation also is from left to right.
(f) The amino end of the peptide is synthesized first (Met) and the carboxyl end last (Asn).

Problem 2: Using the sequence given in Problem 1, what effects would the following mutations have on the peptide produced? Each of the mutations affects the TTTT in the middle of the strand.
(a) TTTT → TCTT (substitution of C for T)
(b) TTTT → TATT (substitution of A for T)

(c) TTTT → TTTTT (one-base insertion)
(d) TTTT → TTT (one-base deletion)

Answer: In problems of this type, it is best to derive the sequence of the mutant mRNA and then the amino acid sequence of the peptide.
(a) DNA 3′-TACGTCTCCAGCGGAGACCTCTTCC-
GGTCGCAACTGAGGTTGATC-5′

mRNA 5′-AUGCAGAGGUCGCCUCUGGAGA-
AGGCCAGCGUUGACUCCAACUAG-3′

Peptide Met-Gln-Arg-Ser-Pro-Leu-Glu-Lys-Ala-Ser-Val-Asp-Ser-Asn
The mutation changes a GAA codon into a GAG codon, which still codes for glutamic acid, and so no change in the peptide results.
(b) DNA 3′-TACGTCTCCAGCGGAGACCTATTCC-
GGTCGCAACTGAGGTTGATC-5′

mRNA 5′-AUGCAGAGGUCGCCUCUGGAUAA-
GGCCAGCGUUGACUCCAACUAG-3′

Peptide Met-Gln-Arg-Ser-Pro-Leu-Asp-Lys-Ala-Ser-Val-Asp-Ser-Asn
The mutation changes a GAA codon into a GAU codon, which changes glutamic acid into aspartic acid in the peptide.
(c) DNA 3′-TACGTCTCCAGCGGAGACCTTTTTC-
CGGTCGCAACTGAGGTTGATC-5′

mRNA 5′-AUGCAGAGGUCGCCUCUGGAAAAA-
GGCCAGCGUUGACUCCAACUAG-3′

Peptide Met-Gln-Arg-Ser-Pro-Leu-Glu-Lys-Gly-Gln-Arg
The insertion introduces a new nucleotide into the mRNA and shifts the reading frame downstream from the site of the mutation. In this case, the normal sequence starting with Lys becomes replaced with Lys-Gly-Gln-Arg, and the UGA stop codon following the Arg results in premature termination of the peptide.
(d) DNA 3′-TACGTCTCCAGCGGAGACCTTTCCG-
GTCGCAACTGAGGTTGATC-5′

mRNA 5′-AUGCAGAGGUCGCCUCUGGAAAG-
GCCAGCGUUGACUCCAACUAG-3′

Peptide   Met-Gln-Arg-Ser-Pro-Leu-Glu-Arg-Pro-Ala-Leu-Thr-Pro-Thr

The deletion again shifts the reading frame downstream from the site of the mutation. In this case, translation continues because a stop codon is not reached immediately, but the entire amino acid sequence downstream from the mutation is altered.

Problem 3: What anticodon sequence would pair with the codon 5'-AUG-3', assuming that no modified bases are present in the anticodon?

Answer: In the absence of modified bases, the base-pairing is the usual A with U and G with C. However, the orientations of the codon and anticodon are antiparallel, and so the anticodon sequence is 3'-UAC-5'.

PROBLEMS

11-1. Which DNA strand serves as the template for RNA polymerase? Is it transcribed in the 5' → 3' or 3' → 5' direction? Which end of an mRNA molecule is translated first? Which end of a polypeptide is synthesized first?

11-2. What are the translation initiation and stop codons in the genetic code? In a random sequence of the four ribonucleotides, what is the probability that any three adjacent nucleotides will be a start codon? A stop codon? In an mRNA of random sequence, what is the average distance between stop codons?

11-3. How do prokaryotes and eukaryotes differ in the mechanism for selecting an AUG site as a start codon?

11-4. Which of the following is the usual cause of chain termination: (1) mRNA synthesis stops at a chain termination codon; (2) the tRNA corresponding to a chain-termination codon cannot be charged with an amino acid; (3) there are no tRNA molecules that have anticodons corresponding to chain-termination codons?

11-5. A part of the coding strand of a DNA molecule that codes for the 5' end of an mRNA has the sequence 3'-TTTTACGGGAATTAGAGTCGCAGGATG-5'. What is the amino acid sequence of the polypeptide encoded by this region, assuming that the normal start codon is needed for initiation of polypeptide synthesis?

11-6. Poly(U) codes for polyphenylalanine. If a G is added to the 5' end of the molecule, the polyphenylalanine has a different amino acid at the amino terminus, and if a G is added to the 3' end, there is a different amino acid at the carboxyl terminus. What are the amino acids?

11-7. The synthetic polymer poly-A is used as an mRNA molecule in an *in vitro* protein-synthesizing system that does not need a special start codon. Polylysine is synthesized. A single guanine nucleotide is added to one end of the poly-A. The resulting polylysine has a glutamic acid at the amino terminus. Was the G added to the 3' or the 5' end of the poly-A?

11-8. What polypeptide products are made when the alternating polymer GUGU . . . is used in an *in vitro* protein-synthesizing system that does not need a start codon?

11-9. What polypeptide products are made when the alternating polymer GUCGUC . . . is used in an *in vitro* protein-synthesizing system that does not need a start codon?

11-10. Some codons in the genetic code were determined by the translation of random polymers. If a ribonucleotide polymer is synthesized that contains 3/4 A and 1/4 C in random order, which amino acids would the resulting polypeptide contain, and in what frequencies?

11-11. How many different sequences of nine ribonucleotides would code for the amino acids Met-His-Thr? For Met-Arg-Thr? Using the symbols Y for any pyrimidine, R for any purine, and N for any nucleotide, state what the sequences are.

11-12. At one time, it was considered that the genetic code might be one in which the codons overlapped. For example, with a one-base overlap, the codons in the mRNA sequence CAUCAU would be translated as CAU AUC UCA CAU rather than as CAU CAU. How is this hypothesis affected by the observation that mutant proteins usually differ from the wild-type protein by a single amino acid?

11-13. What codons could pair with the anticodon 5'-IAU-3'? (I stands for inosine.) What amino acid would be incorporated?

11-14. Two possible anticodons could pair with the codon UGG but only one is actually used. Identify the possible anticodons and explain why one of them is not used.

11-15. Two *E. coli* genes, A and B, are known from mapping experiments to be very close to each other. A deletion mutation is isolated that eliminates the activity of both A and B. Neither the A nor the B protein can be found in the mutant, but a novel protein is isolated in which the amino-terminal 30 amino acids are identical with those of the B gene product and the carboxyl-terminal 30 amino acids are identical with those of the A gene product.
(a) With regard to the 5' → 3' orientation of the noncoding DNA strand, is the order of the

genes *A B* or *B A*?

(b) Can you make any inference about the number of bases deleted?

11-16. The noncoding sequence at the beginning of a gene reads 5'-ATGCATCCGGGCTCATTAGTCT . . . -3'. Two mutations are studied. Mutation X has an insertion of a G immediately after the underlined G, and mutation Y has a deletion of the red A. What is the amino sequence of
(a) the wildtype polypeptide?
(b) mutant X?
(c) mutant Y?
(d) a recombinant molecule containing both mutations?

11-17. The amino terminus of a wildtype enzyme in yeast has the amino acid sequence Met-Leu-His-Tyr-Met-Gly-Asp-Tyr-Pro. A mutant, X, is found that contains an inactive enzyme with the sequence Met-Gly-Asp-Tyr-Pro at the amino terminus and the wildtype sequence at the carboxyl terminus. A second mutant, Y, also lacks enzyme activity, but there is no trace of a full-length protein. Instead, mutant Y makes a short peptide containing just three amino acids. What single-base changes can account for the features of mutation X and mutation Y? What is the sequence of the tripeptide produced by mutant Y?

11-18. Protein synthesis occurs with high fidelity. In prokaryotes, incorrect amino acids are inserted at the rate of approximately 10^{-3} (that is, one incorrect amino acid per 1000 translated). What is the probability that a polypeptide of 300 amino acids has exactly the amino acid sequence specified in the mRNA?

11-19. A DNA fragment containing a particular gene is isolated from a eukaryotic organism. This DNA fragment is mixed with the corresponding mRNA isolated from the organism, denatured, renatured, and observed by electron microscopy. Heteroduplexes of the type shown in the figure are observed. How many introns does this gene contain?

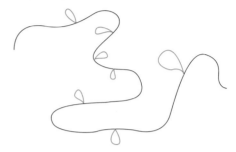

11-20. If the DNA molecule shown here is transcribed from left to right, what is the sequence of the mRNA and the amino acid sequence? What is the sequence of the mRNA and the amino acid sequence if the segment in red is inverted?

5'-AGACTTCAGGCTCAACGTGGT-3'
3'-TCTGAAGTCCGAGTTGCACCA-5'

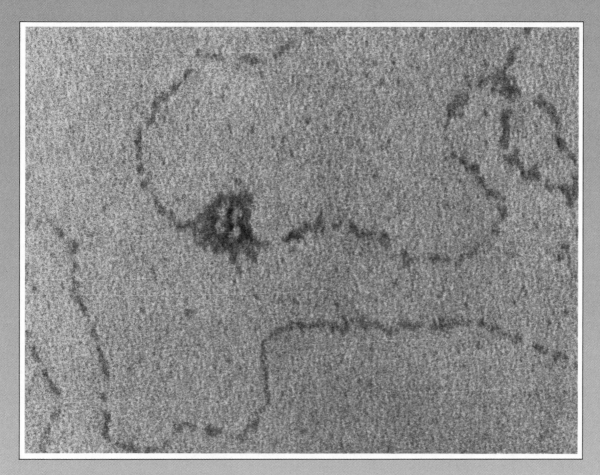

Regulation of Gene Activity

Not all genes are expressed continuously. The level of gene expression may vary from one cell type to the next or according to stage in the cell cycle. For example, the genes for hemoglobin are expressed at high levels only in precursors of the red blood cells. The activity of genes varies according to the functions of the cell. A complex vertebrate animal contains approximately 200 different types of cells with specialized functions. With minor exceptions, all cell types contain the same genetic complement. The cell types differ only in which genes are active. In general, the synthesis of particular gene products is controlled by mechanisms collectively called **gene regulation**.

Above: Electron micrograph of an *E. coli* phage λ repressor molecule bound to the operator region of λ DNA. (Courtesy of Christine Brack.)

In many cases, gene activity is regulated at the level of transcription, either through signals originating within the cell itself or in response to external conditions. For example, many gene products are needed only on occasion, and transcription can be regulated in an on-off manner that enables such products to be present only when demanded by external conditions. However, the flow of genetic information is regulated in other ways also. Control points for gene expression include the following:

1. DNA *rearrangements*, in which gene expression depends on programmed changes in the position of DNA sequences in the genome.
2. *Transcriptional regulation* of the occurrence or rate of synthesis of RNA transcripts.
3. RNA *processing*, or regulation through the occurrence or patterns of RNA splicing.
4. *Translational control* of the occurrence of translation or the rate at which an mRNA is utilized.
5. *Stability of mRNA*, because mRNAs that persist in the cell have longer-lasting effects than those that are degraded rapidly.
6. *Posttranslational control*, which includes a great diversity of mechanisms affecting enzyme inhibition, activation, stability, and so forth.

The regulatory systems of prokaryotes and eukaryotes are somewhat different from each other. Prokaryotes are generally free-living unicellular organisms that grow and divide indefinitely as long as environmental conditions are suitable and the supply of nutrients is adequate. Their regulatory systems are geared to provide the maximum growth rate in a particular environment, except when such growth would be detrimental. Prokaryotes can also use the coupling between transcription and translation (Chapter 11) for regulation, but the absence of introns eliminates RNA splicing as a possible control point.

The requirements of tissue-forming eukaryotes are different from those of prokaryotes. In a developing organism—for example, in an embryo—a cell must not only grow and produce many progeny cells, but also undergo considerable change in morphology and biochemistry and then maintain the changed state. Furthermore, during embryonic development, most eukaryotic cells are challenged less by the environment than are bacteria in that the composition and concentration of the growth medium does not change drastically with time. Finally, in an adult organism, growth and cell division in most cell types have stopped, and each cell needs only to maintain itself and its specialized characteristics.

In this chapter, we consider the basic mechanisms of the regulation of transcription and RNA processing. The examples are those in which the regulation is well understood.

12-1 Transcriptional Regulation in Prokaryotes

In bacteria and phages, on-off regulatory activity is through the control of transcription—that is, synthesis of a particular mRNA is allowed when the gene product is needed and inhibited when the product is not needed. In bacteria, few examples are known of switching a system completely off. When transcription is in the off state, a basal level of gene expression almost always remains, often averaging one transcriptional event or less per cell generation; hence, there is very little synthesis of the gene product. For convenience, in discussing transcription, we use the term "off", but it should be kept in mind that this usually means "very low." Extremely low levels of expression are also found in certain classes of genes in eukaryotes, including many genes that participate in embryonic development. Regulatory mechanisms other than the on-off type are also known in both prokaryotes and eukaryotes; for example, the activity of a system may be modulated from fully on to partly on, rather than to off.

In bacterial systems, when several enzymes act in sequence in a single metabolic pathway, usually either all or none of these enzymes are produced. This phenomenon, which is called **coordinate regulation,** results from control of the synthesis of one or more polycistronic mRNA molecules encoding all of the gene products. This type of regulation is not found in eukaryotes because eukaryotic mRNA is monocistronic, as discussed in Chapter 11.

Several mechanisms of regulation of transcription are common. The particular one used often depends on whether the enzymes being regulated act in degradative or synthetic metabolic pathways. For example, in a multistep degradative system, the availability of the molecule to be degraded helps determine whether the enzymes in the pathway will be synthesized. In contrast, in a biosynthetic pathway the final product is often the regulatory molecule. The molecular mechanisms for each of the regulatory patterns vary quite widely but usually fall into one of two major categories—**negative regulation** and **positive regulation** (Figure 12-1). In a negatively regulated system, a **repressor** protein is present in the cell and prevents transcription. An antagonist of the repressor, generally called an **inducer,** is needed to allow initiation of transcription. In a positively regulated system, an **effector** molecule (usually a protein in combination with a small inducer molecule) activates a promoter; no inhibitor must be overridden. Negative and positive regulation are not mutually exclusive, and some systems are both positively and negatively regulated, utilizing two regulators to respond to different conditions in the cell. In prokaryotes, negative regulation is more common, but, in eukaryotes, positive regulation is more common.

A degradative system may be regulated either positively or negatively. In a biosynthetic pathway, the final product usually negatively regulates its own synthesis; in the simplest type of negative regulation, absence of the product increases its synthesis and presence of the product decreases its synthesis. Even in a system in which a single protein molecule (not necessarily an enzyme) is translated from a monocistronic mRNA molecule, the protein may be **autoregulated**—that is, the protein may regulate its own transcription. In negative autoregulation, the protein inhibits transcription: high concentrations of the protein result in less transcription of the mRNA that codes for the protein.

FIGURE 12-1 The distinction between negative and positive regulation. In negative regulation, a repressor, bound to the DNA molecule, must be removed before transcription can begin. In positive regulation, an effector molecule must bind to the DNA to stimulate transcription. A system may also be regulated both positively and negatively; in such a case, the system is "on" when the positive regulator is bound to the DNA and the negative regulator is not bound to the DNA.

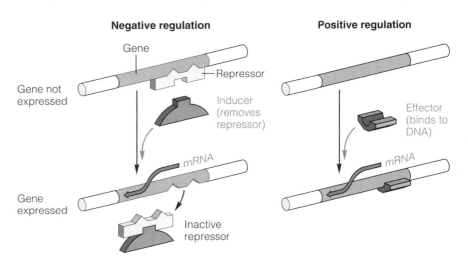

In positive autoregulation, the protein stimulates transcription: as more protein is made, transcription increases to the maximum rate. Positive autoregulation is a common way for weak induction to be amplified. Only a weak signal is necessary to get production of the protein started, but then the positive autoregulation stimulates the production.

The next two sections are concerned with several systems of regulation in prokaryotes. They serve as an introduction to the remainder of the chapter, which deals with regulation in eukaryotes.

12-2 Lactose Metabolism and the Operon

Metabolic regulation was first studied in detail in the system in *E. coli* responsible for degradation of the sugar lactose, and most of the terminology used to describe regulation has come from genetic analysis of this system.

Lac⁻ Mutants

In *E. coli*, two proteins are necessary for the metabolism of lactose—the enzyme β-**galactosidase**, which cleaves lactose (a β-galactoside) to yield galactose and glucose, and a transporter molecule, **lactose permease,** which is required for the entry of lactose into a cell. The existence of two different proteins in the lactose-utilization system was first shown by a combination of genetic experiments and biochemical analysis.

First, hundreds of mutants unable to use lactose as a carbon source— Lac⁻mutants—were isolated. Some of the mutations were in the *E. coli* chromosome and others were in F′ *lac*, a plasmid carrying the genes for lactose utilization. Through F′ × F⁻ matings, partial diploids having the genotypes F′ *lac⁻* / *lac⁺* or F′ *lac⁺* / *lac⁻* were constructed. (The genotype of the plasmid is given at the left of the diagonal line and that of the chromosome at the right.) It was observed that all of these diploids had a Lac⁺ phenotype (that is, they made β-galactosidase); thus, none produced an inhibitor that prevented functioning of the *lac* genes. Other partial diploids were then constructed in which both the F′ *lac* plasmid and the chromosome carried a *lac⁻* allele; these were tested for the Lac⁺ phenotype, with the result that all of the mutants initially isolated could be placed into two complementation groups, *lacZ* and *lacY*. The partial diploids F′ *lacY⁻ lacZ⁺* / *lacY⁺ lacZ⁻* and F′ *lacY⁺ lacZ⁻* / *lacY⁻ lacZ⁺* had a Lac⁺ phenotype, producing both β-galactosidase and permease, but the genotypes F′ *lacY⁻ lacZ⁺* / *lacY⁻ lacZ⁺* and F′ *lacY⁺ lacZ⁻* / *lacY⁺ lacZ⁻* had the Lac⁻ phenotype because they were unable to synthesize either the permease or the β-galactosidase. The existence of two complementation groups was good evidence that the *lac* system consisted of at least two genes. (A third gene participating in lactose metabolism was later discovered; it was not included among the early mutants because it is not essential for growth on lactose.) Biochemical analysis showed that the *lacZ* gene codes for β-galactosidase and the *lacY* gene codes for the permease. A final important result—that the *lacY* and *lacZ* genes are adjacent—was deduced from a high frequency of cotransduction observed in genetic-mapping experiments.

Inducible and Constitutive Synthesis and Repression

The on-off nature of the lactose-utilization system is evident in the following observations:

1. If a culture of Lac$^+$ *E. coli* is growing in a medium lacking lactose or any other β-galactoside, the intracellular concentrations of β-galactosidase and permease are exceedingly low—roughly one or two molecules per bacterial cell. However, if lactose is present in the growth medium, the number of each of these enzymes is about 10^3-fold higher.

2. If lactose is added to a Lac$^+$ culture growing in a lactose-free medium (also lacking glucose, a point that will be discussed shortly), both β-galactosidase and permease are synthesized nearly simultaneously, as shown in Figure 12-2. Analysis of the total mRNA present in the cells before and after addition of lactose shows that almost no *lac* mRNA (the polycistronic mRNA that codes for β-galactosidase and permease) is present before lactose is added and that the addition of lactose triggers synthesis of *lac* mRNA.

These two observations led to the view that transcription of the lactose genes is **inducible** and that lactose is an *inducer* of transcription. Some analogs of lactose also are inducers, such as a sulfur-containing analog denoted IPTG (isopropyl-thiogalactoside), which is convenient for experiments because it induces but is not cleaved by β-galactosidase.

Mutants were also isolated in which *lac* mRNA was synthesized (and hence also β-galactosidase and permease) in *both* the presence and the absence of an inducer. The mutants that eliminated regulation provided the key to

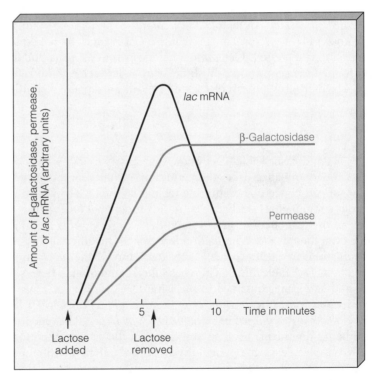

FIGURE 12-2   The "on-off" nature of the *lac* system. The *lac* mRNA appears soon after lactose or another inducer is added; β-galactosidase and permease appear at nearly the same time but are delayed with respect to mRNA synthesis because of the time required for translation. When lactose is removed, no more *lac* mRNA is made, and the amount of *lac* mRNA decreases owing to degradation of mRNA already present. Both β-galactosidase and permease are stable proteins: their amounts remain constant even when synthesis ceases. However, their concentration per cell gradually decreases as a result of cell division.

TABLE 12-1 Characteristics of partial diploids containing several combinations of *lacI* and *lacO* alleles

Genotype	Constitutive or inducible synthesis of lac mRNA
1. F' *lacO*^{c} *lacZ*$^+$ / *lacO*$^+$ *lacZ*$^+$	Constitutive
2. F' *lacO*$^+$ *lacZ*$^+$ / *lacO*^{c} *lacZ*$^+$	Constitutive
3. F' *lacI*$^-$ *lacZ*$^+$ / *lacI*$^+$ *lacZ*$^+$	Inducible
4. F' *lacI*$^+$ *lacZ*$^+$ / *lacI*$^-$ *lacZ*$^+$	Inducible
5. F' *lacO*^{c} *lacZ*$^+$ / *lacI*$^-$ *lacZ*$^+$	Constitutive
6. F' *lacO*^{c} *lacZ*$^-$ / *lacO*$^+$ *lacZ*$^+$	Inducible
7. F' *lacO*^{c} *lacZ*$^+$ / *lacO*$^+$ *lacZ*$^-$	Constitutive

understanding induction; because of their constant synthesis they were termed **constitutive.** Complementation tests using partial diploids showed that the regulatory mutants fall into two groups termed *lacI* and *lacO*^{c}. The characteristics of the mutants are shown in Table 12-1. Genotypes 3 and 4 show that *lacI*$^-$ mutants are recessive. In the absence of inducer, a *lacI*$^+$ cell does not make *lac* mRNA, whereas the mRNA is made by a *lacI*$^-$ mutant. Thus, the *lacI* gene is a regulatory gene *whose product is the repressor protein that keeps the system turned off.* A *lacI*$^-$ mutant lacks the repressor and hence is constitutive. Wildtype copies of the repressor are present in a *lacI*$^+$ / *lacI*$^-$ partial diploid, and so transcription is repressed. Genetic-mapping experiments place the *lacI* gene adjacent to the *lacZ* gene and establish the gene order *lacI lacZ lacY*. How the *lacI* repressor prevents synthesis of *lac* mRNA will be explained shortly.

The Operator Region

Entries 1, 2, and 5 in Table 12-1 show that *lacO*^{c} mutants are dominant. However, the dominance is evident only in certain combinations of *lac* mutations, as can be seen by examining the partial diploids shown in entries 6 and 7. Both combinations are Lac$^+$, because a functional *lacZ* gene is present. However, in the combination shown in entry 6, synthesis of β-galactosidase is inducible even though a *lacO*^{c} mutation is present. The difference between the two combinations in entries 6 and 7 is that, in entry 6, the *lacO*^{c} mutation is contained in a DNA molecule that also has a *lacZ*$^-$ mutation, whereas, in entry 7, *lacO*^{c} and *lacZ*$^+$ are contained in the same DNA molecule. *Thus, a lacO*^{c} *mutation causes constitutive synthesis of* β*-galactosidase only when the lacO*^{c} *and lacZ*$^+$ *alleles are contained in the same DNA molecule.* The *lacO*^{c} mutation is said to be **cis-dominant,** because only genes in the *cis* configuration (in the same DNA molecule as the mutation) are expressed in dominant fashion. Confirmation of this conclusion comes from an important biochemical observation: the mutant enzyme (coded by the *lacZ*$^-$ sequence) is synthesized constitutively in a *lacO*^{c} *lacZ*$^-$ / *lacO*$^+$ *lacZ*$^+$ partial diploid (entry 6), whereas the wildtype enzyme (coded by the *lacZ*$^+$ sequence) is synthesized only if an inducer is added. All *lacO*^{c} mutations are located between the *lacI* and *lacZ* genes; thus, the gene order of the four genetic elements of the *lac* system is

lacI lacO lacZ lacY

An important feature of all *lacO*ᶜ mutations is that they cannot be complemented (a feature of all *cis*-dominant mutations); that is, a *lacO*⁺ allele cannot alter the constitutive activity of a *lacO*ᶜ mutation. Thus, *lacO* does not encode a diffusible product and must define a site that determines whether synthesis of the product of the adjacent *lacZ* gene is inducible or constitutive. The *lacO* region is called the **operator.** In the next section we will see that the operator is a *binding site* in the DNA for the repressor protein.

The Operon Model

The regulatory mechanism of the *lac* system was first explained by the **operon model** of Francois Jacob and Jacques Monod, which has the following features (Figure 12-3):

1. The lactose-utilization system consists of two kinds of components— *structural genes* needed for transport and metabolism of lactose, and *regulatory elements* (the *lacI* gene, the *lacO* operator, and the *lac* promoter).

2. The products of the *lacZ* and *lacY* genes are coded by a single polycistronic mRNA molecule. The mRNA molecule also contains a

FIGURE 12-3   A. A map of the *lac* operon, not drawn to scale; the *p* and *o* sites are actually much smaller than the other genes and together comprise only 83 base pairs. B. A diagram of the *lac* operon in (I) repressed and (II) induced states. The inducer alters the shape of the repressor so that the repressor can no longer bind to the operator. The common abbreviations *i*, *p*, *o*, *z*, *y*, and *a* are used instead of *lacI*, *lacO*, and so forth. The *lacA* gene is not essential for lactose utilization.

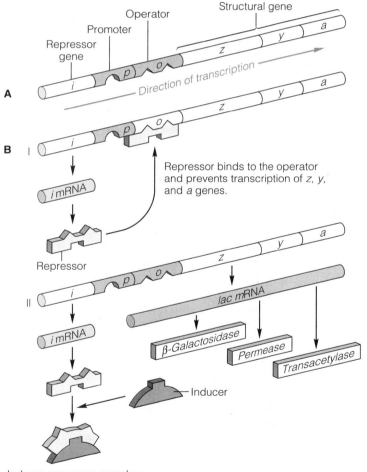

Inducer-repressor complex, which cannot bind to the operator

third gene, denoted *lacA*, which codes for the enzyme transacetylase. This enzyme is used in the metabolism of certain β-galactosides other than lactose and will not be of further concern. The linked structural genes, together with the promoter and *lacO*, constitute the **lac operon.**

3. The promoter for the *lac* mRNA molecule is immediately adjacent to the *lacO* region. This location has been substantiated genetically by the isolation and mapping of promoter mutants ($lacP^-$) that are completely incapable of making *lac* mRNA.

4. The product of the *lacI* gene is a repressor, which binds to a unique sequence of DNA bases constituting the operator.

5. When the repressor is bound to the operator, initiation of transcription of *lac* mRNA by RNA polymerase is prevented.

6. Inducers stimulate mRNA synthesis by binding to and inactivating the repressor. Thus, in the presence of an inducer, the operator is not bound with repressor, and the promoter is available for initiation of mRNA synthesis.

Note that regulation of the operon requires that the *lacO* operator either overlap or be adjacent to the promoter of the structural genes. This results in the proximity of *lacI* to *lacO*, which is not strictly necessary because the *lacI* repressor is a soluble protein and is therefore diffusible throughout the cell.

When the operon is induced, the ratio of the number of copies of β-galactosidase, permease, and transacetylase is $1.0:0.5:0.2$. These differences are partly due to the order of the genes in the mRNA: downstream cistrons are less likely to be translated owing to failure of reinitiation when an upstream cistron has finished translation.

The operon model is supported by a wealth of experimental data and explains many of the features of the *lac* system, as well as numerous other negatively regulated genetic systems in prokaryotes. One aspect of the regulation of the *lac* operon—the effect of glucose—has not yet been discussed. Examination of this feature indicates that the *lac* operon is also subject to positive regulation, as will be seen in the next subsection.

Positive Regulation of the lac Operon

The function of β-galactosidase in lactose metabolism is to form glucose by cleaving lactose. (The other cleavage product, galactose, also is ultimately converted into glucose by the enzymes of the galactose operon.) Thus, if both glucose and lactose are present in the growth medium, activity of the *lac* operon is not needed, and, indeed, no β-galactosidase is formed until virtually all of the glucose in the medium is consumed. The lack of synthesis of β-galactosidase is a result of lack of synthesis of *lac* mRNA. No *lac* mRNA is made in the presence of glucose because, in addition to an inducer to inactivate the *lacI* repressor, another element is needed for initiating *lac* mRNA synthesis; the activity of this element is regulated by the concentration of glucose. However, the inhibitory effect of glucose on expression of the *lac* operon is quite indirect.

The small molecule *cyclic adenosine monophosphate* (**cAMP**) is widely distributed in animal tissues, and in multicellular eukaryotic organisms it is important in mediating the action of many hormones (Figure 12-4). It is also present in *E. coli* and many other bacteria where it has a different function. Cyclic AMP is synthesized by the enzyme *adenyl cyclase*, and the concentration of cAMP is regulated indirectly by glucose metabolism. When bacteria are growing in a medium containing glucose, the cAMP concentration in the cells is quite low. In a medium containing glycerol or any carbon source that cannot

FIGURE 12-4   Structure of cyclic AMP.

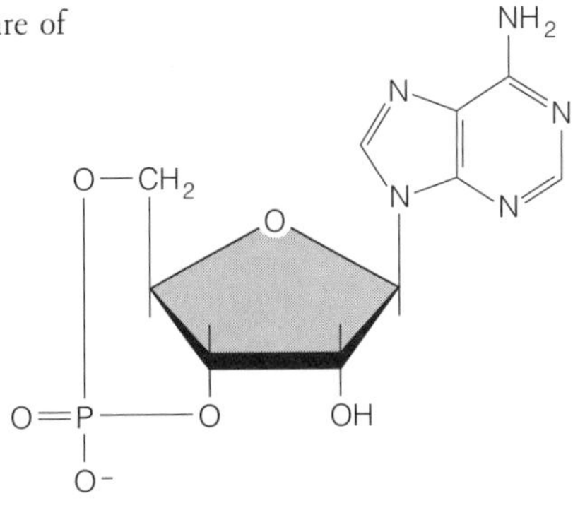

enter the biochemical pathway used to metabolize glucose (the glycolytic pathway) or when the bacteria are otherwise starved of an energy source, the cAMP concentration is high (Table 12-2). Glucose levels help regulate the cAMP concentration in the cell, and *cAMP regulates the activity of the lac operon* (and several other operons that control degradative metabolic pathways).

E. coli (and many other bacterial species) contains a protein called the *cyclic AMP receptor protein* (**CRP**), which is coded by a gene called *crp*. Mutants of either *crp* or the adenyl cyclase gene are unable to synthesize *lac* mRNA, indicating that both CRP function and cAMP are required for *lac* mRNA synthesis. CRP and cAMP bind to one another, forming a complex denoted **cAMP-CRP**, which is an active regulatory element in the *lac* system. The requirement for cAMP-CRP is independent of the *lacI* repression system because *crp* and adenyl cyclase mutants are unable to make *lac* mRNA even if a *lacI⁻* or a *lacOᶜ* mutation is present. The reason is that the cAMP-CRP complex must be bound to a base sequence in the promoter region of the DNA in order for transcription to occur (Figure 12-5). Thus, cAMP-CRP is a *positive* regulator, in contrast with the repressor, which is a *negative* regulator. The positive and negative regulatory systems of the *lac* operon are independent of each other.

Experiments carried out *in vitro* with purified *lac* DNA, *lac* repressor, cAMP-CRP, and RNA polymerase have established two further points:

1. In the absence of the cAMP-CRP complex, RNA polymerase binds only weakly to the promoter, but its binding is stimulated when cAMP-CRP is also bound to the DNA. The weak binding rarely leads to initiation of transcription, because the correct interaction between RNA polymerase and the promoter does not occur.

TABLE 12-2   Concentration of cyclic AMP in cells growing in media with the indicated carbon sources

Carbon source	cAMP concentration
Glucose	Low
Glycerol	High
Lactose	High
Lactose + glucose	Low
Lactose + glycerol	High

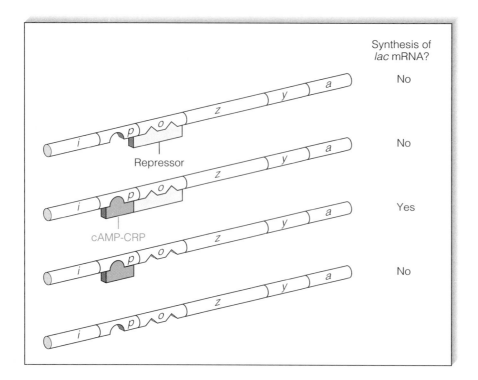

Synthesis of
lac mRNA?

No

No

Yes

No

Repressor

cAMP-CRP

FIGURE 12-5   Four regulatory states of the *lac* operon: the *lac* mRNA is synthesized only if cAMP-CAP is present and repressor is absent.

2. If the repressor is bound to the operator, RNA polymerase cannot stably bind to the promoter.

These results explain how lactose and glucose function together to regulate transcription of the *lac* operon.

12-3   Regulation of the Tryptophan Operon

The tryptophan (*trp*) operon of *E. coli* contains structural genes for enzymes that synthesize the amino acid tryptophan. Regulation of this operon occurs in such a way that, when tryptophan is present in the growth medium, the *trp* operon is not active. That is, when adequate tryptophan is present, transcription of the operon is repressed; however, when the supply is insufficient, transcription occurs. The *trp* operon is quite different from the *lac* operon in that tryptophan acts directly in repression rather than as an inducer. Furthermore, because the *trp* operon codes for a set of biosynthetic rather than degradative enzymes, neither glucose nor cAMP-CRP functions in operon activity.

A simple on-off system, as in the *lac* operon, is not optimal for a biosynthetic pathway. For example, a situation may arise in which some tryptophan is present in the growth medium, but not enough to sustain optimal growth. Under these conditions, it is necessary to synthesize tryptophan, but at less than the maximum possible rate. Cells adjust to this situation by means of a regulatory mechanism in which *the amount of transcription in the derepressed (that is, not repressed) state is determined by the concentration of tryptophan in the cell*. This regulatory mechanism is found in many operons responsible for amino acid biosynthesis.

FIGURE 12-6 The *E. coli trp* operon. For clarity, the regulatory region is enlarged with respect to the coding region. The correct size of each region is indicated by the numbers of base pairs. Region L is the leader. The regulatory elements are shown in red.

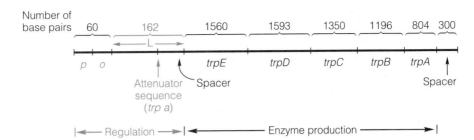

Tryptophan is synthesized in five steps, each requiring a particular enzyme. The genes coding for these enzymes are adjacent in the *E. coli* chromosome, and they are in the same order as that in which the enzymes are used in the biosynthetic pathway. The genes are called *trpE*, *trpD*, *trpC*, *trpB*, and *trpA*, and the enzymes are translated from a single polycistronic mRNA molecule. The *trpE* coding region is the first one translated. Upstream (that is, on the 5′ side) of *trpE* are the promoter, the operator, and two regions called the *leader* and the *attenuator*, which are designated *trpL* and *trp a* (not *trpA*), respectively (Figure 12-6). The repressor gene, *trpR*, is located quite far from this operon.

The regulatory protein of the *trp* operon is the product of the *trpR* gene. Mutations in either this gene or the operator cause constitutive initiation of transcription of *trp* mRNA, as in the *lac* operon. The *trpR* gene product is called the *trp* **aporepressor.** It does not bind to the operator unless it is first bound to tryptophan; that is, the aporepressor and the tryptophan molecule join together to form the active *trp* repressor, which binds to the operator. The reaction scheme is outlined in Figure 12-7. When there is not enough tryptophan, the aporepressor adopts a three-dimensional conformation unable to bind with the *trp* operator, and transcription of the operon occurs (Figure 12-7A). On the other hand, when tryptophan is present at high enough concentration, some molecules bind with the aporepressor and cause it to change conformation into the active repressor. The active repressor binds with the *trp* operator and prevents transcription (Figure 12-7B). Thus, only when tryptophan is present in sufficient amounts is the active repressor molecule formed. This is the basic on-off regulatory mechanism.

FIGURE 12-7 Regulation of the *E. coli trp* operon: (A) by itself, the *trp* aporepressor protein does not bind to the operator, and transcription occurs; (B) in the presence of sufficient tryptophan, the combination of aporepressor and tryptophan forms the active repressor that binds to the operator and transcription is repressed.

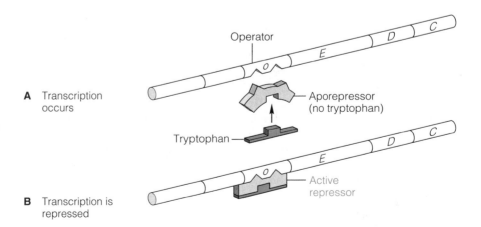

Attenuation

In the *on* state, a finer control, in which the enzyme concentration is adjusted by the amino acid concentration, is effected by (1) premature termination of transcription before the first structural gene is reached and (2) regulation of the frequency of this termination by the internal concentration of tryptophan. This type of regulation is called **attenuation,** and it uses translation to control transcription.

Attenuation occurs by means of interactions between DNA sequences present in the leader region of the *trp* transcript. In wildtype cells, transcription of the *trp* operon is often initiated. However, in the presence of even small amounts of tryptophan, most of the mRNA molecules terminate in a specific 28-base region within the leader sequence. The result of such termination is an RNA molecule that contains only 140 nucleotides and stops short of the genes coding for the *trp* enzymes. The 28-base region in which termination occurs is called the **attenuator.** The base sequence (Figure 12-8) of the region in which termination occurs has the usual features of a termination site—namely, a potential stem-and-loop configuration in the mRNA followed by a sequence of eight AT pairs (see Figure 11-10).

The leader sequence contains several notable features:

1. An AUG codon and a later UGA stop codon in the same reading frame define a region coding for a polypeptide consisting of 14 amino acids—called the **leader polypeptide** (Figure 12-9).

2. Two adjacent tryptophan codons are located in the leader polypeptide at positions 10 and 11. We will see the significance of these repeated codons shortly.

3. Four segments of the leader RNA—denoted 1, 2, 3, and 4—are capable of base-pairing in different combinations (Figure 12-10). They may form two base-paired regions (1–2 and 3–4 in Figure 12-10A), or they may form either a 3–4 pair or a 2–3 pair (Figure 12-10B and C). The 1–2 and 3–4 pairs form in purified *trp* leader mRNA. The paired region 3–4 constitutes the terminator recognition region. Region 1 overlaps the coding region for the leader polypeptide.

This sequence organization allows premature termination in the *trp* leader region by the following mechanism.

Termination of transcription is brought about by translation of the leader peptide region. Because there are two tryptophan codons in this sequence, the translation of the sequence is sensitive to the concentration of charged tRNA^{Trp}.

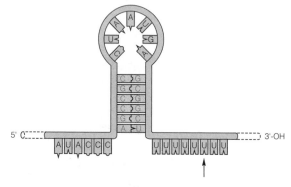

FIGURE 12-8 The terminal region of the *trp* attenuator sequence. The arrow indicates the final uridine in attenuated RNA. Nonattenuated RNA continues past that base. The red bases form the hypothetical stem sequence that is shown.

Leader polypeptide

TrpE protein

⌐Met⌐ ⌐Lys⌐ ⌐Ala⌐ ⌐Ile⌐ ⌐Phe⌐ ⌐Val⌐ ⌐Leu⌐ ⌐Lys⌐ ⌐Gly⌐ ⌐Trp⌐ ⌐Trp⌐ ⌐Arg⌐ ⌐Thr⌐ ⌐Ser—Stop⌐   ⌐Met⌐ ⌐Gln⌐ ⌐Thr⌐ ⌐Gln⌐

pppAAG...(23)... |A|U|G| |A|A|A| |G|C|A| |A|U|U| |U|U|C| |G|U|A| |C|U|G| |A|A|A| |G|G|U| |U|G|G| |U|G|G| |C|G|C| |A|C|U| |U|C|C| |U|G|A| ...(91)... |A|U|G| |C|A|A| |A|C|A| |C|A|A|

FIGURE 12-9 The sequence of bases in the *trp* leader mRNA, showing the leader polypeptide, the two tryptophan codons (shaded red), and the beginning of the TrpE protein. The numbers 23 and 91 are the numbers of bases in sequence that, for clarity, are not shown.

That is, if the supply of tryptophan is inadequate, the amount of charged tRNA^{Trp} will be insufficient and, hence, translation will be stalled at the tryptophan codons.

Figure 12-10 indicates where the end of the *trp* leader peptide occurs in segment 1. Usually, a translating ribosome is in contact with about ten bases in the mRNA past the codons being translated, and all base-pairing is eliminated in the segment of mRNA that is in contact with the ribosome. Thus, when the codons of the leader polypeptide are being translated, segments 1 and 2 are prevented from pairing (Figure 12-10B). Because transcription and translation are coupled in prokaryotes, the leading ribosome is not far behind the RNA polymerase (Figure 12-10C). Thus, if the ribosome is in contact with segment 2 when transcription of segment 4 is in progress (this will occur when there is sufficient tryptophan to read past the tryptophan codons), then segments 3 and 4 will form the duplex region 3–4 (Figure 12-10B). The presence of the 3–4 stem-and-loop configuration allows termination to occur when the terminating sequence of seven uridines is reached. On the other hand, if the concentration of charged tRNA^{Trp} is too low, occasionally a translating ribosome is stalled at the tryptophan codons (Figure 12-10C). These codons are located sixteen bases before the beginning of segment 2. Thus, segment 2 is free before segment 4 has been transcribed, and the 2–3 duplex can form. This prevents the 3–4 stem and loop from forming, and so termination does not occur and the complete mRNA molecule is made, including the coding sequences for the *trp* genes. Hence, if tryptophan is present in excess, termination occurs and little enzyme is synthesized; if tryptophan is absent, termination does not occur and the enzymes are made. At intermediate concentrations, the fraction of initiation

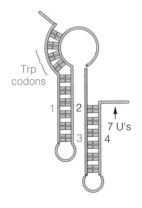

A Free mRNA. Base pairs between 1 and 2 and between 3 and 4.

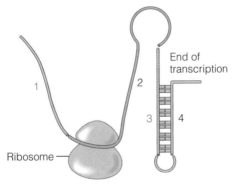

B High concentration of tryptophan. Ribosome reaches region 2 and pairing of 3–4 causes termination of transcription.

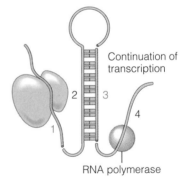

C Low concentration of tryptophan. Ribosome stalled in region 1 permits pairing of 2–3 before transcription of region 4 is completed.

FIGURE 12-10 The explanation for attenuation in the *E. coli trp* operon. The tryptophan codons in part A are those highlighted in red in Figure 12-9.

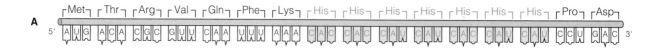

FIGURE 12-11   Amino acid sequence of the leader peptide and base sequence of the corresponding part of mRNA from (A) the histidine operon and (B) the phenylalanine operon. The repetition of these amino acids is emphasized with red shading.

events that result in completion of *trp* mRNA will depend on how often translation is stalled, which in turn depends on the concentration of tryptophan.

Many operons responsible for amino acid biosynthesis (for example, the leucine, isoleucine, phenylalanine, and histidine operons) are regulated by attenuators that function by forming alternative paired regions in the transcript. In the histidine operon, the coding region for the leader polypeptide contains *seven* adjacent histidine codons (Figure 12-11A). In the phenylalanine operon, the coding region for the leader polypeptide contains seven phenylalanine codons divided into three groups (Figure 12-11B). Note that:

> Attenuation cannot occur in eukaryotes because transcription and translation are not coupled; transcription occurs in the nucleus and translation in the cytoplasm.

Regulation of the *lac* and *trp* operons typifies some of the important mechanisms controlling transcription in prokaryotes. In the following sections, we will see that transcriptional regulation in eukaryotes is quite different.

12-4   Regulation in Eukaryotes

Complex eukaryotes with elaborate developmental programs and numerous specialized cell types have different needs for regulation than do prokaryotic cells. For example, the environment of the cells in a complex eukaryote does not usually change drastically in time. During the development of a eukaryotic organism, cells differentiate (1) as a result of sequential changes in gene activity that are programmed in the genome, (2) in response to molecular signals released by other cells, or (3) in response to physical contact with other cells. After the cells have differentiated, they remain genetically quite stable, producing particular substances either at a constant rate or in response to external signals such as hormones, nutrient concentrations, or temperature changes.

The great complexity of multicellular eukaryotes requires a great diversity of genetic regulatory mechanisms. On the whole, the mechanisms are not understood as thoroughly as they are in prokaryotes. However, many important examples of different types of mechanisms have been studied in animals as diverse as mammals (especially the mouse), birds (usually the chicken), amphibians (toads of the genus *Xenopus*), insects (*Drosophila*), nematode worms (*Caenorhabditis elegans*), echinoderms (the sea urchin), and ciliates (*Tetrahymena*), as well as in yeast and other fungi. These examples reveal the general features of eukaryotic gene regulation that are discussed in the following sections.

Important Differences in the Genetic Organization of Prokaryotes and Eukaryotes

Numerous differences exist between prokaryotes and eukaryotes with regard to transcription and translation, and in the spatial organization of DNA, as described in Chapters 5 and 11. Some of those most relevant to regulation are the following:

1. In a eukaryote, usually only a single type of polypeptide chain can be translated from a completed mRNA molecule; thus, polycistronic mRNA of the type seen in prokaryotes is not found in eukaryotes.

2. The DNA of eukaryotes is bound to histones, forming chromatin, and to numerous nonhistone proteins. Only a small fraction of the DNA is bare. In bacteria, some proteins are present in the folded chromosome, but most of the DNA is free.

3. A significant fraction of the DNA of eukaryotes consists of moderately or highly repetitive nucleotide sequences. Some of the repetitive sequences are repeated in tandem copies, but others are not. Other than duplicated rRNA (ribosomal RNA) and tRNA genes and a few specific sequences such as insertion sequences, bacteria contain little repetitive DNA.

4. A large fraction of eukaryotic DNA is untranslated; most of the nucleotide sequences do not code for proteins.

5. Some eukaryotic genes are expressed and regulated by the use of mechanisms for rearranging certain DNA segments in a controlled way and for increasing the number of specific genes when needed.

6. Genes in eukaryotes are split into exons and introns, and the introns must be removed during processing of the RNA transcript before translation begins.

7. In eukaryotes, mRNA is synthesized in the nucleus and must be transported through the nuclear envelope to the cytoplasm where it is utilized. Such extreme compartmentalization does not occur in bacteria.

We shall see in the following sections how some of these features are incorporated into particular modes of regulation.

12-5 Alteration of DNA

Some genes in eukaryotes are regulated by alteration of the DNA. For example, certain sequences may be amplified or rearranged in the genome, or the bases may be chemically modified. Some of the alterations are reversible, but others permanently change the genome of the cells. However, the permanent changes take place only in somatic cells, and so they are not genetically transmitted to the offspring through the germ line.

Gene Dosage and Gene Amplification

Some gene products are required in much larger quantities than others. One means of maintaining particular ratios of certain gene products (other than by differences in transcription and translation efficiency, as discussed earlier) is by **gene dosage**. For example, if two genes, A and B, are transcribed at the same rate and the translation efficiencies are the same, twenty times as much of product A can be made as product B if there are twenty copies of gene A per copy of gene B. The histone genes exemplify a gene-dosage effect: to synthesize the huge amount of histone required to form chromatin, most cells contain hundreds of times as many copies of histone genes as of genes required for DNA replication. In this case, the high expression is automatic because the repeated genes are part of the normal chromosome complement.

In some cases, gene dosage is increased temporarily by a process called **gene amplification,** in which the number of genes increases in response to some signal. The best-understood example of gene amplification is found in the development of the oocytes of the toad *Xenopus laevis*. The formation of an egg from its precursor, the oocyte, is a complex process that requires a huge amount of protein synthesis. To achieve the necessary rate, a very large number of ribosomes is needed. Ribosomes contain molecules of rRNA, and the number of rRNA genes in the genome is insufficient to produce the required number of ribosomes for the oocyte in a reasonable period of time. In the development of the oocyte, the number of rRNA genes increases by about 4000 times. This increase exists only during and for the purpose of development of the egg. The precursor to the oocyte, like all somatic cells of the toad, contains about 600 rRNA-gene (rDNA) units; after amplification, about 2×10^6 copies of each unit are present. This large amount enables the oocyte to synthesize 10^{12} ribosomes, which are required for the protein synthesis that occurs during early development of the embryo, at a time when no ribosomes are being formed.

Before amplification, the 600 rDNA units are arranged in tandem. During amplification, which occurs over a three-week period during which the oocyte develops from a precursor cell, the rDNA no longer consists of a single contiguous DNA segment containing 600 rDNA units but instead forms a large number of small circles and replicating rolling circles. The rolling-circle replication accounts for the increase in the number of copies of the genes. The precise mechanism of excision of the circles from the chromosome and formation of the rolling circles is not known.

When the oocyte is mature, no more rRNA needs to be synthesized until well after fertilization and into early development, at which time 600 copies are sufficient. Thus, the excess rDNA serves no purpose, and it is slowly degraded by intracellular enzymes. Following fertilization, the chromosomal DNA replicates and mitosis ensues, occurring repeatedly as the embryo develops. During this period, the extra chromosomal rDNA does not replicate; degradation continues and, by the time several hundred cells have formed, none of this extra rDNA remains. Amplification of rRNA genes during oogenesis occurs in many organisms including insects, amphibians, and fish.

Amplification of some protein-coding genes also occurs. For example, in *Drosophila* females, the genes that produce chorion proteins (a component of the sac that encloses the egg) are amplified in follicle cells just before maturation of the egg. The amplification enables the cells to produce a large amount of protein in a short time. In some cases, amplification occurs in abnormal regulation. For example, a gene called N-*myc* is frequently amplified in human tumor cells in the disease neuroblastoma, and the degree of amplification is correlated with progress of the disease and tendency of the tumor to spread. N-*myc* is the normal cellular counterpart of a viral oncogene. (Oncogenes are discussed in Section 6-7).

Programmed DNA Rearrangements

Rearrangement of DNA sequences in the genome is an unusual but important mechanism by which some genes are regulated. An example is in the phenomenon known as **mating-type interconversion** in the yeast *Saccharomyces cerevisiae*. As noted in Chapter 3, yeast has two mating types, denoted **a** and α. Mating between haploid **a** and haploid α cells produces the **a**/α diploid, which can undergo meiosis and sporulation to produce four-spored asci containing haploid **a** and α spores in the ratio 2 : 2. If a single yeast spore of either the **a** or α genotype is cultured in isolation from other spores, mating between progeny cells would not be expected because the progeny cells would have the mating type of the original parent. However, *S. cerevisiae* has a mating system called **homothallism,** in which some cells undergo a conversion into the opposite mating type to allow matings to occur in what would otherwise be a pure culture of one mating type or the other.

The outlines of mating-type interconversion are shown in Figure 12-12. An original haploid spore (in this example, α) undergoes germination to produce two progeny cells. Both the mother cell (the original parent) and the daughter cell have mating-type α, as expected from a normal mitotic division. However, in the next cell division a switching (interconversion) of mating type occurs in the mother cell and its *new* progeny cell, in which the original α mating type becomes replaced with the **a** mating type. After the second cell division is complete, the α and **a** cells are able to undergo mating because they now are of opposite mating type. Fusion of the nuclei produces the **a**/α diploid, which undergoes mitotic divisions and later sporulation to again produce **a** and α haploid spores.

FIGURE 12-12 Mating-type switching in the yeast *Saccharomyces cerevisiae*. Germination of a spore (in this example, one of mating type α) forms a mother cell and a bud that grows into a daughter cell. In the next division, the mother cell and its new daughter cell switch to the opposite mating type (in this case, **a**). The result is two α and two **a** cells. Cells of opposite mating type can fuse to form **a** / α diploid zygotes. In a similar fashion, germination of an **a** spore is accompanied by switching to the α mating type.

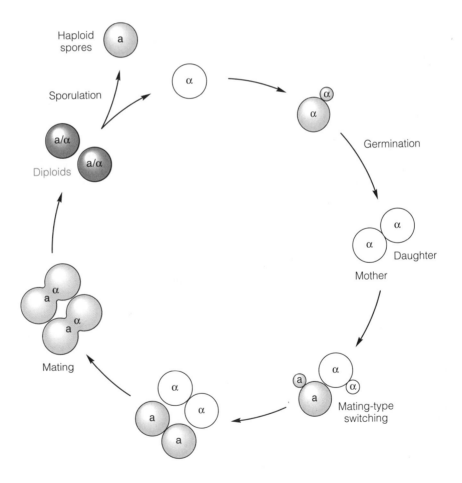

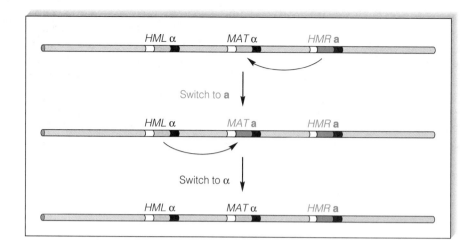

FIGURE 12-13   Genetic basis of mating-type interconversion. The mating type is determined by the DNA sequence present at the *MAT* locus. The *HML* and *HMR* loci are casettes containing unexpressed mating type genes, either α or **a**. In the interconversion from α to **a**, the α genetic information present at *MAT* is replaced with the **a** genetic information from *HMR***a**. In the switch from **a** to α, the **a** genetic information at *MAT* is replaced with the α genetic information from *HML*α.

The genetic basis of mating-type interconversion is outlined in Figure 12-13. The gene controlling mating type is the *MAT* gene in chromosome III, which can have either of two allelic forms, **a** or α. If the allele in a haploid cell is *MAT***a,** the cell has mating-type **a;** and, if the allele is *MAT*α, the cell has mating-type α. However, both genotypes normally contain both **a** and α genetic information in the form of unexpressed **cassettes** present in the same chromosome. The *HML*α cassette contains the α DNA sequence about 200 kb away from one side of the *MAT* gene, and the *HMR***a** cassette contains the **a** DNA sequence about 150 kb away from the other side of *MAT*. When mating-type interconversion occurs, a different yeast gene produces a specific endonuclease that cuts both strands of the DNA in the *MAT* region. The double-stranded break initiates a process in which genetic information in the unexpressed cassette containing the opposite mating type becomes inserted into *MAT*. In this process, the DNA sequence in the donor cassette is duplicated, so that the mating type becomes converted but the same genetic information is retained in unexpressed form in the cassette. The terminal regions of *HML*, *MAT*, and *HMR* are identical (illustrated in white and black), and these regions are critical in enabling recognition of the regions for interconversion. The unique part of the α region is 747 base pairs in length, that of the **a** region is 642 base pairs. The molecular details of the conversion process are similar to those of the double-strand gap mechanism of recombination, which is discussed in Chapter 14.

Figure 12-13 illustrates two sequential mating-type interconversions. In the first, an α cell (containing the *MAT*α allele) undergoes conversion into **a,** using the DNA sequence contained in the *HMR***a** cassette. The converted cell has the genotype *MAT***a.** In a later generation, a descendant **a** cell may become converted into mating-type α, using the unexpressed DNA sequence contained in *HML*α. This cell, again, has the genotype *MAT*α. Mating-type switches can occur again and again in the lineage of any particular cell.

Another important example of programmed DNA rearrangement is found in vertebrates in cells that form the immune system. In this case, the precursor cells contain numerous DNA sequences that can serve as alternatives for various regions in the final gene. In the maturation of each cell, a random combination of the alternatives is created by DNA cutting and rejoining, producing a great variety of possible genes, which enables the immune system to recognize and attack most bacteria and viruses. The process of DNA rearrangement in the immune system is described in Chapter 15.

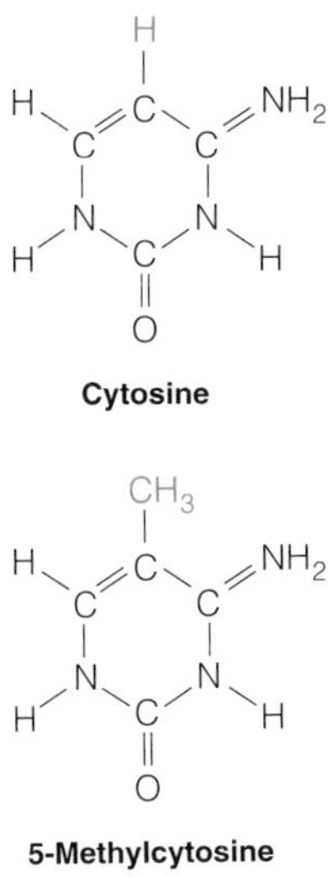

DNA Methylation

In most eukaryotes, a small proportion of the cytosine bases are modified by the addition of a methyl (CH_3) group to the number 5 carbon atom (Figure 12-14). The cytosines are incorporated in their normal, unmodified form in DNA replication, but they are modified later by an enzyme called a **DNA methylase.** Cytosines are modified preferentially in 5'-CG-3' dinucleotides. When a CG dinucleotide that is methylated in both strands undergoes DNA replication, the result is two daughter molecules, each of which contains one parental strand with a methylated CG and one daughter strand with an unmethylated CG. The DNA methylase recognizes the half-methylation in these molecules and methylates the cytocines in the daughter strands. Methylation of CG dinucleotides in the sequence CCGG can easily be detected by the use of the restriction enzymes MspI and HpaII. Both enzymes cleave the sequence CCGG. However, MspI cleaves irrespective of whether the interior C is methylated, whereas HpaII cleaves only unmethylated DNA. Therefore, MspI restriction sites that are not cleaved by HpaII are sites where the interior C is methylated (Figure 12-15).

Many eukaryotic genes have CG-rich regions upstream of the coding region, providing potential sites for methylation that may affect transcription. A number of observations suggest that high levels of methylation are associated with genes that are not actively transcribed. One example is the inactive X chromosome in mammalian cells, which is extensively methylated. Another example is the *Ac* transposable element in maize. With certain *Ac* elements, the activity of the transposase gene can be lost without any change in DNA sequence. These elements prove to have heavy methylation in a region particularly rich in the CG dinucleotides. With regard to the methylated *Ac* elements, a return to normal gene activity is associated with loss of methylation. Although there is a correlation between methylation and gene inactivity, it is possible that heavy methylation is a result of gene inactivity rather than a cause of it. However, treatment of cells with the cytosine analog *azacytidine* reverses methylation and can restore gene activity. For example, some clones of rat pituitary tumor cells express the gene for prolactin, whereas other related clones do not. The gene is methylated in the nonproducing cells but not methylated in the producers. Reversal of methylation in the nonproducing cells with azacytidine results in prolactin expression. On the other hand, not all organisms have methylation. For example, *Drosophila* DNA is not methylated. In organisms in which methylation does occur, it increases susceptibility to certain kinds of mutations, which are discussed in Chapter 14.

FIGURE 12-14 Structures of cytosine and 5-methylcytosine.

Cytosine

5-Methylcytosine

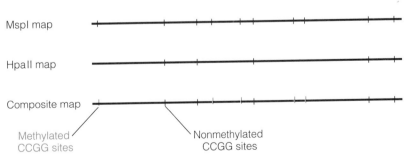

FIGURE 12-15 Detection of methylated cytosines in CCGG sequences by means of restriction enzymes. The enzyme MspI cleaves all CCGG sites irrespective of methylation, whereas HpaII cleaves only nonmethylated sites. The positions of the methylated sites are determined by comparing the restriction maps.

12-6 Regulation of Transcription

Many eukaryotic genes code for essential metabolic enzymes or cellular components and are expressed constitutively at relatively low levels in all cells; these are called **housekeeping genes.** The expression of other genes differs from one cell type to the next or among different stages of the cell cycle; these genes are often regulated at the level of transcription. In prokaryotes, the levels of expression in induced and uninduced cells may differ by a thousandfold or more. Such extreme levels of induction are uncommon in eukaryotes, except for some genes in lower eukaryotes such as yeast. Most eukaryotic genes are induced by factors ranging from 2 to 10.

In this section, we consider some components of transcriptional regulation in eukaryotes.

Yeast Mating Type

As mentioned earlier, the mating type of a cell is controlled by the allele of the *MAT* gene that is present (refer to Figure 12-13 for the genetic basis of mating-type interconversion in yeast). Both *MAT*a (mating-type **a**) and *MAT*α (mating-type α) express a set of haploid-specific genes. They differ in that *MAT*a expresses a set of **a**-specific genes and *MAT*α expresses a set of α-specific genes. The functions of the mating-type-specific genes include (1) secretion of a mating peptide that arrests cells of the opposite mating type before DNA synthesis and prepares them for cell fusion, and (2) production of a receptor for the mating peptide secreted by the opposite mating type. Therefore, when **a** and α cells are in proximity, they prepare each other for mating and undergo fusion.

Regulation of mating type is at the level of transcription according to the regulatory circuits diagrammed in Figure 12-16. The regulatory interactions were originally proposed on the basis of the phenotypes of various types of mutants, and many of the details have since been confirmed by direct molecular studies. The symbols asg, αsg, and hsg represent the **a**-specific genes,

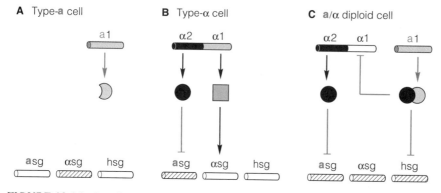

FIGURE 12-16   Regulation of mating type in yeast. The symbols asg, αsg, and hsg denote sets of **a**-specific genes, α-specific genes, and haploid-specific genes, respectively; sets of genes represented with open bars are *on* and those with hatched bars are *off*. A. In an **a** cell, the a1 peptide is inactive and the sets of genes manifest their basal states of activity—namely asg and hsg *on* and αsg *off*, and so the cell is an **a** haploid. B. In an α cell, the α2 peptide turns the asg *off* and the α1 peptide turns the αsg *on*, and so the cell is an α haploid. C. In an **a** / α diploid, the α1 and a1 peptides form a complex that turns the hsg *off*, the α2 peptide turns the asg *off*, and the αsg manifest their basal activity of *off*, and so physiologically the cell is non-**a**, non-α, and nonhaploid (that is, a normal diploid).

the α-specific genes, and the haploid-specific genes, respectively; each set of genes is represented as a single bar, which is crosshatched to indicate when transcription does not occur. In a cell of mating-type **a** (Figure 12-16A), the *MATa* region is transcribed and produces a polypeptide called **a**1 (represented as a kidney-shaped object). By itself, **a**1 has no regulatory activity and, in the absence of any regulatory signal, transcription of asg and hsg but not αsg occurs. In a cell of mating-type α (Figure 12-16B), the *MAT*α region is transcribed and two regulatory proteins, denoted α1 and α2, are produced (represented by the round and square objects): α1 is a *positive regulator* of the α-specific genes, and α2 is a *negative regulator* of the **a**-specific genes. The result is that αsg and hsg are transcribed, but transcription of asg is turned off. Both α1 and α2 bind with particular DNA sequences upstream from the genes that they control.

In the diploid (Figure 12-16C), both *MATa* and *MAT*α are transcribed, but the only polypeptides produced are **a**1 and α2. The reason is that the **a**1 and α2 polypeptides combine as subunits to form a negative regulatory protein that combines with a particular DNA sequence and represses transcription of the α1 gene in *MAT*α and of the haploid-specific genes. The α2 polypeptide acting alone is a negative regulatory protein that turns off asg. Bcause α1 is not produced, transcription of αsg is not turned on. The result is that the **a**/α diploid does not transcribe either mating-type-specific set of genes or the haploid-specific genes.

Transcriptional regulation of the mating-type genes includes negative control (**a**-specific genes and haploid-specific genes) and positive control (α-specific genes). Although both negative and positive regulation of transcription occurs in eukaryotes, positive regulation is the more-usual situation. The type of regulatory proteins required are the subject of the next subsection.

Transcriptional Activator Proteins

The α1 protein is an example of a **transcriptional activator protein,** which must bind with an upstream DNA sequence to prepare a gene for transcription. Transcriptional activator proteins may participate in the assembly of a transcriptional complex including RNA polymerase or they may initiate transcription by an already assembled complex. In either case, the activator proteins are essential for the transcription of genes that are positively regulated.

Many transcriptional activator proteins can be grouped into categories on the basis of shared characteristics of their amino acid sequences. For example, one category has a **helix-turn-helix** motif, which consists of a sequence of amino acids forming a pair of α-helices separated by a bend; the helices are so situated that they can fit neatly into the grooves of a double-stranded DNA molecule. The helix-turn-helix motif is the basis of the DNA's binding ability, although the sequence specificity of the binding results from other parts of the protein. The α2 protein that regulates yeast mating type has a helix-turn-helix motif, as do many other transcriptional activator proteins in both prokaryotes and eukaryotes.

A second large category of transcriptional activator proteins includes a DNA-binding motif called a **zinc finger** because the folded structure incorporates a zinc ion. An example is the product of the *GAL4* gene in yeast, which codes for a transcriptional activator of the genes for utilization of galactose. Part of the GAL4 protein is illustrated in Figure 12-17. A zinc ion (Zn^{2+}, shown in red) is chelated by bonds with four cysteine residues in characteristic positions at the base of a loop that extends for an additional 841 amino acids beyond those shown. The part illustrated constitutes most of the DNA-binding region, and the rest of the molecule determines transcriptional activation of the galactose

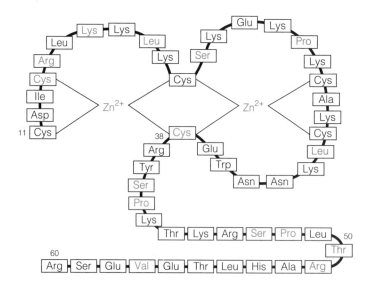

genes. Each amino acid shown in red is the site of a mutation in which the amino acid is replaced by a different one; the mutant proteins are unable to activate transcription. Replacements at amino acid positions 15 (Arg → Gln), 26 (Pro → Ser), and 57 (Val →Met) are particularly interesting because they provide genetic evidence that zinc is necessary for DNA binding. In particular, the mutant phenotypes can be rescued by extra zinc in the growth medium because the molecular defect reduces the ability of the zinc-finger part of the molecule to chelate zinc; extra zinc in the medium is sufficient for chelation in spite of the defect. It has also been shown that the extra zinc restores the ability of the mutant activator protein to attach to its particular binding sites in the DNA.

Hormonal Regulation

Of the known regulators of transcription in higher eukaryotes, the **hormones**—small molecules or polypeptides that are carried from hormone-producing cells to target cells—constitute a group that has been studied in some detail. A large class of hormones consists of small molecules synthesized from cholesterol; these are the steroid hormones, and they include the principal sex hormones. Many of the steroid hormones act by turning on the transcription of specific sets of genes. If a hormone regulates transcription, it must somehow signal the DNA. The mechanism by which this occurs is outlined in Figure 12-18. A steroid hormone penetrates a target cell through diffusion because steroids are hydrophobic (nonpolar) molecules that pass freely through the cell membrane and the nuclear envelope. The nuclei of target cells contain specific receptor proteins that form complexes with the hormone. The receptor usually undergoes some modification in three-dimensional shape after the complex has formed, and this enables the receptor-hormone complex to bind with particular sequences in the DNA and stimulate transcription. Nontarget cells do not contain the nuclear receptors and so are unaffected by the hormone.

A well-studied example of induction of transcription by a hormone is the stimulation of the synthesis of ovalbumin in the chicken oviduct by the steroid sex hormone **estrogen.** When chickens are injected with estrogen, oviduct tissue responds by synthesizing ovalbumin mRNA. This synthesis continues as long as estrogen is administered. Once the hormone is withdrawn, the rate of synthesis decreases. Both before the injection of the hormone and sixty hours

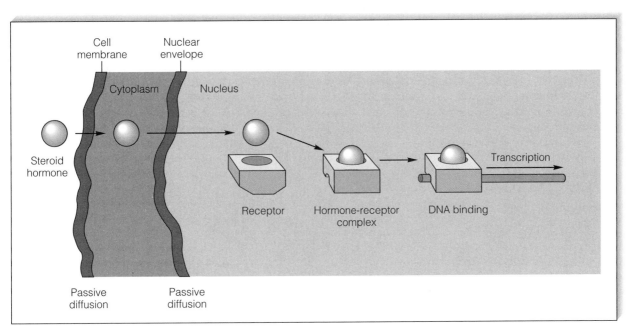

FIGURE 12-18 A schematic diagram showing how a steroid hormone reaches a DNA molecule and triggers transcription by binding with a receptor in the nucleus to form a transcriptional activator. Entry into the cytoplasm and the nucleus is by passive diffusion.

after withdrawal, no ovalbumin mRNA is detectable. When estrogen is given to chickens, only the oviduct synthesizes mRNA because other tissues lack the hormone receptor.

Transcriptional Enhancers

Hormone receptors and other transcriptional activator proteins bind with particular DNA sequences known as **enhancers.** Enhancer sequences are typically rather short (usually fewer than 20 base pairs) and are found at a variety of possible locations around the gene affected. Most enhancers are upstream of the transcriptional start site (sometimes many kilobases away), others are in introns within the coding region, and a few are even located in the 3′ end of genes. One of the most thoroughly studied enhancers is in the mouse mammary tumor virus and determines transcriptional activation by the glucocorticoid steroid hormone. The consensus sequence of the enhancer is AGAQCAGQ, in which Q stands for either A or T. The virus contains five copies of the enhancer positioned throughout its genome (Figure 12-19), which provide five binding sites for the hormone-receptor complex that activates transcription.

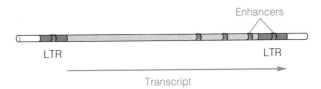

FIGURE 12-19 Positions of enhancers (red) in the mouse mammary tumor virus that allow transcription of the viral sequence to be induced by glucocorticoid steroid hormone. LTR stands for long terminal repeat, a DNA sequence found at both ends of the virus.

Enhancers are essential components of gene organization in eukaryotes because they enable genes to be transcribed only when proper transcriptional activators are present. Some enhancers respond to molecules outside the cell—for example, steroid hormones that form receptor-hormone complexes. Other enhancers respond to molecules that are produced inside the cell—for example, during development—and these enhancers enable the genes under their control to participate in cellular differentiation or to be expressed in a tissue-specific manner. Many genes are under the control of several different enhancers, and so they can respond to a variety of different molecular signals, both external and internal.

Figure 12-20 illustrates several of the genetic elements found in a typical eukaryotic gene. The transcriptional complex binds to the promoter (P) to initiate RNA synthesis. The coding regions of the gene (exons) are interrupted by one or more intervening sequences (introns) that are eliminated in RNA processing. Regulation of transcription is by means of enhancer elements (numbered 1 through 6) that respond to different molecules that serve as induction signals. The enhancers are located both upstream and downstream of the promoter, and some (in this example, enhancer 1) are present in multiple copies.

Many enhancers stimulate transcription by means of **DNA looping,** which refers to physical interactions between relatively distant regions along the DNA. The mechanism is illustrated in Figure 12-21. Part A shows the physical relation between an enhancer and a promoter. In this example, the enhancer is upstream (5′) of the promoter. Transcription requires that a group of protein factors bind with the promoter. These are called **general transcription factors** because they are common to the promoters of many different genes. However, presence of the general transcription factors is not sufficient for transcription to occur (Figure 12-21B). The enhancer must also bind with suitable transcriptional activator proteins. When the enhancer is bound, a DNA loop forms between the promoter and enhancer regions, bringing them into physical contact, and the physical interaction between the bound promoter and enhancer regions activates transcription (Figure 12-21C).

The versatility of some enhancers results from their ability to interact with two different promoters in a competitive fashion; that is, at any one time, the enhancer can stimulate one promoter or the other, but not both. An example of this mechanism is illustrated in Figure 12-22, in which P1 and P2 are alternative promoters that compete for an enhancer located between them (part A). When complexed with a special transcriptional activator protein for promoter P1, the enhancer binds preferentially with promoter P1 and stimulates transcription (part B). When complexed with a different transcriptional activator protein for promoter P2, the enhancer binds preferentially with promoter P2 and stimulates transcription from it (part C). In this way, competition for the enhancer serves as a sort of switch mechanism between the expression of the P1

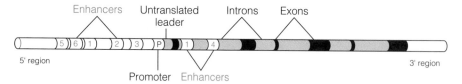

FIGURE 12-20   Schematic diagram of the organization of a typical gene in a higher eukaryote. Scattered throughout the sequence, but tending to be concentrated near the 5′ region, are a number of different enhancer sequences (red numbers). The enhancers are binding sites for transcriptional activator proteins that enable temporal and tissue-specific regulation of the gene.

FIGURE 12-21
Transcriptional activation by means of DNA looping: (A) relation between enhancer and promoter and the protein factors that bind to them; (B) binding of general transcription factors to the promoter is necessary for transcription but not sufficient; (C) transcriptional activator proteins bind with the enhancer, and this complex interacts physically with the bound promoter by forming a loop in the DNA. The physical interaction between the regions stimulates transcription.

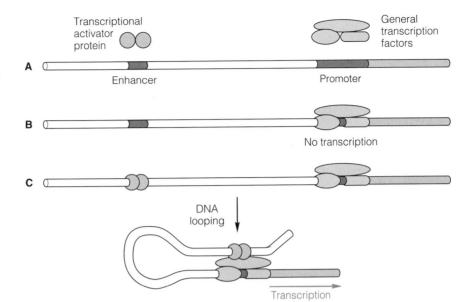

or P2 promoters. This regulatory mechanism is present in chickens in the developmental switch between embryonic and adult β-globin polypeptide chains. In this case, the embryonic globin gene and the adult gene compete for a single enhancer, which in the course of development changes its preferred binding from the embryonic promoter to the adult promoter. In human beings, enhancer competition appears to control the developmental switch from the fetal γ-globin to the adult β-globin polypeptide chains. In persons in whom the β-globin promoter is deleted or altered in sequence and unable to bind with the enhancer, there is no competition for the enhancer molecules, and the γ-globin genes continue to be expressed in adult life when normally they would not be transcribed. Adults who have these types of mutations have fetal hemoglobin instead of the adult forms, but the clinical manifestations of the condition are very mild.

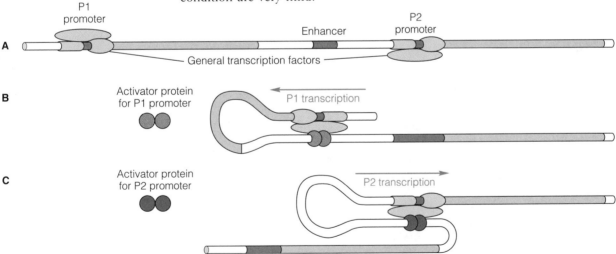

FIGURE 12-22 Genetic switching regulated by competition for an enhancer. A. Promoters P1 and P2 compete for a single enhancer located between them. When complexed with an appropriate transcriptional activator protein, the enhancer binds preferentially with promoter P1 (part B) or promoter P2 (part C). Binding to the promoter stimulates transcription. If either promoter is mutated or deleted, the enhancer binds with the alternative promoter. The location of the enhancer relative to the promoters is not critical.

The Logic of Combinatorial Control

Because the genome of a complex eukaryote contains many possible enhancers that respond to different signals or cellular conditions, in principle each gene could be regulated by its own distinct combination of enhancers. This is called **combinatorial control,** and it is a powerful means of increasing the complexity of possible regulatory states by employing several simple regulatory states in combination.

To consider a simple example, suppose that a gene has a single binding site for only one type of regulatory molecule. Then there are only two regulatory states (call them $+$ or $-$) according to whether the binding site is occupied. If the gene has single binding sites for each of two different regulatory molecules, then there are four possible regulatory states (namely, $+ +$, $+ -$, $- +$, and $- -$). Single binding sites for three different regulatory molecules yield eight combinatorial states, and in general n different types of regulatory molecules yield 2^n states. If transcription occurs according to the particular pattern of which binding sites are occupied, then a small number of types of regulatory molecules can result in a large number of different patterns of regulation. For example, because eukaryotes contain approximately two hundred different cell types, each gene would theoretically need as few as eight $+/-$ types of binding sites to specify whether it should be on or off in each cell type, because $2^8 = 256$.

The real regulatory situation is certainly more complex than the naive calculation based on $+/-$ switches would imply, because genes are not merely on or off; rather, their level of activity is adjusted. For example, a gene may have multiple binding sites for an activator protein, and this allows the level of gene expression to be modulated according to the number of binding sites that are occupied. Furthermore, each cell type is programmed to respond to a variety of conditions both external and internal, and so the total number of regulatory states must be considerably greater than the number of cell types. Nevertheless, the naive example demonstrates that combinatorial control is extremely powerful in multiplying the number of regulatory possibilities, and therefore a large number of regulatory states does not necessarily imply a hopelessly complex regulatory apparatus.

Enhancer-Trap Mutagenesis

The ability of enhancers to regulate transcription is the basis of the **enhancer trap,** a genetically engineered transposable element designed to detect tissue-specific expression resulting from insertion of the element near enhancers. A simplified diagram of an enhancer trap based on the transposable P element in *Drosophila* is shown in Figure 12-23. The element (part A) contains a weak promoter, unable to initiate transcription without the aid of an enhancer, linked to the β-galactosidase gene from *E. coli*; the arrows represent inverted-repeat sequences necessary for transposition. When the enhancer trap transposes and inserts at a random position in the genome, no transcription occurs if the insertion is not near an enhancer (Figure 12-23B). On the other hand, if the insertion is near an enhancer, then the β-galactosidase gene will be transcribed in whatever tissues stimulate the enhancer. Any tissue-specific expression can be detected by the use of staining reagents specific for β-galactosidase; one commonly used stain turns bright blue in the presence of the enzyme.

For example, the enhancer trap has been used to identify genes expressed only in the eye. When the enhancer trap inserts into the genome and comes

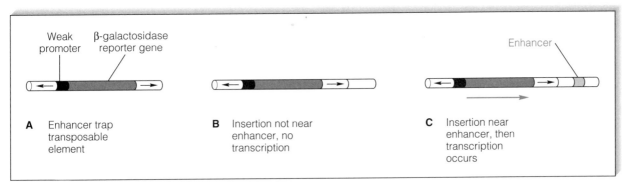

FIGURE 12-23 Identification of enhancers by genetic means: (A) the enhancer trap, consisting of a transposable element containing a gene with a weak promoter called a reporter gene because it "reports" its genetic environment through its level of activity; (B) if insertion is at a site lacking a nearby enhancer, the reporter gene cannot be activated; (C) if insertion is near an enhancer, the reporter gene is transcribed in a temporally regulated or tissue-specific manner determined by the type of enhancer.

under the control of eye-specific enhancers, the β-galactosidase is expressed in the eye and nowhere else. The eyes, and only the eyes, of flies with such insertions stain blue. Insertions of the enhancer trap provide an important method for identifying genes that are expressed in particular cells or tissues because the insertions do not necessarily disrupt the function of the normal gene. Furthermore, the presence of DNA sequences from the *P* element at the site of insertion provides a molecular tag with which to clone the gene, using methods similar to those described in Section 10-7.

Alternative Promoters

Some eukaryotic genes have two or more promoters that are active in different cell types. The different promoters result in different primary transcripts that contain the same protein-coding regions. An example from *Drosophila* is shown in Figure 12-24. The gene codes for alcohol dehydrogenase, and its

FIGURE 12-24 Use of alternative promoters in the gene for alcohol dehydrogenase in *Drosophila*: (A) overall gene organization in which there are two introns within the amino acid coding region; (B) transcription in larvae uses the promoter nearest the 5′ end of the coding region; (C) transcription in adults uses a promoter farther upstream and much of the larval leader sequence is removed by splicing.

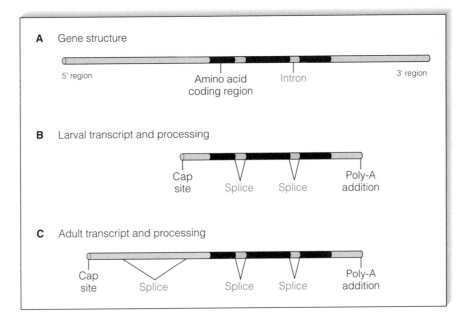

organization in the genome is shown in Figure 12-24A); there are three protein-coding regions interrupted by two introns. Transcription in larvae (part B) uses a different promoter from that used in transcription in adults (part C). The adult transcript has a longer 5′ leader sequence, but most of this is eliminated during splicing. Alternative promoters allow independent regulation of transcription in larvae and adults.

12-7 Alternative Splicing

Even when the same promoter is used to transcribe a gene, different cell types can produce different quantities of the protein—or even different proteins—because of differences in the mRNA produced in processing. The reason is that the same transcript can be spliced differently from one cell type to the next. The different splicing patterns may include exactly the same protein-coding exons, in which case the protein is identical but the rate of synthesis differs because the mRNA molecules are not translated with the same efficiency. In other cases, the protein-coding part of the transcript has a different splicing pattern in each cell type, and the resulting mRNA molecules code for proteins that are not identical even though they share certain exons.

In the synthesis of α-amylase in the rat, different mRNA molecules are produced from the same gene because of different patterns of intron removal in RNA processing. The rat salivary gland produces more of the enzyme than does the liver, although the same coding sequence is transcribed. In each cell type, the same primary transcript is synthesized, but two different splicing mechanisms are used. The initial part of the primary transcript is shown in Figure 12-25. The coding sequence begins 50 base pairs inside exon 2 and is formed by joining exon 3 and subsequent exons. In the salivary gland, the primary transcript is processed such that exon S is joined with exon 2 (that is, exon L is removed as part of introns 1 and 2). In the liver, exon L is joined with exon 2, and exon S is removed along with intron 1 and the leader L. The exons S and L become alternate 5′ leader sequences of the amylase mRNA, and the alternative mRNAs are translated at different rates.

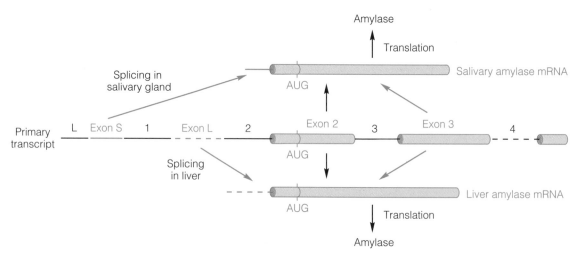

FIGURE 12-25   Production of distinct amylase mRNA molecules by different splicing events in cells of the salivary gland and liver of the mouse. The leader, L, and the introns are in black. The exons are red. The coding sequence begins at the AUG codon in exon 2. Exons S and L form part of the untranslated 5′ end of the mRNA in salivary glands and liver, respectively.

In chicken skeletal muscle, two different forms of the muscle protein myosin are made from the same gene. A different primary transcript is made from each of two promoters, and these transcripts are processed differently to form mRNA molecules encoding distinct forms of the protein. In *Drosophila*, the myosin RNA is processed in four different ways, the precise mode depending on the stage of development of the fly. One class of myosin is found in pupae and another in the later embryo and larval stages. How the mode of processing is varied is not known.

12-8 Translational Control

In bacteria, most mRNA molecules are translated about the same number of times, with little variation from gene to gene. In eukaryotes, translation is sometimes regulated. The principal types of translational control are:

1. Inability of an mRNA molecule to be translated unless a molecular signal is present.
2. Regulation of the lifetime of a particular mRNA molecule.
3. Regulation of the rate of overall protein synthesis.

In this section, examples of each of these modes of regulation will be presented.

An important example of translational regulation is that of **masked mRNA.** Unfertilized eggs are biologically static but, shortly after fertilization, many new proteins must be synthesized—for example, the proteins of the mitotic apparatus, the cell membranes, as well as others. Unfertilized sea-urchin eggs can store large quantities of mRNA for many months in the form of mRNA-protein particles made during formation of the egg. This mRNA is translationally inactive but, within minutes after fertilization, translation of these molecules begins. Here, the timing of translation is regulated; the mechanisms for stabilizing the mRNA, for protecting it against RNases, and for activation are unknown.

Translational regulation of another type also occurs in mature unfertilized eggs. These cells need to maintain themselves but do not have to grow or undergo a change of state. Thus, the rate of protein synthesis in eggs is generally low. This is not a consequence of an inadequate supply of mRNA but apparently results from failure to form the ribosome-mRNA complex.

The synthesis of some proteins is regulated by direct action of the protein on the mRNA. For instance, the concentration of one type of antibody molecule is kept constant by self-inhibition of translation; that is, the antibody molecule itself binds specifically to the mRNA that encodes it and thereby inhibits initiation of translation.

A dramatic example of translational control is the extension of the lifetime of silk fibroin mRNA in the silkworm. During cocoon formation, the silk gland of the silkworm predominantly synthesizes a single type of protein, silk fibroin. Because the worm takes several days to construct its cocoon, it is the *total amount* and not the rate of fibroin synthesis that must be great; the silkworm achieves this by synthesizing a fibroin mRNA molecule that has a very long lifetime. Transcription of the fibroin gene is initiated at a strong promoter and about 10^4 fibroin mRNA molecules are made in a period of several days. (This synthesis is under transcriptional regulation.) A typical eukaryotic mRNA molecule has a lifetime of about three hours before it is degraded. However, fibroin mRNA survives for several days during which each mRNA molecule is translated repeatedly to yield 10^5 fibroin molecules. Thus, each gene is responsible for the synthesis of 10^9 protein molecules in four days. Altogether, the silk gland makes about 10^{15} molecules of fibroin during this period. If the lifetime of the mRNA were not extended, either twenty-five times as many genes would be needed or synthesis of the required fibroin would take about one hundred days.

Another example of an mRNA molecule with an extended lifetime is the mRNA encoding casein, the major protein of milk, in mammary glands. When the hormone prolactin is received by the gland, the lifetime of casein mRNA increases. Synthesis of the mRNA also continues, and so the overall rate of production of casein is markedly increased by the hormone. When the body no longer supplies prolactin, the concentration of casein mRNA decreases because the RNA is degraded more rapidly, and lactation terminates.

Multiple Proteins from a Single Segment of DNA

In prokaryotes, coordinate regulation of the synthesis of several gene products is accomplished by regulating the synthesis of a single polycistronic mRNA molecule encoding all of the products. The analog to this arrangement in eukaryotes is the synthesis of a **polyprotein,** a large polypeptide that is cleaved after translation to yield several different proteins. In such a system, the coding sequences of each gene encoding a protein in the polyprotein unit are not separated by stop and start codons but instead by codons for specific *amino acid sequences* that are recognized as cleavage sites by particular protein-cutting enzymes. Polyproteins with as many as ten cleavage sites have been observed; the cleavage sites are not cut simultaneously, but they are cut in a specific order. Use of a polyprotein serves to maintain an equal molar ratio of the constituent proteins. A temporal sequence in the appearance of the individual proteins can be accomplished by postponing cleavage at certain sites, which is a mechanism of regulation frequently used by animal viruses.

Many examples of polyproteins are known. The synthesis of the RNA precursors uridine triphosphate and cytidine triphosphate proceeds by a biosynthetic pathway that at one stage has the reaction sequence

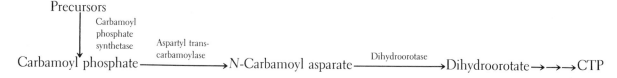

In bacteria, each of the three enzymes is made separately. In yeast and *Neurospora*, only two proteins are made; one is dihydroorotase and other is a large protein that is cleaved to form carbamoyl phosphate synthetase and aspartyl transcarbamoylase. Mammals synthesize a single tripartite protein, which is cleaved to form the three enzymes.

12-9 Is There a General Principle of Regulation?

With microorganisms, whose environment frequently changes drastically and rapidly, a general principle of regulation is that bacteria make what they need when it is required and in appropriate amounts. Although such extraordinary efficiency is common in prokaryotes, it is rare in eukaryotes. For example, efficiency appears to be violated in the large amount of DNA in the genome that has no protein-coding function and in the large amount of intron RNA that is discarded.

What is abundantly clear is that there is no universal regulatory mechanism. Many control points are possible and different genes are regulated in different ways. Furthermore, evolution has not always selected for simplicity in regulatory mechanisms but merely for something that works. If a cumbersome regulatory mechanism were to arise, it would in time evolve, be refined, and become more effective, but it would not necessarily become simpler. On the whole, regulatory mechanisms include a variety of seemingly *ad hoc* processes, each of which has stood the test of time, primarily because it works.

CHAPTER SUMMARY

Most cells do not synthesize molecules that are not needed. There are important opportunities for controlling gene expression in transcription, RNA processing, and translation. Gene expression can also be regulated through the stability of mRNA, and the activity of proteins can be regulated in a variety of ways after translation. The processes of gene regulation generally differ between prokaryotes and eukaryotes.

In bacteria, the synthesis of most proteins is regulated by controlling the rate of transcription of the genes coding for the proteins. The concentration of proteins needed continuously is autoregulated, often by direct binding of the protein to or near its promoter.

The synthesis of degradative enzymes needed only on occasion, such as the enzymes required to metabolize lactose, is typically regulated by an off-on mechanism. When lactose is present, the genes coding for the enzymes required to metabolize lactose are transcribed; when lactose is absent, transcription does not occur. Lactose metabolism is negatively regulated. Two enzymes required to degrade lactose—permease (required for the entry of lactose into bacteria) and β-galactosidase (the actual degrading enzyme)—are coded by a single polycistronic mRNA molecule, *lac* mRNA. Immediately adjacent to the promoter for *lac* mRNA is a regulatory sequence of bases called an operator. A repressor protein is made by still another adjacent gene, and this protein binds tightly to the operator, thereby preventing RNA polymerase from initiating transcription at the promoter. Lactose is an inducer of transcription, because it can bind to the repressor and prevent the repressor from binding to the operator. Therefore, in the presence of lactose, there is no active repressor, and the *lac* promoter is always accessible to RNA polymerase. The repressor gene, the operator, the promoter, and the structural genes are adjacent to one another and together constitute the *lac* operon.

Repressor mutations have been isolated that inactivate the repressor protein, and operator mutations are known that prevent recognition of the operator by an active repressor; such mutations cause continuous production of *lac* mRNA and are said to be constitutive.

When lactose is cleaved by β-galactosidasee, the products are glucose and galactose. Glucose is metabolized by enzymes that are made continually; galactose is broken down by the inducible galactose operon. When glucose is present in the growth medium, the enzymes for degrading lactose and other sugars are unnecessary. The general mechanism for preventing transcription of many sugar-degrading operons is as follows. High concentrations of glucose suppress the synthesis of a small molecule—cyclic AMP (cAMP). Initiation of transcription of many sugar-degrading operons requires binding of a particular protein molecule, called CRP, to a specific region of the promoter. Binding occurs only after CRP has first bound cAMP and formed a cAMP-CRP complex. Only when glucose is absent is the concentration of cAMP sufficient to produce cAMP-CRP and hence to permit transcription of the sugar-degrading operons. Thus, in contrast with a repressor, which must be removed before transcription can begin (negative regulation), cAMP is a positive regulator of transcription.

Biosynthetic enzyme systems exemplify another type of transcriptional regulation. In the synthesis of tryptophan, transcription of the genes encoding the *trp* enzymes is controlled by the concentration of tryptophan in the growth medium. When excess tryptophan is present, it binds with the *trp* aporepressor to form the active repressor that prevents transcription. The *trp* operon is also regulated by attenuation, in which transcription is initiated continually but the transcript forms a hairpin structure resulting in premature termination. The frequency of termination of transcription is determined by the availability of tryptophan; with decreasing concentration of tryptophan, termination occurs less often and the enzymes for tryptophan synthesis are made, thereby increasing the concentration of tryptophan. Attenuators also regulate operons for the synthesis of other amino acids.

Eukaryotes employ a variety of genetic regulatory mechanisms, occasionally including changes in DNA. The number of copies of a gene may be increased by DNA amplification, programmed rearrangements of DNA may occur, and in some cases gene inactivity is associated with methylation of cytosine bases.

Many genes in eukaryotes are regulated at the level of transcription. Although both negative and positive regulation occurs, positive regulation is the usual situation. In the control of yeast mating type by the *MAT* locus, the **a**-specific genes and the haploid-specific genes are negatively regulated and the α-specific genes are positively regulated. In general, positive regulation occurs through transcriptional activator proteins that contain particular structural motifs that bind to DNA—for example, the helix-turn-helix motif or the zinc-finger motif. Hormones also can regulate transcription. Steroid hormones

bind with receptor proteins in the nucleus and form transcriptional activators.

Transcriptional activators bind to DNA sequences known as enhancers, which are usually short sequences that may be present at a variety of positions around the genes they regulate. Enhancers can be identified genetically by the use of transposable elements containing genes that come under the control of enhancers near the sites of insertion. There are many types of enhancers responsive to different transcriptional activators. Combining different types of enhancers provides a large number of possible types of regulation.

Some genes contain alternative promoters used in different tissues, other genes use a single promoter but the transcripts are spliced in different ways. Alternative splicing can result in mRNA molecules that are translated with different efficiencies or even into different proteins if there is alternative splicing of the protein-coding exons.

Regulation can also occur in translation—for example, through masked mRNAs or through factors affecting mRNA stability. Some proteins are regulated coordinately because they are produced by cleavage of a polyprotein coded in a single mRNA.

KEY TERMS

aporepressor
attenuation
attenuator
autoregulation
β-galactosidase
cAMP-CRP complex
cassette
cis-dominant
combinatorial control
constitutive mutant
coordinate regulation
cyclic adenosine monophosphate
  (cAMP)
cyclic AMP receptor protein (CRP)
DNA looping

DNA methylase
effector
enhancer
enhancer trap
estrogen
gene amplification
gene dosage
general transcription factor
gene regulation
helix-turn-helix
homothallism
hormone
housekeeping genes
inducer
inducible transcription

lac operon
lactose permease
leader polypeptide
masked mRNA
mating-type interconversion
methylation
negative regulation
operator
operon model
polyprotein
positive regulation
repressor
transcriptional activator protein
zinc finger

EXAMPLES OF WORKED PROBLEMS

Problem 1: A gene *R* codes for a protein that is a negative regulator of transcription of a gene *S*. Is gene *S* transcribed in an *R*⁻ mutant? How does the situation differ if the product of *R* is a positive regulator of *S* transcription?

Answer: A negative regulator of transcription is needed to turn transcription off; hence, the *R*⁻ mutant has transcription of *S* occurring constitutively. The opposite happens in positive regulation. A positive regulator of transcription is needed to turn transcription on; so, if the product of *R* is a positive regulator, transcription does not occur in *R*⁻ cells.

Problem 2: For each of the following partial diploid genotypes, state whether β-galactosidase is made and whether its synthesis is inducible or constitutive. The convention for writing partial diploid genotypes is to put the plasmid genes at the left of the slash and the chromosomal genes at the right.

(a) *lacZ⁺ lacY⁻ / lacZ⁻ lacY⁺*
(b) *lacOᶜ lacZ⁻ lacY⁺ / lacZ⁺ lacY⁻*
(c) *lacP⁻ lacZ⁺ / lacOᶜ lacZ⁻*
(d) *lacI⁺ lacP⁻ lacZ⁺ / lacI⁻ lacZ⁺*
(e) *lacI⁺ lacP⁻ lacY⁺ / lacI⁻ lacY⁻*

Answer: The location of the genes does not matter because they function in the same manner whether present in a plasmid or in a chromosome. **(a)** In this partial diploid, there are no *cis*-dominant mutations. The plasmid operon is *lacZ⁺* and can make the enzyme, whereas the chromosomal operon is *lacZ⁻* and cannot. The cell will produce the enzyme from the plasmid gene as long as the operon can be turned on. Turning it on requires the presence of functional regulatory elements, which is the case in this example. (Recall that, if a gene is not listed, for example *lacI*, it is assumed to be wildtype.) It is also necessary that the inducer be able to enter the cell, which it can do because the chromosomal genotype can supply the *lacY* permease. Thus, for this partial diploid, β-galactosidase can be made, and its synthesis is inducible. **(b)** This partial diploid has the *cis*-dominant mutation *lacOᶜ* in the plasmid, and so the genes in the plasmid operon are always expressed. However, the *lacZ* gene in the plasmid makes a defective enzyme. The chromosome has a functional *lacZ* gene from which active enzyme can be made, but its synthesis is under the control of a normal operator (because the operator genotype is not indicated, it is wildtype). Hence, enzyme synthesis must be inducible. Thus, the partial diploid makes a defective enzyme constitutively and

a normal enzyme inducibly, and so the overall phenotype is that the cell can be induced to make β-galactosidase. **(c)** A promoter mutation, which is *cis*-dominant, is in the plasmid, which means that no *lac* mRNA can be made from this operon. The chromosome contains a *cis*-dominant *lacOᶜ* mutation that drives constitutive synthesis of an mRNA, but the enzyme is defective. Thus, there is no way for the cell to make active β-galactosidae. **(d)** The plasmid genotype contains a promoter mutation, and so no mRNA can be produced from the *lacZ* structural gene. However, the *lacI* gene in the plasmid has its own promoter, and so *lac* repressor molecules are present in the cell. The chromosomal genotype alone would make the enzyme constitutively because of the *lacI* mutation, but the presence of the functional *lacI* product made by the plasmid means that any synthesis that occurs must be induced. The chromosomal operon can provide both β-galactosidae and permease and so β-galactosidase is inducible in this genotype. **(e)** This genotype differs from that in part **d** by the presence of a *lacY⁻* mutation in the chromosome. Again, the plasmid contributes only *lacI* repressor to the cell, and so any synthesis of the enzyme must be inducible. However, the chromosomal genotype is *lacY⁻*. Because the *lacY⁺* allele in the plasmid cannot be expressed, no inducer can enter the cell, and so the cell is unable to make any enzyme.

Problem 3: With regard to mating type in yeast, what phenotypes would each of the following haploid cells exhibit?

(a) A duplication of the mating-type gene giving the genotype MATa MATα.
(b) A deletion of the *HML*α cassette in a MATa cell.

Answer: **(a)** The haploid cell expresses both **a** and α, and so the **a**-specific genes, the α-specific genes, and the haploid-specific genes are inactive. The phenotype is that of an **a** / α diploid. **(b)** The phenotype is that of a normal **a** haploid, but mating-type switching to α cannot occur.

PROBLEMS

12-1. Distinguish between positive regulation of transcription and negative regulation of transcription.

12-2. Distinguish between a repressor and an aporepressor. Give an example of each.

12-3. What is autoregulation? Distinguish between positive and negative autoregulation. Which would be used to amplify a weak induction signal? Which to prevent overproduction?

12-4. What term describes a gene that is expressed continually?

12-5. The metabolic pathway for glycolysis is responsible for the degradation of glucose and is one of the fundamental energy-producing systems in living

cells. Would you expect the enzymes in this pathway to be regulated? Why or why not?

12-6. Among mammals, the reticulocyte cells in the bone marrow extrude their nuclei in the process of differentiation into red blood cells. Yet the reticulocytes and red blood cells continue to synthesize hemoglobin. Suggest a mechanism by which hemoglobin synthesis can continue for a long time in the absence of the hemoglobin genes.

12-7. Several eukaryotes are known in which a single effector molecule regulates the synthesis of different proteins coded by distinct mRNA molecules X and Y. At what level in the process of gene expression does regulation occur in each of the following situations:

(a) Neither nuclear nor cytoplasmic RNA can be found for either X or Y.

(b) Nuclear but not cytoplasmic RNA can be found for both X and Y.

(c) Both nuclear and cytoplasmic RNA can be found for both X and Y, but none of it is associated with polysomes.

12-8. Is it necessary for the gene that codes for the repressor of a bacterial operon to be near the structural genes? Why or why not?

12-9. Consider a eukaryotic transcriptional activator protein that binds to an enhancer sequence and promotes transcription. What change in regulation would you expect from a duplication in which several copies of the enhancer were present instead of just one?

12-10. The following questions pertain to the *lac* operon in *E. coli*:

(a) Which proteins are regulated by the repressor?

(b) How does binding of the *lac* repressor to the *lac* operator prevent transcription?

(c) Is production of the *lac* repressor constitutive or or induced?

12-11. The permease of *E. coli* that transports the α-galactoside melibiose can also transport lactose, but it is temperature sensitive: lactose can be transported into the cell at 30°C but not at 37°C. In a strain that produces the melibiose permease constitutively, what are the phenotypes of *lacZ⁻* and *lacY⁻* mutants at 30°C and 37°C?

12-12. How do inducers enable transcription to occur in a bacterial operon under negative transcriptional control?

12-13. When glucose is present in an *E. coli* cell, is the concentration of cyclic AMP high or low? Can a mutant with either an inactive adenyl cyclase gene or an inactive *crp* gene synthesize β-galactosidase? Does the binding of cAMP-CRP to DNA affect the binding of a repressor?

12-14. The operon allows a type of coordinate regulation of gene activity in which a group of enzymes with related functions are synthesized from a single polycistronic mRNA. Does this imply that all proteins in the polycistronic mRNA are made in the same quantity? Explain?

12-15. Both repressors and aporepressors bind molecules that are substrates or products of the metabolic pathway encoded by the genes in an operon. Generally speaking, which binds the substrate of a metabolic pathway and which the product?

12-16. Is an attenuator a protein binding site like an operator? Is RNA synthesis ever initiated at an attenuator?

12-17. How do steroid hormones induce transcription of eukaryotic genes?

12-18. With regard to mating-type switching in yeast, what phenotype would you expect from a type-α cell that had a deletion of the *HMR***a** casette?

12-19. What mating-type phenotype would you expect of a haploid cell of genotype *MAT***a**′, in which the prime denotes a mutation that renders the **a**1 protein inactive? What mating type would you expect in the diploid *MAT***a**′/*MAT*α?

12-20. What mating-type phenotype would you expect from a diploid cell of genotype *MAT***a**/*MAT*α with a mutation in α in which the α2 gene product functions normally in turning off the **a**-specific genes but is unable to combine with the **a**1 product?

12-21. If a wildtype *E. coli* strain is grown in a medium without lactose or glucose, how many proteins are bound to the the *lac* operon? How many are bound if the cells are grown in glucose?

12-22. A mutant strain of *E. coli* is found that produces both β-galactosidase and permease constitutively (that is, whether lactose is present or not).

(a) What are two possible genotypes for this mutant?

(b) A second mutant is isolated that produces no active β-galactosidase at any time but produces permease if lactose is present in the medium. What is the genotype of this mutant?

(c) A partial diploid is created from the mutants in (a) and (b): when lactose is absent neither enzyme is made, and when lactose is present both enzymes are made. What is the genotype of the mutant in (a)?

12-23. A *lacI⁺ lacO⁺ lacZ⁺ lacY⁺* Hfr strain is mated with an *F⁻ lacI⁻ lacO⁺ lacZ⁻ lacY⁻* strain. In the absence of any inducer in the medium, β-galactosidase is made for a short time after the Hfr and *F⁻* cells have been mixed. Explain why it is made and why only for a short time.

12-24. An *E. coli* mutant is isolated that is simultaneously unable to utilize a large number of sugars as sources of carbon. However, genetic analysis shows that the operons responsible for the metabolism of each sugar are free of mutations. What genotypes of this mutant are possible?

12-25. The histidine operon is a negatively regulated system containing ten structural genes that code for enzymes needed to synthesize histidine. The repressor protein is also coded within the operon—that is, in the polycistronic mRNA molecule that codes for the structural genes. Synthesis of this mRNA is controlled by a single operator regulating the activity of a single promoter. The corepressor of this operon is tRNA^His, to which histidine is attached. This tRNA is not coded by the operon itself. A collection of mutants with the following defects is isolated. Determine whether the histidine enzymes would be synthesized by each of the mutants and whether each mutant would be dominant, *cis*-dominant only, or recessive to its wildtype allele in a partial diploid.

(a) The promoter cannot bind with RNA polymerase.

(b) The operator cannot bind the repressor protein.

(c) The repressor protein cannot bind with DNA.

(d) The repressor protein cannot bind histidyl-tRNA^{His}.

(e) The uncharged tRNA^{His} (that is, without histidine attached) can bind to the repressor protein.

12-26. The attenuator of the histidine operon contains seven consecutive histidines. The relevant part of the attentuator coding sequence is

5′-AAACACCACCATCATCACCATCATCCT-GAC-3′.

A mutation occurs in which an additional A nucleotide is inserted immediately after the red A. What amino acid sequences are coded by the wildtype and mutant attenuators? What phenotype would you expect of the mutant?

The Genetic Basis of Development

Understanding gene regulation is central to understanding how an organism as complex as a human being develops from a fertilized egg. Genes are expressed according to a prescribed program during development to ensure that the fertilized egg divides repeatedly and the resulting cells become specialized in an orderly way to give rise to the fully differentiated organism. The genotype determines not only the events that occur during development, but also the temporal order in which the events unfold.

Genetic approaches to the study of development often make use of mutations that alter developmental patterns. These mutations interrupt

Above: A wildtype flower of *Antirrhinum majus* (snapdragon), a plant used for genetic and molecular studies of the mechanisms of flower development. (Cover of *Science*, vol. 250, 16 November 1990. From "Genetic Control of Flower Development by Homeotic Genes in *Antirrhinum majus*," by Z. Schwarz-Sommer, P. Huijser, W. Nacken, H. Saedler, and H. Sommer, pp. 931–936. Copyright 1990 by the American Association for the Advancement of Science.)

developmental processes and make it possible to identify factors that control development and to study the interactions among them. This chapter demonstrates how genetics is used in the study of development. Many of the examples come from the nematode *Caenorhabditis elegans* and the fruit fly *Drosophila melanogaster* because these organisms have been studied most intensively from the standpoint of developmental genetics.

13-1 Genetic Determinants of Development

Conceptually, the relation between genotype and development is straightforward:

> The genotype contains a developmental program that unfolds and results in the expression of different sets of genes in different types of cells.

That is, development consists of a program that results in the specific expression of some genes in one cell type and not in another. During development, cell types progressively appear that differ qualitatively in the genes that are expressed. Whether a particular gene is expressed may depend on the presence or absence of particular transcription factors, changes in chromatin structure, or the synthesis of particular receptor molecules. These and other molecular mechanisms define the pattern of interactions by which one gene controls another. Collectively, these regulatory interactions are called the **regulatory circuitry,** and they ultimately determine the fate of the cell. Moreover, many of the regulatory mechanisms that are important in development are widely used in different organisms to control different processes. For example, many important developmental events depend on regional differences in concentration of molecules within a cell or within an embryo, and the activity of enhancers may be modulated by the local concentration of such substances.

The development of an embryo is also affected by the environment in which development takes place. Although genotype and environment work together in development, the genotype determines the developmental potential of the organism. Genes determine whether a developing embryo will become a nematode, a fruit fly, a chicken, or a mouse. However, the expression of the genetically determined developmental potential is also influenced by the environment—in some cases, very dramatically.

The *lac* operon in *E. coli* and the effects of steroid hormones on transcription (Chapter 12) are examples in which genes are selectively turned on or off by the action of regulatory proteins that respond to environmental signals. The patterns of molecular interactions define the genetic regulatory circuits of these genes. A more-complex type of regulatory circuitry is that in Figure 12-16, which summarizes the genetic control of mating type in yeast. The identification of genetic regulatory circuits that operate during development is an important theme in developmental genetics. The implementation of different regulatory circuits means that genetically identical cells can become qualitatively different in the genes that are expressed.

Several key types of regulatory circuits by which genes become activated or repressed at particular stages in development or in particular tissues are illustrated in the hypothetical example in Figure 13-2 (on page 338). In this example, the developmentally regulated genes are denoted *r*, *g*, *b*, and *o*; they are color coded in red, green, blue, and orange, respectively. The initial event is transcription of the *r* gene in the oocyte (Figure 13-2A). Unequal partitioning of the *r* gene product into one region of the egg establishes a polarity or regionalization of the egg with respect to presence of the *r* gene product (part B). When cleavage occurs (part C), the polarized cell produces daughter cells either with (L) or without (R) the *r* gene product. The *r* gene product is a transcriptional activator of the *g* gene, and so the *g* gene product is expressed in cell L but not in R; that is, even at the two-cell stage, the daughter cells may differ in gene expression as a result of the initial polarization of the egg. Also in the two-cell stage, the gene *o* is expressed in both daughter cells. This gene codes for a membrane receptor.

The presence of receptor molecules in the cell membrane provides an important mechanism for signaling between cells during development, as illustrated in this example with the regulation of the *b* gene in Figure 13-2D. In the absence of the receptor, the *b* gene would be expressed in all four cells. However, expression of the *b* gene in one cell inhibits its expression in a neighboring cell because of the receptor. The mechanism is that the *b* gene product in one cell stimulates the *o* gene membrane receptor in the neighboring cell, and stimulation of the receptor represses transcription of *b* in this cell.

At the same time that *b*-gene expression is regulated by lateral inhibition, the initial activation of the *g* gene in cell L is retained in daughter cells 1 and 2 because the *g* gene product has a positive autoregulatory activity and stimulates its own transcription. Positive autoregulation is one way in which cells can amplify weak or transient regulatory signals into permanent changes in gene expression. The result of lateral inhibition of *b* expression and *g* autoregulation is that cells 1 through 4 in Figure 13-2D, though genetically identical, are developmentally different because they express different combinations of the *g* and *b* genes. Specifically, cell 1 has *g* and *b* activity, cell 2 has *g* activity only, cell 3 has *b* activity only, and cell 4 has neither.

Among the most intensively studied developmental genes are those that act early in development—that is, before tissue differentiation—because these genes establish the overall pattern of development.

13-2 Early Embryonic Development in Animals

The early development of the animal embryo establishes the basic developmental plan for the whole organism. Fertilization initiates a series of mitotic **cleavage** divisions in which the embryo becomes multicellular. There is little or no increase in overall size compared with that of the egg because the cleavage divisions are accompanied by little growth and merely partition the egg into progressively smaller cells. The cleavage divisions form the **blastula**, which is essentially a ball of about 10^4 cells containing a cavity (Figure 13-1). Completion of the cleavage divisions is followed by the formation of the **gastrula** through an infolding of part of the blastula wall and extensive cellular migration. In the reorganization of cells in the gastrula, the cells become arranged in several distinct layers. These layers establish the basic body plan of an animal. In higher plants, the developmental processes differ substantially from those in animals. These processes are considered at the end of this chapter.

A fertilized egg has unlimited developmental potential because it contains the genetic program and cytoplasmic information for giving rise to an entire organism. Unlimited developmental potential is maintained in cells produced by the early cleavage divisions. However, the developmental potential is progresssively channeled and limited during early embryonic development. Cells within the blastula usually become committed to particular developmental outcomes, or **fates**, that limit the differentiated states possible among descendant cells.

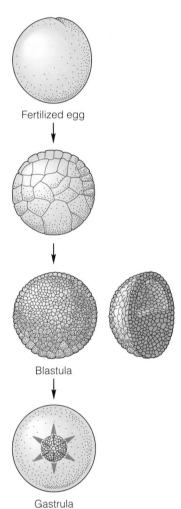

Fertilized egg

Blastula

Gastrula

FIGURE 13-1 Early development of the animal embryo. The cleavage divisions of the fertilized egg result first in a clump of cells and then a hollow ball of cells (the blastula). Extensive cell migrations form the gastrula and establish the basic body plan of the embryo.

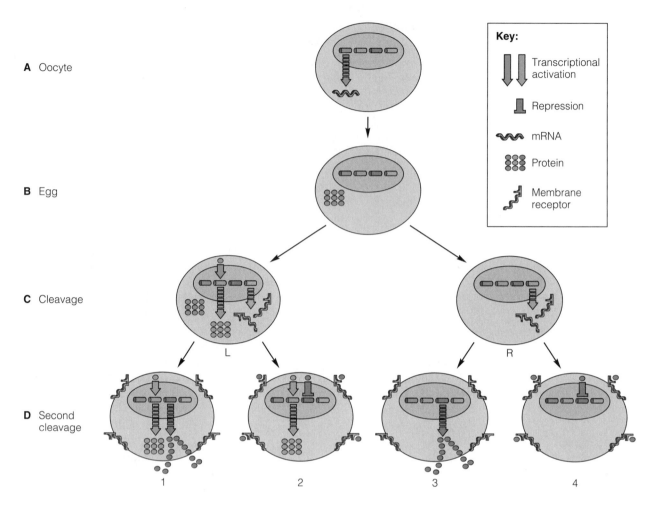

A Oocyte

B Egg

C Cleavage

D Second cleavage

L   R

1   2   3   4

Key:

Transcriptional activation

Repression

mRNA

Protein

Membrane receptor

FIGURE 13-2 Hypothetical example illustrating some regulatory mechanisms that result in differences in gene expression during development. A. The *r* gene is expressed in the oocyte. B. Polarized presence of the *r* gene product in the egg. C. Presence of the *r*-gene product in cell L stimulates transcription of the *g* gene. The *o* gene codes for a membrane receptor that is expressed in both L and R cells. D. The *g* gene has positive autoregulation and so continues being expressed in cells 1 and 2. Expression of the *b* gene in one cell represses its expression in the neighboring cell through stimulation of the *o* membrane receptor. The result of these mechanisms is that cells 1 through 4 differ in their combinations of *b*- and *g*-gene activity.

Two principal mechanisms progressively restrict the developmental potential of cells within a lineage:

1. Developmental restriction may be **autonomous,** which means that it is determined by genetically programmed changes in the cells.
2. Cells may respond to **positional information,** which means that developmental restrictions are imposed by the position of cells within the embryo. Positional information may be mediated by physical interactions between neighboring cells or by gradients in the concentration of molecules called *morphogens* that determine development.

The distinction between autonomous development and development depending on positional information is illustrated in Figure 13-3. In the normal embryo (part A) cells 1 and 2 have different fates, either because of autonomous

FIGURE 13-3 Distinction between autonomous determination and positional information. A. Cells 1 and 2 differentiate normally, as shown. B. Transplantation of the cells to reciprocal locations. If the cells differentiate autonomously, then they differentiate as they would in the normal embryo, but in their new locations. If they differentiate in response to positional information, then their new positions determine their fate.

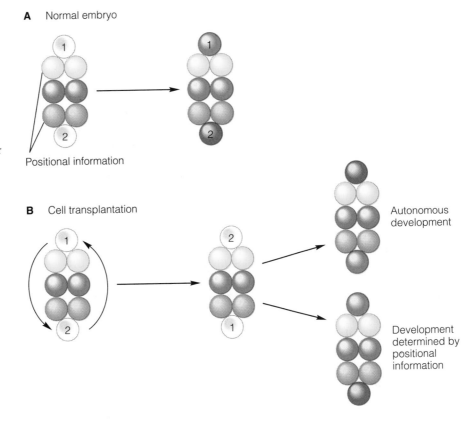

A Normal embryo

Positional information

B Cell transplantation

Autonomous development

Development determined by positional information

developmental programs or because they respond to positional information near the anterior and posterior ends. When the cells are transplanted (part B), development of cells 1 and 2 is autonomous if their developmental fate is unchanged in spite of their new locations; the embryo has anterior and posterior reversed. When development depends on positional information, the transplanted cells respond to their new locations, and the embryo is normal.

Restriction of developmental fate is usually studied by transplanting cells of the embryo to new locations to determine whether they can substitute for the cells they displace. Alternatively, individual cells are isolated from early embryos and cultured in laboratory dishes to study their developmental potential. In some eukaryotes, such as the soil nematode *Caenorhabditis elegans*, many lineages develop autonomously according to genetic programs that are induced by interactions with neighboring cells very early in embryonic development. However, nematode development also includes many examples of regulation by positional information. In *Drosophila* and the mouse, regulation by positional information is more the rule than the exception. Use of positional information is a sort of developmental compensation that helps to overcome the accidental death of individual cells during development.

Early Development Is Independent of the Zygote Genome

In most animals, the earliest events in embryonic development do not depend on genetic information in the zygote. For example, fertilized frog eggs with the nucleus removed are still able to carry out the cleavage divisions and form rudimentary blastulas. Similarly, when transcription in sea-urchin or amphibian embryos is blocked shortly after fertilization, there is no effect on the cleavage divisions or blastula formation; however, gastrulation does not occur.

In mice and probably all mammals, the embryo becomes dependent on the activity of its own genes much earlier than in lower vertebrates. The shift from control by the maternal genome to that by the zygote genome begins in the two-cell stage of development, when proteins coded by paternal genes are first detectable. Inhibitors of transcription stop development of the mouse embryo when applied at any time after the first cleavage division. However, even in mammals, the earliest stages of development are greatly influenced by the cytoplasm of the oocyte, which determines the planes of the initial cleavage divisions and other events that ultimately affect cell fate.

Composition and Organization of Oocytes

The oocyte is a diploid cell during most of oogenesis. The cytoplasm of the oocyte is extensively organized and regionally differentiated (Figure 13-4). This spatial differentiation ultimately determines the different developmental fates of cells in the blastula. The cytoplasm of the animal egg has two essential functions:

1. Storage of the molecules needed to support the cleavage divisions and the rapid RNA and protein synthesis that occurs in early embryogenesis.
2. Organization of the molecules in the cytoplasm to provide the positional information that results in differences between cells in the early embryo.

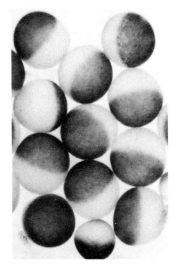

FIGURE 13-4 The animal oocyte is highly organized internally, which is revealed by the visible differences between the dorsal (dark) and ventral (light) parts of these *Xenopus* eggs. The regional organization of the oocyte determines many of the critical events in early development. (Courtesy of Michael W. Klymkowsky.)

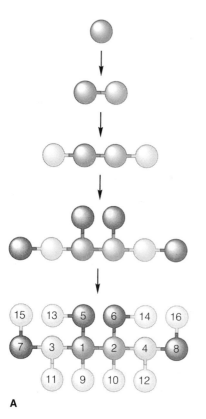

A

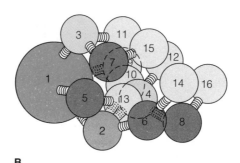

B

FIGURE 13-5 Pattern of cell divisions during the development of the *Drosophila* oocyte (red). A. The cells surrounding the oocyte, which are connected to it and to each other by cytoplasmic bridges, synthesize products transported into the oocyte that contribute to its regional organization. B. Geometric organization of the cells. (After R. C. King. 1965. *Genetics*, Oxford University Press.)

To establish the proper composition and organization of the oocyte requires the participation of many genes within the oocyte itself and in adjacent helper cells of various types (Figure 13-5). Numerous maternal genes are transcribed during oocyte development, and the mature oocyte typically contains an abundance of transcripts coding for proteins needed in the early embryo. Some of the maternal mRNA transcripts are stored complexed with proteins in special ribonucleoprotein particles that cannot be translated, and release of the mRNA, enabling it to be translated, does not occur until after fertilization.

Developmental instructions in oocytes are determined in part by the presence of distinct types of molecules at different positions within the cell and in part by gradients of morphogens that differ in concentration from one position to the next. A **morphogen** is a molecule that participates directly in the control of growth and form during development. The relative importance of distinct molecular signals versus morphogen gradients can be determined from genetic analysis of organisms in which large numbers of mutations with early developmental abnormalities and altered cell fates can be isolated and studied. Some examples are discussed in the following sections.

13-3 Genetic Control of Cell Lineages

In most organisms, it is difficult to trace the lineage of individual cells during development because the embryo is not transparent, the cells are small and numerous, and cell migrations are extensive. The **lineage** of a cell refers to the ancestral-descendant relationships among a group of cells. A cell lineage can be illustrated with a **lineage diagram,** which is a sort of pedigree showing each cell division and indicating the terminal differentiated state of each cell. Figure 13-6 is a lineage diagram of a hypothetical cell A in which the cell fates are programmed for cell death or one of the terminally differentiated cell types designated W, X, and Y. The symbols are the kind normally used for cells in nematodes, in which the name denotes the cell lineage according to ancestry and position in the embryo: for example, the cells A.a and A.p are the anterior and posterior daughters of cell A, respectively; and A.aa and A.ap are the anterior and posterior daughters of cell A.a, respectively.

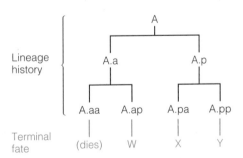

FIGURE 13-6 Hypothetical cell-lineage diagram. Different terminally differentiated cell fates are denoted W, X, and Y. One cell in the lineage (A.aa) undergoes programmed cell death. The lowercase letters (a and p) denote anterior and posterior daughter cells. For example, cell A.ap is the posterior daughter of the anterior daughter of cell A.

Genetic Analysis of Development in the Nematode

The soil nematode *Caenorhabditis elegans* (Figure 13-7) is popular for genetic studies because it is small and easy to culture, and it has a short generation time with a large number of offspring. The worms are grown on agar surfaces in petri dishes and feed on bacterial cells such as *E. coli*. Because they are microscopic in size, as many as 10^5 animals can be contained in a single petri dish. Sexually mature adults of *C. elegans* are capable of laying more than 300 eggs within a few days. At 20°C, about 60 hours are required for the eggs to hatch, undergo four larval molts, and become sexually mature adults.

Nematodes are diploid organisms with two sexes. In *C. elegans*, the two sexes are the hermaphrodite and the male. The hermaphrodite contains two X chromosomes (XX), produces both functional eggs and sperm, and is capable of self-fertilization. The male contains a single X chromosome (X0), produces only sperm, and fertilizes hermaphrodites.

The transparent body wall of the worm has enabled study of the division, migration, and death or differentiation of individual cells during development. Nematode development is unusual in that the pattern of cell division and differentiation is virtually identical from one individual to the next; that is, cell division and differentiation are highly stereotyped. The result is that each sex shows the same geometry in the number and arrangement of somatic cells. The hermaphrodite contains exactly 959 somatic cells, and the male contains exactly 1031 somatic cells. The complete developmental history of each somatic cell is known at the cellular and ultrastructural level, including the *wiring diagram*, which describes all the interconnections among cells in the nervous system.

Nematode development is largely autonomous. Developmental fates are established by interactions among cells in the early embryo, and subsequent division and differentiation of the cells is largely independent of specific interactions among cells or gradients of morphogens. On the other hand, worm development also provides important examples of determination by means of positional information.

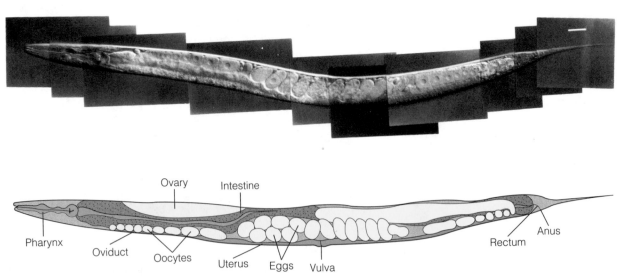

FIGURE 13-7   The soil nematode *Caenorhabditis elegans*. This organism offers several advantages for the genetic analysis of development, including the fact that each individual of each sex exhibits an identical pattern of cell lineages in the development of the somatic cells. (Photograph courtesy of Tim Schedl.)

Mutations Affecting Cell Lineages

Many mutations affecting cell lineages have been studied in nematodes, and they reveal the following general features by which genes control development:

1. The division pattern and fate of a cell is generally affected by more than one gene and can be disrupted by mutations in any of them.
2. Most genes that affect development affect several types of cells.
3. Complex cell lineages often include simpler, genetically determined lineages within them. These components are called *sublineages* because they are expressed as an integrated pattern of cell divisions and terminal differentiations.
4. The lineage of a cell is sometimes triggered autonomously within the cell itself and sometimes by external signals from other cells.
5. Regulation of development is controlled by genes that determine the different sublineages that cells can undergo and the individual steps within the sublineages.

The next subsection deals with some of the types of mutations that affect cell lineages and development.

Types of Lineage Mutations

Mutations can affect cell lineages in two major ways. One is through nonspecific metabolic blocks—for example, in DNA replication—that prevent the cells from undergoing division or differentiation. The other is through specific molecular defects that result in patterns of division or differentiation characteristic of cells normally found elsewhere in the embryo or at a different time in development. From the standpoint of genetic analysis of development, the latter class of mutants is the more informative. *C. elegans* provides several examples of each of the following general types of lineage mutations.

Mutations that cause cells to undergo developmental fates characteristic of other types of cells are called **transformation** mutations. For example, in the wildtype cell lineage shown in Figure 13-8A, the cell designated A.ap differentiates into cell-type W. In the transformation mutation (Figure 13-8B), the cell A.ap differentiates into cell-type Y. Transformation mutations have the consequence that one or more differentiated cell types are missing from the embryo and replaced with extra, supernumerary copies of other cell types. In this example, the missing cell type is W, and the supernumerary cell type is Y. In Section 13-4, we will see that transformation mutations in nematodes are analogous to a class of mutations in *Drosophila* called *homeotic* mutations, in which certain cells undergo developmental fates that are normally characteristic of other cells.

FIGURE 13-8 Transformation mutations cause cells to undergo abnormal terminal differentiation. A. The wildtype lineage. B. A mutant lineage in which the cell A.ap (red) differentiates into a Y-type cell rather than the normal W-type cell.

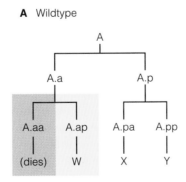 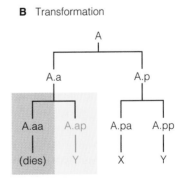

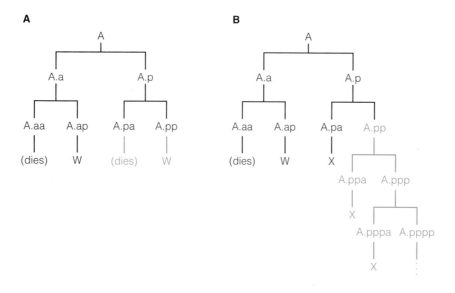

FIGURE 13-9 Two types of segregation mutations, in which related cells fail to become differentiated from each other (see Figure 13-8 for the wildtype lineage). A. Aberrant sister-cell segregation, in which the mutant cell A.p adopts the same lineage as its sister cell A.a instead of undergoing its normal fate. B. Aberrant parent-offspring segregation, in which the mutant cell A.pp (red) behaves like its parent A.p in undergoing cell division, the anterior daughter of which differentiates into an X-type cell. In both parts A and B, the result of abnormal segregation is the absence of certain cell types and the presence of supernumerary copies of other cell types.

Developmental programs often require sister cells or parent and offspring cells to adopt different fates. Mutations in which sister cells or parent-offspring cells fail to become differentiated from each other are called **segregation** mutations because the developmental determinants fail to segregate. Two examples based on the wildtype lineage in Figure 13-8 are illustrated in Figure 13-9. In the mutation in part A, the sister cells A.a and A.p give rise to identical sublineages in which the anterior derivative undergoes programmed cell death and the posterior derivative adopts fate W. This pattern contrasts with the wildtype lineage in Figure 13-8, in which A.a and A.p give rise to different sublineages. In Figure 13-9B, the daughter cell A.pp adopts the fate of its parent cell, A.p, in that A.pp divides, the anterior daughter A.ppa adopts fate X, and the posterior daughter A.ppp divides again. Continuation of this pattern results in a group of supernumerary cells of type X.

Mutations in the *lin-17* gene in C. *elegans* exemplify sister-cell segregation defects. In males, a gonadal precursor cell, Z1, produces daughter cells that are different in that the sister cells Z1.a and Z1.p have different lineages and fates (Figure 13-10A). In *lin-17* mutations, the developmental determinants do not segregate, and the sister cells have the same lineage and fate as the normal Z1.p (Figure 13-10B).

When mutant cells are unable to execute their normal developmental fates, the mutation is an **execution** mutation. Defects in execution may be due to any of a variety of reasons. The *lin-11* gene in C. *elegans* provides an example of an execution mutation. In the normal development of the vulva in the hermaphrodite, a lineage designated the 2° lineage gives rise to four cells that divide in the pattern N-T-L-L (Figure 13-11A), in which N indicates a cell that

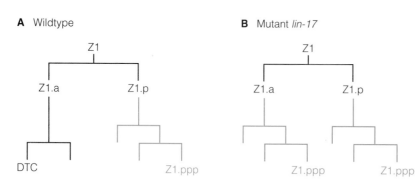

FIGURE 13-10 Aberrant sister-cell segregation in the *lin-17* mutation in C. *elegans*. A. Z1.a and Z1.p normally undergo different fates, and the Z1.a lineage includes a distal tip cell (DTC). B. In the *lin-17* segregation mutation, Z1.a and Z1.p, have the same fate, and the distal tip cell is absent.

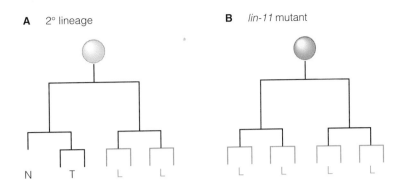

FIGURE 13-11 Example of an execution mutation. A. Wildtype development of the 2° lineage in the vulva of C. elegans. The letter N denotes no further cell division, and T and L denote cell division in either a transverse or a longitudinal plane with respect to the embryo. B. The lin-11 mutation results in failure to execute the N and T sublineage. All cells divide in a longitudinal plane.

does not divide, T one that divides in a transverse plane relative to the orientation of the larva, and L a cell that divides in a longitudinal plane. In lin-11 mutations, the 2° lineage is not executed, and the four cells divide in the abnormal pattern L-L-L-L (Figure 13-11B).

The process of **programmed cell death** is an essential requirement for normal development in many organisms, and specific abnormalities can result from the failure of programmed cell death. For example, compared with the wildtype lineage in Figure 13-12A, the lineage in part B is abnormal in that cell A.aa fails to undergo programmed cell death, but instead differentiates into cell-type V. When failure in programmed cell death occurs, and the surviving cells differentiate into a recognizable cell type, the result is the occurrence of supernumerary cells of that type. For example, in mutations in the ced-3 gene (ced = cell death abnormal) in C. elegans, a particular cell that normally undergoes programmed cell death survives and often differentiates into a supernumerary neuron.

The events of development are coordinated in time, and so mutations that affect the timing of developmental events are of great interest. Mutations that affect timing are called **heterochronic** mutations. An example is shown in Figure 13-12C. In comparison with the wildtype situation in Figure 13-12A, the heterochronic mutant cell A.a delays cell division until the daughter cells

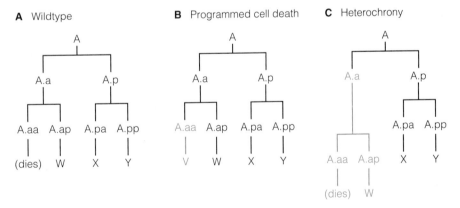

FIGURE 13-12 Programmed cell death and heterochrony. A. A wildtype lineage. B. Failure of programmed cell death, in which the cell A.aa (red) does not die but instead differentiates into a V-type cell. In many cases in which programmed cell death fails to occur, the surviving cells differentiate into identifiable types. C. Heterochronic mutants are abnormal in the timing of events during development. Retarded mutants undergo normal events too late, and precocious mutants undergo normal events too early. The example shown here is a retarded mutant in which cell A.a (red) delays division until its products (A.aa and A.ap) are contemporaneous with the X and Y cells derived from A.p.

A.aa and A.ap become contemporaneous with X and Y. Therefore, the heterochronic mutation in Figure 13-12C is *retarded* in that developmental events are normal but delayed. Heterochronic mutations can also be *precocious* in the expression of developmental events at times earlier than normal. For example, in certain heterochronic mutants in *C. elegans*, specific sublineages that normally develop in males only at the fourth larval molt develop precociously in the mutant in the second or third larval molt.

The lin-12 Developmental-Control Gene

Control genes that cause cells to diverge in developmental fate are not always easy to recognize. For example, an execution mutation may identify a gene that is necessary for the expression of a particular developmental fate, but the gene may not actually control or determine the developmental fate of the cells in which it is expressed. This possibility complicates the search for genes that control major developmental decisions.

Genes that control decisions about cell fate can sometimes be identified by the unusual characteristic that dominant or recessive mutations have opposite effects; that is, if alternative alleles of a gene result in opposite cell fates, then the product of the gene must be both necessary and sufficient for expression of the fate. Identification of possible regulatory genes in this way excludes the large number of genes whose functions are merely necessary, but not sufficient, for the expression of cell fate. Recessive mutations in genes controlling development often result from **loss of function** in that the mRNA or the protein is not produced; loss-of-function mutations are exemplified by nonsense mutations that cause polypeptide chain termination during translation (Chapter 11). Dominant mutations in developmental-control genes often result from **gain of function** in that the gene is overexpressed or expressed at the wrong time.

In *C. elegans*, only a small number of genes have dominant and recessive alleles that affect the same cells in opposite ways. Among them is the *lin-12* gene, which controls developmental decisions in a number of cells. One example concerns the cells denoted Z1.ppp and Z4.aaa in Figure 13-13. These cells lie side by side in the embryo, but they have quite different lineages (the cell P_0 is the zygote). Normally, one of the cells diffentiates into an *anchor cell* (AC), which participates in development of the vulva, and the other one differentiates into a *ventral uterine precursor cell* (VU) (Figure 13-14A). Either Z1.ppp or Z4.aaa may become the anchor cell with equal likelihood.

Direct cell-cell interaction between Z1.ppp and Z4.aaa controls the AC-VU decision. If either cell is burned away (ablated) by a laser microbeam, the remaining cell differentiates into an anchor cell (Figure 13-14B). This result implies that the preprogrammed fate of both Z1.ppp and Z4.aaa is that of an anchor cell. When either cell becomes committed to the anchor-cell fate, its contact with the other cell elicits the VU cell fate. As noted, recessive and dominant mutations of *lin-12* have opposite effects. Mutations in which *lin-12* activity is lacking or greatly reduced are denoted *lin-12(0)*. These mutations are recessive, and in the mutants both Z1.aaa and Z4.aaa become anchor cells (Figure 13-14C). In contrast, *lin-12(d)* mutations are those in which *lin-12* activity is overexpressed. These mutations are dominant or partially dominant, and in the mutants both Z1.aaa and Z4.ppp become ventral uterine precursor cells.

The *lin-12* gene also controls the fate of certain cells in the development of the vulva. Figure 13-15 illustrates five precursor cells, P4.p through P8.p, that participate in development of the vulva. Each precursor cell has the capability of differentiating into one of three alternative fates, called the 1°, 2°,

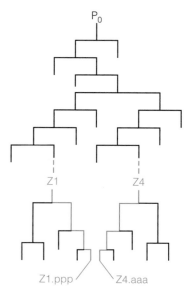

FIGURE 13-13 Complete lineage of Z1.ppp and Z4.aaa in *C. elegans*. P_0 represents the zygote, and the dashed lines indicate three cell divisions not shown. In the normal development of the vulva, there is an equal likelihood that either Z1.ppp or Z4.aaa will differentiate into the anchor cell. The remaining cell differentiates into a ventral uterine precursor cell.

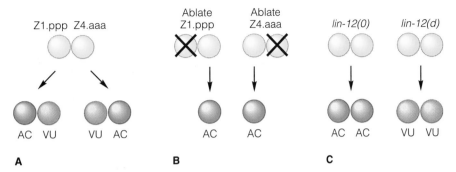

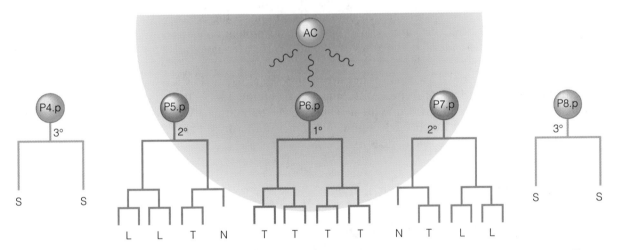

FIGURE 13-14 Control of the fates of Z1.ppp and Z4.aaa in vulval development. For the complete lineage of these cells, see Figure 13-13. A. In wildtype cells, there is an equal chance that either cell will become the anchor cell (AC); the other becomes a ventral uterine precursor cell (VU). B. If either cell is destroyed (ablated) by a laser beam, the other differentiates into the anchor cell. C. Genetic control of cell fate by the *lin-12* gene. In recessive loss-of-function mutations (*lin-12(0)*), both cells become anchor cells. In dominant gain-of-function mutations (*lin-12(d)*), both cells become ventral uterine precursor cells.

and 3° lineages, which differ according to whether descendant cells remain in a syncytium (S) or divide longitudinally (L), transversely (T), or not at all (N). The precursor cells normally differentiate as shown in Figure 13-15, giving five lineages that develop in the order 3°-2°-1°-2°-3°. The vulva itself is formed from the 1° and 2° cell lineages.

The fate adopted by each precursor cell appears to depend on its distance from the anchor cell (AC) in the developing gonad. The anchor cell is apparently responsible for releasing an induction signal (shading), and the precursor cells are exposed to different concentrations of the substance

FIGURE 13-15 Determination of cell lineages by means of a diffusible signal. Cells P4.p through P8.p give rise to lineages in the development of the vulva in the hermaphrodite. The three types of lineages are, designated 1°, 2°, and 3°. The lineage that arises depends on the concentration of an inducing signal released from the anchor cell (AC). P6.p is exposed to the greatest concentration and undergoes differentiation into the 1° lineage, cells P5.p and P7.p are exposed to lower concentrations and differentiate into the 2° lineage, and cells P4.p and P8.p are exposed to the least concentration and differentiate into the 3° lineage. The *lin-12* gene affects the ability of the cells to respond to the inducing signal. Symbols are: S, syncytial hypodermal nucleus; L, longitudinal orientation of cell division; T, transverse orientation of cell division; N, no division. (After P. W. Sternberg and H. R. Horvitz. 1984. *Ann. Rev. Genet.* 18: 489–524.)

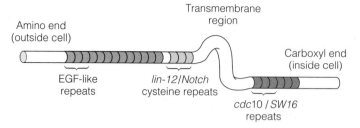

FIGURE 13-16   The structure of the *lin-12* gene product is that of a receptor protein containing a transmembrane region and various types of repeated units resembling those in epidermal growth factor (EGF) and other developmental control genes.

depending on their distance from the anchor cell. In the absence of the anchor cell, all of the precursor cells express the 3° lineage. Thus, development of the 3° lineage is uninduced. If all precursor cells except one are killed with a laser beam, the fate adopted by the remaining cell is still correlated with its position in relation to the anchor cell. The precursor cell (P6.p) closest to the anchor cell takes on the 1° fate, the two that are somewhat farther away (P5.p and P7.p) adopt the 2° fate, and the most distant (P4.p and P8.p) remain uninduced.

The effects of *lin-12* mutations suggest that the wildtype gene product is a receptor of a developmental signal. Indeed, the molecular structure of the *lin-12* gene product is typical of a receptor protein, and it shares domains with other proteins important in developmental control (Figure 13-16). A transmembrane region spans the cell membrane and separates the Lin-12 protein into an extracellular part (the amino end) and an intracellular part (the carboxyl end). The extracellular part contains thirteen repeats of a domain found in a mammalian peptide hormone, epidermal growth factor (EGF), and in the product of the *Notch* gene in *Drosophila*, which controls the decision between epidermal and neural cell fates. Nearer the transmembrane region, the amino end contains three repeats of a cysteine-rich domain also found in the *Notch* gene product. Inside the cell, the carboxyl part of the Lin-12 protein contains six repeats of a domain also found in the genes *cdc10* and *SW16*, which control cell division in yeast.

The function of the Lin-12 receptor is to respond to a signal released by a neighboring cell. In the decision between AC and VU in Figure 13-14, the signal comes from the AC cell and causes the neighboring cell to become a VU cell. In the vulval lineage in Figure 13-15, the signal comes from a 1° cell and causes the neighboring cells to differentiate as 2° cells.

13-4   Development in *Drosophila*

Drosophila also provides excellent opportunities for genetic analysis of development. The developmental cycle of *D. melanogaster*, summarized in Figure 13-17, includes egg, larval, pupal, and adult stages. Early development includes a series of cell divisions, migrations, and infoldings that result in the gastrula. About 24 hours after fertilization, the first-stage larva emerges from the egg. Two larval molts are followed by pupation and a complex metamorphosis that gives rise to the adult fly composed of more than 10^6 cells. In wildtype strains reared at 25°C, development requires from 10 to 12 days.

Early development in *Drosophila* takes place within the egg case (Figure 13-18A). The first nine mitotic divisions are in rapid succession without division of the cytoplasm, producing a cluster of nuclei within the egg (Figure

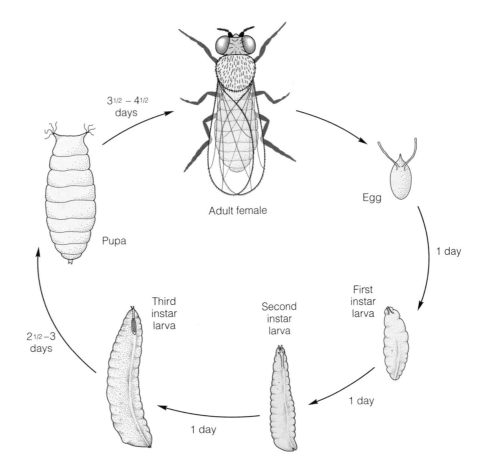

FIGURE 13-17 Developmental program of *Drosophila melanogaster*. The durations of the stages are at 25°C.

3 1/2 – 4 1/2 days

Adult female

Egg

1 day

Pupa

First instar larva

Third instar larva

Second instar larva

1 day

2 1/2 – 3 days

1 day

1 day

FIGURE 13-18 Early development in *Drosophila*. A. The nucleus in the fertilized egg. B. Mitotic divisions occur synchronously within a syncytium. C. Some nuclei migrate to the periphery of the embryo and, at the posterior end, the pole cells (associated with the germ line) become cellularized. D. Additional mitotic divisions occur within the syncytial blastoderm. E. Membranes are formed around the nuclei giving rise to the cellular blastoderm.

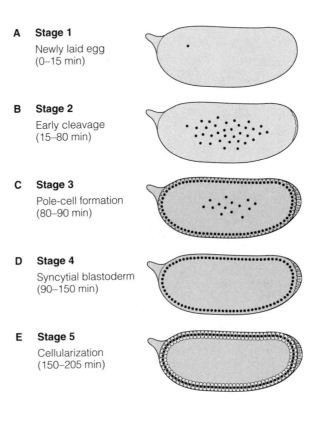

A **Stage 1**
Newly laid egg
(0–15 min)

B **Stage 2**
Early cleavage
(15–80 min)

C **Stage 3**
Pole-cell formation
(80–90 min)

D **Stage 4**
Syncytial blastoderm
(90–150 min)

E **Stage 5**
Cellularization
(150–205 min)

13-18B). The nuclei migrate to the periphery, and the germ line is formed from about 10 **pole cells** set off at the posterior end of the egg (Figure 13-18C); the pole cells undergo two additional divisions and are reincorporated into the embryo by invagination. The nuclei within the embryo proper undergo four more mitotic divisions without division of the cytoplasm, forming the **syncytial blastoderm** containing about 6000 nuclei (Figure 13-18D). Cellularization of the blastoderm occurs about $2\frac{1}{2}$ to 3 hours after fertilization by the synthesis of membranes that separate the nuclei. The **blastoderm** formed by cellularization (Figure 13-18E) is a flattened hollow ball of cells that corresponds to the blastula in other animals.

Elimination of selected cells within a *Drosophila* blastoderm results in localized defects in the larva and adult. Correlating the position of the cells eliminated with the type of defects results in a **fate map** of the blastoderm, which specifies the cells in the blastoderm that give rise to the various larval and adult structures (Figure 13-19). Use of genetic markers in the blastoderm has allowed further refinement of the fate map. Cell lineages can be genetically marked during development by inducing recombination between homologous chromosomes during mitosis, resulting in genetically different daughter cells. Similar to cells in the early blastula of *Caenorhabditis*, cells in the blastoderm of *Drosophila* have predetermined developmental fates, with little ability to substitute in development for other, even adjacent, cells.

Evidence that blastoderm cells in *Drosophila* have predetermined fates comes from experiments in which cells from a genetically marked blastoderm are implanted into host blastoderms. Blastoderm cells implanted into the equivalent regions of the host become part of the normal adult structures. However, blastoderm cells implanted into different regions develop autonomously and are not integrated into host structures.

As with most other insects, the larva and adult of *Drosophila* have a segmented body plan consisting of a head, three thoracic segments, and eight

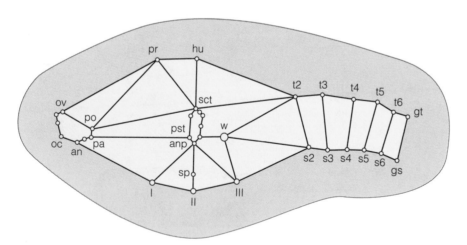

FIGURE 13-19 Fate map of the *Drosophila* blastoderm, which shows the adult structures that derive from various parts of the blastoderm. The map was derived by correlating the expression of genetic markers in different adult structures in genetic mosaics. The abbreviations stand for various body parts in the adult fly. For example, "ov" and "oc" are head structures; "w" the wing; I, II, and III the first, second, and third legs; and "gs" and "gt" are genital structures. Generally speaking, the spatial relations of determined cells in the fate map are similar to those of the structures in the adult fly. (After J. C. Hall, W. M. Gelbart, and D. R. Kankel. 1976. In *The Genetics and Biology of Drosophila*, vol. 1a, M. Ashburner and E. Novitski, eds. Academic Press, pp. 265–314.)

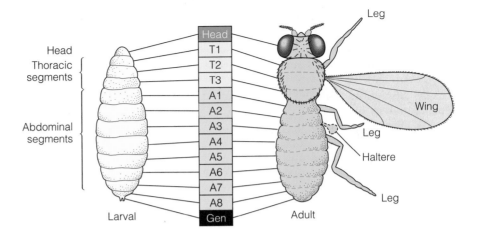

FIGURE 13-20 Relation between larval and adult segmentation in *Drosophila*. Each of the three thoracic segments in the adult carries a pair of legs. The wings develop on the second thoracic segment (T2) and the halteres (flight balancers) on the third thoracic segment (T3).

abdominal segments (Figure 13-20). Metamorphosis makes use of about twenty structures called **imaginal disks** present inside the larva (Figure 13-21). Formed early in development, the imaginal disks ultimately give rise to the principal structures and tissues in the adult organism. Examples of imaginal disks are the pair of wing disks (one on each side of the body), which give rise to the wings and related structures; the pair of eye-antenna disks, which give rise to the eyes, antennae, and related structures; and the genital disk, which gives rise to the reproductive apparatus. During the pupal stage, when many larval tissues and organs break down, the imaginal disks progressively unfold and differentiate into adult structures. The morphogenetic events that take place in the pupa are initiated by the hormone ecdysone, secreted by the larval brain.

Cell determination in *Drosophila* also occurs within bounded units called **compartments.** Cells in the body segments and imaginal disks do not migrate across the boundaries between compartments. For example, the *Drosophila* wing disk includes five compartment boundaries, and most body segments include one boundary that divides the segment into anterior and posterior

FIGURE 13-21 Locations of imaginal disks in a *Drosophila* larva and the adult structures derived from them. A. Larval locations of the nine pairs of disks and one genital disk. B. General morphology of the disks late in larval development. C. Adult structures formed by the disks.

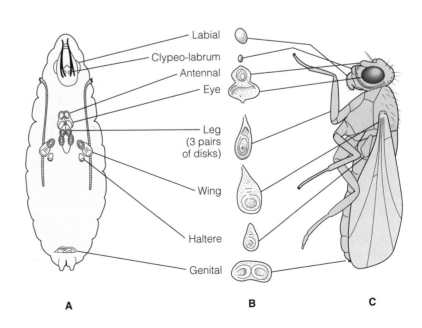

halves. Within each compartment, neighboring groups of cells not necessarily related by ancestry undergo developmental determination together. The evidence for compartments comes from the genetic marking of individual cells by means of mitotic recombination and observation of the positions of their descendants.

Because of the relatively high degree of determination in the blastoderm, genetic analysis of *Drosophila* development has tended to focus on the early stages of development when the basic body plan of the embryo is established and key regulatory processes become activated. The following sections summarize the genetic control of these early events.

Maternal-Effect Genes and Zygotic Genes

Early development in *Drosophila* requires translation of maternal mRNA molecules present in the oocyte. Blockage of protein synthesis during this period arrests the early cleavage divisions. Expression of the zygote genome is also required, but the timing is uncertain. Blockage of transcription of the zygote genome at any time after the ninth cleavage division prevents formation of the blastoderm.

Because the earliest stages of *Drosophila* development are programmed in the oocyte, mutations that affect oocyte composition or structure can upset development of the embryo. Genes that function in the mother that are needed for development of the embryo are called **maternal-effect genes,** and those that need to function in the embryo are called **zygotic genes.** The interplay between the two types of genes is as follows:

> The zygotic genes interpret and respond to the positional information laid out in the egg by the maternal-effect genes.

Mutations in maternal-effect genes are easy to identify because homozygous females produce eggs unable to support normal embryonic development, whereas homozygous males produce normal sperm. Therefore, reciprocal crosses give dramatically different results. For example, a recessive maternal-effect mutation, m, will yield the following results in crosses:

$$m/m \; \male \; \times \; +/+ \; \female \; \to \; +/m \; \text{progeny} \quad \text{(abnormal development)}$$
$$+/+ \; \male \; \times \; m/m \; \female \; \to \; +/m \; \text{progeny} \quad \text{(normal development)}$$

The $+/m$ progeny of the reciprocal crosses are genetically identical, but development is upset when the mother is homozygous m/m.

Maternal-effect genes establish the polarity of the *Drosophila* oocyte before fertilization takes place. They are active during the earliest stages of embryonic development, and they determine the basic body plan of the embryo. Maternal-effect mutations provide a valuable tool for investigating the genetic control of pattern formation and for identifying the molecules important in morphogenesis. Examples are discussed later in this chapter.

Genetic Basis of Pattern Formation in Early Development

Among the genes that control the body plan of the early embryo are the **segmentation genes,** which determine the number, position, and identity of the body parasegments and segments. Each **parasegment** consists of the posterior half of one segment and the anterior half of the next segment in line (Figure

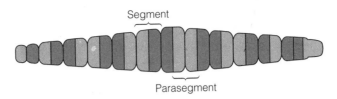

FIGURE 13-22 Segments and parasegments. Each segment (defined by indentations formed by the sites of muscle attachment in the larval cuticle) consists of the anterior half of one parasegment and the posterior half of the next parasegment in line. The distinction is important because the patterns of expression of segmentation genes are more often correlated with parasegment boundaries than with segment boundaries.

13-22). Segments are established from parasegments by muscle attachments that form grooves in the embryo. Each segment thus contains parts of two adjacent parasegments. Parasegments are important because the activity of many developmental-control genes coincides with the parasegment boundaries rather than the segment boundaries. Mutations in segmentation genes result in embryos missing particular regions— for example, the head or thorax.

The early stages of pattern formation are determined by four classes of segmentation genes that differ in their times and patterns of expression in the embryo. They are:

1. The **coordinate genes** determine the principal coordinate axes of the embryo: the anterior-posterior axis, which defines the front and rear; and the dorsal-ventral axis, which defines the top and bottom. Coordinate genes are maternal-effect genes that establish early polarity through the presence of their product at defined positions within the oocyte or by means of gradients of concentration of their products.

2. The **gap genes** are expressed in regions along the embryo, and they establish the next level of spatial organization. Mutations in gap genes result in the absence of contiguous segments, and so gaps appear in the normal pattern of structures in the embryo. Most gap genes bring about their effects by expression in the zygotic genome. An example of a gap gene is *Krüppel*. Extreme alleles of *Krüppel* lack all thoracic and anterior abdominal segments. The wildtype *Krüppel* gene is expressed only in the affected regions of the embryo. The protein is a DNA-binding protein of the zinc-finger class (Chapter 12), and so it regulates the expression of other genes.

3. The **pair-rule genes** determine the separation of the embryo into discrete segments. Approximately ten pair-rule genes have been identified. Mutations in pair-rule genes result in missing pattern elements with a two-segment periodicity (Figure 13-23). For example,

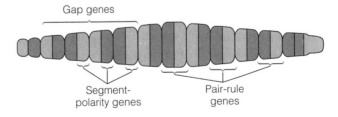

FIGURE 13-23 Groups of segments affected in different segmentation-gene mutants. Mutants in gap genes cause the absence of several contiguous segments, those in pair-rule genes cause the absence of alternating segments or parasegments, and those in segment-polarity genes fail to undergo the normal anterior-posterior differentiation within each parasegment.

even-skipped mutations affect even-numbered segments. The two-segment periodicity of pair-rule genes results from a zebra-stripe pattern of gene expression along the embryo. Another example of a pair-rule gene is *odd skipped*, in which odd-numbered segments are abnormal.

4. The **segment-polarity genes** determine the pattern of development within each segment of the embryo (Figure 13-23). Approximately ten segment-polarity genes have been identified. Mutations in segment-polarity genes affect all segments or parasegments in which the normal gene is active. Many segment-polarity mutations have the normal number of segments, but part of each segment is deleted and the remainder is duplicated in mirror-image symmetry.

Evidence for the existence of the four classes of segmentation genes is presented in the next two subsections.

Coordinate Genes and Gap Genes

Coordinate genes control the major axes of the embryo by the presence of their gene products at localized positions in the oocyte. For example, when *Drosophila* eggs are punctured and small amounts of cytoplasm allowed to escape, loss of cytoplasm from the anterior end results in embryos in which some posterior structures develop in place of the head. Replacement of anterior cytoplasm with posterior cytoplasm by injection yields embryos with two mirror-image abdomens and no head.

Abnormalities resulting from surgical manipulation of *Drosophila* embryos resemble those caused by some maternal-effect mutations. For example, the *bicoid* mutation results in an embryo lacking a head and thorax; this mutation occasionally also causes abdominal segments in reverse polarity to be duplicated at the anterior end. Another mutation, *nanos*, yields an embryo lacking pole cells and an abdomen, which resembles an embryo produced by surgical removal of the posterior cytoplasm. These abnormalities do not result merely from a generalized disruption of development at the posterior end, because a posterior structure called the telson, which normally develops between the pole cells and the abdomen, is not affected in either *nanos* or the surgically manipulated embryos.

The *bicoid* mRNA is produced in nurse cells (the cells surrounding the oocyte in Figure 13-5) and exported to a localized region at the anterior pole of the oocyte. The protein product is less localized and, during the syncytial cleavages (Figure 13-18B), forms an anterior-posterior concentration gradient with the maximum at the anterior tip of the embryo. The bicoid protein is a principal morphogen in determining the blastoderm fate map (Figure 13-19). The protein is a transcriptional activator containing a helix-turn-helix motif for DNA binding (Chapter 12). Genes affected by the bicoid protein contain multiple upstream binding domains consisting of nine nucleotides resembling the consensus sequence 5'-TCTAATCCC-3'. Binding sites that mismatch the consensus sequence in from zero to two base pairs bind the bicoid protein with high affinity, and sites containing four mismatches bind with low affinity. The combination of high- and low-affinity binding sites determines the concentration of bicoid protein needed for gene activation; genes with many high-affinity binding sites can be activated at low concentrations, but those with many low-affinity binding sites need higher concentrations. Such differences in binding affinity mean that the level of gene expression can differ from one regulated gene to the next along the bicoid concentration gradient.

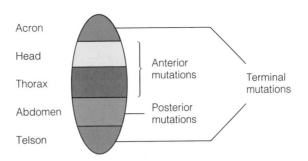

FIGURE 13-24 Regional differentiation of the early *Drosophila* embryo along the anterior-posterior axis. Mutations in any of the classes of genes shown result in elimination of the corresponding region in the embryo.

The coordinate genes like *bicoid* and *nanos*, together with the gap genes, ultimately define the initial anterior-posterior subdivision of the embryo. The early embryo can be divided into five regions (Figure 13-24). These regions ultimately form the most-anterior structures (acron), the most posterior structures (telson), and the head, thorax, and abdomen. Based on the effects of mutations, these regions can be grouped as indicated by color. Mutations in the *anterior* class of genes affect the head and thorax, those in the *posterior* class affect the abdomen, and mutations in the *terminal* class affect the acron and telson. There are from four to eight genes is each class. For example, *bicoid* is a gene in the anterior class and *nanos* a gene in the posterior class. Their functions are to control other genes further along the hierarchy of genes that determine pattern formation. Embryos that are simultaneously mutant for all three classes of genes appear as unorganized blobs.

Another set of about ten coordinate genes determines the dorsal-ventral axis of the embryo. Mutations in these genes eliminate all ventral and lateral pattern elements. In many cases, the mutant embryos can be rescued by injection of wildtype cytoplasm, no matter where the wildtype cytoplasm is taken from or where it is injected. An example is the gene called *snake*, mutations in which eliminate ventral structures. The *snake* gene codes for a protein that is related to a class of enzymes called *serine proteases*. Serine proteases are synthesized as inactive precursors that require a specific cleavage for activation. They often act in a temporal sequence so that activation of one enzyme in the pathway is necessary for activation of the next enzyme in line (Figure 13-25). For example, about half the clotting factors in human blood are serine proteases. The serial activation of the enzymes results in a **cascade** effect that greatly amplifies an initial signal. Each step in the cascade multiplies the signal produced in the preceding step.

FIGURE 13-25 Amplification of a signal by a cascade of activation. The number of activated components at each step increases exponentially. This is a simplified example with a threefold amplification at each step. The primed symbols denote inactive enzyme forms; the unprimed symbols denote active forms.

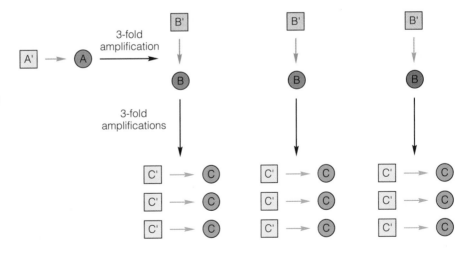

Pair-Rule and Segment-Polarity Genes

The coordinate and gap genes determine the polarity of the embryo and establish broad regions within which subsequent development takes place. The function of the pair-rule and segment-polarity genes is to give the early *Drosophila* larva a segmented body pattern with both repetitiveness and individuality of segments. For example, the eight abdominal segments (Figure 13-20) are repetitive in that they are regularly spaced and have several features in common, but they differ in the details of their differentiation. As development proceeds, the progressively more refined organization of the embryo is correlated with the patterns of expression of the various types of segmentation genes. The pair-rule genes determine the body plan to the level of segments and parasegments, and the segment-polarity genes create a spatial differentiation within each segment.

Together, the segmentation genes determine a unique combinatorial pattern in every segment and parasegment. The evidence for combinatorial regulation comes from loss-of-function and gain-of-function mutations. Loss-of-function mutations in a segmentation gene result in abnormal structures in regions of the embryo in which the gene is normally active. Gain-of-function mutations in a segmentation gene result in abnormal structures in regions of the embryo in which the gene is normally not expressed. Expression of the genes where it normally occurs, and only where it normally occurs, is necessary for normal development.

Regulatory interactions within the hierarchy of segmentation genes are illustrated in Figure 13-26. These interactions govern the activities of the second set of developmental genes, the **homeotic selector genes,** which control the pathways of differentiation in each segment or parasegment. They do this by regulating the important class of developmental-control genes discussed in the next subsection.

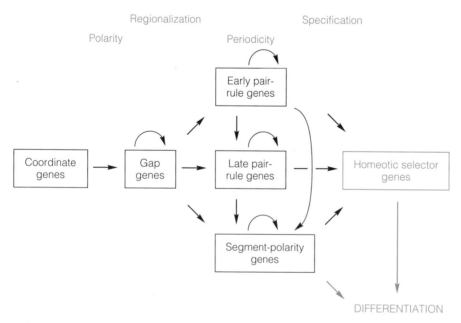

FIGURE 13-26 Regulatory interactions (arrows) among genes controlling early development in *Drosophila*. The terms polarity, regionalization, periodicity, and specification refer to the major developmental determinations that occur during each time interval.

FIGURE 13-27 An adult *Drosophila* with four wings produced by a combination of several mutations in the *bithorax* complex. The mutants convert the third thoracic segment into the second thoracic segment, and the halteres normally present on the third thoracic segment become converted into the posterior pair of wings. (Courtesy of E. B. Lewis.)

Homeotic Genes

Mutations in **homeotic genes** result in a transformation from one body segment to another; that is, homeotic mutations result in structures being replaced by different structures that are normally elsewhere in the embryo. Homeotic genes act within developmental compartments to control other genes concerned with such characteristics as rates of cell division, orientation of mitotic spindles, and the capacity to differentiate bristles, legs, and other features. Homeotic genes are also important in restricting the activities of groups of structural genes to definite spatial patterns.

An example of a homeotic mutation is *bithorax*, which causes transformation of the anterior part of the third thoracic segment into the anterior part of the second thoracic segment, with the result that the halteres (flight balancers) are transformed into an extra pair of wings (Figure 13-27). Another homeotic mutation, *Antennapedia*, results in transformation of the antennae into the second pair of legs. The normal *Antennapedia* gene specifies the second thoracic segment. *Antennapedia* mutations are dominant gain-of-function mutations in which the gene is overexpressed in the dorsal part of the head, transforming it into the second thoracic segment, with the antennae becoming a pair of legs.

Many homeotic genes in *Drosophila* are gene clusters. The cluster containing *bithorax* is designated BX-C (stands for *bithorax*-complex), and that containing *Antennapedia* is called ANT-C (stands for *Antennapedia*-complex). Both gene clusters were initially discovered through their homeotic effects in adults. Later they were shown to affect the identity of larval segments.

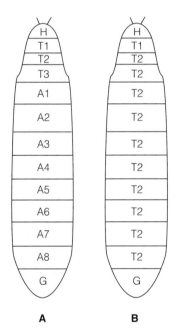

A **B**

FIGURE 13-28 Segmentation patterns in *Drosophila* larvae. A. A wildtype larva. In addition to head (H) and genital (G) segments, there are three thoracic segments (T1–T3) and eight abdominal segments (A1–A8). B. A mutant larva lacking the entire *bithorax* complex. The segments H, T1, and G are normal. All others develop as T2 segments.

Deletion (loss of function) and duplication (gain of function) of genes in the homeotic complexes help to define their functions. For example, deletion of the entire BX-C complex (Figure 13-28) results in a larva with a normal head (H), first thoracic segment (T1), and second thoracic segment (T2), but the remaining segments (T3 and A1–A8) differentiate in the manner of T2. Therefore, the BX-C genes function to shift the development of segments from T2 to progressively more posterior types of segments. Each of the segments T3 through A8 requires the activity of a different BX-C gene, plus all the genes required for the more-anterior segments. For example, the gene *Ultrabithorax* defines one end of the BX-C complex and, if all genes are deleted except for *Ultrabithorax*, all of the segments T3 and A1 through A8 differentiate in the manner of T3. Similarly, when all BX-C genes are deleted except for *Ultrabithorax* and the adjacent gene *bithoraxoid*, all abdominal segments develop in the manner of A1. It is of some interest that the BX-C genes are arranged along the chromosome in the same order that they are expressed during development.

Homeotic genes are transcriptional activators of other genes. Most homeotic genes contain one or more copies of a characteristic sequence of about 180 nucleotides called a **homeobox,** which is also found in key developmental-control genes in organisms as diverse as segmented worms, frogs, chickens, mice, and human beings. Homeobox sequences are present in exons; they code for a protein-folding domain that includes a helix-turn-helix DNA-binding motif (Chapter 12), as well as other transcriptional activation components that are not as well understood.

13-5 Development in Higher Plants

Reproductive and developmental processes in higher plants differ significantly from those in other eukaryotes. For example, plants have an alternation of generations between the diploid sporophyte and the haploid gametophyte, and the plant germ line is not established in a discrete location during embryogenesis but at many locations in the adult organism.

Specialized cells in the pistils and stamens of the flowers undergo meiosis to form haploid male spores (*microspores*) and female spores (*megaspores*). The spores divide mitotically and differentiate into the multicellular male gametophyte (*pollen*) and the female gametophyte (*embryo sac*). Male and female gametes are produced by mitotic divisions within the gametophytes. Transfer of a pollen grain from an anther to the stigma of a pistil in pollination is followed by its germination and the growth of a pollen tube through the style of the pistil and penetration of the embryo sac. A double fertilization occurs, in which one haploid sperm nucleus in the pollen grain unites with the haploid egg nucleus to form the diploid zygote, and another haploid sperm nucleus from the same pollen grain unites with a second (typically diploid) nucleus in the female gametophyte to form the triploid endosperm. The endosperm develops into a tissue that nourishes the differentiating embryo or the germinating seedling. These processes are illustrated as they occur in maize in Figure 2-5.

In animals, most of the major developmental decisions are made during embryogenesis, as we have seen. However, in higher plants, differentiation of the vegetative organs (root, stem, and leaves) and the floral organs (sepal, petal, pistil, and stamen) occurs almost continuously throughout life in regions of actively dividing cells called **meristems.** The shoot and root meristems are formed during embryogenesis and consist of cells that divide in distinctive geometric planes and at different rates to produce the basic morphological pattern of each organ system. The floral meristems are established by a reorganization of the shoot meristem after embryogenesis and eventually differentiate into floral structures characteristic of each particular species. The plastic, or indeterminate, growth patterns of higher plants are the result of continuous production of both vegetative and floral organ systems. These patterns are conditioned largely by day length and the quality and intensity of light.

The mechanisms that regulate development in higher plants are not well understood. Whether maternal factors in the egg contribute to embryonic development in plants, as they do in animals, is not known; nor is it known how meristems are determined or how they produce the distinctive morphological pattern specific for each organ system. However, rapid progress is being made in elucidating molecular aspects of plant development, such as the initiation and differentiation of flowers. For example, one of the molecular signals that initiates flowering by the reorganization of shoot meristems into floral meristems is the diffusible hormone *florigen* produced in leaves. Florigen appears to induce a commitment of cells in the vegetative meristem to the pathway of floral development.

The determination of floral development is of some interest because the process is irreversible. Although many plant cells are **totipotent**, which means capable of regenerating an entire plant, this is not the case for cells in the floral meristem. For example, tissue segments derived from floral meristems of tobacco form only flowers when cultured, but comparable segments not committed to floral development form shoot meristems that differentiate into stems and leaves.

Homeotic mutations also occur in plants and are the basis of a genetic approach to identifying genes that control the reorganization of vegetative meristems into floral meristems. Particularly interesting are homeotic mutations that transform one floral organ system into another. For example, one mutation isolated in the plant *Arabadopsis thaliana* (a member of the mustard family) transforms petals into stamens with anthers that produce pollen. The molecular basis of this transformation has not been identified.

CHAPTER SUMMARY

The genotype determines the developmental potential of the embryo. By means of a developmental program that results in different sets of genes being expressed in different types of cells, the genotype controls the developmental events that occur and their temporal order. Thus, mutations that interrupt developmental processes identify genetic factors that control development.

The early development of the animal embryo establishes the basic developmental plan for the whole organism. The earliest events in embryonic development depend on the correct spatial organization of numerous constituents present in the oocyte. Developmental genes that are needed in the mother for proper oocyte formation and zygotic development are maternal-effect genes. Genes that are required in the zygote nucleus are zygotic genes. Fertilization of the oocyte initiates a series of mitotic cleavage divisions forming the hollow-ball blastula, which rapidly undergoes a restructuring into the gastrula. Accompanying these early morphological events are a series of molecular events that determine the developmental fates that cells undergo. Execution of a developmental state may be autonomous (genetically programmed) or may require positional information supplied by neighboring cells or the local concentration of one or more morphogens.

The soil nematode *Caenorhabditis elegans* is widely used in studies of cell lineages because many lineages in the organism undergo virtually autonomous development and the developmental program is identical from one organism to the next. Most lineages are affected by many genes, including genes that control the sublineages into which the lineage can differentiate. Mutations affecting cell lineages define several types of developmental mutations: (1) execution mutations, which prevent a developmental program from being carried out; (2) transformation mutations, in which a wrong developmental program is executed; (3) segregation mutations, in which developmental determinants fail to segregate normally between sister cells or between mother and daughter cells; (4) programmed cell-death mutations, in which cells fail to undergo a normal developmental program leading to their death; and (5) heterochronic mutations, in which execution of a normal developmental program is either precocious or retarded in time.

Genes that control key points in development can often be identified by the unusual feature that recessive alleles (ideally loss-of-function mutations) and dominant alleles (ideally gain-of-function mutations) result in opposite effects on phenotype. For example, if loss of function results in failure to execute a developmental program in a particular anatomical position, then gain of function should result in execution of the program in an abnormal location. The *lin-12* gene in C. *elegans* is an important developmental-control gene identified by these criteria. The *lin-12* gene, which controls a number of yes-or-no developmental decisions, codes for a receptor protein that shares some domains with the mammalian peptide hormone epidermal growth factor, shares other domains with the *Notch* protein in *Drosophila* (which controls epidermal or neural commitment), and shares still other domains with proteins that control the cell cycle.

Early development in *Drosophila* includes the formation of a syncytial blastoderm by early cleavage divisions without cytoplasmic division, the setting apart of pole cells that form the germ line at the posterior of the embryo, the migration of most nuclei to the periphery of the syncytial blastoderm, cellularization to form the cellular blastoderm, and determination of the blastoderm fate map at or before the cellular blastoderm stage. Metamorphosis into the adult fly makes use of about twenty imaginal disks present in the larva that contain developmentally committed cells that divide and develop into the adult structures. Most imaginal disks include several discrete groups of cells or compartments separated by boundaries that progeny cells do not cross.

Early development in *Drosophila* to the level of segments and parasegments requires four classes of segmentation genes: (1) coordinate genes that establish the basic anterior-posterior and dorsal-ventral aspect of the embryo, (2) gap genes for longitudinal separation of the embryo into regions, (3) pair-rule genes that establish an alternating on/off striped pattern of gene expression along the embryo, and (4) segment-polarity genes that refine the patterns of gene expression within the stripes and determine the basic layout of segments and parasegments. For example, the *bicoid* gene is a maternal-effect gene in the coordinate class that controls anterior-posterior differentiation of the embryo and determines position on the blastoderm fate map. The *bicoid* gene product is a transcriptional activator protein that is present in a concentration gradient from anterior to posterior along the embryo. Genes that are regulated by the bicoid protein have different sensitivities to its concentration, depending on the number and types of high-affinity and low-affinity binding sites that they contain. The segmentation genes can regulate themselves, other members of the same class, and genes of other classes farther along the hierarchy. Together, the segmentation genes control the homeotic selector genes that initiate the final stages of the developmental process.

Mutations in homeotic genes result in the transformation of one body segment into another. For example, *bithorax* causes transformation of the anterior part of the third thoracic segment into the anterior part of the second thoracic segment, and *Antennapedia* results in transformation of the dorsal part of the head into the into the second thoracic segment. Homeotic genes act within developmental compartments to control other genes concerned with such characteristics as rates of cell division, orientation of mitotic spindles, and the capacity to differentiate into bristles, legs, and other features. Most homeotic genes are clusters of genes, each containing homeoboxes that are characteristic of developmental-control genes in many organisms and that code for DNA-binding and other protein domains.

Developmental processes in higher plants differ significantly from those in animals in that developmental decisions continue throughout life—for example, in the periodic reorganization of shoot meristem into floral meristem and the development of flowers. The genetic and molecular mechanisms that control plant development are not as well understood as those in animals.

KEY TERMS

autonomous determination
blastoderm
blastula
cascade
cell fate
cleavage division
compartment
coordinate gene
execution mutation
fate map
gain-of-function mutation
gap gene
gastrula

heterochronic mutation
homeobox
homeotic gene
homeotic selector gene
imaginal disk
lineage
lineage diagram
loss-of-function mutation
maternal-effect gene
meristem
morphogen
pair-rule gene
parasegment

pole cell
positional information
programmed cell death
regulatory circuitry
segmentation gene
segment-polarity gene
segregation mutation
syncytial blastoderm
totipotent cell
transformation mutation
zygotic gene

Problem 1: What is the logic behind the following principle of developmental genetics: "For a gene affecting cell determination, if loss-of-function alleles eliminate a particular cell fate and gain-of-function alleles induce the fate, then the product of the gene must be both necessary and sufficient for expression of the fate."

Answer: A gene product is necessary for cell fate if its absence prevents normal expression of the fate. Hence, the effect of loss-of-function alleles means that the gene product is necessary for the fate. On the other hand, a gene product is sufficient for cell determination if its presence at the wrong time, in the wrong tissues, or in the wrong amount results in expression of the cell fate. The fact that gain-of-function mutations induce a particular cell fate means that the gene product is sufficient to trigger the fate.

Problem 2: A mutation m is called a recessive maternal-effect lethal if eggs from homozygous m/m females are unable to support normal embryonic development, irrespective of the genotype of the zygote. What kinds of crosses are necessary to produce m/m females?

Answer: Although the eggs from m/m females are abnormal, those from $+/m$ females allow normal embryonic development because the wildtype m^+ allele supplies the cytoplasm of the oocyte with sufficient products for embryogenesis. Therefore, homozygous m/m females can be obtained by crossing heterozygous $+/m$ females with either $+/m$ or m/m males. The first cross produces 1/2 m/m females, the second 1/4 m/m.

Problem 3: The *bicoid* mutation in *Drosophila* results in absence of head structures in the early embryo. A similar phenotype results when cytoplasm is removed from the anterior end of a *Drosophila* embryo. What developmental effect would you expect from: **(a)** a *bicoid* embryo that had injected into its anterior end some cytoplasm taken from the anterior end of a wildtype embryo? **(b)** a wildtype *Drosophila* embryo that had injected into its posterior end some cytoplasm taken from the anterior end of another wildtype embryo?

Answer: The experimental results imply that the anterior cytoplasm, including the *bicoid* gene product, induces development of head structures. **(a)** Anterior cytoplasm from a wildtype embryo should be able to rescue *bicoid*, and so head structures should be formed. **(b)** Anterior cytoplasm injected into the posterior of the embryo should induce head structures in that location, and so the embryo should have head structures at both ends.

PROBLEMS

13-1. Distinguish between a developmental fate determined by autonomous development and one determined by positional information. What two types of surgical manipulation are used to distinguish between the processes experimentally?

13-2. What is meant by *polarity* in the mature oocyte?

13-3. Do plants have a germ line in the same sense as animals? Explain.

13-4. What does it mean to say that a cell in an organism is *totipotent*? Are all cells in a mature higher plant totipotent?

13-5. What are lineage mutations and how are they detected?

13-6. What is the *Drosophila* blastoderm, and what is the blastoderm fate map? Does the existence of a fate map imply that *Drosophila* developmental decisions are autonomous?

13-7. What is the result of a maternal-effect lethal gene in *Drosophila*? If an allele is a maternal-effect lethal, how can an individual fly be homozygous for it?

13-8. What is the principal consequence of a failure in programmed cell death?

13-9. With regard to its effects on cell lineages, what kind of mutation is a homeotic mutation in *Drosophila*?

13-10. A cell A in *Caenorhabditis elegans* normally divides and the daughter cells differentiate into cell-types B and C. What developmental pattern would result from a mutation in A that prevents sister-cell differentiation? From a mutation in A that prevents parent-offspring segregation?

13-11. A mutation is found in which the development pattern is normal but slow. Does this qualify as a heterochronic mutation? Explain?

13-12. Why is transcription of the zygote nucleus dispensable in early *Drosophila* development but not in early development of the mouse?

13-13. Distinguish between a loss-of-function mutation and a gain-of-function mutation. Can both types of mutation occur in the same gene? Can both types occur in the same allele?

13-14. A particular gene is necessary, but not sufficient, for a certain developmental fate. What is the expected phenotype of a loss-of-function mutation in the gene? Is the allele expected to be dominant or recessive?

13-15. A particular gene is sufficient for a certain developmental fate. What is the expected phenotype of a gain-of-function mutation in the gene? Is the allele expected to be dominant or recessive?

13-16. What is the phenotype of a *Drosophila* mutation in the *gap* class? Of a mutation in the *pair-rule* class?

13-17. The drug actinomycin D prevents RNA transcription but has little direct effect on protein synthesis. When fertilized sea-urchin eggs are immersed in a solution of the drug, development proceeds to the blastula stage, but gastrulation does not take place. How would you interpret this finding?

13-18. A mutation in the axolotl designated *o* is a maternal-effect lethal because embryos from *o/o* females die at gastrulation, irrespective of their own genotype. However, the embryos can be rescued by injecting oocytes from *o/o* females with an extract of nuclei from either +/+ or +/o eggs. Injection of cytoplasm is not as effective. Suggest an explanation for these results.

13-19. The nuclei of frog brain cells normally do not synthesize DNA or undergo mitosis. However, when transplanted into developing oocytes the brain-cell nuclei behave as follows: (a) In rapidly growing premeiosis oocytes, they synthesize RNA; (b) In more-mature oocytes, they do not synthesize DNA or RNA, but their chromosomes condense and they begin meiosis. How would you explain these results?

13-20. The autosomal gene *rosy* (*ry*) in *Drosophila* is the structural gene for the enzyme xanthine dehydrogenase (XDH), which is necessary for wildtype eye pigmentation. Flies of genotype *ry/ry* lack XDH activity and have rosy eyes. The X-linked gene *maroonlike* (*mal*) also is necessary for XDH activity, and *mal/mal*; ry^+/ry^+ females and *mal*/Y; ry^+/ry^+ males also lack XDH activity; they have maroonlike eyes. The cross mal^+/*mal*; *ry/ry* ♀ × *mal*/Y; ry^+/ry^+ ♂ produces *mal/mal*; ry^+/ry females and *mal*/Y; ry^+/ry males that have wildtype eye color even though they lack active XDH enzyme. Suggest an explanation.

Mutation, Recombination, and DNA Repair

In preceding chapters, numerous examples were presented in which the information contained in the genetic material had been altered by mutation. A *mutation* is any heritable change in the genetic material. In this chapter, we examine the nature of mutations at the molecular level, how they are created, and how they are detected. You will see that mutations can be induced by

Above: Scanning electron micrograph of a mutant *Drosophila melanogaster* in which legs rather than antennae emerge from the head. (Courtesy of F. R. Turner.)

363

radiation and a variety of chemical agents that produce strand breakage and other types of damage to DNA. Most types of damage to DNA can be repaired by enzymes. The breakage and repair of DNA serve as an introduction to the process of recombination at the DNA level.

14-1 General Properties of Mutations

Mutations can occur in any cell and can result in phenotypic effects ranging from minor alterations that are detectable only by biochemical methods to drastic changes in essential processes that cause death of the cell or organism. When genetic phenomena are studied in the laboratory, mutations producing clearly defined effects are usually used, but most mutations are not of this type. The effect of a mutation is determined by the type of cell containing the mutant allele, by the stage in the life cycle or in the development of the organism in which the process affected by the mutation occurs, and, in diploid organisms, by the dominance or recessiveness of the mutant allele. A recessive mutation is usually not detected until a later generation when two heterozygotes happen to mate. Dominance does not complicate the expression of mutations in bacteria and haploid eukaryotes.

Mutations can be classified in a variety of ways. In multicellular organisms, one distinction is based on the type of cell in which the mutation occurs: **germinal mutations** arise in cells that ultimately form gametes, and **somatic mutations** occur in all other cells. A somatic mutation yields an organism that is genotypically, and for many dominant mutations phenotypically, a mixture of normal and mutant tissue. Because reproductive cells are not affected, such a mutant allele will not be transmitted to the progeny and may not be detected or be recoverable for genetic analysis. However, in higher plants, somatic mutations can often be propagated by vegetative means (without going through seed production), such as grafting or the rooting of stem cuttings. This process has been the source of valuable new varieties such as the 'Delicious' apple and the 'Washington' navel orange.

Some of the most-useful mutations for genetic analysis are those whose effects can be turned on or off at will. They are called **conditional mutations,** and they produce changes in phenotype in one set of environmental conditions (the **restrictive** conditions) but not in another (the **permissive** conditions). For example, a **temperature-sensitive** mutation is a conditional mutation whose expression depends on temperature. Usually, the restrictive temperature is high, and the organism exhibits a mutant phenotype above this critical temperature. The permissive temperature is lower, and under permissive conditions the phenotype is wildtype or nearly wildtype. Temperature-sensitive mutations are frequently used to eliminate particular biochemical reactions in laboratory experiments to test the role of the reactions in various cellular processes, such as DNA replication.

Mutations can also be classified by numerous other criteria, such as the kinds of alterations that have occurred in the DNA, the kinds of phenotypic effects produced, and whether the mutational events were **spontaneous** in origin or were **induced** by exposure to a known **mutagen** (a mutation-causing agent). Such classifications are often useful in discussing various aspects of the mutational process. *Spontaneous* usually means that the event that caused a mutation is unknown, any *spontaneous mutations* are those that occur in the absence of any known mutagenic agent. The properties of spontaneous and induced mutations will be described in later sections.

All mutations result from changes in the nucleotide sequence of DNA or from deletions, insertions, or rearrangement of DNA sequences in the genome. Some types of major rearrangements in chromosomes were discussed in Chapter 6. In this section, the discussion concerns common types of mutations.

Base Substitutions

The simplest type of mutation is a **base substitution,** in which a nucleotide in DNA is replaced with a different nucleotide. For example, in an A → G substitution, an A is replaced with a G and, in an A → T substitution, an A is replaced with a T. When the DNA then replicates, the pairing nucleotide in the other strand is also changed, and so an A → G substitution replaces an AT nucleotide pair with a GC pair, and an A → T substitution replaces an AT pair with a TA pair.

Some base substitutions replace one pyrimidine base with the other or one purine base with the other. These are called **transition** mutations. The possible transition mutations are:

$$T \rightarrow C \text{ or } C \rightarrow T \quad (\text{pyrimidine} \rightarrow \text{pyrimidine})$$
$$A \rightarrow G \text{ or } G \rightarrow A \quad (\text{purine} \rightarrow \text{purine})$$

Other base substitutions replace a pyrimidine with a purine or the other way around. These are called **transversion** mutations. The possible transversion mutations are:

$$T \rightarrow A \text{ or } T \rightarrow G \text{ or } C \rightarrow A \text{ or } C \rightarrow G \quad (\text{pyrimidine} \rightarrow \text{purine})$$
$$A \rightarrow T \text{ or } A \rightarrow C \text{ or } G \rightarrow T \text{ or } G \rightarrow C \quad (\text{purine} \rightarrow \text{pyrimidine})$$

Altogether, there are four possible transitions and eight possible transversions. Therefore, if base substitutions occurred strictly at random, one would expect a 1:2 ratio of transitions to transversions. However:

> Spontaneous base substitutions are biased in favor of transitions. Among spontaneous base substitutions, the ratio of transitions to transversions is approximately 2:1.

Examination of the genetic code (Table 11-2) shows that the bias toward transitions has an important consequence for base substitutions that occur in the third position of a codon. In all codons with a pyrimidine in the third position, the particular pyrimidine present does not matter; and, in most codons ending in a purine, either purine will do. This means that most transition mutations occurring in the third codon position do not change the amino acid that is coded. The mutations change the nucleotide sequence without changing the amino acid sequence, and they are called **silent substitutions** because they are not detectable by changes in phenotype. Many mutations that occur outside of coding regions also have no detectable effects on phenotype. Large regions of DNA within introns or in the spacer sequences between genes are nonessential and can undergo base substitutions, small deletions or additions, insertions of transposable elements, and other rearrangements with no detectable effects on phenotype. On the other hand, some noncoding sequences do have essential functions—for example, promoters, transcription termination signals, and intron splice junctions. Mutations in these sequences can have phenotypic effects.

Base substitutions in coding regions that do result in changed amino acids are called **missense** mutations. A change in the amino acid sequence of a protein may alter the biological properties of the protein. A classic example of a phenotypic effect of a single amino acid change is that responsible for the human hereditary disease **sickle cell anemia.** This condition is characterized by a change in the shape of red blood cells into an elongate form that blocks capillaries. It results from the replacement of a glutamic acid for a valine in hemoglobin. The sickle cells also break down more rapidly than do normal red blood cells, causing the anemia. However, an amino acid substitution does not always create a mutant phenotype. For example, substitution of one amino acid for another with the same charge (say, lysine for histidine) may in some cases have no effect on either protein structure or phenotype. Whether there is an effect of substituting a similar amino acid for another depends on the precise role of that particular amino acid in the structure and function of the protein. Any change in the active site of an enzyme will usually decrease enzymatic activity.

Figure 14-1 illustrates the nine possible codons resulting from a single base substitution in the UAU codon for tyrosine. One mutation is silent (box) and six are missense mutations that change the amino acid inserted in the polypeptide at this position. The other two mutations (red) create a stop codon resulting in premature termination of translation and production of a truncated fragment of the polypeptide. A base substitution that creates a new stop codon is called a **nonsense** mutation. Because nonsense mutations cause premature chain termination, the remaining polypeptide fragment is almost always nonfunctional.

Additions and Deletions

DNA sequences consisting of short tandem repeats or runs of a particular nucleotide are susceptible to errors in replication or recombination that change

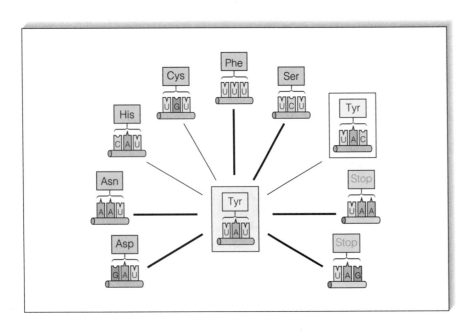

FIGURE 14-1 The nine codons that can result from a single base change of the tyrosine codon UAU. Heavy lines indicate transversions, others transitions. Two possible stop codons are shown in red. Altogether there are six possible missense mutations, two possible nonsense mutations, and one silent mutation.

the number of repeats or the length of the run. For example, a run of consecutive Gs may have extra ones added or a few deleted. Similarly, a short sequence repeated in tandem a number of times may gain or lose a few copies owing to these types of errors. One process that can result in such additions or deletions is unequal crossing-over, discussed in Chapter 6 (see Figure 6-11).

The phenotypic consequences of addition or deletion mutations depend on their location. In nonessential regions, no effects may be seen but, in regulatory or coding regions, their effects may be profound. Small additions or deletions in coding regions will add or delete amino acids from the polypeptide if the number of nucleotides added or deleted is an exact multiple of three (the length of one codon). Otherwise, they will shift the phase in which the ribosome reads the triplet codons and consequently alter all of the amino acids downstream from the site of the mutation. As noted in Chapter 11, such mutations are called *frameshift* mutations because they shift the reading frame of the codons in the mRNA. A common type of frameshift mutation is a single-base addition or deletion. The consequences of a frameshift can be illustrated by the insertion of an adenine in this simple *mRNA* sequence:

Leu Leu Leu Leu
. . . CUG CUG CUG CUG . . .
↓
. . . CUG CAU GCU GCU G . . .
Leu His Ala Ala

An addition or deletion of any number of nucleotides that is not a multiple of three will produce a frameshift. Unless it occurs very near the carboxyl terminus of a protein, a frameshift mutation will result in synthesis of a nonfunctional protein.

Transposable-Element Mutagenesis

All organisms contain multiple copies of several different kinds of transposable elements; that is, DNA sequences capable of readily changing their positions in the genome. The structure and function of transposable elements were examined in Chapter 5 (eukaryotic transposable elements) and Chapter 10 (prokaryotic transposable elements). Transposable elements are also important agents of mutation. For example, in *Drosophila*, approximately half of all spontaneous mutations that have visible phenotypic effects result from insertions of transposable elements.

Transposable elements cause mutations in two principal ways, illustrated in Figure 14-2. Most transposable elements are present in nonessential regions of the genome and usually do little or no harm. However, when an element transposes, it can insert into an essential region and disrupt its function. Figure 14-2A shows the result of transposition into a coding region of DNA (an exon). The insert interrupts the coding region. Because most transposable elements contain coding regions of their own, transcription of the transposable element will interfere with transcription of the gene into which it is inserted or transcription of the gene will terminate within the transposable element. Even if transcription proceeds through the element, the phenotype will be mutant because the coding region then contains incorrect sequences.

Another mechanism of transposable-element mutagenesis results from recombination (Figure 14-2B and C). Transposable elements are present in multiple copies, often with two or more in the same chromosome. If the copies

A Interruption of coding region

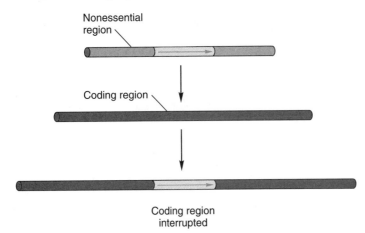

B Intrachromosomal crossing-over

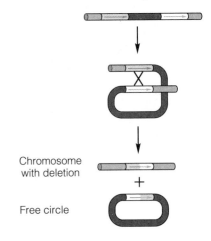

FIGURE 14-2   Some types of mutations resulting from transposable elements. A. Insertion into the coding region of a gene eliminates gene function. B. Crossing-over between homologous elements present in the same orientation in a single chromatid results in the deletion of one of the transposable elements and all of the DNA sequences between them. C. Unequal crossing-over between homologous elements present in the same orientation in different chromatids results in products that contain either a duplication or a deletion of the DNA sequences between the elements. Crossing-over occurs at the four-strand stage of meiosis (Chapter 3) but, in parts B and C, only the two strands participating in the crossover are shown.

C Unequal crossing-over

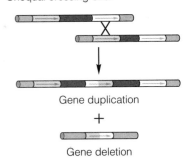

are near enough together, they can pair during synapsis and undergo crossing-over, with the result that the region of DNA between the elements is deleted. Figure 14-2B shows what happens when elements located within the same chromosome undergo pairing and crossing-over. Alternatively, the downstream element in one chromosome can pair with the upstream element in the homologous chromosome, as diagrammed in Figure 14-2C. An exchange within the paired elements results in one chromatid that is missing the region between the elements (bottom) and one chromatid in which the region is duplicated (top).

14-3   Spontaneous Mutations

Mutations are statistically random events, and there is no way of predicting when or in which cell a mutation will occur. However, every gene mutates spontaneously at a characteristic rate, making it possible to assign probabilities to particular mutational events. Thus, there is a definite probability that a given gene will mutate in a particular cell, and likewise a definite probability that a mutant allele of the gene will appear in a population of a particular size. The different kinds of alterations in DNA that lead to mutations differ substantially in complexity, and so their probabilities of occurrence also are quite different. A mutation is also random in the sense that its occurrence is not related to any adaptive advantage that it may confer on the organism in its environment; in the following section, the basis for this conclusion will be presented.

The Nonadaptive Nature of Mutation

The concept that mutations are spontaneous, statistically random events unrelated to adaptation was not widely accepted until the late 1940s. Before that time, it was believed that mutations occur in bacterial populations *in response to* particular selective conditions. The basis for this belief was the observation that, when antibiotic-sensitive bacteria are spread on a solid growth medium containing an antibiotic, some colonies form that consist of cells having an inherited resistance to the drug. The initial interpretation of this observation (and similar ones) was that these adaptive variations were *induced* by the selective agent itself.

Several types of experiments showed that the adaptive mutations in fact occurred spontaneously and were present at low frequency in the bacterial population even *before* it was exposed to the antibiotic. One experiment utilized a technique developed by Joshua and Esther Lederberg called **replica plating** (Figure 14-3). In this procedure, a suspension of bacterial cells is spread on a solid medium. After colonies have formed, a piece of sterile velvet mounted on a solid support is pressed onto the surface of the plate. Some bacteria from each colony stick to the fibers, as shown in Figure 14-3A. Then, the velvet is pressed

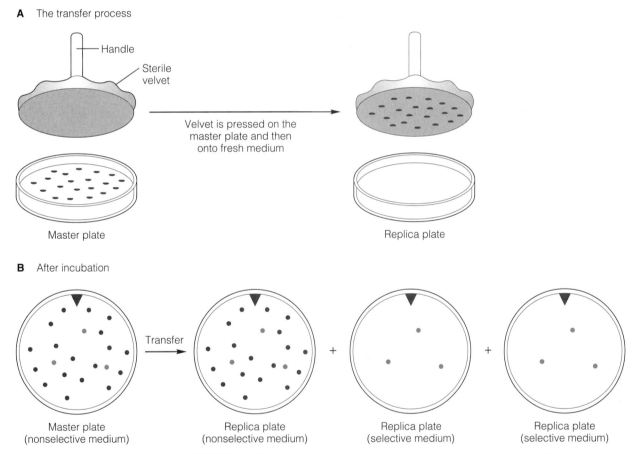

A The transfer process

Handle

Sterile velvet

Velvet is pressed on the master plate and then onto fresh medium

Master plate

Replica plate

B After incubation

Master plate (nonselective medium)

Transfer

Replica plate (nonselective medium)

+

Replica plate (selective medium)

+

Replica plate (selective medium)

FIGURE 14-3 Replica plating. A. The transfer process: a velvet-covered disk is pressed onto the surface of a master plate to transfer cells from colonies on that plate to a second medium. B. Detection of mutants: cells are transferred onto two types of plates containing either a nonselective medium (on which all form colonies) or a selective medium (for example, one spread with T1 phage). Colonies form on the nonselective plate in the same pattern as on the master plate. Only mutant cells (for example, T1-r) can grow on the selective plate; the colonies that form correspond to certain positions on the master plate. Colonies consisting of mutant cells are shown in red.

onto the surface of fresh medium, transferring some of the cells from each colony, which give rise to new colonies having positions identical with those on the first plate. Figure 14-3B shows how this method was used to demonstrate the spontaneous origin of phage T1-r mutants. A master plate containing about 10^7 cells growing on nonselective medium (lacking phage) was replica-plated onto a series of plates that had been spread with about 10^9 T1 phage. After incubation for a time sufficient for colony formation, a few colonies of phage-resistant bacteria appeared in the same positions on each of the selective replica plates. This meant that the T1-r cells that formed the colonies must have been transferred from corresponding positions on the master plate. Because the colonies on the master plate had never been exposed to the phage, the mutations to resistance must have occurred by chance in cells not exposed to the phage.

The replica-plating experiment illustrates the principle that:

Selective techniques merely select mutants that preexist in a population.

This principle is the basis for understanding how natural populations of rodents, insects, and disease-causing bacteria become insensitive to the chemical substances used to control them. A familiar example is the high level of resistance to such insecticides as DDT that now exists in many insect populations, the result of selection for spontaneous mutations affecting behavioral, anatomical, and enzymatic traits that enable the insect to avoid or resist the chemical. Similar problems are encountered in controlling plant pathogens. For example, the introduction of a new variety of a crop plant resistant to a particular strain of disease-causing fungus results in only temporary protection against the disease. The resistance inevitably breaks down because of the occurrence of spontaneous mutations in the fungus that enable it to attack the new plant genotype. Such mutations have a clear selective advantage, and the mutant alleles rapidly become widespread in the fungal population.

Measurement of Mutation Rates

Spontaneous mutations are usually rare events, and the methods used to estimate the frequency with which they arise require large populations and often special techniques. The **mutation rate** is the probability that a gene undergoes mutation in a single generation or in forming a single gamete. Measurement of mutation rates is important in population genetics, studies of evolution, and in analyses of the effects of environmental mutagens.

One of the earliest techniques for measuring mutation rates was Hermann Müller's *ClB* method. *ClB* refers to a special X chromosome of *Drosophila melanogaster*; it has a large inversion (C) that prevents the recovery of crossover chromosomes in the progeny from a female heterozygous for the chromosome (as described in Section 6-6); a recessive lethal (*l*); and the dominant marker Bar (*B*), which reduces the normal round eye to a bar shape. The presence of a recessive lethal in the X chromosome means that males with that chromosome and females homozygous for it cannot survive. The technique is designed to detect mutations arising in a normal X chromosome.

In the *ClB* procedure, females heterozygous for the *ClB* chromosome are mated with males carrying a normal X chromosome (Figure 14-4). From the progeny produced, females with the Bar phenotype are selected and then individually mated with normal males. (The presence of the Bar phenotype

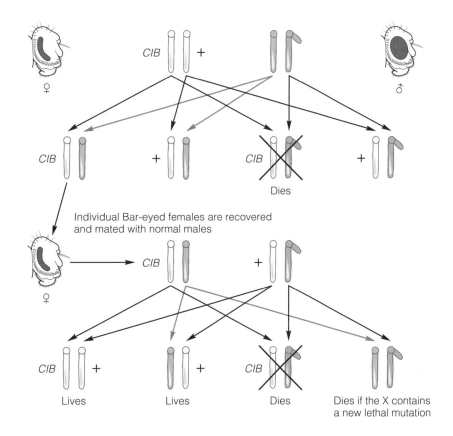

FIGURE 14-4 The *ClB* method for estimating the rate at which spontaneous recessive lethal mutations arise on the *Drosophila* X chromosome.

Individual Bar-eyed females are recovered and mated with normal males

Lives   Lives   Dies   Dies if the X contains a new lethal mutation

indicates that the females are heterozygous for the *ClB* chromosome and the normal X from the male parent.) A ratio of two females to one male is expected among the progeny from such a cross, as shown in the lower part of the illustration. The critical observation in determining the mutation rate is the fraction of males produced in the second cross. Because the *ClB* males die, all of the males in this generation must contain an X chromosome derived from the X of the initial normal male (top row of illustration). Furthermore, it must be a nonrecombinant X chromosome because of the inversion (C) in the *ClB* chromosome. Occasionally, the progeny include no males, and this means that the X chromosome present in the original sperm underwent a mutation somewhere along its length to yield a new recessive lethal. The method provides a quantitative estimate of the rate at which mutation to a recessive lethal allele occurs *at any of the large number of genes* in the X chromosome. About 0.15 percent of the X chromosomes acquire new recessive lethals during spermatogenesis; that is, the mutation rate is 1.5×10^{-3} recessive lethals per X chromosome per generation. Note that the *ClB* method tells us nothing about the mutation rate for a particular gene, because we have no idea how genes on the X chromosome would cause lethality if mutant. Since the development of the *ClB* method, a variety of methods have been developed for determining mutation rates in *Drosophila* and other organisms. Of significance is the fact that mutation rates vary widely from one gene to another. For example, the yellow-body mutation in *Drosophila* occurs at a frequency of 10^{-4} per gamete per generation, whereas mutations to streptomycin resistance in *E. coli* occur at a frequency of 10^{-9} per cell per generation. Furthermore, within a single organism, the frequency can be enormously different, ranging in *E. coli* from 10^{-5} to 10^{-9}.

Hot Spots of Mutation

Certain DNA sequences are called mutational **hot spots** because they are more liable to undergo mutation than others. Mutational hot spots include monotonous runs of a single nucleotide or tandem repeats of short sequences, which may gain or lose copies by any of a variety of mechanisms. Hot spots are found at many sites throughout the genome and within genes. For genetic studies of mutation, hot spots mean that a relatively small number of sites account for a disproportionately large fraction of all mutations. For example, in the analysis of the *rII* gene in bacteriophage T4 discussed in Section 10-5, one extreme hot spot accounted for approximately 20 percent of all the mutations obtained.

Sites of cytosine methylation are usually highly mutable, and the mutations are usually GC → AT transitions. In many organisms, including bacteria, maize, and mammals (but not *Drosophila*), about 1 percent of the cytosine bases are methylated at the 5-carbon position, giving 5-methylcytosine instead of ordinary cytosine (Figure 14-5). Special enzymes add the methyl groups to the cytosines in certain target sequences of DNA. During DNA replication the 5-methylcytosine pairs with guanine and replicates normally. The genetic function of cytosine methylation is not entirely clear, but regions of DNA high in cytosine methylation tend to have reduced gene activity. Examples include the inactive X chromosome in female mammals (Chapter 6) and inactive copies of certain transposable elements in maize.

Cytosine methylation is an important contributor to mutational hot spots, as illustrated in Figure 14-5. Both cytosine and 5-methylcytosine are subject to occasional loss of an amino group. For cytosine, this loss yields uracil. Because uracil pairs with adenine instead of guanine, replication of a molecule containing a GU base pair would ultimately lead to substitution of an AT pair for the original GC pair (by the process GU → AU → AT in successive rounds of replication). However, cells possess a special repair enzyme that specifically removes uracil from DNA, and so the C → U conversion rarely leads to mutation because the U is removed and replaced with C. Unfortunately, loss of the amino group of 5-methylcytosine yields thymine (Figure 14-5B), which is a normal DNA base and hence not removed. Thus, the original GC pair becomes a GT pair and in the next round of replication gives GC (wildtype) and AT (mutant) DNA molecules.

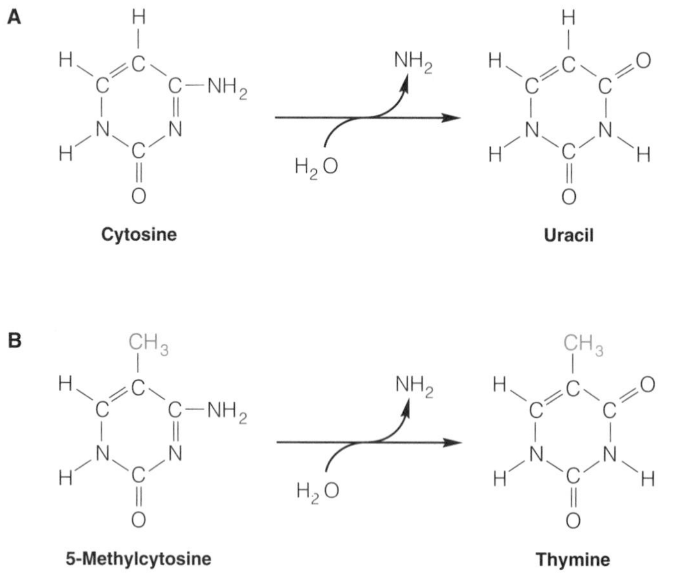

FIGURE 14-5 Spontaneous loss of the amino group of (A) cytosine to yield uracil and (B) 5-methylcytosine to yield thymine.

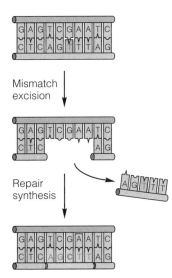

FIGURE 14-6 Mismatch repair consists of the excision of a short segment of a DNA strand containing a base mismatch followed by repair synthesis. Either strand can be excised and corrected but, in newly synthesized DNA, methylated bases in the template strand often direct the excision mechanism to the newly synthesized strand containing the incorrect nucleotide.

Mismatch Repair

Base-substitution mutations usually start as a single mismatched (mispaired) base in a double-stranded DNA molecule—for example, a GT base pair resulting from spontaneous loss of an amino group from 5-methylcytosine (Figure 14-5). However, many of these mismatched bases need not persist through DNA replication to be resolved. The mismatched bases can be detected and corrected by enzymes that carry out **mismatch repair,** a process of excision and resynthesis illustrated in Figure 14-6. When a mismatched base is detected, one of the strands is cut and a region around the mismatch is removed. Then DNA polymerase fills in the gaps, using the uncut strand as template, and thereby eliminates the mismatch. The mismatch-repair system also corrects most single-base insertions or deletions.

If the DNA strand that is removed in mismatch repair is chosen at random, then the repair process will sometimes create a mutant molecule by cutting the strand containing the correct base and using the mutant strand as template. However, in newly synthesized DNA, there is a way to prevent this from happening because the daughter strand is less methylated than the parental strand. The mismatch-repair system recognizes the degree of methylation of a strand and *preferentially excises nucleotides from the undermethylated strand.* This helps ensure that incorrect nucleotides incorporated into the daughter strand during replication will be removed and repaired. The daughter strand is always the undermethylated strand because its methylation lags somewhat behind the moving replication fork, whereas the parental strand was fully methylated in the preceding round of replication.

The mismatch-repair system, like the proofreading function of DNA polymerase (Chapter 4) and all other biochemical systems, is not perfect. Some spontaneous base substitutions still arise as a result of incorporation errors that escape correction. Figure 14-7 summarizes the relevant error rates. The rate of

FIGURE 14-7 Summary of error rates that occur in DNA polymerization, proofreading, and postreplication mismatch repair. The initial rate of nucleotide misincorporation is 10^{-5} per base pair per replication. The proofreading function of DNA polymerase corrects 99 percent of these errors and, of those that remain, postreplication mismatch repair corrects 99.9 percent. The overall rate of misincorporated nucleotides that are not repaired is 10^{-10} per base pair per replication.

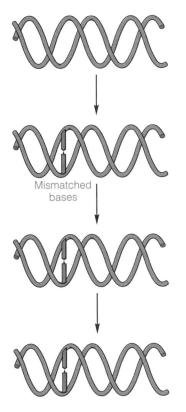

DNA polymerase, probability of mismatch per base pair per replication = 10^{-5}

Mismatched bases

Proofreading function, probability of not correcting mismatch = 10^{-2}

Postreplication mismatch repair, probability of not correcting mismatch = 10^{-3}

Overall probability of nucleotide substitution per base pair per replication = $10^{-5} \times 10^{-2} \times 10^{-3} = 10^{-10}$

initial misincorporation is about 10^{-5} per nucleotide per replication. The chance that a mismatch escapes the proofreading function is approximately 10^{-2}, and the chance that it escapes the postreplication mismatch-repair system is approximately 10^{-3}. Overall, the probability of incorporating a mismatched base that remains uncorrected by either mechanism is $10^{-5} \times 10^{-2} \times 10^{-3} = 10^{-10}$ per base pair per replication.

14-4   Induced Mutations

The first evidence that external agents could increase the mutation rate was presented in 1927 by Hermann Müller, who showed that x rays are mutagenic in *Drosophila*. Since then, a large number of physical agents and chemicals have been shown to increase the mutation rate. The use of these mutagens, several of which will be discussed in this section, provides a means for greatly increasing the number of mutants that can be isolated.

Base-Analog Mutagens

A **base analog** is a compound sufficiently similar to one of the four DNA bases that it can be incorporated into a DNA molecule in normal replication. Such a substance must be able to pair with a base in the template strand. Some base analogs are mutagenic because they are more prone to mispairing than are the normal nucleotides. The molecular basis of the mutagenesis can be illustrated with 5-bromouracil (BU), a commonly used base analog that is efficiently incorporated into the DNA of bacteria and viruses.

The base 5-bromouracil is an analog of thymine, and the bromine atom is about the same size as the methyl group of thymine (Figure 14-8A). If cells are grown in a medium containing 5-bromouracil, during replication of the DNA a thymine is sometimes replaced by a 5-bromouracil, resulting in the formation of an ABU pair at a normal AT site (Figure 14-8B). The subsequent mutagenic activity of the incorporated 5-bromouracil stems from a rare shift in the configuration of the base. This shift is influenced by the bromine atom and occurs in 5-bromouracil more frequently than it does in thymine. In the less-common configuration, 5-bromouracil pairs preferentially with guanine

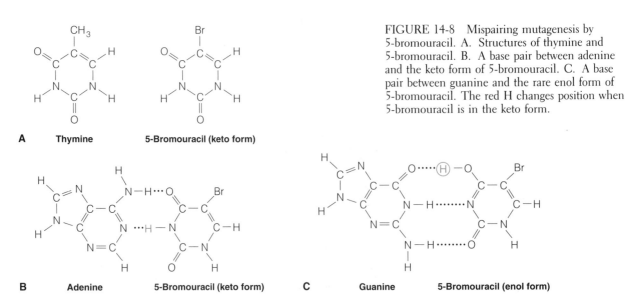

A   Thymine   5-Bromouracil (keto form)

B   Adenine   5-Bromouracil (keto form)   C   Guanine   5-Bromouracil (enol form)

FIGURE 14-8   Mispairing mutagenesis by 5-bromouracil. A. Structures of thymine and 5-bromouracil. B. A base pair between adenine and the keto form of 5-bromouracil. C. A base pair between guanine and the rare enol form of 5-bromouracil. The red H changes position when 5-bromouracil is in the keto form.

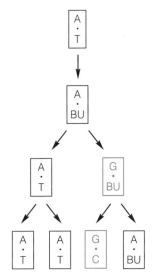

FIGURE 14-9 One mechanism for mutagenesis by 5-bromouracil (BU). An AT → GC transition is produced by the incorporation of 5-bromouracil during DNA replication forming an ABU pair. In the mutagenic round of replication, the BU (in its rare enol form) pairs with G, and in the next round of replication the G pairs with a C, completing the transition mutation.

(Figure 14-8C, and so in the next round of replication one daughter molecule has a GC pair at the altered site, yielding an AT → GC transition (Figure 14-9).

Recent experiments suggest that 5-bromouracil is also mutagenic in another way. The concentration of the nucleoside triphosphates in most cells is regulated by the concentration of deoxythymidine triphosphate (dTTP). This regulation results in appropriate relative amounts of the four triphosphates for DNA synthesis. One part of this complex regulatory process is the inhibition of synthesis of deoxycytidine triphosphate (dCTP) by excess dTTP. The 5-bromouracil nucleoside triphosphate also inhibits production of dCTP. When 5-bromouracil is added to the growth medium, dTTP continues to be synthesized by cells at the normal rate, whereas the synthesis of dCTP is significantly reduced. The ratio of dTTP to dCTP then becomes quite high and the frequency of misincorporation of T opposite G increases. Although repair systems can remove some incorrectly incorporated thymine, in the presence of 5-bromouracil the rate of misincorporation can exceed the rate of correction. An incorrectly incorporated T pairs with A in the next round of DNA replication, yielding a GC → AT transition in one of the daughter molecules. Thus, 5-bromouracil usually induces transitions in both directions—AT → GC by the route in Figure 14-8, and GC → AT by the misincorporation route.

Chemical Agents That Modify DNA

Many mutagens are chemicals that react with DNA and change the hydrogen-bonding properties of the bases. These mutagens are active on both replicating and nonreplicating DNA, in contrast with the base analogs that are mutagenic only when DNA replicates. Several of these chemical mutagens, of which nitrous acid is a well-understood example, are highly specific in the changes they produce. Others, for example, the alkylating agents, react with DNA in a variety of ways and produce a broad spectrum of effects.

Nitrous acid (HNO_2) acts as a mutagen by converting amino (NH_2) groups of the bases adenine, cytosine, and guanine into keto (=O) groups. This process of *deamination* alters the hydrogen-bonding specificity of each base. Deamination of adenine yields a base that pairs with cytosine rather than thymine, resulting in an AT → GC transition (Figure 14-10). As discussed

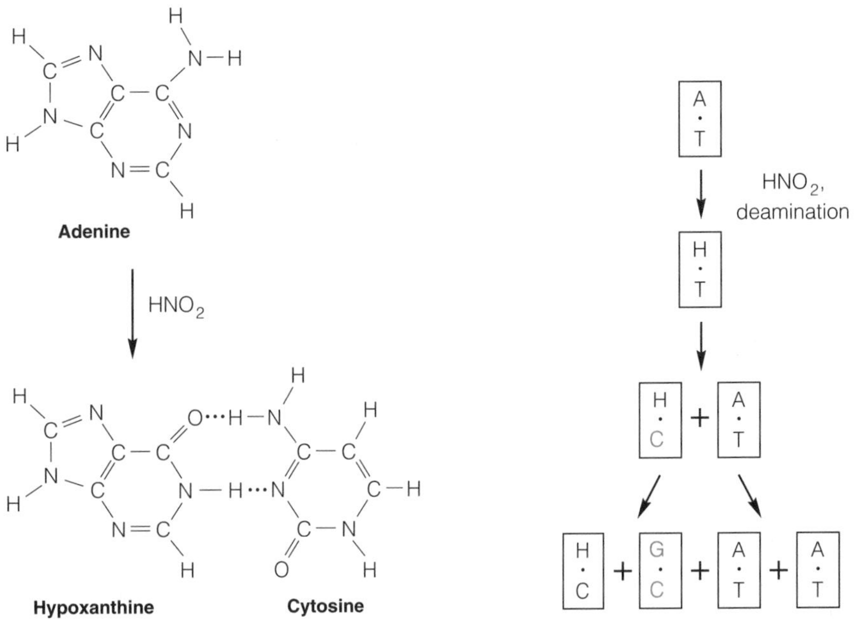

FIGURE 14-10 Nitrous acid mutagenesis. A. Conversion of adenine into hypoxanthine, which pairs with cytosine. The cytosine → uracil conversion is shown in Figure 14-5. B. Production of an AT → GC transition. In the mutagenic round of replication, the hypoxanthine (H) pairs with C. In the next round of replication, the C pairs with G, completing the transition mutation.

earlier, deamination of cytosine yields uracil (Figure 14-5A), which pairs with adenine instead of guanine, producing a GC → AT transition.

Chemical mutagens such as nitrous acid (and many others) are exceedingly useful in prokaryotic systems but are not particularly useful as mutagens in higher eukaryotes because the chemical conditions necessary for reaction are not easily obtained. However, the alkylating agents are highly effective in eukaryotes as well as prokaryotes. The **alkylating agents,** such as ethyl methane sulfonate (EMS) and nitrogen mustard, the structures of which are shown in Figure 14-11, are potent mutagens that have been used extensively in genetic research. These agents add various chemical groups to the DNA bases that either alter their base-pairing properties or cause structural distortion of the DNA molecule. Alkylations of either G or T cause mispairing leading to the transitions AT → GC and GC → AT. EMS reacts less readily with adenine and cytosine. Another phenomenon resulting from alkylation of guanine is **depurination,** or loss of the alkylated base from the DNA molecule by breakage of the bond joining the purine nitrogen and deoxyribose. Depurination is not always mutagenic, because the gap left by loss of the purine can be repaired. However, sometimes the replication fork may reach the gap before repair has occurred. In this case, replication almost always inserts an adenine nucleotide in the daughter strand opposite the apurinic site. After another round of replication, the original GC pair becomes a TA pair, an example of a transversion mutation.

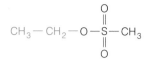

Ethyl methane sulfonate

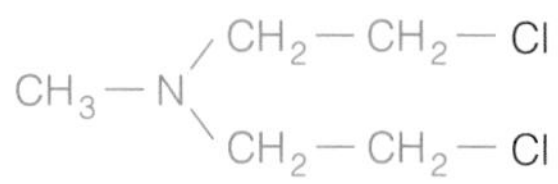

Nitrogen mustard

FIGURE 14-11 The chemical structures of two highly mutagenic alkylating agents with the alkyl groups shown in red.

Misalignment Mutagenesis

The **acridine** molecules, of which proflavin is an example, are planar three-ringed molecules, whose dimensions are roughly the same as those of a purine-pyrimidine pair (Figure 14-12). These substances insert between adjacent base pairs of DNA, a process called **intercalation.** The effect of the intercalation of an acridine molecule is to cause the adjacent base pairs to move apart by a distance roughly equal to the thickness of one pair (Figure 14-13). When DNA containing intercalated acridines is replicated, the template and daughter DNA strands can misalign, particularly in regions where nucleotides are repeated. The result is that a nucleotide is either added or deleted in the daughter strand. In a coding region, the result of a single-base addition or deletion is a frameshift mutation (Section 14-2).

Ultraviolet Irradiation

Ultraviolet (UV) light produces both mutagenic and lethal effects in all viruses and cells. The effects are caused by chemical changes in the bases resulting

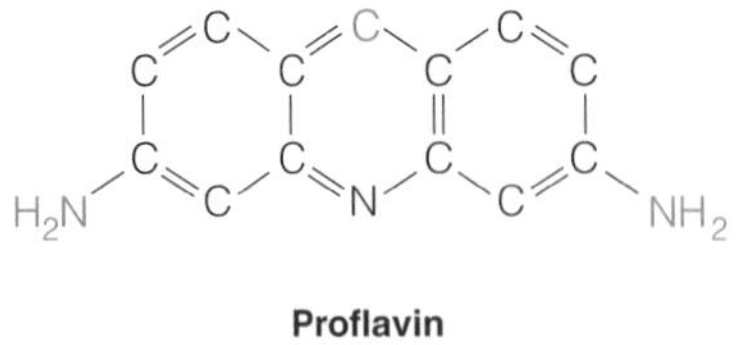

Proflavin

FIGURE 14-12 The structure of proflavin, an acridine derivative. Other mutagenic acridines have additional atoms on the NH₂ groups and on the red C of the central ring. Hydrogen atoms are not shown.

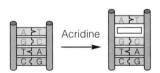

FIGURE 14-13 Separation of two base pairs (red) caused by intercalation of an acridine molecule (open box).

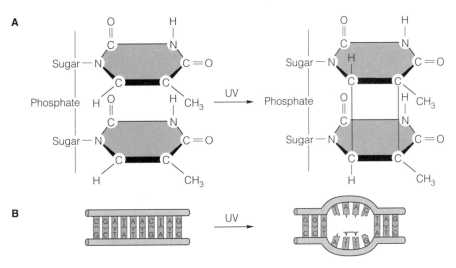

A

B

FIGURE 14-14   A. Structural view of the formation of a thymine dimer. Adjacent thymines in a DNA strand that have been subjected to ultraviolet (UV) irradiation are joined by formation of the bonds shown in red. Other types of bonds between the thymine rings are also possible. Although not drawn to scale, these bonds are considerably shorter than the spacing between the planes of adjacent thymines, so that the double-stranded structure becomes distorted. The shape of each thymine ring also changes. B. The distortion of the DNA helix caused by two thymines moving closer together when joined in a dimer. The dimer is indicated as two thymines joined by a bar.

from absorption of the energy of the light. The major products formed in DNA after UV irradiation are covalently joined pyrimidines (**pyrimidine dimers**), primarily thymine, that are adjacent in the same polynucleotide strand (Figure 14-14A). This chemical linkage brings the bases closer together, causing a distortion of the helix (Figure 14-14B), which blocks transcription and transiently blocks DNA replication. Pyrimidine dimers can be repaired in ways discussed later in this chapter.

In human beings, the inherited autosomal recessive disease **xeroderma pigmentosum** is the result of a defect in a system that repairs ultraviolet-damaged DNA. Persons with this disease are extremely sensitive to sunlight, resulting in excessive skin pigmentation and the development of numerous skin lesions that frequently become cancerous. Examination of cells cultured from patients with the disease shows that the ability to repair pyrimidine dimers is reduced, and the cells are killed by much lower doses of ultraviolet light than are cells from normal persons.

Ionizing Radiation

Ionizing radiation includes x rays and the particles and radiation released by radioactive elements (α and β particles and γ rays). When ionizing radiation interacts with water, highly reactive ions called **free radicals** are formed. The free radicals react with other molecules, including DNA, which results in the mutagenic effects. The intensity of a beam of ionizing radiation can be described quantitatively in several ways. A common measure is the **rad,** which is defined as the amount of radiation that results in the absorption of 100 ergs of energy per gram of matter.

The frequency of mutations induced by x rays is proportional to the radiation dose. Thus, the frequency of X-chromosome recessive lethals in

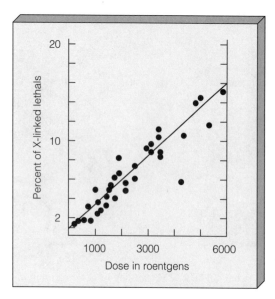

FIGURE 14-15   The relation between the percentage of X-linked recessive lethals in *D. melanogaster* and x-ray dose, obtained from several experiments. The frequency of spontaneous X-linked lethal mutations is 0.15 percent per X chromosome per generation.

Drosophila increases linearly with increasing x-ray dose (Figure 14-15). For example, an exposure of 1000 rads increases the frequency from the spontaneous value of 0.15 percent to about 3 percent.

There appears to be no threshold exposure below which mutations are not induced. Even very low doses of radiation induce some mutations.

The mutagenic and lethal effects of ionizing radiation at low-to-moderate doses result primarily from damage to DNA. Three types of damage in DNA are produced by ionizing radiation—single-strand breakage (in the sugar-phosphate backbone), double-strand breakage, and alterations in nucleotide bases. The single-strand breaks are usually efficiently repaired, but the other damage is responsible for mutation and lethality. In eukaryotes, ionizing radiation results in chromosome breaks. Although systems exist for repairing the breaks, the repair often leads to translocations, inversions, duplications, and deletions.

Ionizing radiation is widely used in tumor therapy. The basis for the treatment is the increased frequency of chromosomal breakage (and the consequent lethality) in cells undergoing mitosis compared with cells in interphase. Tumors usually contain many more mitotic cells than most normal tissues, and so more tumor cells are destroyed than normal cells. Because all tumor cells are not in mitosis at the same time, irradiation is carried out at intervals of several days to allow interphase tumor cells to enter mitosis. The hope is that, over a period of time, most tumor cells will be destroyed.

Table 14-1 gives representative values of doses of ionizing radiation received by human beings, and thus reproductive cells, in the United States in the course of one year. Note that with the exception of diagnostic x rays, which yield important compensating benefits, most of the total radiation exposure comes from natural sources. There are dangers inherent in any exposure to ionizing radiation because of its mutagenic effects, but most geneticists have become less concerned with the potential hazards of ionizing radiation than with those of the many mutagenic and carcinogenic chemicals introduced into the environment from a variety of sources.

TABLE 14-1   Annual exposure of human beings in the United States in 1980 to various forms of ionizing radiation

Source	Dose (in millirems*)
Natural radiation	
Cosmic rays	28
Natural radioisotopes in the body	28
Natural radiosotopes in the soil	26
	82
Other radiation sources	
Diagnostic x rays	20
Radiopharmaceuticals	2–4
Consumer products (x rays from TV, radioisotopes in clock dials) and building materials	4–5
Fallout from weapons tests	4–5
Nuclear power plants	<1
	30–35
Total from all sources	112–117

*For ionizing radiation, 1 millirem is approximately 1/1000 rad.
Source: From National Research Council, Committee on the Biological Effects of Ionizing Radiations. *The Effects on Populations of Exposure to Low Levels of Ionizing Radiation.* National Academy Press. 1980.

14-5   Mechanisms of DNA Repair

Much of the damage done to DNA by chemical mutagens and radiation can be repaired.

1. Various enzymes can recognize and catalyze the direct reversal of specific DNA damage. A classic example is the reversal of UV-induced pyrimidine dimers by **photoreactivation,** in which an enzyme breaks the bonds that join the pyrimidines in the dimer and thereby restores the original bases. The enzyme binds to the dimers in the dark but then utilizes the energy of blue light to cleave the bonds. Another important example is the reversal of guinine methylation in O^6–methyl guanine by a methyl transferase enzyme, which otherwise would pair like adenine during replication.

2. **Excision repair** is a ubiquitous multistep enzymatic process by which a stretch of a damaged DNA strand is removed from a duplex molecule and replaced by resynthesis, using the undamaged strand as a template. Mismatch repair of single mispaired bases (Section 14-3) is an important category of excision repair. The overall process of excision repair is illustrated in Figure 14-16. The DNA damage can be due to anything that produces a distortion in the duplex molecule; for example, a pyrimidine dimer. In excision repair, a repair endonuclease complex recognizes the distortion produced by the DNA damage and makes one to two cuts in the sugar-phosphate backbone, several nucleotides on either side of the damage. A 3′-OH group is produced at the 5′ cut, which DNA polymerase uses as a primer and synthesizes a new strand while displacing the DNA segment that contains the damage. The final step of the repair process is joining of the newly synthesized segment to the contiguous strand by DNA ligase.

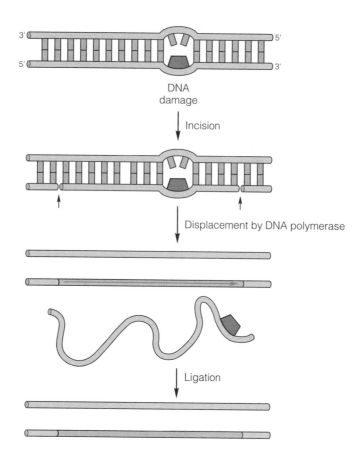

DNA
damage

↓ Incision

↓ Displacement by DNA polymerase

↓ Ligation

FIGURE 14-16   Mechanism of excision repair of DNA damage.

3. Sometimes DNA damage persists rather than being reversed or removed, but its harmful effects may be minimized. This often requires replication across damaged areas, and so the process is called **postreplication repair.** For example, when DNA polymerase reaches a damaged site (such as a pyrimidine dimer), it stops synthesis of the strand. However, after a brief time, synthesis is reinitiated beyond the damage and chain growth continues, producing a gapped strand with the damaged spot in the gap.

The gap can be filled by strand exchange with the parental strand having the same polarity, and the secondary gap produced in that strand can be filled by repair synthesis (Figure 14-17). The products of this exchange and resynthesis are two intact single strands, each of which can then serve in the next round of replication as a template for the synthesis of an undamaged DNA molecule.

4. **SOS repair,** which occurs in *E. coli* and related bacteria, is a complex set of processes that includes a bypass system that allows DNA replication to occur across pyrimidine dimers or other DNA distortions, but at the cost of fidelity of replication. Even though intact DNA strands are formed, the strands are often defective. Thus, SOS repair is said to be *error prone.* A significant feature of the SOS repair system is that it is not always active but is induced by DNA damage. Once activated, the SOS system allows the growing fork to advance across the

FIGURE 14-17 Postreplication repair. A. A molecule with DNA damage in strand 4 is being replicated. B. By reinitiating synthesis beyond the damage, a gap is formed in strand 3. C. A segment of parental strand 1 is excised and inserted in strand 3. D. The gap in strand 1 is then filled in by repair synthesis.

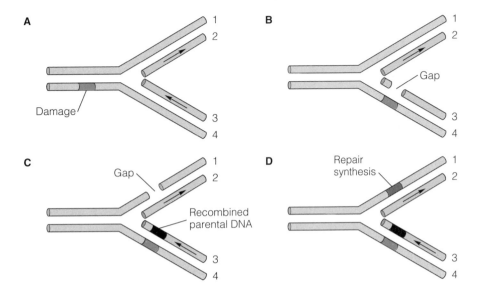

damaged region adding nucleotides that are often incorrect because the damaged template strand cannot be replicated properly. The proofreading system of DNA polymerase is relaxed to allow polymerization to proceed across the damage, despite the distortion of the helix.

14-6 Reverse Mutations and Suppressor Mutations

In most of the mutations considered so far, a wildtype or normal gene is changed into a form that results in a mutant phenotype, an event called a **forward mutation.** Mutations are frequently reversible, and an event that restores the wildtype phenotype is called a **reversion.** A reversion may result from a **reverse mutation,** an exact reversal of the alteration in base sequence that occurred in the original forward mutation, restoring the wildtype DNA sequence. Reversions may also result from the occurrence of a second mutation, at some other site in the genome, which in one of several ways compensates for the effect of the original mutation. Reversion by the exact reversal mechanism is infrequent. The second-site mechanism is much more common, and a mutation of this kind is called a **suppressor mutation.** A suppressor mutation can occur at a different site in the same gene from that of the mutation that it suppresses (*intragenic suppression*) or in a different gene in either the same chromosome or a different chromosome (*intergenic suppression*). Most suppressor mutations do not fully restore the wildtype phenotype for reasons that will be apparent in the following discussion of the two kinds of suppression.

Intragenic Suppression

Reversion of frameshift mutations usually results from intragenic suppression; the mutational effect of the addition (or deletion) of a nucleotide pair in changing the reading frame of the mRNA is rectified by a compensating deletion (or addition) of a second nucleotide pair at a nearby site in the gene (Section 11-6). A second type of intragenic suppression occurs when loss of activity of a protein, caused by one amino acid change, is at least partly restored

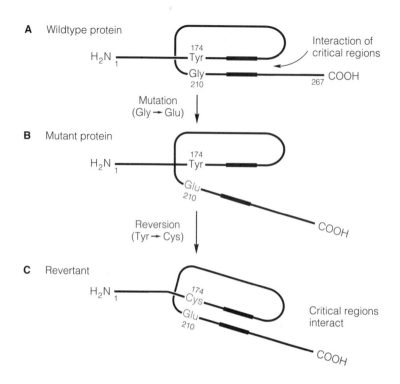

A Wildtype protein

H₂N ──── Tyr ──── ▬▬▬
         174   Interaction of
                               critical regions
         Gly ──── ▬▬▬ ──── COOH
         210   267

Mutation
(Gly → Glu)

B Mutant protein

H₂N ──── Tyr ────
         174

         Glu ──── ▬▬▬
         210
                      ── COOH

Reversion
(Tyr → Cys)

C Revertant

H₂N ──── Cys ──
         174
         Glu ── ▬▬▬   Critical regions
         210   interact
                   COOH

FIGURE 14-18 Simplified model for the effect of mutation and intragenic suppression on the folding and activity of the A protein of tryptophan synthetase in *E. coli*. A. The wildtype protein in which two critical regions (heavy lines) of the polypeptide chain interact. B. Disruption of proper folding of the polypeptide chain by a substitution (red) of amino acid 210 prevents the critical regions from interacting. C. Suppression of the effect of the original forward mutation by a subsequent change in amino acid 174 (red). The structure of the region containing this amino acid is altered, bringing the critical regions together again.

by a change in a second amino acid. An example of this type of suppression in the protein product of the *trpA* gene in *E. coli* is shown in Figure 14-18. The A polypeptide, composed of 268 amino acids, is one of two polypeptides that make up the enzyme tryptophan synthetase in *E. coli*. The mutation shown, one of many that inactivate the enzyme, is a change of amino acid number 210 from glycine in the wildtype protein to glutamic acid. This glycine is not in the enzyme's active site; rather, the inactivation is caused by a change in the folding of the protein, which indirectly affects the active site. The activity of the mutant protein is partly restored by a second mutation, in which amino acid number 174 changes from tyrosine to cysteine. A protein in which only the tyrosine has been replaced by cysteine is inactive, again because of a change in the shape of the protein. This change at the second site, found to occur repeatedly in reversion of the original mutation, restores activity because the two regions containing amino acids 174 and 210 interact fortuitously to produce a protein with the correct folding. This type of reversion can usually be taken to mean that the two regions of the protein interact, and studies of the amino acid sequences of mutants and revertants are often informative in elucidating the structure of proteins.

Intergenic Suppression

Intergenic suppression refers to a mutational change *in a second gene* that eliminates or suppresses the mutant phenotype. Many intergenic suppressors are very specific in being able to suppress the effects of mutations in only one or a few other genes. For example, phenotypes resulting from the accumulation of intermediates in a metabolic pathway due to mutational inactivation of one step in the pathway can sometimes be suppressed by mutations in other genes that function earlier in the pathway, which reduces the concentration of the intermediate. These suppressors act only to suppress mutations further along the pathway, and they suppress all alleles of these genes.

Other intergenic suppressors are more general in being able to suppress mutations in many functionally unrelated genes—and usually only a subset of alleles of these genes. For example, certain intergenic suppressors in *Drosophila* affect transcription factors controlling the transcription of particular families of transposable elements. Alleles that are mutant because they contain a transposable element whose transcription interferes with that of the gene can be suppressed by an intergenic suppressor that prevents transcription of the inserted transposable element.

In prokaryotes, the best-understood type of intergenic suppression is found in some tRNA genes, mutant forms of which can suppress the effects of particular mutant alleles of many other genes. These suppressor mutations change the anticodon sequence in the tRNA and thus the specificity of mRNA codon recognition by the tRNA molecule. Mutations of this type were first detected in certain strains of *E. coli*, which were able to suppress particular phage-T4 mutants that failed to form plaques on standard bacterial strains. These strains were also able to suppress particular alleles of numerous other genes in the bacterial genome. The suppressed mutations were, in each case, nonsense mutations—those in which a stop codon (UAA, UAG, or UGA) had been introduced within the coding sequence of a gene, with the result that polypeptide synthesis was prematurely terminated and only an amino-terminal fragment of the polypeptide was synthesized (Section 11-6).

The reason for suppression of nonsense mutations by a **suppressor tRNA** can be illustrated by examining a chain-termination codon formed by mutation of the tyrosine codon UAC to the stop codon UAG (Figure 14-19A). Such a

A Nonsense mutation

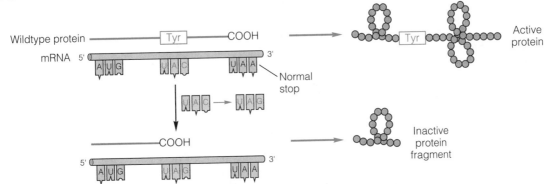

B tRNA suppression

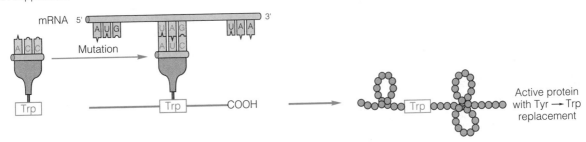

FIGURE 14-19 The mechanism of suppression by a suppressor tRNA molecule. A. A UAC → UAG chain-termination mutation leads to an inactive, prematurely terminated protein. B. A mutation in the tRNA^{Trp} gene produces an altered tRNA molecule, which has a codon complementary to a UAG stop codon but which can still be charged with tryptophan. This tRNA molecule allows the protein to be completed but with a tryptophan at the site of the original tyrosine. Suppression will occur if the substitution restores activity to the protein.

mutation can be suppressed by a mutant tryptophan tRNA molecule. In *E. coli*, tRNA^{Trp} has the anticodon 3'-ACC-5', which pairs with the codon 5'-UGG-3'. A suppressor mutation in the tRNA^{Trp} gene produces an altered tRNA with the anticodon AUC; this tRNA molecule is still charged with tryptophan but responds to the stop codon UAG rather than to the normal tryptophan codon UGG. Thus, in a cell containing this suppressor tRNA, the mutant protein was completed and suppression occurred as long as the mutant protein would tolerate a substitution of tryptophan for tyrosine (Figure 14-19B). Many suppressor tRNA molecules of this type have been observed. Each suppressor is effective against only some nonsense mutations, because the resulting substituted amino acid may not yield a functional protein.

There are three classes of tRNA suppressors: those that suppress only UAG, those that suppress both UAA and UAG, and those that suppress only UGA. They share the following properties:

1. The original mutant gene still contains the mutant base sequence (UAG in Figure 14-19).
2. The suppressor tRNA suppresses all chain-termination mutations with the same stop codon, provided that the amino acid inserted is an acceptable amino acid at the site.
3. A cell can survive the presence of a tRNA suppressor *only* if the original cell contained two or more copies of the same tRNA gene. Taking the example in Figure 14-19, if only one tRNA^{Trp} gene were present in the genome and it was mutated, then the normal tryptophan codon UGG would no longer be read as a sense codon and all polypeptide chains would terminate wherever a UGG codon occurred. However, *multiple copies of most tRNA genes exist*, and so if one copy is mutated to yield a suppressor tRNA, a normal copy nearly always remains.
4. Any chain-termination codon can be translated by a suppressor tRNA mutation that recognizes that codon. For example, translation of UAG by insertion of an amino acid would prevent termination of all wildtype mRNA reading frames terminating in UAG. (However, the anticodon of the suppressor tRNA usually binds rather weakly to the stop codon, and so the stop codon often results in termination anyway.)

Suppressor tRNA mutations are very useful in genetic analysis because they allow nonsense mutations to be identified through their ability to be suppressed. This is important because nonsense alleles usually result in a completely inactive truncated protein and so are considered true loss-of-function alleles. Suppressor tRNAs have been widely used for genetic analysis in prokaryotes, yeast, and even nematodes because their other harmful effects are tolerated by the organism. In higher organisms, suppressor tRNAs have such severe harmful effects that they are of limited usefulness.

Reversion as a Means of Detecting Mutagens and Carcinogens

In view of the increased number of chemicals used and present as environmental contaminants, tests for the mutagenicity of these substances have become important. Furthermore, most carcinogens are also mutagens, and so mutagenicity provides an initial screening for potential hazardous agents. One simple method for screening large numbers of substances for mutagenicity is a reversion test that uses nutritional mutants of bacteria. In the simplest type of reversion test, a compound that is a potential mutagen is added to solid growth medium, known numbers of a mutant bacterium are plated, and the number of revertant colonies is counted. An increase in the reversion frequency significantly greater than that obtained in the absence of the test compound identifies the substance as a mutagen. However, simple tests of this type fail to demonstrate the mutagenicity of a large number of potent carcinogens. The explanation for this failure is that many substances are not directly mutagenic (or carcinogenic), but rather require a conversion into mutagens by enzymatic reactions that take place in the livers of animals and that have no counterpart in bacteria. The normal function of these enzymes is to protect the organism from various harmful naturally occurring substances by converting them into soluble nontoxic substances that can be disposed of in the urine. However, when the enzymes encounter certain man-made and natural compounds, they convert these substances, which may not be harmful in themselves, into mutagens or carcinogens. The enzymes of liver cells, when added to the bacterial growth medium, activate the compounds and allow their mutagenicity to be recognized. The addition of liver extract is one step in the currently used **Ames test** for carcinogens and mutagens.

In the Ames test, histidine-requiring (His^-) mutants of the bacterium *Salmonella typhimurium*, containing either a base substitution or a frameshift mutation, are tested for reversion to His^+. In addition, the bacterial strains have been made more sensitive to mutagenesis by the incorporation of several mutant alleles that inactivate the excision-repair system and that make the cells more permeable to foreign molecules. Because some mutagens act only on replicating DNA, the solid medium used contains enough histidine to support a few rounds of replication but not enough to permit the formation of a visible colony. The medium also contains a potential mutagen to be tested and an extract of rat liver. If the test substance is a mutagen or is converted into a mutagen, some colonies are formed. A quantitative analysis of reversion frequency can also be carried out by incorporating various amounts of the potential mutagen into the medium. The reversion frequency generally depends on the concentration of the substance being tested and, for a known carcinogen or mutagen, correlates roughly with its carcinogenic potency in animals.

The Ames test has been used with thousands of substances and mixtures (such as industrial chemicals, food additives, pesticides, hair dyes, and cosmetics), and numerous unsuspected substances have been found to stimulate reversion in this test. A high frequency of reversion does not necessarily indicate that the substance is definitely a carcinogen but only that it has a high probability of being so. As a result of these tests, many industries have reformulated their products: for example, the cosmetic industry has changed the formulation of many hair dyes and cosmetics to render them nonmutagenic. Ultimate proof of carcinogenicity is determined by testing for tumor formation in laboratory animals. However, only a few percent of the substances known from animal experiments to be carcinogens failed to increase the reversion frequency in the Ames test.

14-7 Oligonucleotide Site-directed Mutagenesis

For many years, the study of mutations was driven in part by the hope of finding agents that could induce mutations of particular types in particular genes. That hope gradually faded as most mutagens were found to produce a variety of effects at random locations throughout the genome. However, a quite different approach using direct manipulation of DNA can be used to create any desired mutation in virtually any gene. The method, called **oligonucleotide site-directed mutagenesis**, is outlined in Figure 14-20 for the production of a

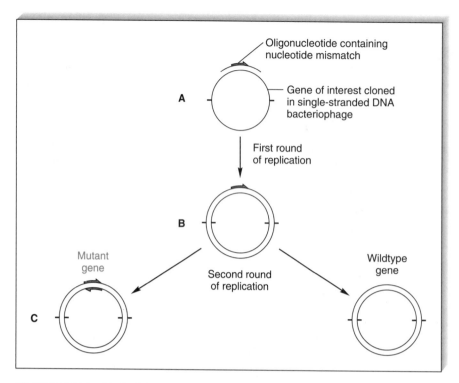

FIGURE 14-20 Method for introducing specific nucleotide substitutions into a DNA molecule. A short oligonucleotide is synthesized that is complementary in sequence to the target DNA but contains a single base substitution. A. This oligonucleotide is annealed with the target DNA, which has been inserted into a single-stranded bacteriophage. B. The first round of DNA replication uses the oligonucleotide as primer and forms a complete double-stranded molecule containing one base-pair mismatch. C. The mismatched bases acquire their correct pairing partners in the second round of replication, and phage containing the mutant gene can be identified by any of a number of methods.

base-substitution mutation. The method uses a clone of the target gene inserted into a single-stranded DNA bacteriophage. This is a type of phage containing single-stranded DNA within the mature infectious particle, although the DNA is double stranded during replication in the bacterial host. The single-stranded phage DNA is annealed with a short (usually from 20 to 50 nucleotides) segment of single-stranded DNA called an **oligonucleotide.** Oligonucleotides of any desired sequence can be obtained from automated DNA synthesizers, and those used in site-directed mutagenesis usually have sequences complementary to the target gene except for an internal nucleotide substitution. When the annealing takes place (Figure 14-20A), the flanking homology is enough to allow the oligonucleotide to pair with the target gene.

The next step is transformation, in which the DNA is introduced into a bacterial cell (Section 10-2). Once inside, the oligonucleotide is used as a primer that is extended by DNA polymerase to create a double-stranded circle (Figure 14-20B). The circle contains a mismatch and is theoretically susceptible to mismatch repair, but the phage DNA usually replicates too rapidly for this repair to occur. After the next round of replication, the mismatched strands segregate (Figure 14-20C), and the cells contain a mixture of phages with either the wildtype gene or the specific mutation. Mature phages are harvested and used to infect other bacterial cells to obtain pure plaques (Section 10-5). Any of a variety of methods can then be used to identify phages containing the newly created mutation.

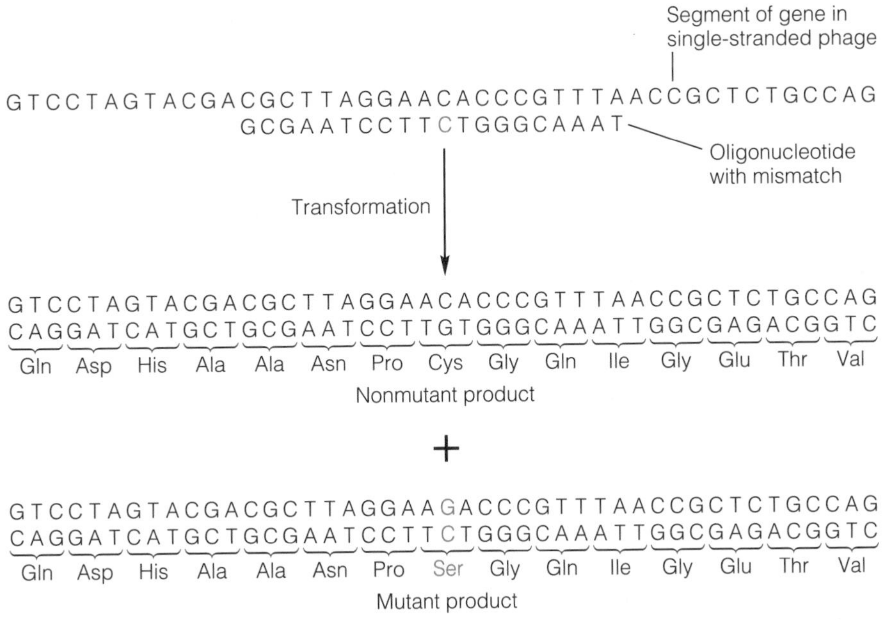

GTCCTAGTACGACGCTTAGGAACACCCGTTTAACCGCTCTGCCAG
GCGAATCCTTCTGGGCAAAT

Oligonucleotide
with mismatch

Transformation

GTCCTAGTACGACGCTTAGGAACACCCGTTTAACCGCTCTGCCAG
CAGGATCATGCTGCGAATCCTTGTGGGCAAATTGGCGAGACGGTC

Gln Asp His Ala Ala Asn Pro Cys Gly Gln Ile Gly Glu Thr Val

Nonmutant product

+

GTCCTAGTACGACGCTTAGGAAGACCCGTTTAACCGCTCTGCCAG
CAGGATCATGCTGCGAATCCTTCTGGGCAAATTGGCGAGACGGTC

Gln Asp His Ala Ala Asn Pro Ser Gly Gln Ile Gly Glu Thr Val

Mutant product

FIGURE 14-21 Production of a specific mutation in the *E. coli* alkaline phosphatase, using oligonucleotide site-directed mutagenesis. The mismatch in the oligonucleotide is in red. The mutant protein contains a cysteine → serine replacement, which renders the enzyme nonfunctional.

As an example, part of the wildtype gene for the *E. coli* enzyme alkaline phosphatase and a synthetic oligonucleotide used for mutagenesis are shown in Figure 14-21. The base substitution in the oligonucleotide is shown in red. The amino acid sequence coded by the wildtype and that coded by the mutant differ in a single replacement; that is, cysteine by serine. This particular cysteine residue is thought to be important in protein folding. As expected, the mutant enzyme is completely nonfunctional.

14-8   Recombination

Genetic recombination may be regarded as a process of breakage and repair between two DNA molecules. In eukaryotes, the process occurs early in meiosis after each molecule has replicated and, with respect to genetic markers, it results in two molecules of the parental type and two recombinants (Chapter 3). For genetic studies of recombination, fungi such as yeast or *Neurospora* are particularly useful because all four products of meiosis are contained in a 4-spore or 8-spore ascus (Section 3-4). Most asci from heterozygous A*a* individuals contain ratios of 2 A : 2 *a* or 4 A : 4 *a* because of normal Mendelian segregation. Occasionally, however, aberrant ratios such as 3 A : 1 *a* or 1 A : 3 *a* in 4-spored asci, or 5 A : 3 *a* or 3 A : 5 *a* in 8-spored asci, are found. Other aberrant ratios also occur. The aberrant asci are said to result from **gene conversion** because it appears as if one allele has "converted" the other allele into a form like itself. Gene conversion is frequently accompanied by recombination between genetic markers on either side of the conversion event, even when the flanking markers are tightly linked. This implies that gene conversion can be one consequence of the recombination process. The accepted explanation of gene conversion is that it results from mismatch repair

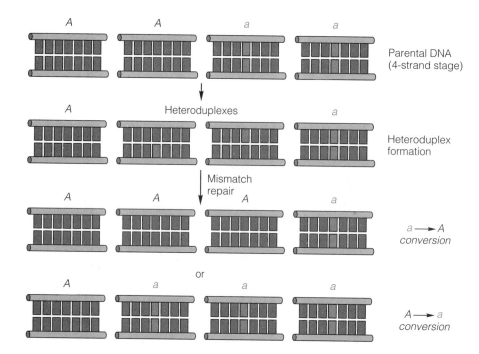

Parental DNA
(4-strand stage)

Heteroduplexes

Heteroduplex
formation

Mismatch repair

$a \longrightarrow A$
conversion

or

$A \longrightarrow a$
conversion

FIGURE 14-22 Mismatch repair resulting in gene conversion. Only a small part of the heteroduplex region is shown. The base pair that differs in the *a* allele is shown in red. Each mismatched heteroduplex can be repaired to give an A allele or an *a* allele. The patterns of repair leading to 3:1 and 1:3 segregation are shown.

in a small segment of duplex DNA containing one strand from each parental molecule (Figure 14-22). The region of hybrid DNA is called a **heteroduplex** region.

Molecular models of recombination are quite complex because heteroduplexes may or may not lead to recombination between flanking genetic markers. The models also strive to account for all the types of aberrant asci that are observed and their relative frequencies. We will consider three models that differ according to the type of DNA breakage that initiates the process.

Symmetric single-strand breaks This model is also called the **Holliday model** after its originator. The process is outlined in Figure 14-23. It begins with the occurrence of nicks in strands of the same polarity (Figure 14-23B). The strands unwind in the region of the nicks, switch pairing partners, and form a heteroduplex region in each molecule (Figure 14-23C). Ligation anchors the ends of the exchanged strands in place, and further unwinding increases the length of the heteroduplex region (termed *branch migration*, Figure 14-23D). Two more nicks and rejoinings are necessary to resolve the structure in part D. As the configuration is drawn, these breaks can occur in the inner strands, resulting in molecules that are *nonrecombinant* for markers flanking the heteroduplex region; or they can occur in the outer strands, resulting in molecules that are *recombinant* for outside markers (Figure 14-23E).

Resolution may lead to recombination of the outside markers or no recombination. These outcomes are equally likely because there is no topological difference between the inner strands and the outer strands. To understand why, note that an end-for-end rotation of the lower duplex in Figure 14-23D gives the configuration in Figure 14-24. When drawn in this manner, it is clear that breakage and rejoining of the east-west strands or the north-south strands are topologically equivalent. However, north-south resolution results in recombination of the outside markers (*Ab* and *aB* gametes), whereas east-west resolution gives nonrecombinants (*AB* and *ab* gametes).

An important feature of the Holliday model is that heteroduplex regions are found in the region of the exchange. Mismatch repair in this region can

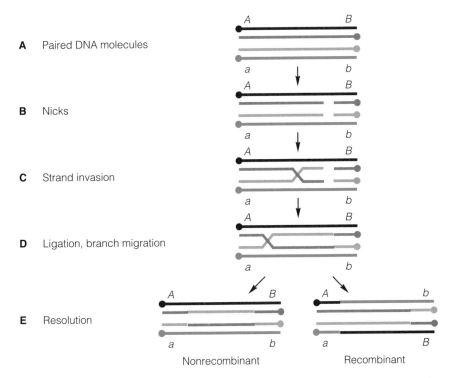

FIGURE 14-23 Holliday model of recombination. A. Paired DNA molecules. B. The exchange process is initiated by single-stranded breaks in strands of the same polarity. C. Strands exchange pairing partners (referred to as "invasion"). D. Ligation and branch migration. E. Exchange is resolved either by breaking and rejoining the inner strands (resulting in molecules that are nonrecombinant for outside markers) or by breaking and rejoining the outer strands (resulting in recombinant molecules). Either type of resolution gives a heteroduplex region of the same length in each participating molecule. The dot at the end of each DNA strand denotes the 5' end. Thus, the left-to-right polarity of the strands is 5' → 3' for the top and bottom and 3' → 5' for the middle two.

FIGURE 14-24 The cross-shaped configuration shows the topological equivalence of outer-strand and inner-strand resolution of the Holliday structure in Figure 14-23D. East-west resolution corresponds to inner-strand breakage and rejoining, north-south resolution corresponds to outer-strand breakage and rejoining. The dot at the end of each DNA strand denotes the 5' end.

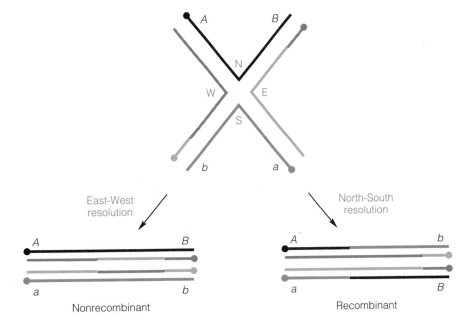

result in gene conversion, and the equal likelihood of north-south and east-west resolution implies that gene conversion will frequently be accompanied by recombination of flanking markers.

Single-strand break and repair Observed patterns of gene conversion indicate that the heteroduplex region is not always the same length in the DNA duplexes participating in an exchange. This is a difficulty for the Holliday model because its heteroduplex regions are symmetric. However, in the single-strand break and repair model (Figure 14-25), the heteroduplex regions can be asymmetric. Starting with a single-strand nick (Figure 14-25A), progressive unwinding and DNA synthesis (dashed pink line) liberates one DNA strand from a duplex. This strand can "invade" the other duplex (Figure 14-25B), resulting in displacement of the original strand and formation of a **D loop** (*displacement loop*). Nucleases degrade the D loop and the invading strand is ligated in place (Figure 14-25C). Branch migration (part D) forms a second heteroduplex region, but the heteroduplex region in the participating duplexes are of different lengths. As with the Holliday model, the configuration in Figure 14-25D can be resolved in either the east-west or north-south direction, yielding nonrecombinant or recombinant products, respectively.

Double-strand break and repair In yeast, duplex DNA molecules with double-stranded breaks are very efficient in initiating recombination with a homologous unbroken duplex. A model for this process is shown in Figure

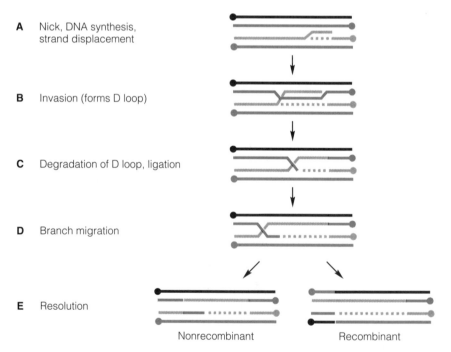

A Nick, DNA synthesis, strand displacement

B Invasion (forms D loop)

C Degradation of D loop, ligation

D Branch migration

E Resolution

Nonrecombinant   Recombinant

FIGURE 14-25   Model of recombination initiated by a break in one strand of a pair of DNA molecules (the layout is like that in Figure 14-23). A. Freed DNA strand is replaced by new synthesis and invades partner molecule. B. Displacement of strand in partner molecule forms D (displacement) loop. C. D loop is degraded and free DNA ends are ligated. D. Branch migration increases length of heteroduplex region. E. Resolution is in either of two topologically equivalent ways. In this model, the heteroduplex regions are not of equal length.

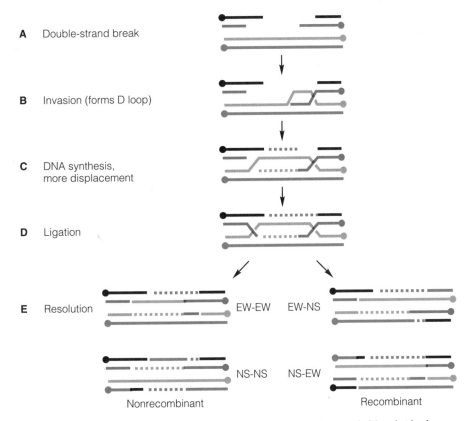

A Double-strand break

B Invasion (forms D loop)

C DNA synthesis,
more displacement

D Ligation

E Resolution

EW-EW   EW-NS

NS-NS   NS-EW

Nonrecombinant   Recombinant

FIGURE 14-26   Model of recombination initiated by a double-stranded break (the layout is like that in Figure 14-23). A. The double-stranded break is trimmed back asymmetrically by nucleases. B. One of the strands invades the partner molecule, forming a displacement loop. C. The invading strand is elongated by new synthesis, and the displacement loop serves as a template for new synthesis in the other duplex. D. Free DNA ends are ligated. E. Two Holliday junctions are formed, which can be resolved in either of four topologically equivalent ways, two of which result in recombination of outside markers. In this model, the heteroduplex regions are unequal in length, and all of them contain some newly synthesized DNA.

14-26. The initiating molecule will typically contain a gap with missing sequences and strands of unequal length at the ends because of exonuclease activity. The ends are free to invade a homologous duplex, forming a D loop (Figure 14-26B), which increases in extent until it bridges the gap (part C). Repair synthesis in both duplexes gives the configuration in part D, which can be resolved in four ways because it contains two Holliday-type junctions, each of which can be resolved in east-west or north-south orientation. Two resolutions yield nonrecombinant duplexes and two yield recombinant duplexes, and so genetic recombination is a frequent outcome.

Although each of the three models of recombination accounts for the principal features of gene conversion and recombination, each falls short of completely explaining all of the details and the aberrations that occasionally occur. For example, the Holliday model predicts an excess of 2-strand double recombination, which is not observed. All three models may represent alternative pathways of recombination whose frequencies depend on whether the initiating event consists of a pair of single-stranded nicks or a double-stranded gap.

Mutations can be classified in a variety of ways: how they come about, the nature of the chemical change, or how they are expressed. Conditional mutations cause a change in phenotype under restrictive but not permissive conditions. For example, temperature-sensitive mutations are expressed only above or below particular temperatures. Spontaneous mutations are of unknown origin, whereas induced mutations result from exposure to chemical reagents or radiation. All mutations ultimately result from changes in the sequence of nucleotides in DNA, including base substitutions, additions, deletions, and rearrangements. Base-substitution mutations are single-base changes that may be silent (for example, in a noncoding region or a synonymous codon position), may change an amino acid in a polypeptide chain (a missense mutation), or may cause chain termination by production of a stop codon (a nonsense mutation). Transitions are base-substitution mutations in which the purine-pyrimidine orientation of the base pair is unaltered; if the orientation is reversed, the mutation is a transversion. A mutation may consist of an addition or deletion of one or more bases; in a coding region, if the number of bases is not a multiple of three, the mutation is a frameshift.

Spontaneous mutations are random and do not occur in response to environmental conditions; rather, mutations with a growth advantage are selected by environmental conditions. Spontaneous mutations often arise by errors in DNA replication that fail to be corrected by either the proofreading system or the mismatch-repair system. The proofreading system removes an incorrectly incorporated base immediately after it has been added to the growing end of a DNA strand. The mismatch-repair system removes incorrect bases at a later time. Methylation of parental DNA strands and delayed methylation of daughter strands allows the mismatch-repair system to operate selectively on the daughter DNA strand.

A variety of systems exist for repairing damage to DNA. Photoreactivation is direct cleavage of the pyrimidine dimers produced by ultraviolet radiation. A methyl transferase deals with O^6–methyl guanine. In excision repair, altered bases that distort the helix are excised by nucleases and the resulting gap is filled. In both of these systems, the correct template is restored. Postreplication repair is an exchange process in which gaps in one daughter strand produced by aberrant replication across damaged sites are filled by nondefective segments from the parental strand of the other branch of the newly replicated DNA. Thus, a new template is produced by this system. The SOS repair system in bacteria is error prone (mutagenic) because it facilitates replication across noninformational (unreadable) stretches in the template DNA.

Mutations can be induced chemically by direct alteration of DNA; for example, by nitrous acid, which deaminates bases. Base analogs are incorporated into DNA during replication. They undergo mispairing more often than do the normal bases, which gives rise to transition mutations. The base analog 5-bromouracil is an example of such a mutagen, although some of its mutagenic effects also result from upsetting the balance of nucleotide precursor pools. Alkylating agents often cause depurination; this mutation is usually repaired, but often an adenine is inserted opposite a depurination site, which results in a transversion. Acridine molecules intercalate (interleaf) between base pairs of DNA and cause misalignment of parental and daughter strands during DNA replication, giving rise to frameshift mutations, usually of one or two bases. Ionizing radiation results in oxidative free radicals that cause a variety of alterations in DNA, including strand breaks.

Mutant organisms sometimes revert to the wild-type phenotype. Reversion normally results from an additional mutation at another site. Reversion by secondary mutations is called suppression, and the secondary mutations are called suppressor mutations. These can be intragenic or intergenic. In intragenic suppression, a mutation in one region of a protein alters the folding of the protein, and a change in another amino acid causes the protein to refold correctly. Intergenic suppression occurs through a variety of mechanisms. For example, in nonsense suppression, a chain-termination mutation (a stop codon) is read by a mutant tRNA molecule with an anticodon that can hydrogen-bond with the stop codon, and an amino acid is inserted at the site of the stop codon. The Ames test measures reversion as an indicator of mutagens and carcinogens; it uses an extract of rat liver, which in mammals occasionally converts intrinsically harmless molecules into mutagens and carcinogens.

Base-substitution mutations of any desired type can be created *in vitro* by the use of oligonucleotide site-directed mutagenesis. In this procedure, a single DNA strand containing the target sequence is hybridized with a short synthetic strand (the oligonucleotide) containing the desired base substitution. DNA replication elongates the oligonucleotide, producing the complete target-gene sequence containing the base mismatch. The DNA strands segregate in the next replication to give a mutant molecule containing the base substitution and a nonmutant molecule containing the original sequence.

Genetic recombination is intimately connected with DNA repair because the process always includes the breakage and rejoining of DNA strands. Three current models of recombination differ in the types of DNA breaks that initiate the process and in predictions about the types and frequencies of gene-conversion events. Gene conversion occurs when one allele becomes converted into a homologous allele, which is detected by aberrant segregation in fungal asci, such as 3:1 or 1:3. Gene conversion is the outcome of mismatch repair in heteroduplexes. All recombination models include the creation of heteroduplexes in the region of the exchange and predict that 50 percent of gene conversions will be accompanied by recombination of genetic markers flanking the conversion event. The Holliday model assumes that single-strand nicks occur at homologous positions in the participating duplexes. Other models assume initiation with a single-strand nick or a double-strand break. All three processes may be operative.

KEY TERMS

acridine
alkylating agent
Ames test
base analog
base-substitution mutation
CIB method
conditional mutation
depurination
displacement loop
excision repair
forward mutation
free radical
gene conversion
germinal mutation
heteroduplex region
Holliday model
hot spot

induced mutation
intercalation
intergenic suppression
intragenic suppression
ionizing radiation
mismatch repair
missense mutation
mutagen
mutation rate
nitrous acid
nonsense mutation
oligonucleotide
oligonucleotide site-directed
   mutagenesis
permissive condition
photoreactivation
postreplicational repair

pyrimidine dimer
rad
replica plating
restrictive condition
reverse mutation
reversion
sickle-cell anemia
silent mutation
somatic mutation
SOS repair
spontaneous mutation
suppressor mutation
suppressor tRNA
temperature-sensitive mutation
transition mutation
transversion mutation
xeroderma pigmentosum

EXAMPLES OF WORKED PROBLEMS

Problem 1: The molecule 2-aminopurine (Ap) is an analog of adenine that pairs with thymine. It also occasionally pairs with cytosine. What types of mutations will be induced by 2-aminopurine?

Answer: In problems of this sort, first note the base that is replaced, and then follow what can happen in subsequent rounds of DNA replication when the base analog pairs normally or abnormally. In this example, the 2-aminopurine (Ap) is incorporated in place of adenine opposite thymine, forming an ApT base pair. In the following rounds of replication, the Ap will usually pair with T but, when it occasionally pairs with C, it forms an ApC pair. In the next round of replication, the Ap again pairs with T (forming an ApT pair), but the C pairs with G (forming a mutant GC pair). Hence, the type of mutation induced by

2-aminopurine is a change from an AT pair to a GC pair, which is a transition mutation.

Problem 2: What amino acids can be present at the site of a UAA codon that is suppressed by a suppressor tRNA created by a mutant base in the anticodon?

Answer: The nonmutant tRNA must be able to pair with the codon at two sites, and the mutant base in the anticodon allows the third site to pair as well. Therefore, the amino acids that can be inserted into the suppressed site are those whose codons differ from UAA in a single base. These codons are AAA (Lys), CAA (Gln), GAA (Glu), UUA (Leu), UCA (Ser), UAC (Tyr), and UAU (Tyr).

Problem 3: Twenty spontaneous white-eye (*w*) mutants of *Drosophila* are examined individually to determine their reversion frequencies. Of these, twelve revert spontaneously at a frequency ranging from 10^{-3} to 10^{-5}. Another six do not revert spontaneously at any detectable frequency but can be induced to revert at frequencies ranging from 10^{-5} to 10^{-6} by treatment with alkylating agents. The remaining ten cannot be made to revert under any conditions. (a) Which class of mutants would you suspect to result from the insertion of transposable elements? (b) Which class of mutants are probably missense mutations? (c) Which class of mutants are probably deletions. Explain your answers.

Answer: (a) Insertions of transposable elements in eukaryotes are often characterized by genetic instability and high reversion rates. Therefore, the twelve mutants with the highest reversion frequencies are probably transposable-element insertions. **(b)** The fact that alkylating agents can revert the second class of mutations suggests that they are missense mutations. Although they do not revert spontaneously, this is probably because the exact reversal of a base substitution is an exceedingly rare spontaneous event. **(c)** Deletion mutations cannot revert under any conditions because there is too much missing genetic information to be restored. Therefore, the class of mutations that does not revert probably consists of deletions.

PROBLEMS

14-1. Occasionally, an individual has one blue eye and one brown eye or a sector of one eye is a different color from the rest. Can such occurrences be explained by new mutations? If so, in what types of cells must the mutations occur?

14-2. A mutation is isolated that cannot be reverted. What types of molecular changes might be responsible?

14-3. There are twelve possible substitutions of one nucleotide pair for another (for example, A → G). Which changes are transitions and which transversions?

14-4. Do all nucleotide substitutions in the second position of a codon necessarily produce an amino acid replacement?

14-5. Do all nucleotide substitutions in the first position of a codon necessarily produce an amino acid replacement? Do all of them necessarily produce a nonfunctional protein?

14-6. In a coding sequence made up of equal proportions of A, U, G, and C, what is the probability that a random nucleotide substitution will result in a chain-termination codon?

14-7. What features of DNA polymerase could qualify it as a repair enzyme?

14-8. If a mutagen produces frameshift mutations, what kinds of mutations will be reverted by it?

14-9. If a deletion occurs that eliminates a single amino acid in a protein, how many base pairs are missing?

14-10. Does the nucleotide sequence of a mutation tell you anything about its dominance or recessiveness?

14-11. What are two ways that pyrimidine dimers are removed from DNA in E. coli?

14-12. This problem illustrates how conditional mutations can be used to determine the order of genetically controlled steps in a developmental pathway. A certain organ undergoes development in the sequence of stages A → B → C, and both genes X and Y are necessary for the sequence to occur. A conditional mutation X′ is sensitive to heat (the gene product is inactivated at high temperatures), and a conditional mutation Y′ is sensitive to cold (the gene product is inactivated at cold temperatures). The double mutation X′/X′; Y′/Y′ is created at either high or low temperatures. How far would development proceed in each of the following cases at the high temperature and at the low temperature?

(a) Both X and Y are necessary for the A → B step.
(b) Both X and Y are necessary for the B → C step.
(c) X is necessary for the A → B step, and Y for the B → C step.
(d) Y is necessary for the A → B step, and X for the B → C step.

14-13. A Lac⁺ culture of E. coli gives rise to a Lac⁻ strain that results from a UGA nonsense mutation in the lacZ gene. The Lac⁻ strain gives rise to Lac⁺ colonies at the rate of 10^{-8} per cell per generation, and 90 percent of the latter are caused by suppressor tRNA molecules. What is the rate of production of suppressor mutations in the original Lac⁺ culture?

14-14. If a gene in a particular chromosome has a probability of mutation of 5×10^{-5} per generation, and if the allele in a particular chromosome is followed through successive generations:
(a) What is the probability that the allele does not undergo a mutation in 10,000 consecutive generations?
(b) What is the average number of generations before the allele undergoes a mutation?

14-15. In the mouse, a dose of approximately 100 rads of x rays produces a rate of induced mutation equal to the rate of spontaneous mutation. Taking the spontaneous mutation rate into consideration, what is the total mutation rate at 100 rads? What dose of x rays will increase the mutation rate by 50 percent? What dose will increase the mutation rate by 10 percent?

14-16. Among several hundred missense mutations in the gene for the A protein of tryptophan synthetase in *E. coli*, fewer than 30 of the 268 amino acid positions are affected by one or more mutations. Explain why the number of positions affected by amino acid replacements is so limited.

14-17. Human hemoglobin C is a variant in which a lysine in the β hemoglobin chain substitutes for a particular glutamic acid. What single-base substitution can account for the hemoglobin-C mutation?

14-18. How many amino acids can substitute for tyrosine by a single-base substitution? (Do not assume that you know which tyrosine codon is being used.)

14-19. Which of the following amino acid replacements would be expected with the highest frequency among mutations induced by 5-bromouracil? (1) Met → Leu, (2) Met → Lys, (3) Leu → Pro, (4) Pro → Thr, (5) Thr → Arg.

14-20. Often dyes are incorporated into a solid medium to determine whether bacterial cells can utilize a particular sugar as a carbon source. For instance, on eosin-methylene blue (EMB) medium containing lactose, a Lac$^+$ cell yields a purple colony and a Lac$^-$ cell yields a pink colony. If a population of Lac$^+$ cells is treated with a mutagen that produces Lac$^-$ mutants and the population is allowed to grow for many generations before the cells are placed on EMB-lactose medium, a few pink colonies are found among a large number of purple ones. However, if the mutagenized cells are plated on the medium immediately after exposure to the mutagen, some colonies appear that are *sectored* (half purple and half pink). Explain how the sectored colonies arise.

14-21. What amino acids can be inserted at the site of a UGA codon that is suppressed by a suppressor tRNA?

14-22. A *Neurospora* strain unable to synthesize arginine (and therefore requiring this amino acid in the growth medium) produces a revertant colony able to grow in the absence of arginine. A cross is made between the revertant and a wildtype strain. What proportion of the progeny from this cross would be arginine-independent if the reversion occurred by:
(a) A precise reversal of the nucleotide change that produced the original *arg*$^-$ mutant allele?
(b) A mutation to a suppressor of the *arg*$^-$ allele in another gene located in a different chromosome?
(c) A suppressor mutation in another gene located 10 map units away from the *arg*$^-$ allele in the same chromosome?

14-23. Mutant alleles of two genes, *a* and *b*, interact as follows: *a b*$^+$ and *a*$^+$ *b* are mutant, but a few *a b* combinations are wildtype. Each *a* allele that is suppressed requires a particular *b* allele, and each *b* allele that is suppressed requires a particular *a* allele. However, most *a* and *b* alleles are not suppressible by any alleles of the other gene. What kind of interaction between the *a* and *b* gene products can explain these results?

14-24. The following technique has been used to obtain mutations in bacterial genes that are otherwise difficult to isolate. Suppose that a mutation is desired in a gene, *a*, and a mutation, *b*$^-$, in a closely linked gene is available. The method is to grow phage P1 on a *b*$^+$ strain to obtain *a*$^+$*b*$^+$ transducing particles. The P1 phages containing the transducing particles are exposed to a mutagen, such as hydroxylamine, and the *b*$^-$ strain is infected with the phages. What step or steps would you take next to obtain an *a*$^-$ mutation? On what genetic phenomenon does this procedure depend?

14-25. In the oligonucleotide mutagenesis outlined in Figure 14-20, what oligonucleotide would you use to produce a specific Cys → Tyr replacement?

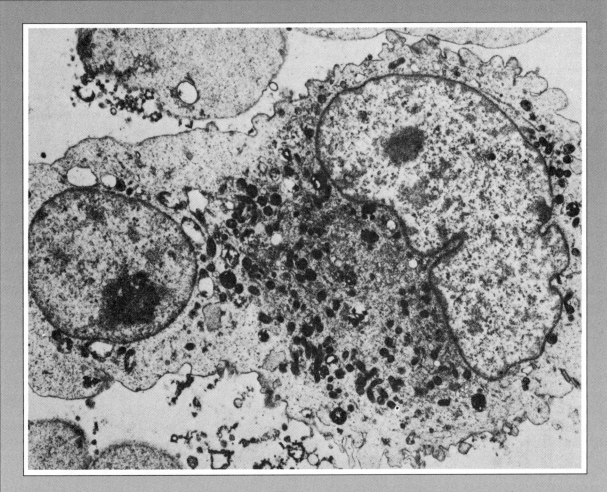

Somatic-Cell Genetics and Immunogenetics

Genetic processes that take place in somatic cells—the cells in the body other than gametes or gamete-forming cells—constitute **somatic-cell genetics.** Certain methods of somatic-cell genetics have been highly developed for the study of mammalian cells grown in laboratory cultures, and these methods,

Above: Cell fusion. An immature red blood cell, from a five-day-old chick embryo, in the process of fusing with a mouse cell. (From H. Harris, 1970. *Cell Fusion.* Harvard University Press. Photograph taken by Gutta Schoefl.)

which are a main topic of this chapter, are important in the genetic analysis of human chromosomes.

Some genetic events take place only in somatic cells and not in germ cells. One of the best-known examples is in the programming of the immune system to make antibodies, which is dependent upon genetic changes in somatic cells; in these cells, gene-splicing events occur in which new functional genes are created by joining together pieces of other genes. Such unusual genetic processes form the second major subject of this chapter.

15-1   Somatic-Cell Genetics

Figure 15-1 shows the mitotic chromosomes of an unusual type of cell—a **cell hybrid** formed by the fusion of a human cell and a mouse cell in a laboratory culture. Such hybrids are initially complete hybrids in that the nucleus contains a complete set of 40 mouse chromosomes and a complete set of 46 human chromosomes. However, the hybrid cells usually lose some chromosomes in the course of the mitotic divisions following the cell fusion. The chromosomes that are lost are usually human chromosomes. Studying a series of cell hybrids containing single human chromosomes or various combinations of human chromosomes enables identification of the particular chromosome containing any human gene. In this way, restriction fragment length polymorphisms (Section 8-1) and other genetic markers can be assigned their physical positions in the genome. The hybrid-cell method has been most widely used for human genome mapping, but similar methods can be used for almost any mammal.

Somatic-Cell Hybrids and Gene Mapping

Although occasionally lost, the chromosomes in human-mouse hybrids are sufficiently stable to permit the establishment of **cell lines,** or clones, containing particular human chromosomes, and within these cell lines most cells will have the same combination of human and mouse chromosomes. The particular human chromosomes that are occasionally lost from hybrid cells vary, which

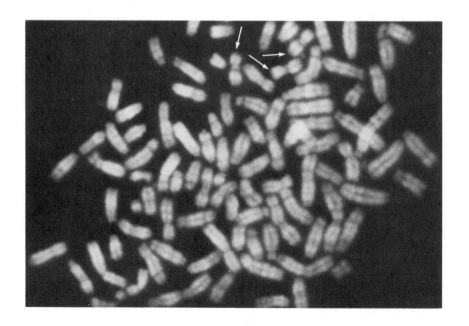

FIGURE 15-1   Metaphase chromosomes in a human-mouse cell hybrid. The chromosomes have been stained by a procedure that produces fluorescence. Each chromosome has a distinctive pattern of fluorescence, allowing identification. Several human chromosomes are indicated by arrows. (Courtesy of Raju Kucherlapati.)

allows a large number of cell lines to be established, each containing a *different* group of human chromosomes.

Loss of chromosomes by hybrid cells is illustrated in Figure 15-2, in which the human chromosomes are indicated in black and the mouse chromosomes in red. In this drawing, the cell at the top is assumed to have four human chromosomes—one copy each of chromosomes 1, 2, 3, and 4. After several mitotic divisions (arrows), a daughter cell is produced that has lost chromosome 1, and this cell is cultured individually, producing cell line A. Other mitotic divisions of the original hybrid cell may result in the loss of chromosomes 3 and 4 (cell line B) or the loss of chromosomes 1, 2, and 4 (cell line C). The particular human chromosome present in a hybrid cell line is identified by its pattern of chromosome bands produced by the Giemsa staining procedure discussed in Chapter 6.

The homologous human and mouse genes can be distinguished by electrophoresis of restriction fragments and hybridization with appropriate probe DNA (Section 4-10). *If the human gene is found only in hybrid cells containing a particular chromosome (or perhaps fragment of a chromosome), then the particular chromosome (or fragment) must contain the gene.* For example, a gene contained in chromosome 1 will be present in cell line B in Figure 15-2 but not in lines A or C. In this way, the chromosomal location of any gene can be identified if a cell line is available lacking that chromosome, because for each gene there is a unique pattern of presence or absence of the gene among the cell lines. Specifically, if + indicates the presence of the gene

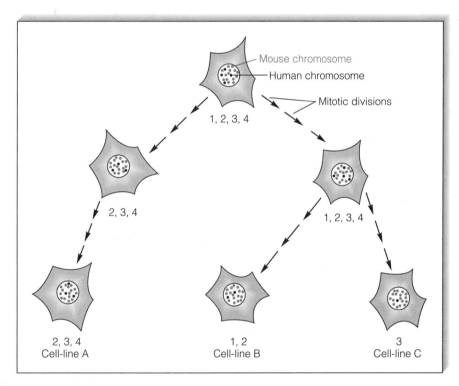

FIGURE 15-2   Three lines of hybrid cells that are chromosomally distinct; each line is formed by the loss of human chromosomes. The uppermost cell is a human-mouse hybrid having four human chromosomes (1, 2, 3, 4). After several mitotic divisions, one daughter cell (left) is produced that has lost chromosome 1. The chromosome complement of this cell line is stable, yielding cell-line A. Other mitotic divisions of the original cell result in loss of chromosomes 3 and 4 (line B) and chromosomes 1, 2, and 4 (line C).

product and − indicates its absence, then the patterns corresponding to genes in chromosomes 1, 2, 3, and 4 in Figure 15-2 are

	Line A	*Line B*	*Line C*
Chromosome 1	−	+	−
Chromosome 2	+	+	−
Chromosome 3	+	−	+
Chromosome 4	+	−	−

Data from actual human-mouse hybrid cell lines are analyzed in essentially the same way as that used in the hypothetical example in Figure 15-2, but the actual data are somewhat more challenging because there are 23 pairs of human chromosomes to consider. In some cases, probe DNA needed to identify the genes directly is not available. However, the method can still be used, provided the gene is expressed in cultured cells and the human gene product (for example, an enzyme) can be distinguished from the mouse gene product (for example, by electrophoresis).

One set of eight human-mouse hybrid cell lines used in chromosome assignments is shown in Table 15-1. Each of the clones, designated by letters A through H, carries a different group of human chromosomes, and, conversely, each human chromosome has a different pattern of presence (+) or absence (−) among the clones. (Note that all eight clones carry the human X chromosome. This is not accidental but is a result of the particular method by which these clones were produced, which will be discussed shortly.) One application of these clones has been in locating the gene for uridine monophosphate kinase (UMPK), which among the clones has the pattern

A B C D E F G H
− + + + − − − +

Comparison of these results and the columns in Table 15-1 reveals perfect agreement between the presence or absence of the human gene and the presence or absence of chromosome 1 among the clones (red column). On the basis of this correspondence, the gene for UMPK is assigned to chromosome 1.

TABLE 15-1   Human chromosomes among human-mouse hybrid cell lines

Clone*	\\	\\	\\	\\	\\	\\	\\	\\	\\	Chromosome													
	1	*2*	*3*	*4*	*5*	*6*	*7*	*8*	*9*	*10*	*11*	*12*	*13*	*14*	*15*	*16*	*17*	*18*	*19*	*20*	*21*	*22*	*X*
A	−	+	+	+	−	−	−	−	−	−	+	+	−	+	−	+	−	−	−	−	−	−	+
B	+	+	−	−	+	−	−	+	−	−	−	−	+	−	−	+	+	+	−	−	−	−	+
C	+	+	−	+	−	−	+	+	−	+	−	+	+	−	+	+	−	+	+	+	+	+	+
D	+	−	+	−	−	+	+	−	−	+	+	+	−	+	+	+	+	+	−	−	+	−	+
E	−	+	−	+	+	−	+	+	−	+	+	+	+	+	−	−	+	+	−	+	−	−	+
F	−	−	−	−	−	+	+	−	+	−	+	+	+	+	+	+	+	+	+	+	+	+	+
G	−	+	+	−	−	−	+	−	−	+	−	+	−	−	−	−	−	−	−	−	−	−	+
H	+	+	+	+	+	−	+	+	−	+	−	+	−	+	+	+	−	+	−	−	+	+	+

*Each clone has a unique combination of human chromosomes.
Source: Data from A. Satlin, R. Kucherlapati, and F. H. Ruddle, *Cytogenet. Cell Genet.* 15(1975):146−152

Similarly, the gene coding for the enzyme β-galactosidase in these clones has the pattern

A B C D E F G H
+ – – + – – + +

which is that of a gene located in chromosome 3.

Several extensions of the hybrid-cell method can be used to identify the location of genes with even greater precision. For example, human-mouse hybrid cells produced with human chromosomes carrying a translocation (Section 6-6) can be used to assign a gene to one or the other segments of the translocation. In the simplest cases, the translocation breakpoint is near the centromere of one chromosome, and the presence of a gene on the long or short arm of the chromosome is indicated by its expression in hybrids carrying only the long or short arm. Finer resolution of the position can be obtained by the use of cell lines containing small fragments of the X chromosome (obtained by heavy x irradiation) or by direct *in situ* hybridization of probe DNA to metaphase chromosomes, as illustrated in Figure 15-3.

Production of Hybrid Cells

Cells occasionally fuse spontaneously. With most cell types, the frequency of spontaneous fusion is so low that fusion must be enhanced. This is usually done by exposing a mixture of two cell types to polyethylene glycol. Once fusion has been achieved, a method is needed to select fused cells from a population of individual cells. The most-common method is based on the use of two enzymes that participate in the synthesis of DNA precursors. Most eukaryotic cells have two metabolic pathways for producing the dTTP needed for DNA synthesis—

FIGURE 15-3 *In situ* hybridization of DNA to human metaphase chromosomes. In this case, the fluorescence results from hybridization to sequences surrounding the centromeres of most chromosomes. (Courtesy of Terry Featherstone.)

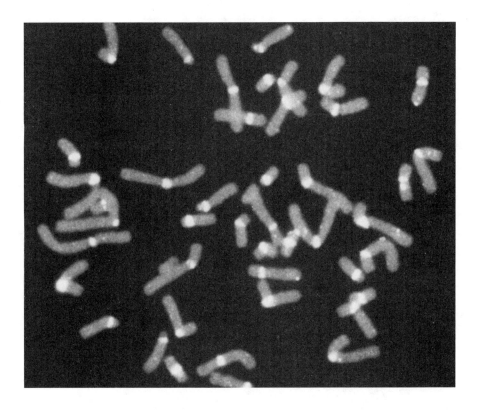

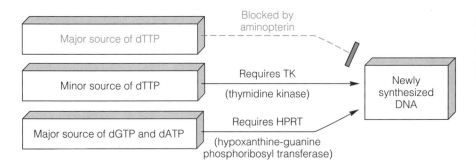

FIGURE 15-4   Major pathway and salvage pathway for DNA synthesis. (The pathway that supplies dCTP is not shown here.)

the major pathway, which yields most of the dTTP, and a **salvage pathway,** a normally minor route. The drug *aminopterin* inhibits the major pathway of dTTP synthesis; in the presence of aminopterin, cells must rely exclusively on the salvage pathway for dTTP precursors (Figure 15-4). The salvage pathway for dTTP synthesis includes the enzyme *thymidine kinase,* or TK, which phosphorylates the deoxyribonucleoside dT to the monophosphate nucleotide dTMP. In the presence of aminopterin, cells that are defective in TK (TK^- cells) are unable to synthesize new DNA.

The enzyme *hypoxanthine-guanine phosphoribosyl transferase,* or HPRT, participates in the formation of the purine deoxynucleotides used in DNA synthesis. Cells that are defective in HPRT ($HPRT^-$ cells) cannot synthesize DNA for lack of sufficient purine nucleotide precursors.

The necessity for both HPRT and TK for cell growth in the presence of aminopterin is the basis of the method for selecting fused cells (Figure 15-5).

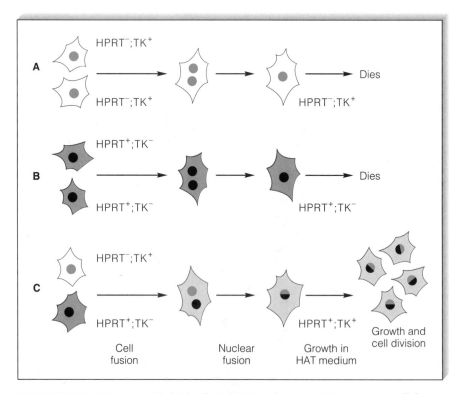

FIGURE 15-5   Selection of hybrid cells in HAT medium. A. Mouse-mouse cell fusions fail to survive because they are $HPRT^-$. B. Human-human cell fusions die because they are TK^-. C. The only cells that can grow and divide are the mouse-human hybrids because they are $HPRT^+$ TK^+ hybrids.

Human TK^- cells are mixed with mouse $HPRT^-$ cells in a medium containing aminopterin. Neither type of cell will be able to grow because the human cells lack dTTP and the mouse cells lack sufficient dGTP and dATP. Among the cells that undergo fusion, the mouse-mouse hybrids are still $HPRT^-$ (Figure 15-5A), and the human-human hybrids are still TK^- (part B); the human-mouse hybrids alone have genes coding for active HPRT and TK and thus are able to grow (part C). In other words, the $HPRT^-$ and TK^- mutations complement one another in the hybrids (which they must, because the genes are nonallelic); therefore, the hybrid cells are the only ones that survive. The medium in which fused cells are selected is called **HAT medium,** because it contains hypoxanthine (a substrate of HPRT), aminopterin (to block the major pathway of dTTP synthesis), and thymidine (the substrate of TK).

In the discussion of Table 15-1, it was pointed out that all of the hybrid clones in the table retained the human X chromosome. This occurs whenever the clones are obtained by fusing human TK^- and mouse $HPRT^-$ cells. The human cell must provide the normal HPRT gene, which is present in the human X chromosome, and so all clones must carry the human X.

15-2 The Immune Response

For centuries, it has been recognized that people who have recovered from an infectious disease, such as smallpox or influenza, are less likely to succumb to the same disease a second time. As a result of prior exposure, these people acquire the ability to resist the infectious agent; they are said to be **immune** to the agent. The process of becoming immune is called the **immune response.** The recognition that immunity can be acquired has led to the development of vaccines, harmless in themselves, that produce immunity to many diseases.

Immunity results from the ability of the body to recognize viruses, bacteria, or other foreign substances that may invade the body and to attack and destroy them. Most large molecules, and almost all viruses and cells, have the ability to elicit an immune response. Such immunity-evoking substances are called **antigens.**

When blood transfusions first began to be administered about 1900, it was quickly recognized that the blood of the donor and that of the recipient had to match in some way in order for the transfusion to be successful. In many cases, transfusions could be carried out successfully with no complications; when this occurred, the donor and the recipient were said to be *compatible*. In other cases, upon transfusion, recipients went into severe shock and many died; in these cases, the donor and the recipient were obviously *incompatible*. However, blood from the same donor might be compatible with some recipients and at the same time incompatible with other recipients. Additional studies revealed that the basis of incompatibility was the presence of certain genetically determined antigens present on the surface of the red blood cells of the donor, and laboratory tests for compatibility were soon developed. These studies led to the first rule of immunogenetics:

> A recipient individual organism produces an immune response against any antigen that the individual itself does not possess.

That is, if the antigens present on the red blood cells of the donor are also present on the red blood cells of the recipient, then the donor and the recipient are compatible for transfusion.

Shortly after blood transfusions were first attempted in human beings, skin transplants were undertaken in highly inbred, and therefore virtually homozygous, strains of mice. Usually, transplants between individual mice of the same inbred strain were accepted. The small transplanted piece of skin would grow on the recipient mouse just as well as if the transplanted skin had been taken from another place on the recipient's own body. However, with individual mice of different inbred strains, the skin around the transplant became inflamed and the transplanted skin would not heal; this phenomenon was called **rejection.** Such rejection reactions also were shown to result from the presence of genetically determined antigens on the transplanted skin—but these were antigens of a different type from those implicated in blood transfusion. When two different inbred strains were crossed to produce an F_1, and the F_1 was used as donor or recipient of skin transplants, the following results were obtained: (1) transplants using either parental strain as the donor and the F_1 as the recipient were almost always accepted; (2) transplants using the F_1 as the donor and either parental strain as the recipient were almost always rejected (Figure 15-6). Such results led to a second rule of immunogenetics:

> Antigens are genetically determined by dominant or codominant alleles, and so the cells of an individual organism possess all antigens corresponding to the alleles that the individual has inherited from its parents.

The two rules of immunogenetics account for the observations about transplantation. Because the F_1 must inherit one allele from each homozygous parent, the F_1 will possess the antigens of *both* parental strains. Consequently, transplants from an inbred parent to the F_1 will be accepted, because the parental strain possesses no antigens not also present in the F_1. However, transplants from the F_1 to either parental strain will be rejected, because the F_1 possesses antigens inherited from the other parental strain, which are not present in the parental recipient of the transplant.

Transfusion incompatibility and skin-graft rejection exemplify a division of functions of the immune system arising from different characteristics of two classes of white blood cells—B cells and T cells. As will be discussed later in this chapter, **B cells** secrete proteins called **antibodies** capable of combining with antigens. Stimulation of a B cell by a suitable antigen causes secretion of

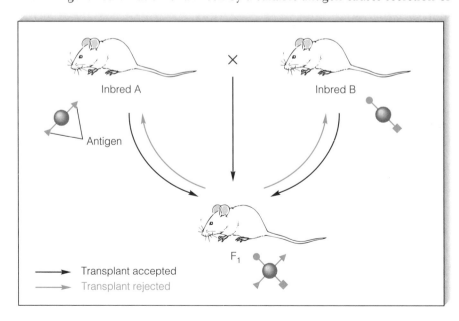

FIGURE 15-6 Results of skin grafts. Skin transplants in mice are usually accepted when the donor animal is from a homozygous inbred strain and the recipient is one of the F_1 progeny, because all antigens in the donor tissue are also present in the recipient. However, transplants in the reverse direction are usually rejected.

an antibody, which circulates in the blood and lymph and combines with the antigen, clumping the antigen molecules together and marking them for destruction by other classes of white blood cells.

Unlike B cells, which attack foreign antigens indirectly by producing circulating antibodies, **T cells** attack foreign antigens directly, releasing substances that cause a local inflammation, and attract other types of white blood cells to aid in the destruction of the antigens. Individuals with T-cell deficiency will accept skin transplants and are prone to viral infections. T-cell deficiency is a defect in several rare genetic disorders.

Two types of T cells may be distinguished:

1. **Helper T cells** are required for antigens to stimulate B cells to produce antibody. Helper T cells must be assisted by another type of white blood cell called a **macrophage.** Antigens invading an organism are initially engulfed by macrophages, and small parts of the antigen (typically, peptides of 15–20 amino acids) are brought to the cell surface in a form suitable for **antigen presentation** to the helper T cell. Effective antigen presentation requires that the macrophage possess certain proteins coded by genes in the *major histocompatibility complex*, or MHC. The antigen-binding region of an MHC protein is illustrated in Figure 15-7. Resembling a clam shell with a β-sheet hinge and α-helical edges, the structure is able to bind any antigen having a shape and charge that fits into the cleft; that is, the structure "clamps" the segment of antigen during antigen presentation. More information about the major histocompatibility complex is presented in Section 15-5.

2. **Cytotoxic T cells** attack antigens directly. Cytotoxic T cells also require antigen presentation. For example, a cell infected by a virus will bring some viral antigens to the cell surface, where they combine with the MHC product (shown in Figure 15-7) and enable the antigen to be presented to a cytotoxic T cell for activation of the T cell. Antigen presentation may be carried out by a macrophage, but any cell carrying an appropriate MHC product can function in presentation. Cells

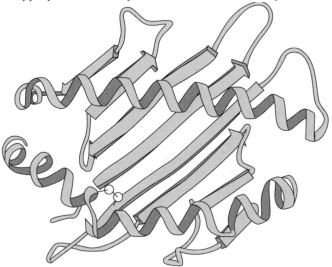

FIGURE 15-7   Top view of the antigen-binding site in a molecule coded by the major histocompatibility complex. The "pocket" can be said to resemble a clam shell, with the β-sheet structure (pink) forming the hinge. Binding of a peptide fragment in the pocket is necessary for antigen presentation and stimulation of the immune response. (Reprinted by permission from *Nature*, vol. 329, pp. 506–512. Copyright © 1987 Macmillan Magazines Ltd.)

capable of carrying out the presentation function are called **antigen-presenting cells.**

The remainder of this chapter deals with three aspects of immunogenetics: (1) the genetic determination of the red-blood-cell antigens that are important in transfusions, (2) the genetic determination of antigens implicated in tissue compatibility, and (3) the genetic basis of B-cell and T-cell responses.

15-3 Blood-Group Systems

More than thirty genes that determine surface antigens of red blood cells have been identified. Each gene defines a **blood-group system** in accord with the alleles that may be present and the corresponding red-blood-cell antigens. The most-important blood-group systems are the familiar ABO and Rh.

The ABO Blood Groups

The human ABO blood-group system, which has already been described in Chapter 1, is the most-significant system in blood transfusions. For the purpose of the present discussion, let us review the basic phenomena:

1. Genotypes $I^A I^A$ and $I^A I^O$ produce the A antigen on red cells. This blood group is A.
2. Genotypes $I^B I^B$ and $I^B I^O$ produce the B antigen on red cells. This blood group is B.
3. Genotype $I^A I^B$ produces both A and B antigens on red cells. The cells are type AB.
4. Genotype $I^O I^O$ has neither A nor B antigen on the red cells. The cells are type O.

Because of similarities between A and B antigens and particular antigens present on certain bacteria, the immune system is stimulated quite early in life to produce anti-A and anti-B antibodies against A and B antigens that the persons do not themselves possess. Hence,

1. People of blood-group A produce anti-B antibody.
2. Those of blood-group B produce anti-A antibody.
3. Those of blood-group AB produce neither anti-A nor anti-B antibody.
4. Those of blood-group O produce both anti-A and anti-B antibody.

Anti-A and anti-B antibodies are large molecules capable of the agglutination (clumping) of red blood cells that carry the corresponding antigens. Thus, when a person is transfused with red blood cells carrying the wrong antigen, antibodies present in the recipient will attack the foreign red blood cells, causing agglutination and destruction of the cells. The destruction is potentially very harmful for the recipient because of the disruption of normal circulation. Problems can also arise in the transfusion of a large volume of donor blood containing antibodies that can attack the recipient's own red blood cells, because significant amounts of the recipient's blood may be destroyed. Consequently, modern medical practice requires that donor and recipient have the same ABO blood type.

Rh Blood Groups

The antigens forming the Rh blood groups are determined by the alleles of three genes, which with their alternative alleles are designated C, c, D, d, E, and e.

A chromosome may carry any combination of the alleles; so twenty-seven genotypes are possible. Each allele, with the exception of *d*, determines the presence of a distinct red-blood-cell antigen that can be detected with the appropriate antibody. These antigens are designated by the same letters as the alleles, and so the antigens are C, c, D, E, and e. All three genotypes determined by the C and c alleles and the three determined by the E and e alleles are distinguishable. For example, genotype *CC* produces C antigen only, *Cc* produces both C and c antigens, and *cc* produces c only. The E and e alleles similarly determine three distinguishable phenotypes. However, only two phenotypes associated with the D gene are distinguishable: (1) the presence of antigen D (these people have genotype *DD* or *Dd* and are said to be **Rh-positive**), and (2) absence of antigen D (these people have genotype *dd* and are said to be **Rh-negative**). Altogether, the five common antigens define eighteenth Rh phenotypes.

Unlike the ABO system, for which the antigens are common in environmental substances, Rh-like substances are rare. Consequently, most people do not produce antibodies against the Rh antigens that they do not themselves possess. However, exposure to an appropriate "external" Rh antigen stimulates a person's immune system, and antibodies will be produced. For example, when a person having a *dd* gentotype is exposed to D antigen, anti-D antibodies will be produced.

A severe blood disorder affecting fetuses and newborns is caused when Rh antibodies present in the mother cross the placenta and attack the red blood cells of the fetus. Anti-A and anti-B antibodies normally cannot cross the placental barrier, but antibodies against Rh antigens can do so because they are smaller molecules than the ABO antibodies. Rh blood disease can occur whenever a *dd* female carries consecutive *Dd* fetuses (Figure 15-8A). The first *Dd* fetus is usually unaffected because the mother does not yet produce anti-D antibody. However, red cells from the fetus carry the D antigen, and a few of these cells can enter the maternal circulation during birth trauma and stimulate production of anti-D antibody in the mother. Once the mother has been stimulated by Rh-positive cells from the first fetus, subsequent *Dd* fetuses

FIGURE 15-8   Rh incompatibility between mother and fetus. A. A small number of red blood cells (red) from an Rh⁺ fetus may enter the blood stream of the mother. If the mother is Rh⁻, these cells stimulate the production of antibodies against the Rh⁺ antigen. In a subsequent pregnancy, these antibodies can cross the placenta and, if the fetus is again Rh⁺, attack the fetal red blood cells. B. After the birth of an Rh⁺ child, stimulation of the immune system in the mother can be prevented by injection of anti-Rh⁺ antibody, which rapidly removes the Rh⁺ red blood cells from the circulation.

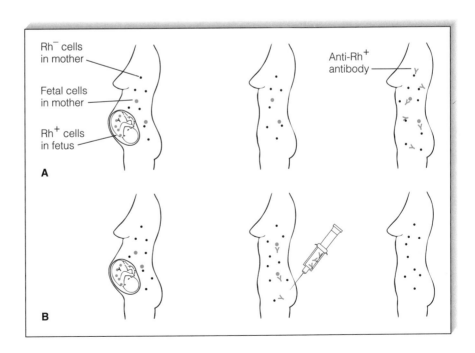

(which make D antigen) are at risk of the Rh blood disease, *hemolytic disease of the newborn*. Note, however, that *dd* fetuses are never at risk. The D antigen is almost always responsible for the maternal-fetal incompatibility that results in hemolytic disease of the newborn. Antibodies directed against other Rh antigens rarely cause severe problems.

Among Caucasians, about 13 percent of all matings are between *dd* females and *DD* or *Dd* males and so might result in maternal-fetal incompatibility. However, the risk of hemolytic disease of the newborn is greatly reduced by a preventive treatment (Figure 15-8B). In this treatment, the *dd* mother is injected with anti-D antibody immediately after the birth of each *Dd* child. The injected antibody attacks any fetal cells that may have entered into the blood of the mother, and the mother escapes being stimulated by the cells because they are destroyed too rapidly to evoke antibody production. The injected antibodies gradually disappear. Because the mother is not stimulated to produce anti-D antibodies, each pregnancy is just like the first with respect to antibody production, which means there is little risk of hemolytic disease. The only cases in which antibody treatment is ineffective are when the mother's immune system has already been stimulated by D antigen to produce anti-D antibodies. For this reason, *dd* females should not be transfused with red cells carrying the D antigen.

Applications of Blood Groups

In addition to the importance of certain blood-group systems in transfusions and maternal-fetal incompatibility, blood groups have applications in studies of genetic linkage, population studies, parentage exclusion, and criminology. Blood groups are useful in identification because, in many cases, all the alternative blood groups are reasonably common and any two unrelated persons can be expected to differ in one or more blood groups. Indeed, the probability that any two randomly chosen Europeans are identical for all known blood groups is less than 0.0003. Moreover, blood-group frequencies may vary among populations. For example, among black Africans, the chromosome with the highest frequency with respect to the Rh genes is *c D e*, whereas, among Chinese, it is *C D e*. Such observations are the basis of the use of blood groups in anthropology and population genetics to infer the genetic ancestry of populations.

Evidence based on blood groups is admissible in many courts in cases depending on the identification of parentage. In this application, the blood groups are interpreted according to two rules: (1) an allele present in a child must be present in one or both parents; (2) if a parent is homozygous for an allele, the allele must be present in the child. Suppose, for example, that the blood groups of mother, child, and two possible fathers are

Mother	O	*Cc D– ee*
Child	O	*Cc dd Ee*
Male 1	O	*CC D– Ee*
Male 2	A	*cc dd ee*

Because of the phenotypes of the mother and child, the sperm from the father must have carried I^O, either *C* or *c* (depending on whether the egg carried *c* or *C*, respectively), *d*, and *E*. With respect to ABO, either male could have contributed such a sperm, because male 2 could have the genotype $I^A I^O$. However, male 2 cannot contribute a sperm carrying *E*, and so he is excluded as the possible father. It is important to note that this analysis does not imply that male 1 *is* the father, but merely that he *could* be.

Various technical difficulties lie in the way of widespread use of blood-group systems for personal identification. Significant quantities of material are needed for blood-group determination, some blood-group phenotypes are difficult to analyze, the antigens are usually destroyed in old or dried samples, and the level of genetic polymorphism is often inadequate for discriminating among individuals. Many of these difficulties are overcome by the use of highly polymorphic restriction fragment length polymorphisms (RFLPs), of which there are many in the human genome. RFLPs provide a powerful method for the identification of individuals and their parentage through direct studies of DNA (Section 8-1). The DNA may be extracted from fresh or dried blood or semen, hair roots, and many other types of material. With respect to one highly variable RFLP sequence that is present in numerous copies in each human genome, the chance that any two unrelated people would possess sets of restriction fragments giving electrophoretic bands of identical size has been estimated as less than 10^{-11}. In other words, except for identical twins, each person has a unique pattern of DNA bands. The high degree of genetic variation has led to the name **DNA fingerprinting** for RFLP-based identification procedures. For example, the DNA fingerprint of a rapist can be determined from semen, and any suspect can either be proved innocent or be positively identified by examination of his DNA fingerprint. DNA fingerprinting is the method of choice in settling parentage disputes and in criminal investigations.

15-4 Antibodies and Antibody Variability

It has been estimated that a normal mammal is capable of producing more than 10^8 different antibodies, each having the ability to combine specifically with a particular antigen. Antibodies are proteins, and each unique antibody has a different amino acid sequence. If antibody genes were conventional in the sense that each gene codes for a single polypeptide, then mammals would need more than 10^8 genes for the production of antibodies—more genes than are present in the genome. Actually, mammals use only a few hundred genes for antibody production, and the huge number of different antibodies derives from remarkable events that take place in the DNA of certain somatic cells. These events will be discussed in this section.

Although an individual organism is capable of producing a vast number of different antibodies, only a fraction of them are synthesized at any one time. Each B cell can produce a single type of antibody, but antibody is not secreted until the cell has been stimulated by the appropriate antigen. Once stimulated, the B cell undergoes successive mitoses and eventually produces a clone of identical cells that secrete the antibody. Moreover, antibody secretion may continue even if the antigen is no longer present. In this manner, organisms produce antibodies only to the antigens to which they were have been exposed.

Five distinct classes of antibodies are known, which are designated IgG, IgM, IgA, IgD, and IgE (Ig stands for **immunoglobulin**). These classes serve specialized functions in the immune response and exhibit small structural differences. However, each contains two types of polypeptide chains differing in size: a large one called the **heavy (H) chain** and a small one called the **light (L) chain.**

Immunoglobulin G is the most-abundant antibody class and has the simplest molecular structure. The molecular organization is illustrated in

Figure 15-9. An IgG molecule consists of two heavy and two light chains held together by disulfide bridges (two joined sulfur atoms) and has the overall shape of the letter Y. The sites on the antibody that carry its specificity and combine with the antigen are located in the upper half of the arms above the fork of the Y. Each IgG molecule with a different antigen specificity has a different amino acid sequence for the heavy and light chains in this part of the molecule. These specificity regions are called the **variable regions** (denoted V in Figure 15-9) of the heavy and light chains. The remaining regions of the polypeptide are the **constant regions,** denoted C, which have virtually the same amino acid sequence in all IgG molecules. In this section, we will see how the antibody polypeptide chains are formed.

Initial understanding of the genetic mechanisms responsible for antibody variability came from cloning the IgG gene. The critical observation was made by comparing the nucleotide sequence of the gene in embryonic cells or germ cells with that in mature antibody-producing cells. In the genome of a B cell that was actively producing an antibody, the DNA segments corresponding to the constant and variable regions of the light chains were found to be very close together, as expected of DNA that codes for different parts of the same polypeptide. However, in embryonic cells, these same DNA sequences were located far apart. Similar results were obtained for the variable and constant regions of the heavy chains: segments encoding these regions were close together in B cells but widely separated in embryonic cells.

Extensive DNA sequencing of the genes revealed not only the reason for the different gene locations in B cells and germ cells, but also the mechanism for the origin of antibody variability. Cells in the germ line contain a small number of genes corresponding to the constant region of the light chain, and these genes are close together along the DNA. Separated from them but on the same chromosome is another cluster consisting of a much larger number of genes that correspond to the variable region of the light chains. In the differentiation of a B cell, one gene for the constant region is spliced (cut and joined) to one gene for the variable region, and this splicing produces a complete light-chain antibody gene. A similar splicing mechanism yields the constant and variable regions of the heavy chains.

The formation of a finished antibody gene is slightly more complicated than this description implies because light-chain genes consist of three parts and heavy-chain genes consist of four parts. Gene splicing in the origin of a mouse

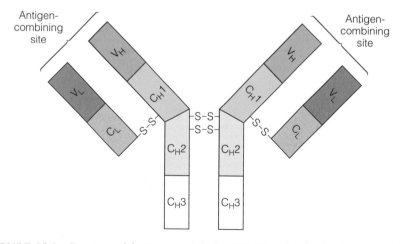

FIGURE 15-9   Structure of the immunoglobulin G (IgG) molecule showing the light chains (L, gray) and heavy chains (H, white and various shades of pink). V and C refer to variable and constant regions, respectively.

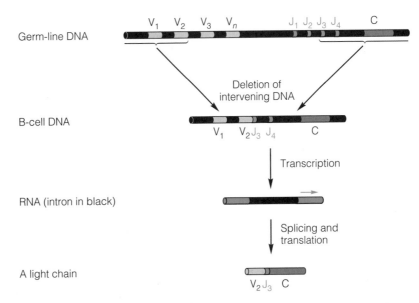

FIGURE 15-10 Formation of a gene for the light chain of an antibody molecule. One variable (V) region is randomly joined with one J region by deletion of the intervening DNA. The remaining J regions are eliminated from the RNA transcript during RNA processing.

antibody light chain is illustrated in Figure 15-10. For each of two parts of the variable region, the germ line contains multiple coding sequences called the **V** and **J (joining) regions.** In the differentiation of a B cell, a deletion (which is variable in length) enables the joining of one of the V regions with one of the J regions. The DNA joining process is called **combinatorial joining,** because it can create many combinations of the V and J regions. When transcribed, this joined V-J sequence forms the 5′ end of the light-chain RNA transcript. Transcription continues through the DNA region coding for the constant (C) part of the gene. RNA splicing subsequently attaches the C region, creating the light-chain mRNA.

Combinatorial joining also occurs in the genes for the antibody heavy chains. In this case, the DNA splicing joins the heavy-chain counterparts of V and J with a third set of sequences, called D (for diversity), located between the V and J clusters.

The amount of antibody variability that can be created by combinatorial joining is calculated as follows. In mice, the light chains are formed from combinations of about 250 V and 4 J regions, giving $250 \times 4 = 1000$ different chains. For the heavy chains, there are approximately 250 V, 10 D, and 4 J regions, producing $250 \times 10 \times 4 = 10,000$ combinations. Because any light chain can combine with any heavy chain, there are at least $1000 \times 10,000 = 10^7$ possible types of antibody. Thus, the number of DNA sequences used for antibody production is quite small (about 500), but the number of possible antibodies is very large.

The value of 10^7 different antibody types is an underestimation because there are two additional sources of antibody variability:

1. The junction for V-J (or V-D-J) splicing in combinatorial joining can be between different nucleotides of a particular V-J combination in light chains or a particular V-D or D-J combination in heavy chains). The different splice junctions can generate different codons in the spliced gene. For example, a particular combination of V and J sequences can be spliced in five different ways. At the splice junction,

the V sequence contains the nucleotides CATTTC and the J sequence contains CTGGGTG. The splicing event determines the codons for amino acids 97, 98, and 99 in the completed antibody light chain, and it can occur in any of the following ways, the last two of which result in altered amino acid sequences:

Spliced sequence	97	98	99
CACTGGGTG	His	Trp	Val
CATTGGGTG	His	Trp	Val
CATTGGGTG	His	Trp	Val
CATTTGGTG	His	Leu	Val
CATTTCGTG	His	Phe	Val

In this manner, variability in the junction of V-J joining can result in polypeptides differing in amino acid sequence.

2. The V regions are susceptible to a high rate of *somatic mutation*, which occurs in B-cell development. These mutations allow different B-cell clones to produce different polypeptide sequences, even if they have undergone exactly the same V-J joining. The mechanism for this high mutation rate is unknown.

Gene Splicing in the Origin of T-Cell Receptors

The T cells carry antigen receptors on their surfaces that combine with presented antigens, stimulating the T-cell response. Like the antibody molecules produced in B cells, the T-cell receptors are highly variable in amino acid sequence, enabling the T cells to respond to many different antigens. Although the polypeptide chains in T-cell receptors are different from those in antibody molecules, they have a similar organization in that they are formed from the aggregation of two pairs of polypeptide chains. A T cell may carry either of two types of receptors. The majority carry the αβ receptor, composed of polypeptide chains designated α and β, and the rest carry the γδ receptor, composed of chains designated γ and δ. Each receptor polypeptide includes a variable and a constant region. As suggested by their variability and similarity in organization, the T-cell receptor genes are formed by somatic rearrangement of components analogous to those of the V, D, J, and C regions in the B cells. For example, in the mouse, the gene encoding the β chain of the T-cell receptor is spliced together from one each of approximately 20 V regions, 2 D regions, 12 J regions, and 2 C regions. Note that there are far fewer V regions for T-cell receptor genes than there are for antibody genes; yet T-cell receptors seem able to recognize just as many foreign antigens as do B cells. The extra variation results from a higher rate of somatic mutation in the T-cell receptor genes.

15-5 Histocompatibility Antigens

Skin grafts and other transplanted tissues between unrelated individual organisms are usually rejected, except for a few tissues such as bone or the cornea of the eye. Rejection results from cell-surface antigens, present in the graft, that are recognized as foreign and attacked by a class of T cells in the recipient. The antigens responsible for rejection are genetically determined by the alleles of more than forty genes, and the alleles of each gene are codominantly expressed. These cellular antigens are called **histocompatibility antigens,** and the genes that code for them are **histocompatibility genes.** Paradoxically, the histocompatibility antigens formed by one's own tissues do not elicit a rejection reaction, a phenomenon called **tolerance,** which is associated with the programmed cell death of potentially responding T cells during the development of the immune system. On the other hand, in the mature immune system, certain genes in the major histocompatibility complex (MHC) are required for foreign antigen presentation in order for cytotoxic T cells to mount a rejection reaction against histocompatibility antigens present in transplanted tissues.

The genetics of histocompatibility has been extensively studied in laboratory mice because of the availability of many unrelated inbred strains derived from different ancestors. These unrelated inbred strains have been used in crosses to produce still other inbred strains that differ in only one histocompatibility gene, and these strains provide an opportunity to evaluate the effects of individual genes. The strength of individual genes can be assessed in accord with the speed and severity of the graft rejection. Histocompatibility genes are often divided qualitatively into two groups — "weak" and "strong" — though the boundary between the groups is not sharp. Among the strong histocompatibility genes, the *H-2* gene elicits a stronger and more-rapid rejection than any other. For this reason, the *H-2* gene is the most important in histocompatibility. The *H-2* gene is not a single gene; rather, it is the set of genes constituting the *major histocompatibility complex,* or MHC, which extends for at least 0.5 map units on mouse chromosome 17. Analysis of *H-2* has shown that it contains four major regions, each of which consists of several genes with many alleles. Studies with human beings are more difficult because of the lack of inbred lines and because mating cannot be controlled. However, the work with mice has enabled geneticists to conclude from human pedigree analysis and from tissue typing that the *MHC* in human beings also consists of four regions located in chromosome 6—*HLA-A, HLA-B, HLA-C,* and *HLA-D*—each having many alleles. As in mice, numerous minor genes are scattered throughout the genome. The set of *HLA* alleles in one chromosome defines the **haplotype** of the chromosome. Thus, each person, unlikely to be homozygous for such a complex genetic region, has two haplotypes, corresponding to the chromosomes inherited through egg and sperm. Many *HLA* antigens can be detected on the surface of white blood cells, and so identification of the haplotypes is often possible. A significant observation is that the haplotypes are usually inherited as an indivisible unit because the genes are sufficiently close that genetic recombination in the *HLA* region is rare. Thus, haplotypes are usually inherited as if they were alternative alleles of a single gene.

For any histocompatibility gene at which there may be many possible alleles, the maximum number of alleles that can be represented in a mating pair is four (that is, one each in both homologous chromosomes in both parents). In this case, the probability that two siblings will have an identical genotype is 1/4 (Figure 15-11). For this reason, transplant success in human beings decreases in the order: identical twin donor > sibling donor > parental donor > unrelated donor, in which the symbol ">" means "more often successful than." With an unrelated donor, the probability of a successful transplant is miniscule; the numbers of alleles of the *HLA-A, B, C, D* genes are at least 30, 8,)6, and 12, respectively, and so the number of possible haplotypes is the product of these values, or 103,680. Matching within the family does not guarantee complete success (except for identical twins) because of the large number of minor histocompatibility genes. Nonetheless, HLA matching improves the survival of the transplant considerably.

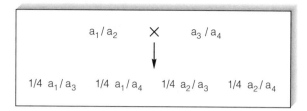

FIGURE 15-11   Possible types of offspring from parents that differ in genotype for a gene with many alleles. Ignoring the small chance of recombination within the major histocompatibility complex, the probability that two siblings will have identical genotypes is 1/4.

In recent years, certain transplants have become routine (kidney transplants), and even heart and liver transplants are commonplace. HLA matching is essential, but the major breakthrough has been the use of the immunosuppresive agent **cyclosporin,** which reduces the overall activity of the immune system. Some organ recipients must be maintained on cyclosporin indefinitely, which makes them especially susceptible to infection, but this problem can be controlled with appropriate antibiotic therapy.

Blood Groups, HLA, and Disease

Among people with particular diseases, certain blood groups and *HLA* haplotypes are found more frequently than would be expected by chance. Examples include stomach cancer, which is found with elevated frequency in blood type A, and duodenal ulcer, which is found with elevated frequency in blood type O. The reason for the disease association is unclear.

HLA haplotypes also have been studied extensively with regard to disease, and there are many significant associations. The strongest association is between ankylosing spondylitis, an inflammatory disease of the spinal column, and haplotypes carrying *B27*. Although most people carrying *B27* do not develop the disease, the frequency of *B27* among patients with the disorder is 90 percent, whereas among unaffected persons it is only 8 percent. Other associations between *HLA* and disease include *B13* with psoriasis (a familial skin disease) and *Dw3* with gluten-sensitive enteropathy (an immune disorder resulting in destruction of intestinal cells triggered by eating wheat products, which contain gluten proteins). In the last example, the disease association has been shown to result from a single amino acid replacement within the histocompatibility antigen.

CHAPTER SUMMARY

Somatic-cell genetics is the study of genetic processes that take place in cells that are neither gametes nor precursors to gametes. An example of these processes are the genetic events that occur in the development of the immune system. Hybrid somatic cells created by cell fusion have been indispensible in the study of mammalian cells in the laboratory.

A variety of techniques are available for fusing somatic cells of the same or different species. Hybrid cells can be selected easily if each of two parental cell lines carries a different mutation in the salvage pathway for DNA synthesis. The mutations complement in the hybrids and permit growth in HAT selective medium. Interspecies cell fusion is used in genetics to assign genes to particular chromosomes or parts of chromosomes. Fusion of a mouse cell with a human cell produces a hybrid cell with the full complements of 40 mouse chromosomes and of 46 human chromosomes. Human chromosomes in the newly formed hybrids are frequently lost in the course of cell division. When only a few human chromosomes remain in a cell, a hybrid becomes fairly stable, with very little loss of human chromosomes. Study of the phenotypes of these more-or-less stable hybrid cell lines allows the assignment of a given human gene to a particular chromosome, because the gene product will be present only in those hybrid cells that have retained the particular human chromosome containing the gene.

Persons who have recovered from an infectious disease are less likely to become ill again with the same disease, because the initial exposure to the agent that causes the disease has made them immune to the organism. Immunity is an acquired ability of the body to recognize and destroy invading viruses, bacteria, and large molecules. A substance able to elicit an immune response is called an antigen. The immune response relies on two types of white blood cells—B cells and T cells. B cells produce antibodies that circulate in the blood and lymph and combine with specific antigens; T cells attack specific antigens directly.

The antigens responsible for most types of transfusion incompatibility are the A and B antigens, whose presence on the surface of red blood cells is determined by codominant alleles. Adverse transfusion reactions result when antibodies in the blood of the recipient attack antigens on the transfused red blood cells. Hemolytic disease of the newborn results from the incompatibility of Rh blood groups between mother and fetus if anti-Rh$^+$ antibodies in the mother cross the placenta and destroy red blood cells of the fetus. Various blood groups are used in anthropological and other types of population studies, but their use in parentage disputes and criminology has been superceded by DNA fingerprinting.

A normal mammal is able to produce more than 10^8 different types of antibody molecules encoded by only a few hundred antibody genes. This variety results primarily from DNA splicing in the development of B cells, which joins together one each of numerous alternative coding sequences for three parts of the light chain of the antibody molecule. The heavy chain of the antibody molecule is formed by the splicing of four segments. Additional antibody variability originates from somatic mutations. After splicing and mutation, the cell containing the new gene sequence multiplies, forming a clone of cells, each of which is able to synthesize the same antibody. Each antibody type made by an organism is the result of a unique splicing event and is the product of a particular clone. At a later time, if the organism is exposed to the same antigen, the clone responds by making large quantities of the specific anti-antigen antibody; this response is the cause of immunity. The T-cell response depends on receptor molecules on the cell surface; they, too, are highly variable because of a combinatorial splicing mechanism analogous to that for antibodies in B cells. Stimulation of T cells and B cells requires antigen presentation by other cells and is mediated by certain products of the major histocompatibility complex.

In the rejection of transplanted tissues or organs, T cells attack the antigens present in cells of the transplanted tissue. These antigens are determined by dominant or codominant alleles, and an individual organism possesses all antigens corresponding to the alleles in its genotype. Although many genes produce tissue antigens that are subject to rejection, the chromosomal region resulting in the most-rapid and severe rejection reaction is the major histocompatibility complex (MHC). The MHC is a complex genetic region containing several genes that participate in overall regulation of the immune response. There are many alleles of each of these genes, and the particular combination of alleles in a chromosome is called the haplotype of the chromosome. Some MHC alleles are associated with particular diseases because of immunity problems resulting from amino acid replacements in the MHC antigens.

KEY TERMS

antibody	cyclosporin	immune response
antigen	cytotoxic T cell	immunity
antigen presentation	DNA fingerprinting	immunoglobulin
B cell	haplotype	J (joining) region
blood-group system	HAT medium	L (light) chain
C (constant) region	helper T cell	macrophage
cell hybrid	H (heavy chain)	major histocompatibility complex
cell line	histocompatibility antigen	(MHC)
combinatorial joining	histocompatibility gene	rejection

Rh-negative
Rh-positive
salvage pathway

somatic-cell genetics
T cell

tolerance
V (variable) region

EXAMPLES OF WORKED PROBLEMS

Problem 1: Suppose that four mouse-human cell lines (A, B, C, and D) have been isolated. Each contains one human chromosome, either 1, 2, 3, or 4, respectively. Cell-line B possesses a human enzymatic activity not present in the other three cell lines. What information does this give you?

Answer: The gene encoding the enzyme is located in human chromosome 2.

Problem: A series of mouse-human hybrid cell lines are examined and found to have the following human chromosomes:

(1) 3, 4, 9, 13, 17, X
(2) 1, 3, 4, 9, 21, X
(3) 9, 13, 19, 21, X
(4) 1, 4, 13, 17, 19, X
(5) 2, 4, 13, 15, 17, X

Southern blots from the cell lines are hybridized with probes for two human genes, P and Q. DNA homologous to gene P is present in cell-lines 3 and 4 only, and DNA homologous to gene Q is absent from these lines but present in all the others. What do these results tell you about the chromosomal locations of: **(a)** gene P? **(b)** gene Q?

Answer: The chromosomal location of the gene is deduced by identifying the chromosome that is common to all cell lines containing the homologous DNA but not present in cell lines lacking the DNA. **(a)** Gene P could be present in chromosome 9, 21, or X, because these are common to cell-lines 2 and 3. However, chromosome 9 is also present in line 1, and the X chromosome is present in all lines; so the likely location of gene P is chromosome 21. **(b)** Candidate chromosomes for gene Q are 4, 13, 17, and X, because all are common to cell-lines 1, 4, and 5. However, chromosome 4 is present in line 2, chromosome 13 in line 3, and the X chromosome in all lines; hence the probable location of gene Q is chromosome 17.

Problem 3: A woman accused of abandoning a baby claims that she never had a baby. The blood types of the woman and the baby are as follows:

Woman   O *cc* D– *EE*
Baby   A *Cc* *dd* *ee*

Could it be her baby?

Answer: The thing to check is whether the baby's genotype for any of the genes implies that the mother must have contributed an allele that the accused woman does not possess. Let us start with the ABO group and see if she could have a type A baby. The ABO data is compatible with the baby being hers, because the baby could have received an I^A allele from the father and an I^O allele from the mother. In the Rh system, the mother could have contributed the baby's c allele, and the mother could be heterozygous Dd, which is compatible with having a dd baby. However, because the baby is ee, it must have received an e allele from both its father and its mother, but this woman is EE. Hence, the blood-group data are consistent with the woman's story that the baby is not hers.

Problem 4: A cow is artificially inseminated with the sperm of her own father and a single calf results.
(a) Ignoring recombination, what is the probability that the calf will be homozygous for a bovine major histocompatibility complex (MHC) haplotype?
(b) What is the probability that the calf and the bull will be a perfect match at the bovine MHC?

Answer: **(a)** The father and daughter must share one MHC haplotype. The probability that the sperm carries the haplotype present in the daughter is 1/2, and the probability that the egg carries the haplotype present in the father is 1/2. Hence, a homozygote for the MHC can arise only by union of egg and sperm with the identical haplotype, which occurs at a frequency of $1/2 \times 1/2 = 1/4$. **(b)** For the calf and bull to have a perfect match at the MHC, the calf must receive the haplotype shared by father and daughter from the daughter, which is transmitted with a probability of 1/2, and the unshared haplotype from the father, which is also transmitted with a probability of 1/2. Thus, when father and daughter mate, the probability of a calf that matches the father perfectly is $1/2 \times 1/2 = 1/4$.

PROBLEMS

15-1. What are somatic-cell hybrids and how are they produced?

15-2. How are the enzymes HPRT and TK used in cell fusion? What genotypes cannot grow in HAT medium?

15-3. What is an antigen? An antibody? Antigen presentation?

15-4. How is it possible for a molecule to be antigenic in one species but not antigenic in another?

15-5. Distinguish between a germ-line cell, a B cell, and a T cell. Are their genetic constitutions identical?

15-6. Distinguish between helper T cells and cytotoxic T cells.

15-7. What is the purpose of matching ABO blood types when transfusing blood? In an emergency, is it more important for all donor antigens or for all donor antibodies to also be present in the recipient? Explain.

15-8. Is the rejection of tissue transplants due to preexisting antibodies in the recipient?

15-9. Which combination of Rh blood groups in mother and child can cause hemolytic disease of the newborn? Does a firstborn child normally have the disease? Can a firstborn child ever have it? Explain.

15-10. What blood types could a child have whose mother is type O and whose father is type B?

15-11. A husband is suspicious of his wife's fidelity because he has blood type A, she has blood type B, and their newborn child has blood type O. Are the husband's suspicions justified from the standpoint of genetics?

15-12. How many polypeptide chains, and how many different types of polypeptide chains, are present in an IgG molecule? What is meant by the terms variable region and constant region when referring to the individual chains of an IgG molecule?

15-13. Do the V and J segments of an antibody gene join to form the variable or the constant region of the light chain of an IgG molecule? What segments join to form the variable portion of the heavy chain?

15-14. Rh antibodies are of the IgG class and can cross the placenta, but naturally occurring anti-A and anti-B antibodies are of the IgM class. What does this imply about the ability (or inability) of IgM to cross the placenta?

Problems 15-15 through 15-17 use data from the set of eight hypothetical human-mouse clones described in the adjoining table. Each clone may carry an intact (numbered) chromosome (+), or only its long arm (q), or only its short arm (p), or it may lack the chromosome (−).

	Chromosome								
Clone	1	2	6	9	12	13	17	21	X
A	+	+	−	q	−	p	+	+	+
B	+	−	p	+	−	+	+	−	−
C	−	+	+	+	p	−	+	−	+
D	+	+	−	+	+	−	q	−	+
E	p	−	+	−	q	−	+	+	q
F	−	p	−	−	q	−	+	+	p
G	q	+	−	+	+	+	+	−	−
H	+	q	+	−	−	q	+	−	+

15-15. Suppose that the clones were obtained by the HAT selection method.
(a) Were the human cells used in the cell fusions unable (TK^-) or able (TK^+) to utilize thymidine as a DNA precursor? How can you tell?
(b) On which chromosome arm is the TK gene located?

15-16. The following human enzymes were tested for their presence (+) or absence (−) in clones A through H. Identify the chromosome carrying each enzyme locus, and, where possible, identify the chromosome arm.

	Cell line							
Enzyme	A	B	C	D	E	F	G	H
Steroid sulfatase	+	−	+	+	−	+	−	+
Phosphoglucomutase-3	−	−	+	−	+	−	−	+
Esterase D	−	+	−	−	−	−	+	+
Phosphofructokinase	+	−	−	−	+	+	−	−
Amylase	+	+	−	+	+	−	−	+
Galactokinase	+	+	+	+	+	+	+	+

15-17. A cloned human DNA sequence is used as a hybridization probe with DNA from each of cell lines A through H. Lines A, C, D, E, and H test positive for presence of the sequence, and lines B, F, and G test negative. What chromosomal locations for the cross-hybridizing DNA sequences can definitely be excluded on the basis of these results? Where are the possible chromosomal locations of the cross-hybridizing DNA sequences? Could there be more than one location? Explain.

15-18. Persons of genotype Se/Se or Se/se secrete the ABO antigens into their saliva and other body fluids, and those of genotype se/se are nonsecretors. A nonsecretor woman of blood-type A has a secretor child of blood-type B.
(a) What are the possible genotypes of the father?
(b) What are the possible phenotypes of the father?
(c) What is the genotype of the child?

15-19. A woman has a child with hemolytic disease of the newborn.
(a) What is the genotype of the mother with regard to the Dd pair of Rh alleles?
(b) Will antibody treatment in subsequent pregnancies be effective? Why or why not?
(c) If the woman's husband has the genotype Dd, what is the probability that the next child will not be at risk for hemolytic disease?

15-20. A woman has a baby and thinks that either of two men could be its father. The individual blood-group phenotypes are as follows.

Mother	O	CC D– ee	M
Child	O	CC dd Ee	M
Male 1	A	cc D– EE	MN
Male 2	B	Cc dd Ee	N

Based on these phenotypes, can either male be excluded from paternity? Explain your reasoning. (*Note:* The MN blood groups are determined by a pair of codominant alleles *M* and *N*. Genotypes *MM* and *NN* have phenotypes M and N, respectively, and genotype *MN* has phenotype MN.)

15-21. What familial relationship is the best possible for blood transfusions or transplants? Why?

15-22. A woman claims to have had a child by spontaneous embryonic development of an unfertilized diploid egg cell (parthenogenesis). How would one use skin transplants to determine the validity of the claim? What tests are needed if the child is a normal male?

15-23. Each of a pair of identical twins of blood-type B produces an anti-A antibody of the IgM class. Would these antibodies be expected to have identical amino acid sequences? Explain.

15-24. What is the maximum number of cellular antigens that can be coded by the alleles of a single histocompatibility gene in a diploid, triploid, or trisomic individual?

15-25. A female mouse of inbred strain CH is mated with a male of inbred strain C57. The F_1 hybrid is called CH57. From which of the following mice will a CH57 mouse accept skin grafts: CH, C57, CH57, a wild mouse, a mouse of a different inbred strain?

Genetic Engineering

Genetic studies continue to be instrumental in advancing the understanding of biological processes. An equally important role of genetics is the deliberate manipulation of biological systems for scientific or economic reasons, which has resulted in the creation of organisms with novel genotypes and phenotypes. The traditional foundations of genetic manipulation have been mutation and

Above: Genetic engineering of a plant. *Left panel.* Young shoots of a regenerating tobacco plant, growing on a solid medium containing hormones that prevent root formation and stimulate leaf and stem formation. The cluster developed from callus tissue obtained from tobacco stems infected with the bacterium *Agrobacterium tumefaciens* containing the hybrid plasmid pGV3850. *Right panel.* A plant formed from one of the shoots seen in the left panel. A shoot was cut from the cluster, placed on a medium that allows roots to form, and finally placed in soil. The regenerated plant synthesizes a substance, nopaline, that is encoded in the plasmid. (From P. Zambryski, H. Joos, C. Genetello, J. Leemans, M. Van Montague, and J. Schell. 1983. *EMBO Journal,* 2: 2143.)

recombination. These processes are essentially random, and selective procedures, often quite complex, are required to identify organisms with the desired characteristics among the many other types of organisms produced. Since the 1970s, techniques have been developed in which the genotype of an organism can instead be modified in a directed and predetermined way. This is alternately called **recombinant DNA technology, genetic engineering,** or **gene cloning.** It entails isolating DNA fragments, joining them in new combinations, and introducing the recombined molecules back into a living organism. Selection of the desired genotype is still necessary, but the probability of success is usually many orders of magnitude greater than with traditional procedures. The basic technique is quite simple: two DNA molecules are isolated and cut into fragments by one or more specialized enzymes and then the fragments are joined together *in any desired combination* and introduced back into a cell for replication and reproduction. Recombinant DNA procedures complement the traditional methods of plant and animal breeding in supplying new types of genetic variation for improvement of the organisms.

Current interest in genetic engineering centers on its many practical applications, a few examples of which are: (1) isolation of a particular gene, part of a gene, or region of a genome; (2) production of particular RNA and protein molecules in quantities formerly unobtainable; (3) more-efficient production of biochemicals (such as enzymes and drugs) and commercially important organic chemicals; (4) creation of varieties of plants with particular desirable characteristics (for example, requiring less fertilizer or resistant to disease); and (5) correction of genetic defects in higher organisms. Specific examples will be considered later in the chapter.

16-1 Cloning Strategies

In genetic engineering, the immediate goal of an experiment is usually to insert a *particular* fragment of chromosomal DNA into a plasmid or a viral DNA molecule. This is accomplished by techniques for breaking DNA molecules at specific sites and for isolating particular DNA fragments. These techniques are described in the following subsection.

Production of Defined DNA Fragments

DNA fragments are usually obtained by treatment of DNA samples with restriction enzymes. *Restriction enzymes* are nucleases that cleave DNA wherever it contains a particular short sequence of nucleotides matching the *restriction site* of the enzyme (see Section 4-10). Most restriction sites consist of four or six nucleotides, within which the restriction enzyme makes two single-strand breaks, one in each strand, generating 3'-OH and 5'-P groups at each position. Several hundred restriction enzymes with different restriction sites have been isolated from microorganisms.

The sequences recognized by restriction enzymes are usually *palindromes*, which means that the sequence is identical in both strands of the DNA molecule. For example, the restriction enzyme EcoRI isolated from *Escherichia coli* has the restriction site 5'-GAATTC-3' and cuts the strand between G and A. The sequence of the other strand is 3'-CTTAAG-5', which is identical but written with the 3' end at the left.

Soon after restriction enzymes were discovered, observations with the electron microscope indicated that the fragments produced by many restriction enzymes could spontaneously form circles. The circles could be made linear again by heating; however, if after circularization they were treated with *E. coli* DNA ligase, which joins 3′-OH and 5′-P groups (Section 4-6), the ends became covalently joined. This observation was the first evidence for three important features of restriction enzymes:

1. Restriction enzymes make breaks in palindromic sequences.
2. The breaks need not be directly opposite one another in the two DNA strands.
3. Enzymes that cleave the DNA strands asymmetrically generate DNA fragments with complementary ends.

These properties are illustrated for EcoRI in Figure 16-1.

Most restriction enzymes are like EcoRI in that the cuts in the DNA strands are staggered, producing single-stranded ends called **sticky ends** or **cohesive ends** that can adhere to each other because they contain complementary nucleotide sequences. Some restriction enzymes (such as EcoRI) leave a single-stranded overhang at the 5′ end (Figure 16-2A), others leave a 3′ overhand. A number of restriction enzymes cleave both DNA strands at the center of symmetry, forming **blunt ends**. Figure 16-2B shows the blunt ends produced by the enzyme BalI. Blunt ends can also be ligated by DNA ligase. However, whereas ligation of sticky ends recreates the original restriction site, any blunt end can join with any other blunt end and not necessarily create a restriction site.

Most restriction enzymes recognize their restriction sequence without regard to the source of the DNA. Thus,

> DNA fragments obtained from one organism will have the same sticky ends as the fragments from another organism if they have been produced by the same restriction enzyme.

FIGURE 16-1 Circularization of DNA fragments produced by a restriction enzyme. Arrows indicate cleavage sites.

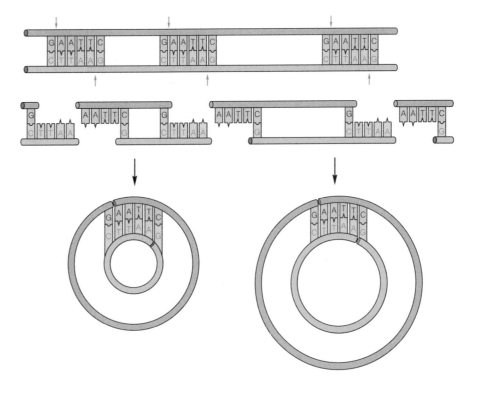

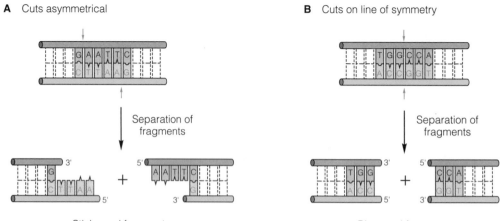

A Cuts asymmetrical

B Cuts on line of symmetry

Separation of fragments

Separation of fragments

Sticky-end fragments

Blunt-end fragments

FIGURE 16-2   Two types of cuts made by restriction enzymes. The arrows indicate the cleavage sites. A. Cuts made in each strand at an equal distance from the center of symmetry of the restriction site. In this example with the enzyme EcoRI, the resulting molecules have complementary 5' overhangs. Other restriction enzymes produce fragments with complementary 3' overhangs. B. Cuts made in each strand at the center of symmetry of the restriction site. The products have blunt ends. The specific enzyme in this example is BalI.

This principle will be seen to be one of the foundations of recombinant DNA technolology.

Because most restriction enzymes recognize a unique sequence, the number of cuts made in the DNA from an organism by a particular enzyme is limited. For example, the DNA molecule from *E. coli* contains 4.7×10^6 base pairs, and any restriction enzyme that cleaves at a six-base restriction site will cut the molecule into about a thousand fragments. The nuclear DNA of mammals would be cut into about a million fragments. These numbers are large but still small compared with the number that would be produced by random cuts. Of special interest are the smaller DNA molecules, such as viral or plasmid DNA, which may have only from one to ten sites of cutting (or even none) for particular enzymes. Plasmids containing a single site for a particular enzyme are especially valuable, as we will see shortly.

Recombinant DNA Molecules

In genetic engineering, a particular DNA segment of interest is joined to a small, but essentially complete, DNA molecule that is able to replicate––a *vector*. By a transformation procedure, described later in this section, this recombinant molecule is placed in a cell in which the molecule can replicate (Figure 16-3). When a stable transformant has been isolated, the genes or DNA sequences linked to the vector are said to be **cloned**. In the following subsection, several types of vectors are described.

Vectors

A **vector** is a DNA molecule into which a DNA fragment can be cloned. The most generally useful vectors have three properties:

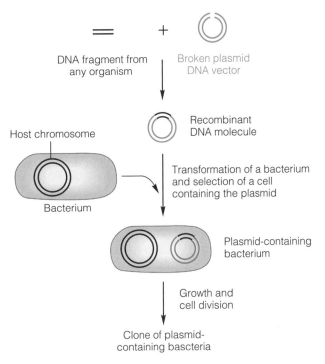

DNA fragment from
any organism

Broken plasmid
DNA vector

Recombinant
DNA molecule

Host chromosome

Bacterium

Transformation of a bacterium
and selection of a cell
containing the plasmid

Plasmid-containing
bacterium

Growth and
cell division

Clone of plasmid-
containing bascteria

FIGURE 16-3   An example of cloning. A fragment of DNA from any organism is joined to a cleaved plasmid. The recombinant plasmid is then used to transform a bacterial cell, and, thereafter, the foreign DNA is present in all progeny bacteria.

1. The vector DNA can be introduced into a host cell.
2. The vector contains a replication origin and so can replicate inside the host cell.
3. Cells containing the vector can usually be selected by a straightforward assay, most conveniently by allowing growth of the host cell on a solid selective medium.

At present, the most commonly used vectors are *E. coli* plasmids and derivatives of the bacteriophages λ and M13. Other plasmids and viruses also have been developed as vectors for cloning into the cells of animals and plants. Recombinant DNA can be detected in host cells by means of genetic features or particular markers made evident in the formation of colonies or plaques. Plasmid and phage DNA can be introduced into cells by a **transformation** procedure in which cells gain the ability to take up free DNA by exposure to a $CaCl_2$ solution. Recombinant DNA can also be introduced into cells by a kind of electrophoretic procedure called **electrophoration.** After introduction of the DNA, the cells containing the recombinant DNA are plated on a solid medium. If the added DNA is a plasmid, colonies consisting of bacterial cells containing the recombinant plasmid are formed, and the transformants can usually be detected by the phenotype that the plasmid confers on the host cell. For example, plasmid vectors typically include one or more genes for resistance to antibiotics, and plating the transformed cells on a selective medium with antibiotic prevents all but the plasmid-containing cells from growing (Section 10-1). Alternately, if the vector is phage DNA, the infected cells are plated in the usual way to yield plaques. Variants of these procedures are used to transform animal or plant cells with suitable vectors, but the technical details may differ considerably.

Three types of vectors commonly used for cloning into *E. coli* are illustrated in Figure 16-4. Plasmids (part A) are most convenient for cloning relatively small DNA fragments (5–10 kb). Somewhat larger fragments can be cloned with bacteriophage λ (part B). The wildtype phage is approximately 50 kb in length, but the central part of the genome is nonessential for lytic growth and can be removed and replaced with donor DNA. After the donor DNA has been ligated in place, the recombinant DNA is packaged into mature phage *in vitro*, and the phage is used to infect bacterial cells. However, to be packaged into a phage head, the recombinant DNA must be neither too large nor too small, which means that the donor DNA must be roughly the same size as the part of the λ genome that was removed. Most λ cloning vectors accept inserts ranging in size from 15 to 20 kb. Still larger DNA fragments can be inserted into cosmid vectors (Figure 16-4C). These vectors can exist as plasmids but also contain the cohesive ends of phage λ (Section 10-7), allowing them to be packaged into mature phages. The size limitation on cosmid inserts usually ranges from 30 to 40 kb.

Joining DNA Fragments

Circularization of restriction fragments having terminal single-stranded regions with complementary bases was described in Section 16-1. Because a particular restriction enzyme produces fragments with *identical* sticky ends, without regard for the source of the DNA, fragments from DNA molecules isolated from two different organisms can be joined, as shown in Figure 16-5. In this example, the restriction enzyme EcoRI is used to digest DNA from any organism of interest and to cleave a bacterial plasmid that contains only one EcoRI restriction site. The donor DNA is digested into many fragments (one of which is shown) and the plasmid into a single linear fragment. When the donor fragments and the linearized plasmid are mixed, recombinant molecules can form by base-pairing between the complementary single-stranded ends. At this point, the DNA is treated with DNA ligase to seal the joints, and the fragments become permanently joined in a combination that may never have existed before. The ability to join a donor DNA fragment of interest to a vector is the basis of the recombinant DNA technology.

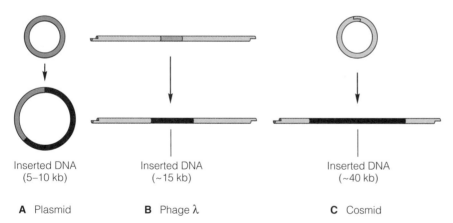

Inserted DNA
(5–10 kb)

Inserted DNA
(~15 kb)

Inserted DNA
(~40 kb)

A Plasmid   **B** Phage λ   **C** Cosmid

FIGURE 16-4   Common cloning vectors for use with *E. coli*. Plasmid vectors (A) are ideal for cloning relatively small fragments of DNA. Bacteriophage λ vectors (B) contain convenient restriction sites for removing the middle section of the phage and replacing it with the DNA of interest. Cosmid vectors (C) are useful for cloning DNA fragments as long as about 40 kb. They can replicate as plasmids but contain the cohesive ends of phage λ and so can be packaged in phage particles.

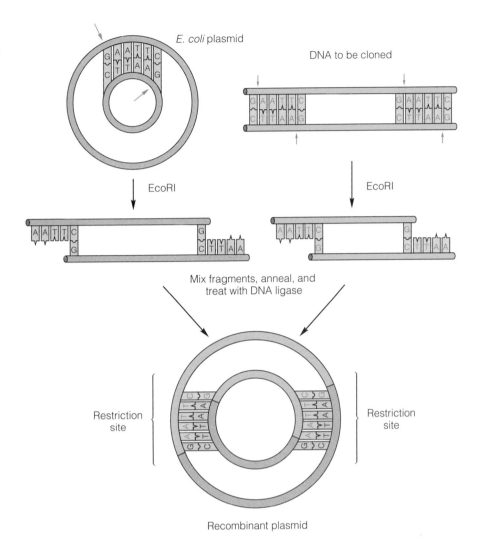

FIGURE 16-5 Construction of a recombinant DNA plasmid containing fragments derived from a donor organism, using a restriction enzyme (in this example, EcoRI) and cohesive-end joining. Short red arrows indicate cleavage sites.

E. coli plasmid

DNA to be cloned

EcoRI

EcoRI

Mix fragments, anneal, and treat with DNA ligase

Restriction site

Restriction site

Recombinant plasmid

Joining sticky ends does not always produce a DNA sequence that has functional genes. For example, consider a linear DNA molecule that is cleaved into four fragments—A, B, C, and D—whose sequence in the original molecule was ABCD. Reassembly of the fragments can yield the original molecule but, because B and C have the same pair of sticky ends, molecules with the fragment arrangements ACBD or BADC also can form with the same probability as ABCD. The problem of the scrambling of vector fragments is minimized by the use of a vector having only one cleavage site for a particular restriction enzyme. Plasmids of this type are available (most have been created by genetic engineering). Many vectors contain unique sites for several different restriction enzymes, but often only one enzyme is used at a time.

DNA molecules lacking sticky ends also can be joined. A direct method uses the DNA ligase made by E. coli phage T4. This enzyme differs from other DNA ligases in that it not only seals single-stranded breaks in double-stranded DNA, but it can also join molecules with blunt ends (Figure 16-2B).

Insertion of a Particular DNA Molecule into a Vector

In the cloning procedure described so far, a collection of fragments obtained by digestion with a restriction enzyme can be made to anneal with a cleaved vector molecule, yielding a large number of recombinant molecules containing different fragments of donor DNA. However, if a particular gene or segment of DNA is to be cloned, the recombinant molecule possessing that particular segment must be isolated from among all the recombinant molecules containing donor DNA. The kinds of selections used will be described shortly.

For many genes, simple selection techniques are adequate for recovery of recombinant molecules containing the gene. For example, if the gene to be cloned were a bacterial *leu* gene, one would use a Leu$^-$ host and select a Leu$^+$ colony on a medium lacking leucine. However, often the clone of interest is either so rare or so difficult to detect that it is preferable if, before joining, the fragment containing the DNA segment can be purified. One method has already been discussed: if the DNA of interest (for example, a viral gene) is known to be contained in a particular restriction fragment, that fragment can be isolated from a gel after electrophoresis and joined to an appropriate vector. However, eukaryotic cells contain about a million cleavage sites for a typical restriction enzyme; so direct isolation of a eukaryotic gene from a mixture of fragments separated by electrophoresis is not feasible. In the following subsections, two other procedures for cloning a particular DNA molecule are described.

The Use of cDNA

In both prokaryotes and eukaryotes, selection of a recombinant DNA molecule containing a particular gene from a collection of recombinant vectors containing other DNA from the donor is not always straightforward. In theory, a gene is most easily detected by its expression—that is, selection for the phenotype conferred by the gene as exemplified in the selection of the *E. coli leu* gene mentioned in the preceding subsection. However, not all prokaryotic genes produce such easily detected phenotypes, and for eukaryotes a particular gene that has been inserted into a bacterial DNA vector will, for a variety of reasons, usually fail either to be transcribed or to be correctly translated into a functional protein. However, the method described in this section allows direct cloning of any coding sequence whose mRNA can be isolated in almost pure form, and this is possible for many eukaryotic genes.

To use a specific example, let us assume that the chicken gene for the eggwhite protein ovalbumin is to be cloned into a bacterial plasmid. Such a plasmid could be used as a source of the ovalbumin gene for the purpose of DNA sequencing. A recombinant plasmid could be formed by treating both chicken cellular DNA and the *E. coli* plasmid DNA with the same restriction enzyme, mixing the fragments, annealing, and finally transforming *E. coli* with the DNA mixture containing the recombinant plasmid. However, a bacterial colony containing this plasmid could not be identified by a test for the presence of ovalbumin because a bacterial cell containing the intact ovalbumin gene will not synthesize ovalbumin. Introns are present in the gene (Section 11-4), and bacteria lack the enzymes needed to remove these noncoding sequences. Other tests might be used to find the desired colony, such as hybridization of bacterial DNA with either the primary transcript or the mRNA of the ovalbumin gene isolated from chicken cells but, because only about 1 plasmid-containing colony in 10^5 would contain the ovalbumin gene, this method would be very tedious. A more-convenient procedure is clearly desirable.

Some specialized animal cells, such as those producing ovalbumin, make only one or a very small number of proteins in large amount. In these cells, the cytoplasm contains specific mRNA molecules whose introns have been removed in RNA processing in the nucleus. Indeed, the specific mRNA molecules constitute a large fraction of the total mRNA synthesized in the cell. Consequently, mRNA samples can usually be obtained that consist predominantly of a single mRNA species—in this case, ovalbumin mRNA. In cloning genes whose products are the major cellular proteins, the purified mRNA can serve as a starting point for creating a collection of recombinant plasmids, many of which contain coding sequences of the gene of interest.

The technique depends on an unusual polymerase, **reverse transcriptase,** which can use a single-stranded RNA molecule (such as an mRNA) as a template and synthesize a double-stranded DNA copy, called **complementary DNA,** or **cDNA** (Figure 16-6). If the template RNA molecule is an mRNA molecule (that is, if the introns have been removed from the primary transcript), the corresponding full-length cDNA will contain an uninterrupted coding sequence. Because the introns are absent, this sequence is not identical with that in the genome of the original donor organism. However, if the purpose of forming the recombinant DNA molecule is to identify the coding sequence or to synthesize the gene product in a bacterial cell, cDNA formed from processed mRNA is the material of choice for cloning. Joining of cDNA to a vector can be accomplished by procedures for joining blunt-ended molecules (Figure 16-6).

Detection of Recombinant Molecules

When a vector is cleaved by a restriction enzyme and renatured in the presence of many different restriction fragments from a particular organism, many types of molecules result, including such examples as a self-joined vector that has not acquired any fragments, a vector containing one or more fragments, and a molecule consisting only of many joined fragments. To facilitate the isolation of a vector containing a particular gene, some means is needed to ensure (1) that the vector does indeed possess an inserted DNA fragment, and (2) that the fragment is in fact the DNA segment of interest. In this section, several useful procedures for detecting recombinant vectors will be described.

In using transformation to introduce recombinant plasmids into bacterial cells, the initial goal is to isolate bacteria that contain the plasmid from a mixture of plasmid-free and plasmid-containing cells. A common procedure is to use a plasmid possessing an antibiotic-resistance marker and to grow the transformed bacteria on a medium containing the antibiotic: only cells in which

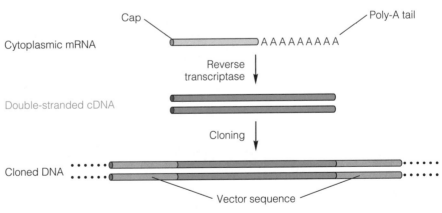

FIGURE 16-6 Cloning of cDNA molecules derived from messenger RNA by the enzyme reverse transcriptase. Analysis of cDNA is usually necessary to identify the positions and lengths of introns present in genomic DNA that are removed from the primary transcript in RNA processing.

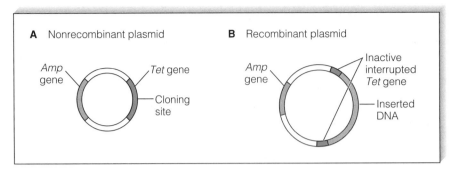

A Nonrecombinant plasmid   **B** Recombinant plasmid

FIGURE 16-7 Detection of recombinant plasmids by means of insertional inactivation of an antibiotic-resistance gene. A. Nonrecombinant plasmid containining genes for resistance to ampicillin (*Amp*) and tetracycline (*Tet*). The tetracycline gene contains a unique restriction site for inserting donor DNA. B. Recombinant plasmid with donor DNA inserted into the *Tet* cloning site. The plasmid now confers ampicillin resistance but not tetracycline resistance because the *Tet* gene is interrupted and rendered nonfunctional by the inserted DNA.

a plasmid is present can form a colony. A good example is the plasmid pBR322. It is small (4362 nucleotide pairs) and contains two genes for resistance to different antibiotics—tetracycline (*tet-r*) and ampicillin (*amp-r*). Therefore, transformed bacterial cells that contain pBR322 are easily selected by growth of the cells on medium containing either tetracycline or ampicillin. Also, pBR322 is very useful because it contains only one cleavage site for each of seven different restriction enzymes, and when donor DNA is inserted into any of these sites the exact position of the insertion is always known.

In addition to a screening procedure for identifying plasmid-containing cells, a method is needed to identify plasmids in which donor DNA has been inserted. The presence of two antibiotic-resistance markers, such as are available in pBR322, allows the use of a procedure for detecting insertion called **insertional inactivation**. In pBR322, the *tet* gene contains sites for cleavage by the restriction enzymes BamHI and SalI. Thus, insertion of donor DNA at either of these sites yields a plasmid that is *tet-s* because the insertion interrupts and hence inactivates the *tet* gene (Figure 16-7). However, the recombinant plasmid remains *amp-r*. If bacterial cells are transformed with a DNA sample in which the cleaved pBR322 and restriction fragments have been joined, and the cells are plated on a medium containing ampicillin, all surviving colonies must be Amp-r and hence must possess the plasmid. Some of these colonies will be Tet-r and others Tet-s, and these can be identified by replica plating onto a medium containing tetracycline (Section 14-3). Because the unaltered pBR322 contains the *tet-r* allele, the Amp-r colonies will also be Tet-r unless the *tet-r* allele was inactivated by insertion of donor DNA. Thus, the bacterial cells that are Amp-r Tet-s must contain recombinant pBR322 plasmids that have donor DNA inserted into the *tet-r* gene.

Screening for Particular Recombinants

The simplest procedure for detecting a cloned gene is complementation of a bacterial gene, as described earlier for the cloning of a *leu* gene by complementation in a Leu⁻ host grown on a medium lacking leucine. Some eukaryotic genes can also be detected in this way. For example, a cloned *his* gene for histidine synthesis in yeast was first detected by transforming a particular His⁻ *E. coli* mutant and selecting for growth on a medium lacking

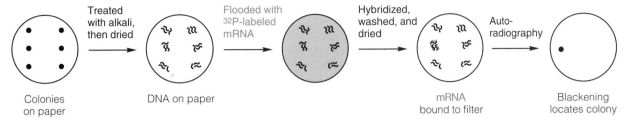

FIGURE 16-8   Colony hybridization. The reference plate, from which the colonies on paper were obtained, is not shown.

histidine. However, this method is not generally successful with animal or plant DNA cloned in bacteria, because either the genes are not expressed or there is no corresponding bacterial gene. The following two procedures are more generally applicable and are commonly used for eukaryotic genes.

The procedure of **colony hybridization** allows detection of the presence of any gene for which DNA or RNA labeled with radioactivity or some other means is available (Figure 16-8). Colonies to be tested are transferred (*lifted*) from a solid medium onto a nitrocellulose or nylon filter by gently pressing the filter onto the surface of the solid medium. A part of each colony remains on the agar medium, which constitutes the reference plate. The filter is treated with sodium hydroxide (NaOH), which simultaneously breaks open the cells and denatures the DNA. The filter is then saturated with labeled DNA or RNA, complementary to the gene being sought, and the cellular DNA is renatured. The labeled nucleic acid used in the hybridization is called the *probe*. After washing to remove unbound probe, the positions of the bound probe identify the desired colonies. For example, with radioactively labeled probe, the desired colonies are located by means of autoradiography. A similar assay is done with phage vectors, but in this case plaques are lifted onto the filters.

If transformed cells can synthesize the protein product of a cloned gene or cDNA, immunological techniques may allow the protein-producing colony to be identified. In one method, the colonies are transferred as in colony hybridization, and the transferred copies are exposed to a labeled antibody directed against the particular protein. Colonies to which the antibody adheres are those containing the gene of interest.

16-2   Polymerase Chain Reaction

The great value of recombinant DNA technology is that large quantities of the cloned DNA can be obtained for use in DNA sequencing and other applications. In some cases, it is possible to obtain large quantities of a particular DNA sequence without any cloning. The method is called the **polymerase chain reaction (PCR)**, and it uses DNA polymerase and short, synthetic oligonucletides, usually about twenty nucleotides in length, that are complementary in sequence to the ends of any DNA fragment of interest. Starting with a mixture containing as little as one molecule of the fragment of interest, repeated rounds of DNA replication increase the number of molecules exponentially. For example, starting with a single molecule, thirty rounds of DNA replication will result in $2^{30} = 10^9$ molecules. This number of molecules of the amplified fragment is so much greater than that of the other unamplified molecules in the original mixture that the amplified DNA can often be used without further purification for DNA sequencing, as probe DNA in hybridizations, or even for cloning.

An outline of the polymerase chain reaction is shown in Figure 16-9. The DNA sequence to be amplified is shown in black and the oligonucleotide sequences in red. The oligonucleotides are called **primer** sequences because they anneal to the ends of the sequence to be amplified and are used as primers for chain elongation by DNA polymerase. In the first cycle of PCR amplification, the DNA is denatured to separate the strands and then renatured in the presence of a vast excess of the primer oligonucleotides. Then DNA polymerase is added, and elongation of the primers produces double-stranded molecules; the first cycle in PCR produces two copies of each molecule containing sequences complementary to the primers. The second cycle of PCR is similar to the first. The DNA is denatured and renatured in the presence of an excess of primer oligonucleotides, and DNA polymerase elongates the primers; after this cycle there are four copies of each molecule in the original mixture. The steps of denaturation, renaturation, and replication are usually repeated from twenty to thirty times, and, in each cycle, the number of molecules of the amplified sequence is doubled. The theoretical result of thirty rounds of amplification is 2^{30} copies of the DNA sequence of interest.

Figure 16-10A presents a closer look at the primer oligonucleotide sequences and their relation to the DNA sequence to be amplified. The primers are designed to anneal to opposite DNA strands at the extreme ends of the region to be amplified. The primers are oriented with their 3' ends facing the sequence to be amplified, so that, in DNA synthesis, the primers grow toward each other but along opposite template strands (Figure 16-10B). In this way, each newly synthesized strand terminates in a sequence that can anneal with the complementary primer and so can be used for further amplification (Figure 16-10C).

PCR amplification is very useful in producing large quantities of a specific DNA sequence without the need for cloning. The limitations of the technique are that (1) the DNA sequences at the ends of the region to be amplified must

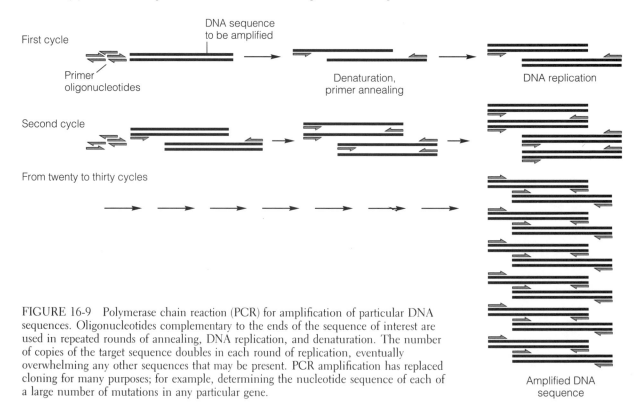

FIGURE 16-9 Polymerase chain reaction (PCR) for amplification of particular DNA sequences. Oligonucleotides complementary to the ends of the sequence of interest are used in repeated rounds of annealing, DNA replication, and denaturation. The number of copies of the target sequence doubles in each round of replication, eventually overwhelming any other sequences that may be present. PCR amplification has replaced cloning for many purposes; for example, determining the nucleotide sequence of each of a large number of mutations in any particular gene.

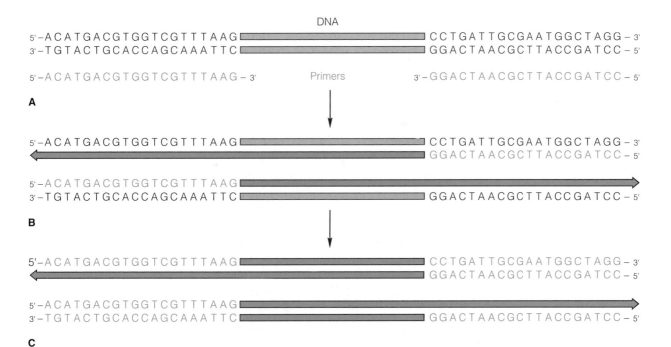

FIGURE 16-10 Example of primer sequences used in PCR amplification. Newly synthesized DNA is shown in red. After the first round of replication, each newly synthesized DNA molecule forms a new template, and so the number of copies increases exponentially. Sequences of as many as about 3 kb can be amplified using PCR.

be known so that primer oligonucleotides can be synthesized, and (2) the sequence to be amplified must be smaller than about 3000 nucleotides to allow the replication steps to occur efficiently. However, there are many applications in which these requirements are met. For example, it is often desirable to study many different mutant alleles of a gene whose wildtype sequence is known in order to identify the molecular basis of the mutations. Similarly, DNA sequence variation among alleles present in natural populations can easily be determined using PCR.

16-3 Reverse Genetics

Genetics has traditionally relied on mutation to provide the raw material needed for analysis. The usual procedure has been to use a mutant gene and phenotype to identify the wildtype allele of the gene and its normal function. This approach has proved highly successful as evidenced by numerous examples throughout this book. The approach also has its limitations. For example, it may prove to be difficult or impossible to isolate mutations in genes that duplicate the functions of other genes or that are essential for the viability of the organism.

Recombinant DNA technology has enabled another approach in genetic analysis in which wildtype genes are cloned, deliberately mutated in specific ways, and introduced back into the organism to facilitate the study of the phenotypic effects of the mutations. Because the mutations are predetermined, a very fine level of resolution is possible in defining promoter and enhancer sequences that are necessary for transcription, the sequences necessary for normal RNA splicing, particular amino acids that are essential for protein function, and so forth. The procedure is often called **reverse genetics** because it reverses the usual flow of study: instead of starting with a mutant phenotype and trying to identify the wildtype gene, reverse genetics starts by making a mutant gene and studying the resulting phenotype. The following subsections describe some techniques of reverse genetics.

Germ-Line Transformation in Animals

Reverse genetics can be carried out in most animals that have been extensively studied genetically, including the nematode *Caenorhabditis elegans*, *Drosophila*, the mouse, and many domesticated animals. In nematodes, the basic procedure is to manipulate the DNA of interest in a plasmid that also contains a selectable genetic marker that will alter the phenotype of the transformed animal. This DNA is injected directly into the germ line of the hermaphrodite. The injected DNA is often maintained in the germ line nuclei and sometimes spontaneously becomes incorporated into the chromosomes. The result of transformation is observed and can be selected in the progeny of the injected animals because of the phenotype conferred by the selectable marker.

A somewhat more elaborate procedure is necessary for germ-line transformation in *Drosophila*. The usual method makes use of a 2.9 kb transposable element (Section 5-8) called the **P element**, which consists of a central region coding for transposase flanked by 31 base-pair inverted repeats (Figure 16-11A). A genetically engineered derivative of this element, called *wings clipped*, can make functional transposase but cannot itself transpose because of deletions introduced at the ends of the inverted repeats (Figure 16-11B). For germ-line transformation, the vector is a plasmid containing a P element that includes, within the inverted repeats, a selectable genetic marker (usually one affecting eye color), as well as a large internal deletion that removes much of the transposase coding region. By itself, this P element cannot transpose because it makes no transposase, but it can be mobilized by the transposase produced by the wings-clipped or other intact P elements.

In *Drosophila* transformation, any DNA fragment of interest is introduced between the ends of the deleted P element. The resulting plasmid and a

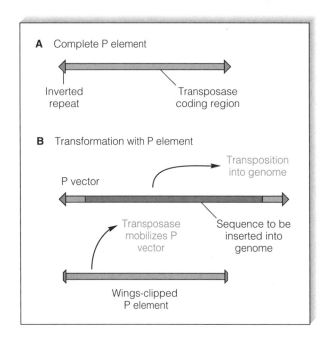

FIGURE 16-11 Transformation in *Drosophila* mediated by the transposable element P: (A) complete P element containing inverted repeats at the ends and an internal transposase-coding region; (B) two-component transformation system. The vector component contains the DNA of interest flanked by the recognition sequences needed for transposition. The wings-clipped component is a modified P element that codes for transposase but cannot transpose itself because critical recognition sequences are deleted.

different plasmid containing the wings-clipped element are injected into the pole cell region of the early embryo (Section 13-4). The DNA is taken up by the germ cells and the wings-clipped element produces functional transposase (Figure 16-11B). This mobilizes the engineered P vector and results in its transposition into an essentially random location in the genome. Transformants are detected among the progeny of the injected flies because of the eye color or other genetic marker included in the P vector.

Transformation of the germ line in mammals can be carried out in several ways. The most common is direct injection of vector DNA into the nucleus of a fertilized egg, which is then transferred to the uterus of a foster mother for development. The vector is usually a modified retrovirus. **Retroviruses** have RNA as their genetic material and code for a reverse transcriptase that converts the retrovirus genome into double-stranded DNA that becomes inserted into the genome in infected cells (Section 6-7). Genetically engineered retroviruses containing inserted genes undergo the same process. Animals that have had new genes inserted into the germ line in this manner are called **transgenic** animals.

Another method of transforming mammals uses **embryonic stem cells,** which are cells present in the blastocyst that give rise to the embryo proper (other cells give rise to various extraembryonic membranes). Although embryonic stem cells are not very hardy, they can be isolated and then grown and manipulated in culture; mutations in them can be selected or introduced by the use of recombinant DNA vectors. These stem cells can then be injected into the cavity of another blastocyst where they become incorporated into the developing embryo and often participate in forming the germ line. In this way, any mutations introduced into the embryonic stem cells while they are in culture may become incorporated into the germ line of living animals. The method was first used to create mice with a deficiency of the enzyme HPRT used in purine biosynthesis (Section 15-1) in order to have an animal model for the human disease. Surprisingly, the HPRT-deficient mice are virtually normal, whereas the enzyme deficiency in human beings results in a very serious disease.

Genetic Engineering in Plants

A procedure for the transformation of plant cells makes use of a plasmid found in the soil bacterium *Agrobacterium tumefaciens* and related species. Infection of susceptible plants with this bacterium results in the growth of tumors at the entry site, usually a wound. The bacterium contains a large (200 kb) plasmid that includes a smaller (30 kb) transposable element of 30 kb flanked by 25 base-pair inverted repeats. The transposable element is called **T DNA,** and it codes for proteins that stimulate the division of infected cells (hence the tumors) and that convert the amino acid arginine into a molecule that the bacterium needs to grow. The transposase functions are not present in the T DNA itself but in the plasmid region outside of the T DNA. In infected cells, the transposase functions mobilize the T DNA and cause its insertion into the plant genome. Susceptible plants comprise about 160,000 species of flowering plants known as the dicots, including the great majority of the most-common flowering plants.

Use of T DNA in plant transformation is illustrated in Figure 16-12. Two kinds of plasmids are introduced into the plant cells to be transformed. One contains the transposase genes that can mobilize the T DNA. The other contains the inverted repeats of T DNA flanking a selectable marker and whatever DNA sequence is to be introduced into the plant genome. Inside the cell, the transposase functions mobilize the T DNA and transposition occurs

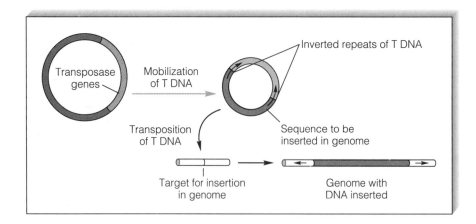

FIGURE 16-12 Transformation system for plants using the transposable element T, which is derived from a plasmid occurring naturally in the bacterium *Agrobacterium tumefacians*. Two genetically engineered plasmids are used. One contains sequences coding for the T DNA transposase functions. The other contains the DNA of interest flanked by the inverted repeats of T needed for recognition.

into the chromosomes. Transformed cells are selected in culture by the use of the selectable marker and then grown into mature plants using the methods described in Section 6-3.

16-4   Applications of Genetic Engineering

Recombinant DNA technology has revolutionized modern biology. Its main uses at the present time are (1) efficient production of useful proteins, (2) creation of cells capable of synthesizing economically important molecules, (3) generation of DNA and RNA sequences as research tools or in medical diagnosis, (4) manipulation of the genotype of organisms such as plants, and (5) potential correction of genetic defects in animals, including human patients (gene therapy). Some examples of these applications follow.

Commercial Possibilities

Cells with novel phenotypes can be produced by genetic engineering, sometimes by combining the features of several different organisms. For example, genes from different bacterial species that can metabolize some components of petroleum have been inserted into a plasmid and used to transform a species of marine bacteria, yielding an organism capable of metabolizing petroleum for use in cleaning up oil spills in the oceans. Many biotechnology companies are at work designing bacteria that can synthesize industrial chemicals or degrade industrial wastes. Bacteria have been designed that are able to compost waste more efficiently and to fix nitrogen (to improve the fertility of soil), and a great deal of effort is currently being expended to create organisms that can convert biological waste into alcohol. A number of medicinally or commercially important molecules are routinely produced in genetically engineered cells. For example, human insulin and growth hormone are produced in bacteria.

Altering the genotypes of plants (Figure 16-12) is an important application of recombinant DNA technology. By such means, it is possible to transfer genes from one plant species to another. These include genes affecting yield, hardiness, or disease resistance. There are also attempts to alter the surface structure of the roots of grains, such as wheat, by introducing certain genes from legumes (peas, beans), to give the grains the ability that legumes possess to establish root nodules of nitrogen-fixing bacteria. If successful, this would eliminate the need for the addition of nitrogenous fertilizers to soils where grains are grown.

The first engineered recombinant plant of commercial value was developed in 1985. An economically important herbicide is glyphosate, a weed killer that inhibits a particular essential enzyme in many plants. However, most herbicides cannot be applied to fields growing crops because both the crop and the weeds would be killed. (The chlorinated acids that selectively kill dicot plants but not the grasses, such as maize and the cereals, are out of favor because of their persistence in soil, toxicity to animals, and possible carcinogenicity in human beings.) The target gene of glyphosate is also present in the bacterium *Salmonella typhimurium*. A resistant form of the gene was obtained by mutagenesis and growth of *Salmonella* in the presence of glyphosate. Then the gene was cloned in *E. coli* and recloned in the T DNA of *Agrobacterium*. Transformation of plants with the glyphosate-resistance gene has yielded varieties of maize, cotton, and tobacco that are resistant to the herbicide. Thus, fields of these crops can be sprayed with glyphosate at any stage of growth of the crop. The weeds are killed but the crop is unharmed.

Genetic engineering can also be used to control insect pests. For example, the black cutworm causes extensive crop damage and is usually combatted with noxious insecticides. The bacterium *Bacillus thuringiensis* produces a protein that is lethal to the black cutworm, but the bacterium does not normally grow in association with the plants that are damaged by the worm. However, the gene coding for the lethal protein has been introduced into the soil bacterium *Pseudomonas fluorescens*, which lives in association with maize and soybean roots. Inoculation of soil with the engineered *P. fluorescens* helps control the black cutworm and reduces crop damage.

Uses in Research

Recombinant DNA technology has become an essential research tool in modern molecular genetics. Reverse genetics makes it possible to isolate and alter a gene at will and introduce it back into a living cell or even the germ line. Besides saving time and labor, reverse genetics enables the construction of mutants that cannot in practice be formed in any other way. An example is the formation of double mutants of animal viruses, which undergo crossing-over at such a low frequency that mutations can rarely be recombined by genetic crosses. The greatest effect of recombinant DNA on basic research has been in the study of eukaryotic gene regulation and development, and many of the principles of regulation and development summarized in Chapters 12 and 13 have been derived through the use of these methods.

Production of Useful Proteins

Among the most-important applications of genetic engineering is the production of large quantities of particular proteins that are otherwise difficult to obtain (for example, proteins of which there are only a few molecules per cell or that are present in only a small number of cells or only in human cells). The method is simple in principle. A DNA sequence coding for the desired protein is cloned in a vector adjacent to a bacterial promoter. This step is usually done with cDNA because cDNA has all the coding sequences spliced together in the right order. Using a vector with a high copy number ensures that many copies of the coding sequence will be present in each bacterial cell, which can result in synthesis of the gene product at concentrations ranging from 1 to 5 percent of the total cellular protein. In practice, production of large quantities of a protein in bacterial cells is straightforward, but there are often problems because many eukaryotic proteins are unstable, do not fold properly, or fail to undergo necessary chemical modification in bacterial cells.

Several approaches to solving these problems have been taken with some degree of success. One approach uses a plasmid containing the *E. coli lac* region, cleaved in the *lacZ* gene by a restriction enzyme that makes blunt ends. Either cDNA or a synthetic DNA molecule that codes for the desired amino acid sequence is inserted into the *lacZ* gene, so that the eukaryotic protein is synthesized as the terminal region of β-galactosidase, from which it can be cleaved. The first example of this approach resulted in a synthetic gene capable of yielding a 14-residue polypeptide hormone—somatostatin—which is synthesized *in vivo* in the mammalian hypothalamus. The procedure, applicable to most short polypeptides, was the following.

By chemical techniques, a double-stranded DNA molecule was synthesized containing 51 base pairs; the base sequence of the coding strand was

TAC-(42 bases encoding somatostatin)-ACTATC

The vector was the plasmid pBR322 modified to contain the *lac* promoter-operator region and a part of the *lacZ* gene coding for only the amino-terminal segment of β-galactosidase. The vector was cleaved at a site in the *lacZ* segment (Figure 16-13) by a restriction enzyme that leaves blunt ends, and the synthetic DNA molecule was ligated to the cleaved plasmid.

When transcription driven by the *lac* promoter occurred, the base sequence of the corresponding mRNA molecule was

(part of *lacZ*)-AUG-(42 bases encoding somatostatin)-UGAUAG

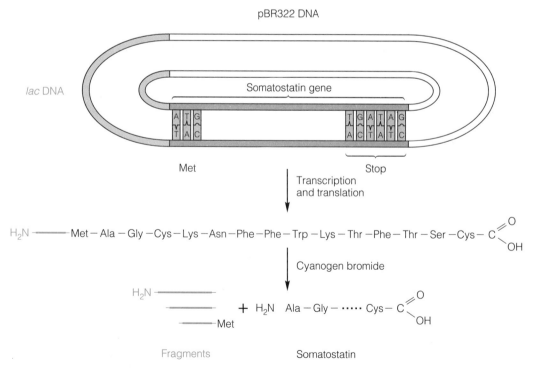

FIGURE 16-13 Synthesis of somatostatin from a chemically synthesized gene joined to the plasmid pBR322 *lac*.

The AUG linking the *lacZ* coding sequence with that of somatostatin was not used for initiation; rather, it specified an internal methionine. The mRNA terminated with two stop codons, UGA and UAG. The protein produced from this mRNA consists of the amino-terminal segment of β-galactosidase *coupled by methionine* to somatostatin. The protein was purified and treated with cyanogen bromide, a reagent that cleaves proteins only at the carboxyl side of methionine. In this way, the methionine linker remained attached to the β-galactosidase fragment and the somatostatin was released and could be purified. Use of a methionine linker followed by cleavage with cyanogen bromide is a useful technique for separating any short polypeptide that does not itself contain methionine. Other linkers and cleavage reagents can be used for polypeptides that do contain methionine.

Proteins that do not fold or become modified properly in bacterial cells can also be produced in genetically engineered cells of yeast or any other suitable eukaryote. Taking yeast as an example, one procedure uses a genetically engineered plasmid called a **shuttle vector** containing sequences from both *E. coli* and yeast and capable of undergoing replication and segregation in both organisms. Cloning is done in *E. coli* because it is easier than in yeast for a variety of technical reasons, and then the plasmid is isolated and transferred to yeast for expression of the cloned gene. In yeast or mammalian cells, the protein-folding and processing problems are sometimes substantially reduced or eliminated.

Genetic Engineering with Animal Viruses

Genetic engineering of animal cells often makes use of retroviruses because their reverse transcriptase makes a double-stranded DNA copy of the viral RNA genome, which then becomes inserted into the chromosomes of the cell. DNA-to-RNA transcription occurs only after the DNA copy is inserted. The infected host cell survives the infection, retaining the retroviral DNA in its genome. These features of retroviruses make them convenient vectors for the genetic manipulation of animal cells, including those of birds, rodents, monkeys, and human beings.

Genetic engineering with retroviruses allows the possibility of altering the genotype of animal cells. Because a wide variety of retroviruses are known, including several that infect human cells, genetic defects may be corrected by these procedures in the future. However, many retroviruses contain a gene that results in uncontrolled growth of the infected cell, thereby causing a tumor. When retrovirus vectors are used for genetic engineering, the tumor-causing gene is first deleted. The deletion also provides the space needed for incorporation of foreign DNA sequences. The recombinant DNA procedure employed with retroviruses consists of synthesis in the laboratory of double-stranded DNA from the viral RNA, through use of reverse transcriptase. The DNA is then cleaved with a restriction enzyme and, by the techniques already described, foreign DNA is inserted. Transformation yields cells with the recombinant retroviral DNA permanently inserted into the genome. In this way, the genotype of the cells can be altered.

For example, human cells from patients with Lesch-Nyhan syndrome are deficient in the synthesis of purines because they lack the enzyme HPRT (Chapter 15). These cells can be converted to produce HPRT *in vitro* by transformation with recombinant DNA with the use of retrovirus vectors. The exciting potential of this technique lies in the possibility of correcting genetic defects—for example, restoring the ability of a diabetic person to make insulin or correcting immunological deficiencies. This technique has been termed

gene therapy. However, a number of major problems stand in the way of gene therapy becoming a practical technique in medicine; for example, there is no reliable way to ensure that a gene will be inserted in the appropriate target cell or target tissue, and there is as yet no reliable means to regulate the expression of the inserted genes.

A major breakthrough in disease prevention has been the development of synthetic vaccines. Production of certain vaccines, such as antihepatitis B, has been difficult because of the extreme hazards of working with large quantities of the active virus. The danger is prevented by cloning and producing the viral antigen in a nonpathogenic organism. Vaccinia virus (the antismallpox agent) has been very useful for this purpose. Viral antigens are often on the surface of virus particles and some of these antigens can be engineered into the coat of vaccinia. For example, engineered vaccinia with surface antigens of hepatitis B, influenza virus, and vesicular stomatitis virus (which kills cattle, horses, and pigs) have produced useful vaccines in animal tests. A surface antigen of *Plasmodium falciparum*, the parasite that causes malaria, has also been placed in the vaccinia coat, and this may lead to an antimalaria vaccine.

Diagnosis of Hereditary Diseases

Probes derived by recombinant DNA methods are widely used in prenatal detection of disease; for example, in detecting cystic fibrosis, Huntington disease, sickle-cell anemia, and many other genetic disorders. In some cases, probes derived from the gene itself are used and, in other cases, restriction fragment polymorphisms (Section 8-1) genetically linked to the disease gene are employed. If the disease gene itself, or a region close to it in the chromosome, differs from the normal chromosome in the positions of one or more cleavage sites for restriction enzymes, then these differences can be detected with Southern blots (Section 4-10), with the use of cloned DNA from the region as the probe. The genotype of the fetus can be determined from the restriction fragments present in its DNA. These techniques are very sensitive and can be carried out as soon as tissue from the fetus—or even from the placenta—can be obtained.

16-5 Analysis of Complex Genomes

Gene structure and function is understood in considerable detail because of the manipulative capability of recombinant DNA technology combined with genetic analysis. However, most recombinant DNA techniques are limited by the size of the DNA molecule that can be cloned and analyzed. The methods are best suited for DNA fragments smaller than about 50 kb. Although this range includes most cDNAs and many genes, some genes are much larger than 50 kb and so must be analyzed in smaller pieces.

Recent advances in recombinant DNA technology enable very large DNA molecules to be cloned and manipulated. These methods not only are suitable for the study of very large genes, but have also made the analysis of entire genomes a possibility. The following subsections deal with complex genomes and the recombinant DNA technology that has made their analysis possible.

Magnitude of Complex Genomes

The simplest genomes are those of small bacterial viruses whose DNA is smaller than 10 kb; for example, the *E. coli* phage ϕX174 whose genome contains 5386 base pairs (Chapter 11). Toward the larger end of the spectrum is the human genome, consisting of 3×10^9 base pairs. The range is so large that comparisons are often made with regard to information content, with the base pairs of DNA analogous to letters in a book. The analogy is apt because the sizes range from a few pages of text to an entire library.

Genome sizes are compared with reference to the book analogy in Figure 16-14. The volumes are about the size of a big-city telephone book: 1500 pages per volume with 25,000 characters (nucleotides) per page. In these terms, the 50 kb of bacteriophage λ fills two pages and the 4.7 Mb genome of *E. coli* needs about 200 pages. (The abbreviation Mb stands for megabase pairs, or one million base pairs; 4.7 Mb equals 4700 kb.)

Turning to eukaryotic genomes, we find that the size range among organisms of genetic interest varies considerably. Yeast would take up one-third volume (genome size, 14 Mb), *Caenorhabditis elegans* three volumes (100 Mb), *Drosophila* five (165 Mb), and human beings eighty (3000 Mb). To put the matter in practical terms, consider how many cosmids would be necessary for cloning an entire genome in a cosmid library in which the average insert size is 40 kb. To have a good chance of containing most sequences of interest, a genomic library must contain several copies of the genome. For example, a library large enough to contain three complete genomes of DNA fragments (three *genome equivalents*) would have a 95 percent chance of containing any given sequence. In 40-kb cosmids, three genome equivalents for yeast would require a little more than 1000 clones; for nematodes, 7500; for *Drosophila*, 12,000; and for human beings, 225,000. Clearly, the larger genomes require a formidable number of clones. An alternative to larger numbers of clones is larger fragments within the clones, and this can be done with the cloning system described in the following subsection.

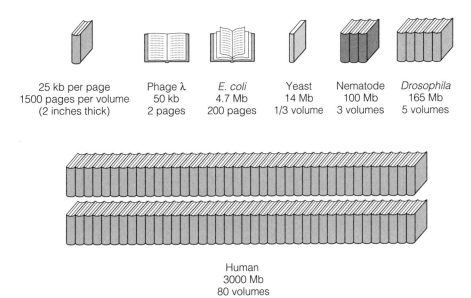

25 kb per page 1500 pages per volume (2 inches thick)	Phage λ 50 kb 2 pages	*E. coli* 4.7 Mb 200 pages	Yeast 14 Mb 1/3 volume	Nematode 100 Mb 3 volumes	*Drosophila* 165 Mb 5 volumes

Human
3000 Mb
80 volumes

FIGURE 16-14 Relative size of genomes if they were they printed at 25,000 characters per page and bound in 1500-page volumes. One volume would contain about as many characters as a telephone book 2½ inches thick. The *E. coli* genome would require about 200 pages, yeast 500 pages, the nematode one volume, and so forth. (Mb = megabase.)

Cloning in Yeast Artificial Chromosomes

Large DNA molecules can be cloned intact in yeast cells with the use of special vectors for creating **yeast artificial chromosomes,** or **YACs.** The general structure of a YAC vector is diagrammed in Figure 16-15. The YAC vector contains four types of genetic elements: (1) a cloning site, (2) a yeast centromere and genetic markers that are selectable in yeast, (3) an *E. coli* origin of replication and genetic markers that are selectable in *E. coli,* and (4) a pair of telomere sequences from *Tetrahymena.* Therefore, a YAC vector is a shuttle vector that can replicate and be selected in both *E. coli* and yeast.

Use of the YAC vector in cloning is illustrated in Figure 16-16. The circular vector is isolated after growth in *E. coli* and cleaved with two different restriction enzymes—one that cuts only at the cloning site (denoted A) and one that cuts near the tip of each of the telomeres (denoted B). Discarding the segment between the telomeres results in the two telomere-bearing fragments shown, which form the arms of the yeast artificial chromosomes.

In the meantime, the DNA to be cloned is very gently isolated and broken into large fragments either by physical means, by treatment with restriction enzymes that have infrequent cleavage sites (for example, enzymes that cleave at eight-base restriction sites), or by treatment with ordinary restriction enzymes under conditions in which only a fraction of the restriction sites are cleaved **partial digestion.** Mixing the large fragments of source DNA with the YAC arms and ligation results in a number of possible products. However, transformation of yeast cells and selection for the genetic markers in the YAC arms yield only the ligation product shown in Figure 16-16, in which a fragment of source DNA is sandwiched between the left and right arms of the YACs. This product is recovered because it is the only true chromosome possessing a single centromere and two telomeres. Products containing two YAC left arms are dicentric, those with two YAC right arms are acentric, and both of these types of products are genetically unstable (Section 6-1). YACs that have donor DNA inserted at the cloning site can be identified because the inserted DNA interrupts a yeast gene present at the cloning site and renders it nonfunctional. Yeast cells containing YACs with inserts of particular sequences of donor DNA can be identified in a number of ways, including colony

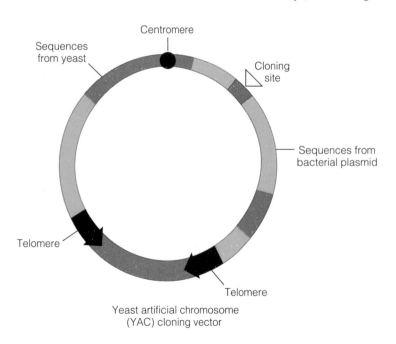

Centromere

Sequences from yeast

Cloning site

Sequences from bacterial plasmid

Telomere

Telomere

Yeast artificial chromosome (YAC) cloning vector

FIGURE 16-15   Yeast artificial chromosome cloning vector. It contains sequences that allow replication and selection in both *E. coli* (pink) and yeast (gray), a yeast centromere, and telomeres from *Tetrahymena.*

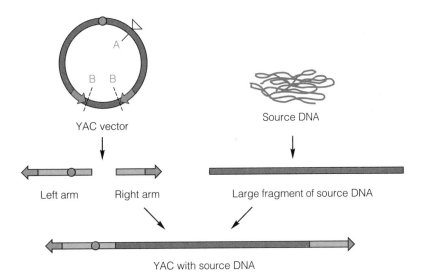

YAC vector

Source DNA

Left arm   Right arm   Large fragment of source DNA

YAC with source DNA

FIGURE 16-16 Cloning in yeast artificial chromosomes. The vector is cut with two restriction enzymes (A and B) to free the chromosome arms. These arms are ligated to the ends of large fragments of source DNA, and yeast cells are transformed. Many ligation products are possible, but only those consisting of source DNA flanked by the left and right vector arms form stable artificial chromosomes in yeast.

hybridization (Figure 16-8), and these cells can be grown and manipulated in order to isolate and study the YAC insert.

The great advantage of using YACs for the study of complex genomes is illustrated in the comparison of cloning vector capacities in Figure 16-17. There is no theoretical size limit to a YAC insert. An average size may be taken as 300 kb, but many YACs contain much larger inserts (as large as 1 Mb). Figure 16-18 shows a large region of the *Drosophila* salivary-gland chromosomes that hybridizes with a 300-kb YAC. Using the 300-kb average, Figure 16-17 demonstrates that a single YAC clone contains as much inserted DNA as

FIGURE 16-17 Relative sizes of fragments that can be cloned in various vectors. The bar at the top represents 1 kb. Generally speaking, 40 kb is about the maximum size of fragments that can be cloned into cosmids, but many YACs are greater than 300 kb (some are as many as 1000 kb).

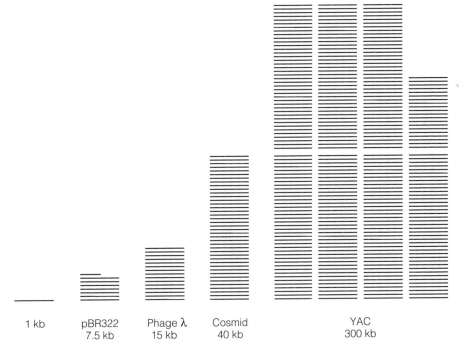

| 1 kb | pBR322 7.5 kb | Phage λ 15 kb | Cosmid 40 kb | YAC 300 kb |

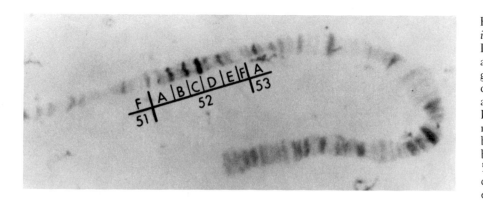

FIGURE 16-18 Hybridization *in situ* between *Drosophila* DNA cloned into a yeast artificial chromosome and the giant salivary-gland chromosomes. The yeast artificial chromosome contains DNA sequences derived from numerous adjacent salivary bands—in this example, the bands in regions 52B through 52E in the right arm of chromosome 2. On average, each salivary band contains about 20 kb of DNA, but there is wide variation among bands.

7 or 8 cosmid clones, 20 λ clones, or 40 pBR322 clones. The increased insert size simplifies the mapping problem considerably. For example, YAC libraries containing three genome equivalents of nematode or *Drosophila* DNA would need only from 1000 to 2000 clones. The human genome would require 30,000 YAC clones for three genome equivalents. Though large, this number is manageable, and therefore the YAC cloning technology has stimulated a major effort to map and sequence the genomes of human beings and model organisms such as yeast, nematodes, and *Drosophila*. With this effort and the data coming from it, the field of genetics embarks on an exciting new phase of its development.

CHAPTER SUMMARY

Recombinant DNA technology allows the genotype of an organism to be modified in a directed, predetermined way by enabling different DNA molecules to be joined into novel genetic units, altered as desired, and reintroduced into the organism. Restriction enzymes play a key role in the technique because they can cleave DNA molecules within particular base sequences. Many restriction enzymes produce DNA fragments with complementary single-stranded ends, which can anneal and be ligated with similar fragments from other DNA molecules. The carrier DNA molecule used to propagate a desired DNA fragment is called a vector. The most-common vectors are plasmids, phages, viruses, and yeast artificial chromosomes (YACs). Transformation is an essential step in the propagation of recombinant molecules because it enables the recombinant DNA molecules to enter host cells, such as those of bacteria, yeast, or mammals. If the recombinant molecule has its own replication system or can use the host replication system, it can replicate. Plasmid vectors become permanently established in the host cell; phages can multiply and produce a stable population of phages carrying source DNA; retroviruses can be used to establish a gene in an animal cell; and YACs include

source DNA within an artificial chromosome containing a functional centromere and telomeres. Particular DNA sequences can also be amplified without cloning by means of the polymerase chain reaction (PCR), in which short synthetic oligonucleotides are used as primers to repeatedly replicate and amplify the sequence between them.

Recombinant DNA can also be used to transform the germ line of animals or to genetically engineer plants. These techniques form the basis of reverse genetics, in which genes are deliberately mutated in specified ways and introduced back into the organism to determine the effects on phenotype. Reverse genetics is routine in genetic analysis in bacteria, yeast, nematodes, *Drosophila*, the mouse, and other organisms. In *Drosophila*, transformation employs a system of two vectors based on the transposable P element. One vector contains sequences that produce the P transposase; the other contains the DNA of interest between the inverted repeats of P and other sequences needed for mobilization by transposase and insertion into the genome. Germ-line transformation in the mouse makes use of retrovirus vectors or embryonic stem cells. Dicot plants are transformed with transposable T DNA derived from a plasmid found in a species of *Agrobacterium*, using a two-component system consisting of transposase producer and transforming vectors analogous to P element transformation in *Drosophila*.

Practical applications of recombinant DNA technology include efficient production of useful proteins, creation of novel genotypes for the synthesis of economically important molecules, generation of DNA and RNA sequences for use in medical diagnosis, manipulation of the genotypes of domesticated animals and plants, development of new types of vaccines, and potential correction of genetic defects in animals (gene therapy). Production of eukaryotic proteins in bacterial cells is sometimes hampered by the instability of the proteins, their inability to fold properly, or their failure to undergo necessary chemical modification. These problems are often eliminated by the use of yeast or mammalian cells.

Cloning in yeast artificial chromosomes (YACs) allows very large DNA molecules to be isolated and manipulated and has stimulated a major effort to analyze complex genomes such as those in nematodes, *Drosophila*, the mouse, and human beings. These efforts constitue a new phase in the development of genetics and will further the understanding of gene structure and function, genome organization, and evolution.

KEY TERMS

blunt end
cDNA
cloning
cohesive end
colony hybridization assay
complementary DNA
electrophoration
embryonic stem cell
gene cloning

gene therapy
genetic engineering
insertional inactivation
partial digestion
polymerase chain reaction (PCR)
primer oligonucleotide
P transposable element
recombinant DNA technology
retrovirus

reverse genetics
reverse transcriptase
shuttle vector
sticky end
T DNA
transformation
transgenic animal
vector
yeast artificial chromosome (YAC)

EXAMPLES OF WORKED PROBLEMS

Problem 1: In genetic engineering, with the aim of expressing eukaryotic gene products in bacterial cells, why is it necessary **(a)** to use cDNA instead of genomic DNA, and **(b)** to hook the cDNA to a bacterial promoter?

Answer: (a) Most eukaryotic genes have introns, which cannot be removed by bacterial cells. **(b)** Eukaryotic promoters have a different sequence from that of prokaryotic promoters and are not normally recognized in bacterial cells.

Problem 2: What is the average distance between restriction sites for each of the following restriction enzymes? Assume that the DNA substrate has a random sequence with equal amounts of each base. The symbol N stands for any nucleotide, R for any purine (A or G), and Y for any pyrimidine (T or C).

(a) TCGA   (TaqI)
(b) GGTACC   (KpnI)
(c) GTNAC   (MaeIII)
(d) GGNNCC   (NlaIV)
(e) GRCGYC   (AcyI)

Answer: The average distance between restriction sites equals the reciprocal of the probability of occurrence of the restriction site. You must therefore calculate the probability of occurrence of each restriction site in a random DNA sequence. **(a)** The probability of occurrence of the sequence TCGA is $1/4 \times 1/4 \times 1/4 \times 1/4 = (1/4)^4 = 1/256$, and so 256 bases is the average distance between TaqI sites. **(b)** By the same reasoning, the probability of a KpnI site is $(1/4)^6 = 1/4096$, and so 4096 bases is the average distance between KpnI sites. **(c)** The probability of N (any nucleotide at a site) is 1, and so the probability of the sequence GTNAC equals $1/4 \times 1/4 \times 1 \times 1/4 \times 1/4 = (1/4)^4 = 1/256$, giving 256 as the average distance between MaeIII sites. **(d)** The same reasoning gives the average distance between NlaIV sites as $(1/4 \times 1/4 \times 1 \times 1 \times 1/4 \times 1/4)^{-1} = 256$ bases. **(e)** The probability of an R (A or G) at a site is 1/2 and the probability of a Y (T or C) at a site is 1/2. Hence the probability of the sequence GRCGYC is $1/4 \times 1/2 \times 1/4 \times 1/4 \times 1/2 \times 1/4 = 1/1024$, and so there is an average of 1024 bases between AcyI sites.

Problem 3: What fundamental structural components are

necessary for yeast artificial chromosomes to be stable in yeast cells?

Answer: Yeast artificial chromosomes are true chromosomes needing a centromere for segregation in cell division and a telomere at each end to stabilize the tips.

PROBLEMS

16-1. What is meant by the term recombinant DNA?

16-2. How does the polymerase chain reaction work and what is it used for? To use the polymerase chain reaction, what information about the target sequence must be known in advance?

16-3. To what process does the term reverse genetics refer?

16-4. In the context of recombinant DNA, what is a YAC? What feature makes YACs useful in the analysis of complex genomes?

16-5. The part of the *Drosophila* genome highly replicated in the banded salivary-gland chromosomes is approximately 100 Mb (million base pairs) in size. The salivary-gland chromosomes contain approximately 5000 bands. For ease of reference, the salivary chromosomes are divided into about 100 approximately equal numbered sections (1–100), each of which consists of six lettered subdivisions (A–F). On average, how much DNA is in a salivary-gland band? In a lettered subdivision? In a numbered section? How do these compare with the size of the DNA insert in a 200-kb (kilobase pair) YAC?

16-6. What features are essential in a bacterial cloning vector?

16-7. Restriction enzymes generate one of three possible types of ends on the DNA molecules that they cleave. What are the three possibilities?

16-8. Are the ends of different restriction fragments produced by a particular restriction enzyme always the same? Must opposite ends of each restriction fragment be the same? Why?

16-9. Will the sequences 5'-GGCC and 3'-GGCC in a double-stranded DNA molecule be cut by the same restriction enzyme?

16-10. In cloning into bacterial vectors, why is it useful to insert DNA fragments to be cloned into a restriction site inside an antibiotic-resistance gene? Why is another gene for resistance to a second antibiotic also required?

16-11. What does the enzyme reverse transcriptase do and how is it used in recombinant DNA technology?

16-12. If the genomic and cDNA sequences of a gene are compared, what information does the cDNA sequence give you that is not obvious from the genomic sequence? What information does the genomic sequence contain that is not in the cDNA?

16-13. What might prevent a cloned eukaryotic gene from yielding a functional mRNA in a bacterial host? Assuming that these problems are overcome, why might the desired protein still not be produced?

16-14. When DNA isolated from phage J2 is treated with the enzyme SalI, eight fragments are produced with sizes of 1.3, 2.8, 3.6, 5.3, 7.4, 7.6, 8.1, and 11.4 kilobase pairs. However, if J2 DNA is isolated from infected cells, only seven fragments are found, with sizes of 1.3, 2.8, 7.4, 7.6, 8.1, 8.9, and 11.4 kb. What form of the intracellular DNA can account for these results?

16-15. Phage X82 DNA is cleaved into six fragments by the enzyme BglI. A mutant is isolated with plaques that look quite different from the wildtype plaque. DNA isolated from the mutant is cleaved into only five fragments. What genetic changes can account for this change?

16-16. The wildtype allele of a bacterial gene is easily selected by growth on a special medium. Repeated attempts to clone the gene by digestion of cellular DNA with the enzyme EcoRI are unsuccessful. However, if the enzyme HindIII is used, clones are easily found. Explain.

16-17. A Kan-r Tet-r plasmid is treated with the restriction enzyme BglI, which cleaves the *kan* (kanamycin) gene. The DNA is annealed with a BglI digest of *Neurospora* DNA and then used to transform *E. coli.*

(a) What antibiotic would you put in the growth medium to ensure that each colony has the plasmid?

(b) What antibiotic-resistance phenotypes would be found among the resulting colonies?

(c) Which phenotype will contain *Neurospora* DNA inserts?

16-18. A Lac⁺ Tet-r plasmid is cleaved in the *lac* gene with a restriction enzyme. The enzyme cleaves at four-base restriction sites and produces fragments with a two-base single-stranded overhang. The cutting site in the *lac* gene is in the codon for the second amino acid in the chain, a site that can tolerate any amino acid without loss of enzyme function. After cleavage, the single-stranded ends are converted into blunt ends by DNA polymerase I, and then the ends are joined by blunt-end ligation to recreate a circle. A Lac⁻Tet-s bacterial strain is transformed with the DNA, and Tet-r bacteria are selected. What is the Lac phenotype of the colonies?

16-19. Plasmid pBR607 DNA is a double-stranded circle of 4 kilobase pairs. This plasmid carries two genes whose protein products confer resistance to tetracycline (Tet-r) and ampicillin (Amp-r) in host bacteria. The DNA has a single site for each of the following restriction enzymes: EcoRI, BamHI, HindIII, PstI, and SalI. Cloning DNA into the EcoRI site does not affect resistance to either drug. Cloning DNA into the BamHI, HindIII, and Sal sites abolishes tetracycline resistance. Cloning into the PstI site abolishes ampicillin resistance. Digestion with the following mixtures of restriction enzymes yields fragments with the sizes listed below. Indicate the positions of the PstI, BamHI, HindIII, and SalI cleavage sites on a restriction map, relative to the EcoRI cleavage site.

Enzyme mixture	Fragment size (kb)
EcoRI + PstI	0.70, 3.30
EcoRI + BamHI	0.30, 3.70
EcoRI + HindIII	0.08, 3.92
EcoRI + SalI	0.85, 3.15
EcoRI + BamHI + PstI	0.30, 0.70, 3.00

aberrant 4:4 segregation   The presence of equal numbers of alleles among the spores in a single ascus, yet with a spore pair in which the two spores have different genotypes.

acentric chromosome   A chromosome without a centromere.

acridine   A class of small organic molecules capable of inserting between adjacent bases in DNA and causing small additions or deletions.

acrocentric chromosome   A chromosome with the centromere near one end.

active site   The part of an enzyme at which substrate molecules bind and are converted into their reaction products.

acylated tRNA   A tRNA molecule to which an amino acid is linked.

adaptation   Any characteristic of an organism that improves its chances of survival and reproduction in its environment; the evolutionary process by which organisms undergo modification favoring their survival and reproduction in a given environment.

addition rule   The principle that the probability of occurrence of any one of a set of mutually exclusive events equals the sum of the individual probabilities of the events.

additive variance   The magnitude of the genetic variance that results from the additive action of genes; the value that the genetic variance would assume if there were no dominance or interaction of alleles affecting the trait.

adenine (A)   A nitrogenous purine base found in DNA and RNA.

adenyl cyclase   The enzyme that catalyzes synthesis of cyclic AMP.

adjacent segregation   Segregation of a heterozygous reciprocal translocation in which a translocated chromosome and a normal chromosome segregate together, producing an aneuploid gamete.

agarose   A component of agar used as a gelling agent in gel electrophoresis; its value is that few molecules bind to it, and so it does not interfere with electrophoretic movement.

agglutination   The clumping or aggregation, as of viruses or blood cells, caused by an antigen-antibody interaction.

albinism   Absence of melanin pigment in the iris, skin, and hair of an animal; absence of chlorophyll in plants.

alkaptonuria   A recessively inherited metabolic disorder in which a defect in the breakdown of tyrosine leads to excretion of homogentisic acid (alkapton) in the urine.

alkylating agent   An organic compound capable of transferring an alkyl group to other molecules.

allele   One of an array of different forms of a given gene.

allele frequency   Relative proportion of all alleles of a gene that are of a designated type.

allopolyploid   A polyploid formed by doubling the chromosome number of a hybrid between two different species.

allosteric protein   Any protein whose activity is altered by a change of shape induced by binding a small molecule.

allozyme   One of several alternative electrophoretic forms of a protein coded by different alleles of a single gene.

alternate segregation   Segregation of a heterozygous reciprocal translocation in which both translocated chromosomes separate from both normal homologs.

amber codon   Common jargon for the UAG stop codon; an amber mutation is a mutation in which a sense codon has been altered to UAG.

Ames test   A bacterial test for mutagenicity that is used to screen for potential carcinogens.

amino acid   Any one of a class of organic molecules having an amino group and a carboxyl group; twenty different amino acids are the usual components of proteins.

amino acid attachment site   The 3′ terminus of a tRNA molecule at which an amino acid is attached.

aminoacyl tRNA synthetase   The enzyme that attaches the correct amino acid to a specific tRNA molecule.

aminopterin   An organic molecule that inhibits the major pathway of DNA synthesis.

amino terminus   The end of a polypeptide chain at which the amino acid bears a free amino group ($-NH_2$).

amniocentesis   A procedure for obtaining fetal cells from the amniotic fluid for the diagnosis of genetic abnormalities.

anaphase   The stage of mitosis or meiosis in which chromosomes move to opposite ends of the spindle; follows metaphase and precedes telophase.

aneuploid   A cell or organism in which the chromosome number is not an exact multiple of the haploid number; more generally, aneuploidy refers to a condition in which particular genes or chromosomal regions are present in extra or fewer copies compared with wildtype.

annealing   A laboratory technique by which single strands of DNA or RNA having complementary base sequences become paired to form a double-stranded molecule.

antibody   A protein, found in blood serum, produced in animals in response to a specific antigen and capable of binding to the antigen.

anticodon   The three bases in a tRNA molecule that are complementary to three bases of a specific codon in mRNA.

antigen   Any substance capable of stimulating the production of specific antibodies.

antigen presentation   An immune process in which a cell carrying an antigen bound to an MHC protein stimulates a responsive cell.

antigen-presenting cell Any cell, such as a macrophage, expressing certain products of the major histocompatibility complex and capable of stimulating T cells having appropriate antigen receptors.

antiparallel A term used to describe the chemical orientation of the two strands of a double-stranded nucleic acid molecule; the $5' \rightarrow 3'$ orientations of the two strands are opposite one another.

aporepressor A protein that is converted into a repressor when another (specific) molecule binds to it.

artificial selection Selection imposed by a breeder, in which individual organisms of only certain phenotypes are allowed to breed.

ascospore *See* ascus.

ascus A sac containing the spores (ascospores) produced by meiosis in certain groups of fungi, which include *Neurospora* and yeast.

assortative mating Nonrandom selection of mating partners with respect to one or more traits; it is positive when like phenotypes mate more frequently than would be expected by chance, and negative when the reverse occurs.

ATP Adenosine triphosphate, the primary molecule for storing chemical energy in a living cell.

attached-X chromosome A chromosome in which two X chromosomes are joined to a common centromere.

attachment site The base sequence in a bacterial chromosome at which bacteriophage DNA can integrate to form a prophage; either of the two attachment sites that flank an integrated prophage.

attenuation *See* attenuator.

attenuator A regulatory base sequence near the beginning of an mRNA molecule at which transcription can be terminated; when an attenuator is present, it precedes coding sequences.

AUG Usual initiation codon for polypeptide synthesis, coding for methionine.

autogamy Reproductive process in ciliates in which the genotype is reconstituted from the fusion of two identical haploid nuclei resulting from mitosis of a single product of meiosis; the result is homozygosity for all genes.

autonomous determination Cellular differentiation not dependent on external signals or interactions with other cells.

autopolyploidy The condition of being polyploid with more than two sets of homologous chromosomes.

autoradiography A process for the production of a photographic image of the distribution of a radioactive substance in a cell or large cellular molecule; the image is produced on a photographic emulsion by decay emission from the radioactive material.

autoregulation Regulation of synthesis of the product of a gene by the product itself.

autosomes All chromosomes other than the sex chromosomes.

auxotroph A mutant microorganism unable to synthesize a compound required for its growth but able to grow if the compound is supplied.

backcross The cross of an F_1 heterozygote with an individual organism having the same genotype as one of the parents of the F_1.

bacteriophage A virus that infects bacterial cells; commonly called a phage.

Barr body A condensed, inactivated mammalian X chromosome found in certain cells that stains darkly during interphase.

base Single-ring (pyrimidine) or double-ring (purine) component of a nucleic acid.

base analog A purine or pyrimidine that is chemically so similar to one of the normal bases that it can be incorporated into DNA.

base composition Relative proportions of the bases in a nucleic acid or of A + T versus G + C in duplex DNA.

base pair A pair of nitrogenous bases, most commonly one purine and one pyrimidine, held together by hydrogen bonds in a double-stranded region of a nucleic acid molecule; commonly abbreviated bp, the term is often used interchangeably with the term nucleotide pair.

base substitution Incorporation of an incorrect base into a DNA duplex.

B cells A class of white blood cells, derived from bone marrow, with the potential for producing antibodies.

bidirectional replication DNA replication proceeding in opposite directions from an origin of replication.

binding site A DNA or RNA base sequence that serves as the target for binding by a protein; the site on the protein that does the binding.

bivalent A pair of homologous chromosomes, each consisting of two chromatids, associated in meiosis I.

blastoderm Structure formed in the early development of an insect larva; the syncytial blastoderm is formed from repeated cleavage of the zygote nucleus without cytoplasmic division; the cellular blastoderm is formed by migration of the nuclei to the surface and their inclusion in individual cell membranes.

blastula A hollow sphere of cells formed early in development.

blood-group system A set of antigens on red blood cells resulting from the action of a series of alleles of a single gene, such as the ABO, MN, or Rh blood-group systems.

blunt ends Ends of a DNA molecule in which all terminal bases are paired; the term usually refers to termini formed by a restriction enzyme that does not produce single-stranded ends.

branched pathway A metabolic pathway in which an intermediate serves as a precursor for more than one product.

branch migration In a DNA duplex invaded by a single strand from another molecule, the process in which the size of the heteroduplex region increases by progressive displacement of the original strand.

broad-sense heritability The ratio of genotypic variance to total phenotypic variance.

cAMP *See* cyclic AMP.

cAMP-CRP complex The regulatory complex consisting of

cyclic AMP (cAMP) and the CAP protein, needed for transcription of certain operons.

cap A complex structure at the 5′ termini of most eukaryotic mRNA molecules, having a 5′-5′ linkage.

CAP protein Acronym for catabolite activator protein, the protein that binds cAMP and regulates the activity of inducible operons in prokaryotes; also called CRP, for cAMP receptor protein.

carboxyl terminus The end of a polypeptide chain at which the amino acid has a free carboxyl group (–COOH).

carcinogen A physical agent or chemical reagent that causes cancer.

carrier A heterozygote for a recessive allele that is not expressed because of the presence of the dominant allele.

cascade Series of enzyme activations serving to amplify a weak chemical signal.

cassette In yeast, each set of inactive mating-type genes that can become active by relocating to the *MAT* locus.

catabolite activator protein *See* **CAP protein.**

cDNA *See* **complementary DNA.**

cell cycle The growth cycle of an individual cell; in eukaryotes, it is subdivided into G_1 (gap 1), S (DNA synthesis), G_2 (gap 2), and M (mitosis).

cell fate The pathway of differentiation that a cell normally undergoes.

cell hybrid Product of fusion of two cells in culture, often from different species.

cell line Clone of cells derived by mitosis from a single parental cell.

cellular oncogene A gene coding for a cellular growth factor whose abnormal expression predisposes to malignancy. *See also* **oncogene.**

centromere The region of a chromosome associated with spindle fibers and participating in normal chromosome movement in mitosis and meiosis.

chain elongation The addition of successive amino acids to the growing end of a polypeptide chain.

chain initiation Process by which polypeptide synthesis is begun.

chain termination Process of ending polypeptide synthesis and releasing the polypeptide from the ribosome; a chain-termination mutation creates a new stop codon, resulting in premature termination of synthesis of a polypeptide chain.

charged tRNA A tRNA molecule to which an amino acid is linked; acylated tRNA.

chiasma The cytological manifestation of crossing-over; the cross-shaped exchange configuration between nonsister chromatids of homologous chromosomes visible in prophase I of meiosis.

chiasmata More than one chiasma. The plural *chiasmas* is also accepted.

chi-square (χ^2) A statistical quantity calculated to assess the goodness of fit between a set of observed numbers and their theoretical expectations.

chromatids The longitudinal subunits produced by chromosome replication and joined at the centromere.

chromatin The aggregate of DNA and histone proteins that makes up a eukaryotic chromosome.

chromocenter The aggregate of centromeres and adjacent heterochromatin in nuclei of *Drosophila* larval salivary-gland cells.

chromomere A tightly coiled, beadlike region of a chromosome most readily seen during pachytene of meiosis; the beads are in register in a polytene chromosome, resulting in the banded appearance of the chromosome.

chromosome In eukaryotes, a DNA molecule containing genes in linear sequence to which numerous proteins are bound and having a telomere at each end and a centromere; in prokaryotes, the DNA is associated with fewer proteins, lacks telomeres and a centromere, and is often circular; in viruses, the chromosome may be DNA or RNA, single- or double-stranded, and linear or circular, and it is usually free of bound proteins.

chromosome complement The set of chromosomes in a cell or individual organism. The haploid chromosome complement is that in a gamete formed by a diploid organism.

chromosome map A diagram showing the locations and relative spacing of genes along a chromosome.

***cis* configuration** The arrangement of linked genes in a double heterozygote in which both mutations are present in the same chromosome—for example, $a_1 a_2 / + +$; also called coupling.

***cis*-dominant mutation** A mutation that affects the expression of genes only on the same chromosome as the mutation.

***cis*-heterozygote** *See cis* configuration.

***cis-trans* test** *See* complementation test.

cistron A DNA sequence specifying a single genetic function as defined by a complementation test; a nucleotide sequence coding for a single polypeptide.

***ClB* method** A genetic procedure used to detect X-linked recessive lethal mutations in *Drosophila melanogaster*; so named because one X chromosome in the female parent is marked with an inversion (*C*), a recessive lethal allele (*l*), and the dominant allele for Bar eyes (*B*).

cleavage Mitosis in the early embryo.

clone A collection of organisms derived from a single parent and, except for acquired mutations, genetically identical with that parent; also, in genetic engineering, the linking of a specific gene or DNA fragment to a replicable DNA molecule, such as a plasmid or phage DNA.

cloned gene A DNA sequence incorporated into a vector molecule capable of replication in the same or a different organism.

cloning The process of producing cloned genes.

coding sequence Nucleotide sequence in DNA or RNA that specifies a particular sequence of amino acids in a polypeptide.

coding strand In a particular gene, the DNA strand that is transcribed.

codominance The expression of both alleles in a heterozygote.

codon A sequence of three nucleotides in an mRNA molecule specifying either an amino acid or a stop signal in protein synthesis.

coefficient of coincidence An experimental value obtained by dividing the observed number of double crossovers by the expected number calculated by assuming that the two exchanges occur independently.

cohesive end A single-stranded region at the end of a double-stranded DNA molecule that is complementary in sequence to a single-stranded region at the other end of the same molecule or at the end of a different molecule.

colchicine A chemical that prevents formation of the spindle in nuclear division.

colinearity The linear correspondence between the sequence of amino acids in a polypeptide chain and the corresponding nucleotide sequence in the DNA molecule.

colony A visible cluster of cells formed on a solid growth medium by repeated division of a single parental cell and its daughter cells.

colony hybridization assay A technique for identifying colonies containing a particular cloned gene; many colonies are transferred to a filter, lysed, and exposed to radioactive DNA or RNA complementary to the DNA sequence of interest, and then bound radioactivity, which identifies colonies containing a complementary sequence, is located by autoradiography.

combinatorial control Strategy of gene regulation in which a relatively small number of time- and tissue-specific positive and negative regulatory elements are used in various combinations to control the expression of a much larger number of genes.

combinatorial joining The DNA splicing mechanism by which antibody variability is produced, resulting in many possible combinations of V and J regions in the formation of a light-chain gene and many possible combinations of V, D, and J regions in the formation of a heavy-chain gene.

common ancestor An ancestor of both the father and the mother of an individual organism.

compartment In *Drosophila* development, a group of descendants of a small number of founder cells with a determined pattern of development.

compatible Refers to blood or tissue that can be transfused or transplanted without rejection.

complementary DNA A DNA molecule made by copying RNA with reverse transcriptase; usually abbreviated cDNA.

complementation The phenomenon in which two recessive mutations with similar phenotypes result in a wildtype phenotype when both are heterozygous in the same individual organism; complementation means that the mutations are in different genes.

complementation group A group of mutations that fail to complement one another.

complementation test A genetic test to determine whether two mutations are alleles (are present in the same functional gene).

complex A term used to refer to an ordered aggregate of molecules, as in an enzyme-substrate complex or a DNA-histone complex.

compound-X chromosome *See* **attached-X chromosome**.

conditional mutation A mutation resulting in a mutant phenotype under certain (restrictive) environmental conditions but in a wildtype phenotype in other (permissive) conditions.

congenital Present at birth.

conjugation A process of DNA transfer in sexual reproduction in certain bacteria; in *E. coli*, the transfer is unidirectional, from donor cell to recipient cell.

consanguineous mating A mating between related individual organisms.

consensus sequence A generalized base sequence derived from closely related sequences that occur in many locations in a genome or in many organisms, consisting of the most commonly occurring base at each position.

conserved sequence A base or amino acid sequence that changes very slowly in the course of evolution.

constant region The part of the heavy and light chains of an antibody molecule that has the same amino acid sequence among all antibodies derived from the same heavy-chain and light-chain genes.

constitutive Refers to synthesis of a particular mRNA molecule (and the protein it codes) at a constant rate, independent of the presence or absence of any inducer or repressor molecule.

continuous trait A trait in which the possible phenotypes occur in a continuous range from one extreme to the other rather than in discrete classes.

coordinate genes Genes that establish the basic anterior-posterior and dorsal-ventral axes of the early embryo.

coordinate regulation Control of synthesis of several proteins by a single regulatory element; in prokaryotes, the proteins are usually translated from a single mRNA molecule.

core particle The aggregate of histones and DNA in a nucleosome, without the linking DNA.

correlated response Change of the mean in one trait in a population accompanying selection for another trait.

correlation coefficient A measure of association between pairs of numbers, equaling the covariance divided by the product of the standard deviations.

cotransduction Transduction of two or more linked genetic markers by one transducing particle.

cotransformation Transformation in bacteria of two genetic markers carried on a single DNA fragment.

counterselected marker A mutation used to prevent growth of a donor cell in an Hfr $\times$ F^- bacterial mating.

coupled transcription-translation In prokaryotes, the translation of an mRNA molecule before its synthesis is completed.

coupling *See cis* configuration.

covalent bond A chemical bond in which electrons are shared.

covalently closed circle A circular double-stranded DNA molecule in which each polynucleotide strand is an uninterrupted circle.

covariance A measure of association between pairs of numbers defined as the average product of the deviations from the respective means.

C region *See* **constant region**.

crossing-over A process of exchange between nonsister chromatids of a pair of homologous chromosomes that results in the recombination of linked genes.

CRP protein *See* **CAP protein**.

cryptic splice site A potential splice site not normally used in

RNA processing unless a normal splice site is blocked or mutated.

cyclic AMP A molecule used in the regulation of cellular processes; its synthesis is regulated by glucose metabolism; its action is mediated by the CAP protein in prokaryotes and protein kinases (enzymes that phosphorylate proteins) in eukaryotes; its technical name is cyclic adenosine monophosphate, which is usually abbreviated cAMP.

cytogenetics The discipline concerned with the genetic implications of chromosome structure and behavior.

cytological map Diagrammatic representation of a chromosome.

cytoplasmic inheritance Transmission of hereditary traits through self-replicating factors in the cytoplasm; for example, mitochondria and chloroplasts.

cytoplasmic male sterility Type of pollen abortion in maize with cytoplasmic inheritance through the mitochondria.

cytosine (C) A nitrogenous pyrimidine base found in DNA and RNA.

cytotoxic T cell Type of white blood cell that participates directly in attacking cells marked for destruction by the immune system.

daughter strand The DNA, or chromosome, strand newly synthesized.

deamination Removal of an amino ($-NH_2$) group from a molecule.

deficiency *See* deletion.

degeneracy The feature of the genetic code in which an amino acid corresponds to more than one codon; also called redundancy.

degrees of freedom An integer that determines the significance level of a particular statistical test. In the goodness-of-fit type of chi-square test used in this book, in which the expected numbers are not based on any quantities estimated from the data themselves, the number of degrees of freedom equals one less than the number of classes of data.

deletion Loss of a segment of the genetic material from a chromosome; also called deficiency.

deletion mapping The use of overlapping deletions to locate a gene on a chromosome or a genetic map.

deme *See* local population.

denaturation Loss of the normal three-dimensional shape of a macromolecule without breaking covalent bonds, usually accompanied by loss of its biological activity; conversion of DNA from the double-stranded into the single-stranded form; unfolding of a polypeptide chain.

denaturation mapping An electron-microscopic technique for localizing regions of a double-stranded DNA molecule by noting the positions at which the individual strands separate in the early stages of denaturation.

deoxyribonuclease An enzyme that breaks sugar-phosphate bonds in DNA, forming either fragments or the component nucleotides; abbreviated DNase.

deoxyribonucleic acid *See* DNA.

deoxyribose The five-carbon sugar present in DNA.

depurination Removal of purine bases from DNA.

derepression Activation of a gene or set of genes by inactivation of a repressor.

deviation In statistics, a difference from an expected value.

diakinesis The substage of meiotic prophase I, immediately preceding metaphase I, in which the bivalents attain maximum shortening and condensation.

dicentric chromosome A chromosome with two centromeres.

dideoxy sequencing method Procedure for DNA sequencing in which a template strand is replicated from a particular primer sequence and terminated by the incorporation of a nucleotide containing dideoxyribose instead of deoxyribose; the resulting fragments are separated by size using electrophoresis.

differentiation The complex changes that occur in progressive diversification in cellular structure and function in the development of an organism; for a particular line of cells, this results in a continual restriction in the types of transcription and synthesis of which each cell is capable.

diploid Refers to a cell or organism with two complete sets of homologous chromosomes.

diplotene The substage of meiotic prophase I that immediately follows pachytene and precedes diakinesis, in which pairs of sister chromatids that make up a bivalent (tetrad) begin to separate from each other and chiasmata become visible.

direct repeat Copies of an identical or very similar DNA or RNA base sequence in the same molecule and in the same orientation.

discontinuous variation Variation in which the phenotypic differences for a trait fall into two or more discrete classes.

disjunction Separation of homologous chromosomes to opposite poles of a division spindle during anaphase of a mitotic or meiotic nuclear division.

displacement loop Loop formed when part of a DNA strand is dislodged from a duplex molecule because of the partner strand's pairing with another molecule.

distribution The relation giving the proportion of individuals that exhibit each possible phenotype in a population.

disulfide bond Two sulfur atoms covalently linked; found in proteins when the sulfur atoms in two different cysteines are joined.

dizygotic twins Twins that result from the fertilization of separate ova; genetically related as siblings; also called fraternal twins.

DNA The macromolecule, usually composed of two polynucleotide chains in a double helix, that is the carrier of the genetic information in all cells and many viruses.

DNA cloning *See* cloned gene.

DNA fingerprinting Electrophoretic identification of individuals using DNA probes for highly polymorphic regions of the genome, such that the genome of virtually every individual exhibits a unique pattern of bands.

DNA gyrase One of a class of enzymes called topoisomerases, which function during DNA replication to relax positive supercoiling of the DNA molecule and which introduce negative supercoiling into nonsupercoiled molecules early in the life cycle of many phages.

DNA ligase An enzyme that catalyzes formation of a covalent bond between adjacent 5′-P and 3′-OH termini in a broken polynucleotide strand of double-stranded DNA; a few DNA ligases can join two double-stranded DNA molecules.

DNA looping A mechanism by which enhancers that are removed from the immediate proximity of a promoter can still regulate transcription; the enhancer and promoter, both bound with suitable protein factors, come into physical contact by means of the looping out of the DNA between them, and the physical interaction stimulates transcription.

DNA methylase An enzyme that adds —CH₃ groups to certain bases, particularly cytosine.

DNA polymerase Any enzyme that catalyzes synthesis of DNA from deoxynucleoside 5′-triphosphates under the direction of a DNA template strand.

DNA repair Any one of several different processes for restoration of the correct base sequence of a DNA molecule into which incorrect bases have been incorporated or whose bases have been modified in some way.

DNA replication The copying of a DNA molecule.

DNase *See* **deoxyribonuclease.**

domain A folded region of a polypeptide chain that is spatially somewhat isolated from other folded regions.

dominance variance The part of the genotypic variance that results from the dominance effects of alleles affecting the trait.

dominant Refers to an allele, or the corresponding phenotypic trait, that is expressed in a heterozygote.

dosage compensation A mechanism regulating X-linked genes so that their activities are equal in males and females; in mammals, random inactivation of one X chromosome in females results in equal amounts of the products of X-linked genes in males and females.

double-Y syndrome The clinical features of the karyotype 47,XYZ.

Down syndrome The clinical features of the karyotype 47,+21 (trisomy 21).

drug-resistance plasmid A plasmid containing genes whose products inactivate certain antibiotics.

duplex DNA A double-stranded DNA molecule.

duplication A chromosome aberration in which a chromosome segment occurs more than once in the haploid genome; if the two segments are adjacent, the duplication is a tandem duplication.

editing function The activity of DNA polymerases that removes incorrectly incorporated nucleotides; also called the proofreading function.

electrophoresis A technique used to separate molecules on the basis of their different rates of movement in response to an applied electric field through a liquid or a gel; the movement is often called electrophoretic migration.

electroporation Introduction of DNA fragments into a cell by means of an electric field.

elongation Addition of amino acids to a growing polypeptide chain.

embryo An organism in the early stages of development; for human beings, the second through seventh week.

embryoid A small mass of dividing cells formed from haploid cells in anthers that can give rise to a mature haploid plant.

embryonic stem cells Cells in the blastocyst that give rise to the embryo itself, not the embryonic membranes.

endonuclease An enzyme that breaks internal phosphodiester bonds in a single- or double-stranded nucleic acid molecule; usually, specific for either DNA or RNA.

endosperm Nutritive tissues formed adjacent to the embryo in most flowering plants: in most diploid plants, the endosperm is triploid.

endosymbiont theory Theory that mitochondria and chloroplasts were originally free-living prokaryotes that invaded eukaryotes, first perhaps as parasites, later becoming symbionts.

enhancer A base sequence in eukaryotes and eukaryotic viruses that increases the rate of transcription of nearby genes; its defining characteristics are that it need not be adjacent to the transcribed gene and that its enhancing activity is independent of its orientation with respect to the gene.

enhancer trap Mutagenesis strategy in which a reporter gene with a weak promoter is introduced at many random locations in the genome; tissue-specific expression of the reporter gene identifies an insertion near a tissue-specific enhancer.

environmental variance The part of the phenotypic variance among individual organisms that is attributable to differences in environment.

enzyme A protein or ordered aggregate of proteins that catalyzes a specific biochemical reaction and is not itself altered in the process.

epistasis Interaction between nonallelic genes such that one gene interferes with or prevents expression of the other.

equilibrium centrifugation Centrifugation of molecules until there is no longer any net movement of the molecules; in a density gradient, each molecule comes to rest when its density equals that of the solution.

erythroblastosis fetalis Hemolytic disease of the newborn; blood-cell destruction when anti-Rh⁺ antibodies in a mother cross the placenta and attack Rh⁺ cells in a fetus.

estrogen A female sex hormone; of great interest in the study of cellular regulation in eukaryotes because the number of types of target cells is large.

ethidium bromide A fluorescent molecule that binds to DNA and changes its density; used to purify supercoiled DNA molecules and to localize DNA in gel electrophoresis.

euchromatin A region of a chromosome having normal staining properties and undergoing the normal cycle of condensation; relatively uncoiled in the interphase nucleus (compared with condensed chromosomes), and apparently containing most of the genes.

eukaryote A cell or an organism composed of cells with true nuclei (DNA enclosed in nuclear envelopes) and membrane-bounded cytoplasmic organelles, in which nuclear division is by mitosis or meiosis.

euploid A cell or an organism having a chromosome number that is an exact multiple of the haploid number.

evolution Cumulative change in the genetic characteristics of a species through time, resulting in greater adaptation.

excision Removal of a DNA fragment from a chromosome; prophage excision.

excisionase An enzyme that is needed for prophage excision; works together with an integrase.

excision repair Type of DNA repair in which segments of a DNA strand that are damaged are removed enzymatically and then resynthesized, using the other strand as a template.

execution mutation A developmental mutation in which affected cells are unable to carry out their normal fate.

exon The sequences in a gene that are retained in the messenger RNA after the introns are removed from the primary transcript.

exon shuffle Theory in which new genes can evolve by the assembly of individual exons from preexisting genes, each coding for a discrete functional domain in the new protein.

exonuclease An enzyme that removes a terminal nucleotide in a polynucleotide chain by cleavage of the terminal phosphodiester bond; nucleotides are removed successively, one by one; usually specific for either DNA or RNA and for either single-stranded or double-stranded nucleic acids; a $5' \rightarrow 3'$ exonuclease cleaves successive nucleotides from the 5' acids; a $5' \rightarrow 3'$ exonuclease cleaves successive nucleotides from the 5' end of the molecule, a $3' \rightarrow 5'$ cleaves successive nucleotides from the 3' end.

expressivity The degree of phenotypic expression of a gene.

extranuclear inheritance Inheritance mediated by self-replicating cellular factors located outside the nucleus; for example, in mitochondria or chloroplasts.

fate map A diagram of the insect blastoderm identifying the regions from which particular adult structures derive.

feedback inhibition Inhibition of an enzyme by the product of the enzyme or, in a metabolic pathway, by a product of the pathway.

fertility restorer A suppressor of cytoplasmic male sterility.

F_1 generation The first generation of descent from a given mating.

F_2 generation The second generation produced by inter-crossing or self-fertilizing F_1 progeny.

first-division segregation Separation of a pair of alleles into different nuclei in the first meiotic division; associated with the absence of crossing-over between the gene and the centromere in a particular cell.

first meiotic division The meiotic division that reduces the chromosome number; sometimes called the reduction division.

fitness A measure of the average ability of organisms with a given genotype to survive and reproduce.

fixed allele An allele whose allele frequency equals 1.0.

fluctuation test A statistical test used to determine whether bacterial mutations occur at random or are produced in response to selective agents.

folded chromosome The form of DNA in a bacterial cell in which the circular DNA is folded to have a compact structure; contains protein.

forward mutation A change from a wildtype allele to a mutant allele.

founder effect Random genetic drift resulting when a group of founders of a population are not genetically representative of the population from which they were derived.

F plasmid A bacterial plasmid—often called the F factor, fertility factor, or sex plasmid—capable of transferring itself from a host (F^+) cell to a cell not carrying an F factor (F^- cell); when an F factor is integrated into the bacterial chromosome (in an Hfr cell), the chromosome becomes transferrable to an F^- cell during conjugation.

F′ plasmid An F plasmid that contains genes obtained from the bacterial chromosome in addition to plasmid genes; formed by aberrant excision of an integrated F factor, taking along adjacent bacterial DNA.

frameshift mutation A mutational event caused by either the insertion or the deletion of one or more nucleotide pairs in a gene, resulting in a shift in the reading frame of all codons following the mutational site.

fraternal twins Twins that result from the fertilization of separate ova; genetically related as siblings; also called dizygotic twins.

free radical A highly reactive molecule produced when ionizing radiation interacts with water; free radicals are potent oxidizing agents.

frequency of recombination Proportion of gametes containing recombinant chromosomes.

G_1 See cell cycle.

G_2 See cell cycle.

gain-of-function mutation Mutation in which a gene is overexpressed or inappropriately expressed.

β-galactosidase The enzyme produced by a gene in the *lac* operon responsible for cleavage of lactose.

gamete A mature reproductive cell, such as sperm or egg in animals.

gametophyte The haploid, gamete-producing generation in plants (reduced in higher plants to the embryo sac and the pollen grain), which alternates with the diploid, spore-producing generation (sporophyte).

gap genes Genes that control the development of contiguous segments or parasegments in *Drosophila*, so that mutations result in gaps in the pattern of segmentation.

gastrula Stage in early animal development marked by extensive cell migration.

gel electrophoresis See electrophoresis.

gene The hereditary unit that contains genetic information that is transcribed into RNA and processed into an RNA molecule that functions directly or is translated into a polypeptide chain; a gene can mutate to various forms.

gene amplification A process in which certain genes undergo differential replication either within the chromosome or extrachromosomally, increasing the number of copies of the gene.

gene cloning See cloned gene.

gene conversion The phenomenon in which the products of a meiotic division in an *Aa* heterozygous individual are in some ratio other than the expected 1A : 1a—for example, 3A : 1a, 1A : 3a, 5A : 3a, or 3A : 5a.

gene dosage   Number of gene copies.

gene expression   The multistep process by which a gene is regulated and its product synthesized.

gene flow   Exchange of genes among populations, as a result of either dispersal of gametes or migration of individual organisms; also called migration.

gene library   A large collection of cloning vectors containing a complete (or nearly complete) set of fragments of the genome of an organism.

gene pool   The totality of genetic information in a population of organisms.

gene product   A term used for the polypeptide chain translated from an mRNA molecule transcribed from a gene; if the RNA is not translated (for example, ribosomal RNA), the RNA molecule is the gene product.

general transcription factor   A protein molecule needed to bind with a promoter before transcription can occur; transcription factors are necessary, but not sufficient, for transcription, and they are shared among many different promoters.

generalized transduction   *See* **transducing phage.**

gene regulation   Processes by which gene expression is controlled in response to external or internal signals.

gene therapy   Deliberate alteration of the human genome for alleviation of disease.

genetic code   The set of sixty-four triplets of bases (codons) corresponding to each amino acid and to signals for initiation and termination of polypeptide synthesis.

genetic differentiation   Accumulation of differences in allele frequency between isolated or semiisolated populations.

genetic divergence   *See* **genetic differentiation.**

genetic engineering   Linking two DNA molecules by *in vitro* manipulations for the purpose of creating a novel organism with desired characteristics.

genetic equilibrium   The condition in which the frequencies of particular alleles in a population remain constant in successive generations; also called equilibrium.

genetic map   *See* **linkage map.**

genome   The total complement of genes contained in a cell or virus; commonly used to refer to all genes present in one complete haploid set of chromosomes in eukaryotes.

genotype   The genetic constitution of an organism or virus, as distinguished from its appearance or phenotype; often used to refer to the allelic composition of one or a few genes of interest.

genotype-environment association   The condition in which genotypes and environments are not in random combinations.

genotype-environment interaction   The condition in which genetic and environmental effects on a trait are not additive.

genotype frequency   The proportion of individual members of a population that are of a prescribed genotype.

genotypic variance   The part of the phenotypic variance attributable to differences in genotype among individuals.

germ cell   Cell that gives rise to reproductive cells.

germinal mutation   A mutation occurring in a cell from which gametes are derived, as distinguished from a somatic mutation.

germ line   Cell lineage consisting of germ cells.

goodness of fit   Extent to which observed numbers agree with the numbers expected based on some specified genetic hypothesis.

guanine (G)   A nitrogenous purine base found in DNA and RNA.

gynandromorph   A sexual mosaic; an individual organism exhibiting both male and female sexual differentiation.

H1, H2A, H2B, H3, H4   The five major histones in chromatin.

haploid   A cell or organism having only one set of chromosomes.

haplotype   The allelic form of each of a set of linked genes present in a single chromosome, particularly genes of the major histocompatibility complex.

Hardy-Weinberg principle   The genotype frequencies expected with random mating.

HAT medium   A growth medium for animal cells, containing hypoxanthine, aminopterin, and thymidine, used in the selection of hybrid cells.

H chain   *See* **heavy chain.**

heavy (H) chain   One of the larger polypeptide chains in an antibody molecule.

helicase   An enzyme that participates in DNA replication by unwinding double-stranded DNA in or near the replication fork.

α helix   A fundamental unit of protein folding in which successive amino acids form a right-handed helical structure held together by hydrogen-bonding between the amino and carboxyl components of the peptide bonds in successive loops of the helix.

helix-turn-helix   A structural motif found in many DNA-binding proteins.

helper T cell   A type of white blood cell that participates in activation of other cells in the immune system.

hemizygous gene   A gene present in only one dose, such as the genes on the X chromosome in XY males.

hemophilia A   One of two X-linked forms of hemophilia; deficient in blood-clotting factor VIII.

heritability   A measure of the degree to which a phenotypic trait can be modified by selection. *See also* **broad-sense heritability** and **narrow-sense heritability.**

heterochromatin   Chromatin that remains condensed and can be heavily stained during interphase; commonly present adjacent to the centromere and in the telomeres of chromosomes; some chromosomes are composed primarily of heterochromatin.

heterochronic mutation   A mutation in which an otherwise normal gene is expressed at the wrong time.

heteroduplex region   Region of a double-stranded nucleic acid molecule in which the two strands have different hereditary origins; produced either as an intermediate in recombination or by the *in vitro* annealing of single-stranded complementary molecules.

heterogametic   Refers to the production of dissimilar gametes with respect to the sex chromosomes; in most animals, the male is the heterogametic sex but, in birds, moths, butterflies, and some reptiles, it is the female.

heterogeneous nuclear RNA (hnRNA) The collection of primary RNA transcripts and incompletely processed products found in the nucleus of a eukaryotic cell.

heterokaryon A cell or individual organism having nuclei from genetically different sources, the result of cell fusion not accompanied by nuclear fusion.

heterosis Superiority of hybrids over either inbred parent with respect to one or more traits; also called hybrid vigor.

heterozygote superiority The condition in which a heterozygous genotype has a greater fitness than either of the homozygotes.

heterozygous Refers to having dissimilar alleles of one or more genes; not homozygous.

hexaploid A cell or organism with six complete sets of chromosomes.

Hfr An *E. coli* cell in which an F plasmid is integrated into the chromosome, enabling transfer of part or all of the chromosome to an F^- cell.

high-copy-number plasmid A plasmid for which there are usually considerably more than two copies (often more than twenty) per cell.

histocompatability Acceptance by a recipient of transplanted tissue from a donor.

histocompatibility antigens Tissue antigens that determine transplant compatibility or incompatibility.

histocompatibility genes Genes coding for histocompatibility antigens.

histone Any of the small basic proteins bound to DNA in chromatin; the five major histones are designated H1, H2A, H2B, H3, and H4.

Holliday model A molecular model of genetic recombination in which the participating duplexes contain heteroduplex regions of the same length.

homeobox A DNA sequence motif found in the coding region of many genes important in early development in animals. The amino acid sequence corresponding to the homeobox has a helix-turn-helix structure.

homeotic genes Genes that determine fundamental patterns of development. *See also* **homeotic mutation.**

homeotic mutation A mutation that results in the replacement of one body structure by another body structure in the course of development.

homeotic selector genes Genes that regulate the expression of homeotic genes.

homogametic Producing only one kind of gamete with respect to the sex chromosomes. *See also* **heterogametic.**

homologous In reference to DNA, having the same or nearly the same nucleotide sequence.

homologous chromosomes Chromosomes that pair in meiosis and have the same genetic loci and structure; also called homologs.

homologous recombination Genetic exchange between identical or nearly identical DNA sequences.

homothallism The capacity of cells in certain fungi to undergo a conversion in mating type to enable mating between cells produced by the same parental organism.

homozygous Refers to having the same allele of a gene in homologous chromosomes.

hormone A small molecule in higher eukaryotes, synthesized in specialized tissue, that regulates the activity of other specialized cells; in animals, hormones are transported from their source to target tissue by the blood stream.

hot spot A site in a DNA molecule at which the mutation rate is much higher than the rate for most other sites.

housekeeping genes Genes expressed at the same level in virtually all cells whose products participate in basic metabolic processes.

H substance The carbohydrate precursor of the A and B red-blood-cell antigens.

Huntington disease Dominantly inherited degeneration of the neuromuscular system; onset is in middle age.

hybrid An individual organism produced by the mating of genetically unlike parents; a duplex nucleic acid molecule produced from strands derived from different sources.

hybridoma A cell hybrid between an antibody-producing B cell and certain types of tumor cells; hybridomas survive well in laboratory culture and continue to produce monoclonal antibody.

hybrid vigor *See* **heterosis.**

hydrogen bond A weak noncovalent linkage in which a hydrogen atom is shared by two atoms.

hydrophobic interaction A noncovalent interaction between nonpolar molecules or nonpolar groups, causing the molecules or groups to cluster when water is present.

identical twins *See* **monozygotic twins.**

imaginal disk Structures present in the body of insect larvae from which the adult structures develop during pupation.

immune response Tendency to develop resistance to a disease-causing agent after an initial infection.

immunity A general term for resistance of an organism to specific substances, particularly agents of disease.

immunoglobulin One of several classes of antibody protein.

immunoglobulin class The category in which an immunoglobulin is placed based on its chemical characteristics, including the identity of its heavy chain.

inborn error of metabolism A genetically determined biochemical disorder, usually in the form of an enzyme defect that produces a metabolic block.

inbreeding Mating between genetically related individual organisms.

inbreeding coefficient A measure of the genetic effects of inbreeding in terms of the proportionate reduction in heterozygosity in an inbred individual organism compared with the heterozygosity expected from random mating.

inbreeding depression Deterioration in fitness or performance of a population accompanying inbreeding.

incompatible Refers to blood or tissue whose transfusion or transplantation results in rejection.

incomplete penetrance Condition in which a mutant phenotype is not expressed in individual organisms having the mutant genotype.

independent assortment Random distribution of unlinked genes into gametes, as with genes in different (nonhomologous) chromosomes.

individual selection Selection based on each individual organism's own phenotype.

inducer A small molecule that inactivates a repressor, usually by binding to it and thereby altering the ability of the repressor to bind to an operator.

inducible Refers to a gene that is expressed, or an enzyme that is synthesized, only in the presence of an inducer molecule. *See also* **constitutive.**

induction Activation of an inducible gene; prophage induction is the derepression of a prophage that initiates a lytic cycle of phage development.

initiation The beginning of protein synthesis.

inosine One of a number of unusual bases found in transfer RNA.

insertional inactivation Inactivation of a gene by interruption of its coding sequence; used in genetic engineering as a means of detecting insertion of a foreign DNA sequence into the coding region of a gene.

insertion sequence A DNA sequence capable of transposition in a prokaryotic genome; such sequences usually code for their own transposase.

in situ **hybridization** Renaturation of a radioactive probe nucleic acid to the DNA or RNA in a cell; used to localize particular DNA molecules to chromosomes or parts of chromosomes and to identify the time and tissue distribution of RNA transcripts; sometimes called cytological hybridization.

integrase An enzyme that catalyzes the site-specific exchange when a prophage is inserted into or excised from a bacterial chromosome; in the excision process, an accessory protein, excisionase, also is needed.

integration The process by which one DNA molecule is inserted intact into another replicable DNA molecule, as in prophage integration and integration of plasmid or tumor viral DNA into a chromosome.

intercalation Insertion of a planar molecule between the stacked bases in duplex DNA.

interference The tendency for a crossover to inhibit the occurrence of other crossovers nearby.

intergenic complementation Complementation between mutations in different genes. *See also* **complementation test.**

intergenic suppression Suppression of mutant phenotype because of a mutation in a different gene; often refers specifically to a tRNA molecule able to recognize a stop codon or two different sense codons.

interphase The interval between nuclear divisions in the cell cycle, extending from the end of telophase of one division to the beginning of prophase of the next division.

interrupted-mating technique In an Hfr × F⁻ cross, a technique by which donor and recipient cells are broken apart at specific times, allowing only a particular amount of DNA to be transferred.

intervening sequence *See* **intron.**

intragenic complementation Apparent complementation between different mutations in the same gene. *See also* **complementation test.**

intragenic suppression Suppression of a mutant phenotype because of a second mutation in the same gene.

intron A transcribed noncoding DNA sequence in a gene that is excised from the primary transcript in forming a mature mRNA molecule; found primarily in eukaryotic cells. *See also* **exon.**

inversion A structural aberration in a chromosome in which the order of several genes is reversed from the normal order; a pericentric inversion includes the centromere within the inverted region, and a paracentric inversion does not include the centromere.

inverted repeat One of a pair of base sequences present in the same molecule that are identical or nearly identical but oriented in opposite directions; often found at the ends of transposable elements.

in vitro **experiment** An experiment carried out with components isolated from cells.

in vivo **experiment** An experiment performed with intact cells.

ionizing radiation Electromagnetic or particulate radiation that produces ion pairs when dissipating its energy in matter.

IS element *See* **insertion sequence.**

isochromosome A chromosome with two identical arms containing homologous genes.

isotopes The forms of a chemical element having the same number of electrons and protons but differing in the number of neutrons in the atomic nucleus; unstable isotopes undergo transitions to a more-stable state and, in so doing, emit radioactivity.

J (joining) region One of multiple DNA sequences coding for alternative amino acid sequences of part of the variable region of an antibody molecule. The J regions of heavy and light chains are different.

kappa particle An intracellular parasite in *Paramecium* that releases a substance capable of killing sensitive cells.

karyotype The chromosome complement of a cell or an individual organism; often represented by an arrangement of metaphase chromosomes according to their lengths and to the positions of their centromeres.

kindred A group of related individuals.

Klinefelter syndrome The clinical features of human males with the karyotype 47,XXY.

lac **operon** The set of genes required to metabolize lactose in bacteria.

lac **repressor** A protein coded by the *lacI* gene that in the absence of lactose binds with the operator and prevents transcription.

lactose A twelve-carbon sugar consisting of the simple sugars glucose and galactose covalently linked.

lactose permease An enzyme responsible for transport of lactose from the environment into bacteria.

lagging strand The single DNA strand synthesized in short fragments that are ultimately joined together.

lariat structure Structure of an intron immediately after excision in which the 5′ end loops back and forms a 5′−2′ linkage with another nucleotide.

L chain *See* **light chain.**

leader sequence   The region of an mRNA molecule from the 5′ end to the beginning of the coding sequence, sometimes containing regulatory sequences; in prokaryotic mRNA, it contains the ribosomal binding site.

leading strand   The single DNA strand that is synthesized as a continuous unit.

leptotene   The initial substage of meiotic prophase I, during which the chromosomes become visible in the light microscope as unpaired threadlike structures.

lethal mutation   A mutation that results in the death of an affected individual organism.

liability   Risk, particularly toward a threshold type of quantitative trait.

light (L) chain   One of the small polypeptide chains in an antibody molecule.

lineage   The ancestry of a cell during development or of a species during evolution.

lineage diagram   A diagram of cell lineages and their developmental fates.

linkage   The tendency of genes located in the same chromosome to be associated in inheritance more frequently than expected from their independent assortment in meiosis.

linkage group   The set of genes present in a single chromosome of a species.

linkage map   A chromosome map showing the relative locations of the known genes on the chromosomes of a given species; also called a genetic map.

linker   In genetic engineering, synthetic DNA fragments that contain restriction-enzyme cleavage sites and that are used to join two DNA molecules.

local population   A group of individual organisms of the same species occupying an area within which most individuals find their mates; synonymous terms are deme and Mendelian population.

locus   The site or position of a particular gene on a chromosome.

loss-of-function mutation   A mutation that eliminates gene function; also called a null mutation.

lysis   Breakage of a cell caused by rupture of its cell membrane and cell wall.

lysogen   Clone of bacterial cells that have acquired a prophage.

lysogenic cycle   In temperate bacteriophage, the phenomenon in which the DNA of an infecting phage is repressed and stably becomes part of the genetic material of the cells.

lysozyme   One of a class of enzymes that dissolve the cell wall of bacteria; found in chicken egg white and human tears, and coded by many phages.

lytic cycle   The life cycle of a phage in which progeny phages are produced and the host bacterial cell is lysed.

M   *See* cell cycle.

macrophage   One of a class of white blood cells that processes antigens for presentation as a necessary step in stimulation of the immune response.

major histocompatibility complex (MHC)   The group of closely linked genes coding for antigens that play a major role in tissue incompatibility and that function in regulation and other aspects of the immune response.

map unit   A unit of distance in a linkage map that corresponds to a recombination frequency of 1 percent.

masked mRNA   Messenger RNA present in eukaryotic cells, particularly eggs, that cannot be translated until specific regulatory substances are available; storage mRNA.

maternal effect   A phenomenon in which the genotype of a mother affects the phenotype of the offspring through substances present in the cytoplasm of the egg.

maternal-effect genes   Genes that influence early development through their expression in the mother and presence of their products in the oocyte.

maternal inheritance   Extranuclear inheritance of a trait through cytoplasmic factors or organelles contributed by the female gamete.

maternal sex ratio   Aberrant sex ratio in certain *Drosophila* species caused by cytoplasmic transmission of a parasite that is lethal to male embryos.

mating system   The norms by which individual members of a population choose their mates; important systems of mating include random mating, assortative mating, and inbreeding.

mating-type interconversion   Phenomenon in homothallic yeast in which cells switch mating type as a result of transposition of genetic information from an unexpressed cassette into the active mating-type locus.

Maxam-Gilbert method   A technique for determining the nucleotide sequence of DNA by means of strand cleavage at the positions of particular nucleotides.

mean   The arithmetic average.

meiocyte   A germ cell that undergoes meiosis to yield gametes in animals or spores in plants.

meiosis   The process of nuclear division in gametogenesis or sporogenesis in which one replication of the chromosomes is followed by two successive divisions of the nucleus to produce four haploid nuclei.

Mendelian population   *See* local population.

meristem   The mitotically active growing point of plant tissue.

meristic trait   A trait in which the phenotype is determined by counting, such as number of ears on a stalk of corn or number of eggs laid by a hen.

messenger RNA (MRNA)   An RNA molecule transcribed from a DNA sequence and translated into the amino acid sequence of a polypeptide. In eukaryotes, the primary transcript undergoes elaborate processing to become the mRNA.

metabolic pathway   A set of chemical reactions taking place in a definite order to convert a particular starting molecule into one or more specific products.

metacentric chromosome   A chromosome with its centromere about in the middle so that the arms are equal or almost equal in length.

metaphase   The state of nuclear division in mitosis, meiosis I, or meiosis II, during which the centromeres of the condensed chromosomes are arranged in a plane between the two poles of the spindle.

methylation The modification of a DNA or RNA base by the addition of a methyl (—CH$_3$) group.

MHC *See* **major histocompatibility complex.**

migration Movement of individual organisms among subpopulations; also, the movement of molecules in electrophoresis. *See also* **branch migration** for another use of the term.

minimal medium A growth medium consisting of simple inorganic salts, a carbohydrate, vitamins, organic bases, essential amino acids, and other essential compounds; its composition is precisely known; the term minimal medium contrasts with complex medium or broth, which is an extract of biological material (vegetables, milk, meat) containing a large number of compounds, the composition of which is unknown.

mismatch An arrangement in which two nucleotides opposite one another in double-stranded DNA are unable to form hydrogen bonds.

mismatch repair Removal of one nucleotide from a pair that cannot properly hydrogen-bond, followed by replacement with a nucleotide that can hydrogen-bond.

missense mutation An alteration in a coding sequence of DNA that results in an amino acid replacement in the polypeptide.

mitosis The process of nuclear division in which the replicated chromosomes divide and the daughter nuclei have the same chromosome number and genetic composition as the parent nucleus.

mitotic spindle The set of fibers arching through the cell during mitosis to which the chromosomes are attached. *See also* **spindle.**

monoclonal antibody Antibody directed against a single antigen, produced by a single clone of B cells or a single cell line of hybridoma cells.

monohybrid An individual organism heterozygous for one pair of alleles; the offspring of a cross between individual organisms homozygous for different alleles of a gene.

monomorphic gene A gene for which the most-common allele has a frequency greater than 0.95.

monoploid The chromosome complement of a gamete.

monosomic Condition of an otherwise diploid organism in which one member of a pair of chromosomes is missing.

monozygotic twins Twins developed from a single fertilized egg that splits into two embryos at an early division; also called identical twins.

morphogen Substance that induces differentiation.

mosaic An individual organism composed of two or more genetically different types of cells.

mRNA *See* **messenger RNA.**

multifactorial trait A trait determined by the combined action of many factors, typically some genetic and some environmental.

multiple alleles The occurrence in a population of more than two alleles of a gene.

multiplication rule The principle that the probability of simultaneous occurrence of each of a set of independent events equals the product of the individual probabilities of the events.

multivalent An association of more than two homologous chromosomes resulting from synapsis in meiosis in a polysomic or polyploid individual.

mutagen An agent that is capable of increasing the rate of mutation.

mutagenesis The process by which a gene undergoes a heritable alteration; also called mutation.

mutant An allele or heritable phenotype that is different from wildtype; also the individual organism in which a mutant allele is expressed.

mutation A heritable alteration in a gene; the process by which such changes occur.

mutation pressure The generally very weak tendency of mutation to change allele frequency.

mutation rate The probability of occurrence of a new mutation in a particular gene, either per gamete or per generation.

narrow-sense heritability The fraction of the phenotypic variance revealed as resemblance between parents and offspring; technically, the ratio of the additive genetic variance to the total phenotypic variance.

natural selection The process of evolutionary adaptation in which individual organisms genetically best suited to survive and reproduce in a particular environment give rise to a disproportionate share of the offspring and so gradually increase the overall ability of the population to survive and reproduce in that environment.

negatively supercoiled Refers to having DNA supercoils that counteract underwinding of the DNA.

negative regulation Regulation of gene expression in which mRNA is not synthesized until a repressor is removed from the DNA of the gene.

neutral allele An allele that has a negligible effect on the ability of an organism to survive and reproduce.

neutral petites A cytoplasmically inherited deficiency in yeast mitochondrial DNA resulting in small colonies.

nick A single-strand break in a DNA molecule.

nicked circle A circular DNA molecule containing one or more single-strand breaks.

nondisjunction Failure of chromosomes to separate (disjoin) and move to opposite poles of the division spindle; the result is loss or gain of a chromosome.

nonhistone chromosomal proteins A large class of proteins, not of the histone class, found in isolated chromosomes.

nonparental ditype An ascus containing two pairs of recombinant spores.

nonpermissive conditions Environmental conditions that result in expression of the phenotype of a conditional mutation.

nonselective medium A growth medium that allows growth of wildtype and one or more mutant genotypes.

nonsense mutation A mutation that changes a codon specifying an amino acid into a stop codon, resulting in premature polypeptide-chain termination; also called chain-termination mutation.

normal distribution A symmetric bell-shaped distribution characterized by the mean and the variance; in a normal distribution, approximately 68 percent of the observations are within one standard deviation from the mean and approximately 95 percent within two standard deviations.

nuclease An enzyme that breaks phosphodiester bonds in nucleic acid molecules.

nucleic acid A polymer composed of repeating units of phosphate-linked five-carbon sugars to which nitrogenous bases are attached. *See also* **DNA** and **RNA.**

nucleoid A DNA mass, not bounded by a membrane, within the cytoplasm of a prokaryotic cell, chloroplast, or mitochondrion; often refers to the major DNA unit in a bacterium.

nucleolar organizer region A chromosome region containing the genes for ribosomal RNA; abbreviated NOR.

nucleoli Nuclear organelles in which ribosomal RNA is made and ribosomes are partly synthesized; usually associated with the nucleolar organizer region; several can fuse to form a single nucleolus.

nucleoside A purine or pyrimidine base covalently linked to a sugar.

nucleosome The basic repeating subunit of chromatin, consisting of a core particle composed of two molecules each of four different histones, around which a length of DNA containing about 145 nucleotide pairs is wound, joined to an adjacent core particle by about 55 nucleotide pairs of linker DNA associated with a fifth type of histone.

nucleotide A nucleoside phosphate.

nucleus The organelle bounded by a membranous envelope that contains the chromosomes in a eukaryotic cell.

nullisomic Refers to a cell or individual organism containing no copies of a particular chromosome.

ochre codon Jargon for the UAA stop codon; an ochre mutation is a UAA codon formed from a sense codon.

octoploid A cell or an organism with eight complete sets of chromosomes.

Okazaki fragment One of the short strands of DNA produced during discontinuous replication of the lagging strand; also called precursor fragment.

oligonucleotide A short, single-stranded nucleic acid, usually synthesized for use in DNA sequencing as a primer in the polymerase chain reaction, as a probe, or for oligonucleotide site-directed mutagenesis.

oligonucleotide site-directed mutagenesis Replacement of a small region in a gene with an oligonucleotide containing a specified mutation.

oncogene A gene that can initiate tumor formation.

operator A regulatory region in DNA that interacts with a specific repressor protein in controlling the transcription of adjacent structural genes.

operon A collection of adjacent structural genes regulated by an operator and a repressor.

organelle A membrane-bounded cytoplasmic structure having a specialized function, such as a nucleus, chloroplast, or mitochondrion.

origin A DNA base sequence at which replication of a molecule is initiated.

overdominance A condition in which the fitness of a heterozygote is greater than the fitness of both homozygotes.

overlapping genes Genes that share part of their coding sequences.

ovule The structure in seed plants that contains the embryo sac (female gametophyte) and develops into a seed after fertilization of the egg.

pachytene The middle substage of meiotic prophase I in which the homologous chromosomes are closely synapsed and the synaptonemal complex is fully formed.

pair-rule genes Genes active early in *Drosophila* development that specify the fates of alternating segments or parasegments. Mutations in pair-rule genes result in loss of even-numbered or odd-numbered segments or parasegments.

palindrome In nucleic acids, a segment of DNA in which the sequence of bases on complementary strands reads the same from a central point of symmetry—for example, GAATTC; frequently, the sites of recognition and cleavage by restriction endonucleases are palindromic.

paracentric inversion An inversion that does not include the centromere.

parasegment Developmental unit in *Drosophila* consisting of the posterior part of one segment and the anterior part of the next segment in line.

parental ditype An ascus containing two pairs of nonrecombinant spores.

partial digestion Condition in which a restriction enzyme cleaves some, but not all, of the restriction sites present in a DNA molecule.

partial diploid A cell in which a segment of the genome is duplicated, usually in a plasmid.

partial dominance A condition in which the phenotype of the heterozygote is intermediate between the corresponding homozygotes but more closely resembles one than the other.

Pascal's triangle Triangular configuration of integers in which the nth row gives the binomial coefficients in the expansion of $(x + y)^{(n-1)}$. The first and last numbers in each row equal 1, and the others equal the sum of the adjacent numbers in the row immediately above.

paternal inheritance Extranuclear inheritance of a trait through cytoplasmic factors or organelles contributed by the male gamete.

pathogen An organism that causes disease.

PCR *See* polymerase chain reaction.

pedigree A diagram representing the familial relationships of individuals.

penetrance The proportion of individual organisms having a particular genotype that express the corresponding phenotype; if all individuals express the phenotype, penetrance is complete, otherwise, it is incomplete.

peptide bond A covalent bond between the amino ($-NH_2$) group of one amino acid and the carboxyl ($-COOH$) group of another.

peptidyl transferase The enzymatic activity of ribosomes responsible for forming a peptide bond; the active site is formed from various regions of several ribosomal proteins.

pericentric inversion An inversion that includes the centromere.

permissive condition An environmental condition in which the phenotype of a conditional mutation is not expressed; the term contrasts with nonpermissive conditions.

petite mutants Mutations in yeast resulting in slow growth and small colonies.

5′-P group The end of a DNA or RNA strand that terminates in a free phosphate group not connected to a sugar farther along.

phage *See* bacteriophage.

phage-attachment site The base sequence in a bacterial chromosome at which bacteriophage DNA can integrate to form a prophage.

phage repressor Regulatory protein that prevents transcription of genes in a prophage.

phenotype The observable properties of a cell or an organism, resulting from the interaction of the genotype and the environment.

phenotypic variance The total variance in a phenotypic trait among individual members of a population.

phenylketonuria A hereditary human condition resulting from inability to convert phenylalanine into tyrosine; causes severe mental retardation unless treated in childhood by a low-phenylalanine diet; abbreviated PKU.

Philadelphia chromosome Abnormal human chromosome 22, resulting from reciprocal translocation and often associated with a certain type of leukemia.

phosphodiester bond In nucleic acids, the covalent bond formed between the phosphate group of one nucleotide and the 3′-OH group of the next nucleotide in line; these bonds form the backbone of a nucleic acid molecule.

photoreactivation The enzymatic splitting of pyrimidine dimers produced in DNA by ultraviolet light; requires visible light and the photoreactivation enzyme.

plaque A clear area in an otherwise turbid layer of bacteria growing on a solid medium, caused by the infection and killing of the cells by a phage; because each plaque is a result of the growth of one phage, plaque counting is a way of counting viable phage particles; the term is used occasionally for animal viruses that cause clear areas in layers of animal cells grown in culture.

plasmid An extrachromosomal genetic element that replicates independently of the host chromosome; it may exist in one or many copies per cell and may segregate in cell division to daughter cells in either a controlled or random fashion; some plasmids, such as the F factor, may be integrated into the host chromosome.

pleiotropy The condition in which a single mutant gene affects two or more distinct and seemingly unrelated traits.

point mutation A mutation caused by the substitution, deletion, or addition of a single nucleotide pair; sometimes used to designate a mutation that can be mapped to a single specific locus.

pole cells Cells set off at the posterior end of the *Drosophila* embryo from which the germ cells are derived.

poly-A tail The sequence of adenines added to the 3′ end of many eukaryotic mRNA molecules in processing.

polycistronic mRNA An mRNA molecule from which two or more polypeptides are translated; found primarily in prokaryotes.

polygenic inheritance Determination of a trait by alleles of two or more genes.

polymer A regular, covalently bonded arrangement of basic subunits or monomers into a large molecule, such as a polynucleotide or polypeptide chain.

polymerase An enzyme that catalyzes covalent joining of nucleotides—for example, DNA polymerase and RNA polymerase.

polymerase chain reaction Repeated cycles of DNA denaturation, renaturation with primer oligonucleotide sequences, and replication, resulting in exponential growth in the number of copies of the DNA sequence located between the primers and replicated.

polymorphic gene A gene for which the most-common allele in a population has a frequency smaller than 0.95.

polymorphism The presence in a population of two or more relatively common forms of a gene, chromosome, or genetically determined trait.

polynucleotide chain A single-stranded molecule consisting of nucleotides covalently linked end to end.

polypeptide chain A polymer of amino acids linked together by peptide bonds.

polyploidy The condition of a cell or organism with more than two complete sets of chromosomes.

polyprotein A protein molecule that can be cleaved to form two or more finished protein molecules.

polyribosome *See* polysome.

polysome A complex of two or more ribosomes associated with an mRNA molecule and actively engaged in polypeptide synthesis; a polyribosome.

polysomy The condition of a diploid cell or organism having three or more copies of a particular chromosome.

polytene chromosome A giant chromosome consisting of many identical strands laterally apposed and in register, exhibiting a specific pattern of transverse banding.

population A group of organisms of the same species.

population genetics Application of Mendel's laws and other principles of genetics to entire populations of organisms.

population structure Description of the manner in which a population is subdivided into local populations and the patterns and rates of migration among them.

population subdivision Organization of a population into smaller breeding groups between which migration is restricted.

positional information Developmental signals transmitted to a cell by virtue of its position in the embryo.

position effect A change in the expression of a gene depending on its position within the genome.

positive regulation Mechanism of gene regulation in which an element must be bound to DNA in an active form in order for transcription to occur; positive regulation contrasts with negative regulation, in which a regulatory element must be removed from DNA.

postmeiotic segregation Segregation of genetically different products in a mitotic division following meiosis, as in the formation of a pair of ascospores having different genotypes in *Neurospora*.

postreplication repair Any DNA repair process that takes place in a region of a DNA molecule after the replication fork has moved some distance past that region or in nonreplicating DNA.

precursor fragment *See* **Okazaki fragment.**

prediction equation In quantitative genetics, an equation used to predict the improvement in mean performance of a population by means of artificial selection; always includes heritability as one component.

Pribnow box A base sequence in prokaryotic promoters to which RNA polymerase binds in an early step of initiating transcription.

primary transcript An RNA copy of a gene; in eukaryotes, the transcript must be processed to form a translatable mRNA molecule.

primase The enzyme responsible for synthesizing the RNA primer needed for initiating DNA synthesis.

primer In nucleic acids, a short RNA or single-stranded DNA segment that functions as a growing point in polymerization.

primer oligonucleotides Single-stranded DNA molecules, typically from eighteen to twenty-two nucleotides in length, that can hybridize with a longer DNA strand and serve as a primer for replication; in the polymerase chain reaction, two primers anneal to opposite strands of a DNA duplex with their 3′-OH ends facing. *See also* **polymerase chain reaction.**

probability A mathematical expression of the degree of confidence that certain events will or will not occur.

probe A radioactive DNA or RNA molecule used in DNA-RNA or DNA-DNA hybridization assays.

processing A series of chemical reactions in which primary RNA transcripts are converted into mature mRNA, rRNA, or tRNA molecules or in which polypeptide chains become converted into finished proteins.

programmed cell death Cell death that occurs as part of the normal developmental process.

prokaryote An organism in which a nuclear envelope is lacking and in which the genetic material does not divide by mitosis or meiosis; bacteria and blue-green algae.

promoter A specific DNA sequence at which RNA polymerase binds and initiates transcription.

promoter mutation A mutation that increases or decreases the ability of a promoter to initiate transcription.

promoter recognition The first step in transcription.

proofreading *See* **editing function.**

prophage The form of phage DNA in a lysogenic bacterium; the phage DNA is repressed and usually integrated in the bacterial chromosome, but some prophages are in plasmid form.

prophage induction Activation of a prophage to undergo the lytic cycle.

prophase The initial stage of mitosis or meiosis, occurring after DNA replication and terminating with the alignment of the chromosomes at metaphase; often absent or abbreviated between meiosis I and meiosis II.

protein A molecule composed of one or more polypeptide chains.

prototroph Microbial strain capable of growth in a defined minimal medium, ideally containing only a carbon source and inorganic compounds; the wildtype genotype is usually regarded as a prototroph.

pseudogene A DNA sequence that is not expressed because of one or more mutations but that has a functional counterpart in the same organism; pseudogenes are regarded as a mutated forms of ancient gene duplications.

P transposable element A DNA sequence capable of changing position within a DNA molecule or moving between molecules.

Punnett square A cross-multiplication square used for determining the expected genetic outcome of matings.

purine An organic base found in nucleic acids; the predominant purines are adenine and guanine.

pyrimidine An organic base found in nucleic acids; the predominant pyrimidines are cytosine, uracil (in RNA only), and thymine (in DNA only).

pyrimidine dimer Two adjacent pyrimidine bases, typically a pair of thymines, in the same polynucleotide strand, between which chemical bonds have formed; the most-common lesion formed in DNA by exposure to ultraviolet light.

quantitative trait A trait, typically measured on a continuous scale, such as height or weight, that results from the combined action of several or many genes in conjunction with environmental factors.

race A genetically or geographically distinct subgroup of a species.

rad A unit of ionizing radiation; the amount resulting in the dissipation of 100 ergs of energy in one gram of matter; rad stands for radiation absorbed dose.

random genetic drift Fluctuation in allele frequency from generation to generation resulting from restricted population size.

random mating System of mating in which mating pairs are formed independent of genotype and phenotype.

reading frame One of three ways to translate an mRNA base sequence linearly into an amino acid sequence to form a polypeptide; the particular reading frame is defined by the AUG codon that is selected for chain initiation.

reannealing Reassociation of dissociated single strands of DNA to form a duplex molecule.

recessive Refers to an allele, or the corresponding phenotypic trait, expressed only in homozygotes.

reciprocal cross A cross in which the sexes of the parents are reversed compared with another cross.

reciprocal translocation Interchange of parts between non-homologous chromosomes.

recombinant A chromosome resulting from crossing-over that carries a combination of alleles different from either chromosome participating in the crossover; the cell or individual organism containing a recombinant chromosome.

recombinant DNA technology Procedures for creating DNA molecules composed of one or more segments from other DNA molecules.

recombination Exchange of parts between DNA molecules

or chromosomes; recombination in eukaryotes usually entails a reciprocal exchange of parts, but in prokaryotes it is often nonreciprocal.

recombination repair Repair of damaged DNA by exchange of good for bad segments between two damaged molecules.

redundancy The feature of the genetic code in which an amino acid corresponds to more than one codon; also called degeneracy.

regulatory circuitry The set of interactions between regulatory genes.

regulatory gene A gene with the primary function of controlling the rate of synthesis of the products of one or more other genes.

rejection An immune response against transfused blood or transplanted tissue.

relative fitness The fitness of one genotype expressed as a proportion of the fitness of another genotype.

relaxed DNA A DNA circle whose supercoiling has been removed either by introduction of a single-strand break or by the activity of a topoisomerase.

rem The quantity of any kind of ionizing radiation that has the same biological effect as one rad of high-energy gamma rays; rem stands for roentgen equivalent man.

renaturation Restoration of the normal three-dimensional structure of a macromolecule; in reference to nucleic acids, the term means the formation of a double-stranded molecule by complementary base-pairing between two single-stranded molecules.

repair *See* **DNA repair.**

repair synthesis The enzymatic filling of a gap in a DNA molecule at the site of excision of a damaged DNA segment.

repetitive sequence A DNA sequence present more than once per haploid genome.

replica plating Procedure in which a particular spatial pattern of colonies on an agar surface is reproduced on a series of agar surfaces by stamping them with a template containing an image of the pattern; the template is often produced by pressing a piece of sterile velvet upon the original surface, which transfers cells from each colony to the cloth.

θ replication Bidirectional replication of a circular DNA molecule starting from a single origin of replication.

replication *See* **DNA replication.**

replication fork In a replicating DNA molecule, the region in which nucleotides are added to growing strands.

replication origin The base sequence at which DNA synthesis begins.

replicon A DNA molecule that has a replication origin.

repressor A protein that binds specifically to a regulatory sequence adjacent to a gene and blocks transcription of the gene.

repulsion *See* ***trans* configuration.**

restriction endonuclease A nuclease that recognizes a short nucleotide sequence (restriction site) in a DNA molecule and cleaves the molecule at that site; also called restriction enzyme.

restriction enzyme *See* **restriction endonuclease.**

restriction fragment A segment of duplex DNA produced by cleavage of a larger molecule with a restriction enzyme.

restriction fragment length polymorphism (RFLP) Genetic variation in a population associated with the size of restriction fragments containing sequences homologous to a particular probe DNA; the polymorphisms result from the positions of restriction sites flanking the probe, and each variant is essentially a different allele.

restriction map A diagram of a DNA molecule showing the positions of cleavage by one or more restriction endonucleases.

restriction site The base sequence at which a particular restriction endonuclease makes a cut.

restrictive condition A growth condition in which the phenotype of a conditional mutation is expressed.

retinoblastoma Gene for a tumor-suppressor located in chromosome band 13q14; inheritance of one copy of the mutation results in multiple malignancies in the retina of the eyes; the malignancies are in retinal cells in which the mutation becomes homozygous—for example, through a new mutation or mitotic recombination.

retrovirus One of a class of RNA animal viruses that cause the synthesis of DNA complementary to their RNA genomes on infection.

reverse genetics Procedure in which mutations are deliberately produced in cloned genes and introduced back into cells or the germ line of an organism.

reverse mutation A mutation that undoes the effect of a preceding one.

reverse transcriptase An enzyme contained within the coats of retroviruses that makes complementary DNA from a single-stranded RNA template.

reversion Restoration of a mutant phenotype to the wildtype phenotype by a second mutation.

RFLP *See* **restriction fragment length polymorphism.**

R group *See* **side chain.**

Rh Rhesus blood-group system in human beings; maternal-fetal incompatibility of this system may result in hemolytic disease of the newborn.

Rh-negative Refers to phenotype in which the red blood cells lack the D antigen of the Rh blood-group system.

Rh-positive Refers to phenotype in which the red blood cells possess the D antigen of the Rh blood-group system.

ribonuclease Any enzyme that cleaves phosphodiester bonds in RNA; abbreviated RNase.

ribonucleic acid *See* **RNA.**

ribose The five-carbon sugar in RNA.

ribosomal RNA (rRNA) RNA molecules that are structural components of the ribosomal subunits; in eukaryotes, there are four rRNA molecules—5S, 5.8S, 18S, and 28S; in prokaryotes, there are three—5S, 16S, and 23S.

ribosome The cellular organelle, consisting of two subunits, each composed of RNA and proteins, on which the codons of mRNA are translated into amino acids in protein synthesis; in prokaryotes, the subunits are 30S and 50S particles and, in eukaryotes, they are 40S and 60S particles.

ribosome-binding site The base sequence in a prokaryotic mRNA molecule to which a ribosome can bind to initiate protein synthesis; also called the Shine-Dalgarno sequence.

RNA Ribonucleic acid; a nucleic acid in which the sugar constituent is ribose; typically, RNA is single-stranded and contains the four bases adenine, cytosine, guanine, and uracil.

RNA polymerase An enzyme that makes RNA by copying the base sequence of a DNA strand.

RNA processing The conversion of a primary transcript into an mRNA, rRNA, or tRNA molecule; includes splicing, cleavage, modification of termini, and, in tRNA, modification of internal bases.

RNase *See* ribonuclease.

RNA splicing Excision of introns and joining of exons.

Robertsonian translocation A chromosomal aberration in which the long arms of two acrocentric chromosomes become joined to a common centromere.

roentgen A unit of ionizing radiation, defined as the amount of radiation resulting in 2.083×10^9 ion pairs per cm^3 of dry air at 0°C and 1 atm pressure; abbreviated R.

rolling-circle replication A mode of replication in which a circular parent molecule produces a linear branch of newly formed DNA.

Rous sarcoma virus A well-studied avian retrovirus that infects chickens.

R plasmid A bacterial plasmid that carries drug-resistance genes; commonly used in genetic engineering.

rRNA *See* ribosomal RNA.

S *See* cell cycle.

salvage pathway A minor source of deoxythymidine triphosphate (dTTP) for DNA synthesis that uses the enzyme thymidine kinase (TK).

satellite DNA Eukaryotic DNA that forms a minor band at a different density from that of most of the cellular DNA upon equilibrium density gradient centrifugation; consists of short sequences repeated many times in the genome (highly repetitive DNA) or of mitochondrial or chloroplast DNA.

scaffold A protein-containing material in chromosomes, believed to be responsible in part for the compaction of chromosomes.

second-division segregation Segregation of a pair of alleles into different nuclei in the second meiotic division, the result of crossing-over between the gene and the centromere of the pair of homologous chromosomes.

second meiotic division The meiotic division in which the centromeres split and the chromosome number is not reduced; also called the equational division.

segmentation genes Genes that determine the spatial pattern of segments and parasegments in *Drosophila* development.

segment-polarity genes Genes that determine the spatial pattern of development within the segments of *Drosophila* larvae.

segregation Separation of the members of a pair of alleles into different gametes in meiosis.

segregational petites Slow-growing yeast colonies deficient in respiration due to mutation in a chromosomal gene; inheritance is Mendelian.

segregation mutation Developmental mutation in which sister cells express the same fate when they should express different fates.

selected marker A genetic mutation that allows growth in selective medium.

selection In evolution, intrinsic differences in the ability of genotypes to survive and reproduce; in plant and animal breeding, the choosing of individuals with certain phenotypes to be parents of the next generation; in mutation studies, a procedure designed in such a way that only a desired type of cell can survive, as in selection for resistance to an antibiotic.

selection coefficient The amount by which relative fitness is reduced or increased.

selection differential In artificial selection, the difference between the mean of the selected individuals and the mean of the population from which they were chosen.

selection limit The mean phenotype of a trait in a population when the trait no longer responds to artificial selection.

selection-mutation balance Equilibrium determined by the opposing effects of mutation tending to increase the frequency of a deleterious allele and of selection tending to decrease the frequency.

selection pressure The tendency of natural or artificial selection to change allele frequency.

selectively neutral mutation A mutation having no effects, or only negligible ones, on fitness.

selective medium Medium that allows growth only of cells with particular genotypes.

self-fertilization The union of male and female gametes produced by the same individual organism.

selfish DNA DNA sequences that do not contribute to the fitness of an organism but are maintained in the genome through their ability to replicate and transpose.

semiconservative replication The usual mode of DNA replication, in which each strand of a double-stranded molecule serves as a template for the synthesis of a new complementary strand and the daughter molecules are composed of one old (parental) and one newly synthesized strand.

semisterility A condition in which a significant proportion of the gametophytes produced by a plant or the zygotes produced by an animal are inviable, as in a translocation heterozygote.

sense strand The DNA strand that serves as the template for transcription of a given gene.

sex chromosome A chromosome, such as the human X or Y, that participates in the determination of sex.

sex-influenced trait A trait whose expression depends on the sex of the individual organism.

sex-limited trait A trait expressed in one sex and not in the other.

sex-linked Refers to a trait determined by a gene on a sex chromosome, usually the X.

β sheet A fundamental structure formed in protein folding in which two polypeptide segments are held together in an antiparallel array by hydrogen bonds.

Shine-Dalgarno sequence *See* **ribosome-binding site.**

shuttle vector A vector capable of replication in two or more organisms; for example, yeast and *E. coli.*

sib *See* **sibling.**

sibling A brother or sister, each having the same parents.

sickle-cell anemia A severe anemia in human beings inherited as an autosomal recessive and caused by an amino acid substitution in the β-globin chain; heterozygotes tend to be more resistant to falciparum malaria than are normal homozygotes.

side chain In protein structure, the chemical group attached to the α-carbon atom of an amino acid; amino acids are differentiated on the basis of side-chain differences; also called R group.

sigma (σ) subunit The subunit of RNA polymerase needed for promoter recognition.

silent mutation Any mutation having no phenotypic effect.

single-copy sequence A DNA sequence present only once in each haploid genome.

sister chromatids Chromatids produced by replication of a single chromosome.

site-specific exchange Genetic exchange that occurs only between particular base sequences.

small ribonucleoprotein particles Small nuclear particles containing short RNA molecules and several proteins; responsible for intron excision and splicing and other aspects of RNA processing.

somatic cell Any cell of a multicellular organism other than the gametes and the germ cells from which they develop.

somatic-cell genetics Study of somatic cells in culture.

somatic mutation A mutation arising in a somatic cell.

SOS repair An inducible, error-prone system for repair of DNA damage in *E. coli.*

Southern blot A nucleic acid hybridization method in which, following electrophoretic separation, denatured DNA is transferred from a gel to a membrane filter and then exposed to radioactive DNA or RNA under conditions of renaturation; the radioactive regions locate the homologous DNA fragments on the filter.

spacer sequence A noncoding base sequence between the coding segments of polycistronic mRNA or between genes in DNA.

specialized transduction *See* **transducing phage.**

species Genetically, a group of actually or potentially inbreeding organisms that is reproductively isolated from other such groups.

spindle A structure composed of fibrous proteins on which chromosomes align during metaphase and move during anaphase.

splice acceptor The 3′ end of an intron.

splice donor The 5′ end of an intron.

spliceosomes The RNA-protein particles in the nucleus responsible for removing the introns from RNA transcripts.

spontaneous mutation A mutation occurring in the absence of any known mutagenic agent.

spore A unicellular reproductive entity that becomes detached from the parent and can develop into a new individual upon germination; in plants, spores are the haploid products of meiosis.

sporophyte The diploid, spore-forming generation in plants, which alternates with the haploid, gamete-producing generation (the gametophyte).

standard deviation The square root of the variance.

start codon An mRNA codon, usually AUG, at which polypeptide synthesis begins.

statistically significant Refers to the result of an experiment or study that has only a small probability of occurring by chance on the assumption that some hypothesis is true; conventionally, if results as bad or worse would be expected less than 5 percent of the time, the result is said to be statistically significant, or if less than 1 percent of the time, it is called statistically highly significant; both outcomes cast the hypothesis into serious doubt.

sticky ends Single-stranded ends of DNA fragments produced by certain restriction enzymes that are capable of reannealing.

stop codon One of the three mRNA codons—UAG, UAA, and UGA—at which polypeptide synthesis stops.

structural gene A gene that encodes the amino acid sequence of a polypeptide chain.

submetacentric chromosome A chromosome whose centromere divides it into arms of unequal length.

subpopulations Breeding groups within a larger population between which migration is restricted.

subspecies A relatively isolated population or group of populations distinguishable from other populations in the same species by allele frequencies or chromosomal arrangements and sometimes exhibiting incipient reproductive isolation.

substrate A specific substance acted on by an enzyme.

subunit A macromolecule that is a component of an ordered aggregate of macromolecules—for example, a single polypeptide chain in a protein containing several chains.

supercoiling Coiling of double-stranded DNA in which strain caused by overwinding or underwinding of the duplex makes the circle twist; a supercoiled circle is also called a twisted circle or a superhelix.

suppressive petites A class of cytoplasmically inherited, small-colony mutations in yeast that are not corrected by the addition of normal mitochondria.

suppressor mutation A mutation that partly or completely restores the function impaired by another mutation at a different site in the same gene (intragenic suppression) or in a different gene (intergenic suppression).

suppressor tRNA Usually, a tRNA molecule capable of translating a stop codon by inserting an amino acid in its place, but a few suppressor tRNAs replace one amino acid with another.

synapsis The pairing of homologous chromosomes or chromosome regions during zygotene of the first meiotic prophase.

synaptonemal complex A complex protein structure that forms between synapsed homologous chromosomes in the pachytene substage of the first meiotic prophase.

syncytial blastoderm Stage in early *Drosophila* development formed by successive nuclear divisions without division of the cytoplasm.

syndrome A group of symptoms appearing together with sufficient regularity to warrant designation by a special name; also, a disorder, disease, or anomaly.

synteny The presence of two different genes on the same chromosome.

tandem duplication A pair of identical or closely related DNA sequences that are adjacent and in the same orientation.

TATA box The base sequence (5′-TATA-3′) in the DNA of a promoter to which RNA polymerase binds.

tautomeric shift A reversible change in the location of a hydrogen atom in a molecule, altering the molecule from one isomeric form to another; in nucleic acids, the shift is typically between a keto group (keto form) and a hydroxyl group (enol form).

T cells A class of white blood cells instrumental in various aspects of the immune response.

T DNA Transposable element found in *Agrobacterium tumefaciens*, which produces crown gall tumors in a wide variety of dicotyledonous plants.

telomerase An enzyme that adds specific nucleotides to the tips of the chromosomes to form the telomeres.

telomeres The tips of a chromosome that contain DNA sequences required for stability of chromosome ends.

telophase The final stage of mitotic or meiotic nuclear division.

temperate phage A phage that is capable of both a lysogenic and a lytic cycle.

temperature-sensitive mutation A conditional mutation that causes a phenotypic change at certain temperatures and not at others.

template A nucleic acid strand whose base sequence is copied in a polymerization reaction.

testcross A cross between a heterozygote and a recessive homozygote, resulting in progeny in which each phenotypic class represents a different genotype.

testis-determining factor (TDF) Genetic element on the mammalian Y chromosome that determines maleness.

tetrad The four chromatids that make up a pair of homologous chromosomes in meiotic prophase I and metaphase I; also, the four haploid products of a single meiosis.

tetrad analysis A method for the analysis of linkage and recombination using the four haploid products of single meiotic divisions.

tetraploid A cell or organism with four complete sets of chromosomes; in an autotetraploid, the chromosome sets are homologous; in an allotetraploid, the chromosome sets consist of a complete diploid complement from each of two distinct ancestral species.

tetratype An ascus containing spores of four different genotypes—one each of the four genotypes possible with two alleles of each of two genes.

30-nm fiber Level of compaction of eukaryotic chromatin resulting from coiling of extended, nucleosome-bound DNA fiber.

three-point cross Cross in which three genes are segregating; used to obtain unambiguous evidence of gene order.

3′-OH group The end of a DNA or RNA strand that terminates in a sugar and so has a free hydroxyl group on the number 3′ carbon.

threshold trait A trait with a continuously distributed liability or risk in which individuals with a liability greater than a critical value (the threshold) exhibit the phenotype of interest, such as a disorder.

thymine (T) A nitrogenous pyrimidine base found in DNA.

thymine dimer *See pyrimidine dimer.*

time of entry In an Hfr × F^- bacterial mating, the earliest time that a particular gene in the Hfr parent is transferred to the F^- recipient.

tolerance The process that normally prevents one's immune system from attacking oneself.

topoisomerases Enzymes that introduce or remove either underwinding or overwinding of double-stranded DNA; they act by introducing a single-strand break, changing the relative positions of the strands, and sealing the break.

totipotent cells Cells capable of differentiation into a complete organism; the zygote is totipotent.

trans **configuration** The arrangement in linked inheritance in which an individual organism heterozygous for two mutant sites has received one of the mutant sites from each parent—that is $a_1 +/+ a_2$.

transcription The process by which the information contained in the coding strand of DNA is copied into a single-stranded RNA molecule of complementary base sequence.

transcriptional activator protein Positive control element that stimulates transcription by binding with particular sites in DNA.

transducing phage A phage type capable of producing particles containing bacterial DNA (transducing particles); a specialized transducing phage produces particles carrying only specific regions of chromosomal DNA; a generalized transducing phage produces particles that may carry any region of the genome.

transduction The carrying of genetic information from one bacterium to another by a phage.

transfer RNA (tRNA) A small RNA molecule that translates a codon into an amino acid in protein synthesis; it has a three-base sequence, called an anticodon, complementary to a specific codon in mRNA and a site to which a specific amino acid is bound.

transformation The conversion of the genotype of a bacterial cell by exposure of the cell to DNA isolated from bacteria with a different genotype; also, the conversion of an animal cell, whose growth is limited in culture, into a tumorlike cell whose pattern of growth is different from that of a normal cell.

transformation mutation A mutation in which affected cells undergo a developmental fate characteristic of other cells.

transgenic animals Animals in which novel DNA has been incorporated into the germ line.

trans-**heterozygote** *See trans configuration.*

transition A mutation resulting from the substitution of one purine for another purine or one pyrimidine for another pyrimidine.

translation The process by which the amino acid sequence of a polypeptide is derived from the nucleotide sequence of an mRNA molecule associated with a ribosome.

translocation Interchange of parts between nonhomologous chromosomes; also, the movement of mRNA with respect to a ribosome in protein synthesis. *See also* **reciprocal translocation.**

transposable element A DNA sequence capable of moving (transposing) from one location to another in a genome.

transposase Protein necessary for transposition.

transposition The movement of a transposable element.

transposon A transposable element that contains bacterial genes—for example, for antibiotic resistance; also loosely used as a synonym for transposable element.

transposon tagging Insertion of a transposable element containing a genetic marker into a gene of interest.

transversion A mutation resulting from the substitution of a purine for a pyrimidine or a pyrimidine for a purine.

triplet code A code in which each codon consists of three bases.

triplication Presence of three copies of a DNA sequence ordinarily present only once.

triploid A cell or individual organism with three complete sets of chromosomes.

trisomic A diploid organism with an extra copy of one of its chromosomes.

trisomy-X syndrome The clinical features of the karyotype 47,XXX.

tRNA *See* **transfer RNA.**

true breeding Refers to a strain, breed, or variety of organisms that gives progeny like itself; homozygous.

truncation point In artificial selection, the critical phenotype determining which organisms will be retained for breeding and which will be culled.

tumor-suppressor gene A gene whose absence predisposes to malignancy; also called an anti-oncogene.

Turner syndrome The clinical features of human females with the karyotype 45, X.

uncharged tRNA A tRNA molecule lacking an amino acid.

underwound DNA A DNA molecule whose strands are untwisted somewhat; hence, some of its bases are unpaired.

unequal crossing-over Crossing-over between nonallelic copies of duplicated or other repetitive sequences—for example, in a tandem duplication, between the upstream copy in one chromosome and the downstream copy in the homologous chromosome.

uniparental inheritance Extranuclear inheritance of a trait through cytoplasmic factors or organelles contributed by only one parent. *See also* **maternal inheritance, paternal inheritance.**

unique sequence A DNA sequence that is present in only one copy in a haploid genome, in contrast with repetitive sequences.

uracil (U) A nitrogenous pyrimidine base found in RNA.

variable expressivity Differences in the severity of expression of a particular genotype.

variable region The part of an immunoglobulin molecule that varies greatly in amino acid sequence among antibodies in the same subclass. *See also* **V region.**

variance A measure of the spread of a statistical distribution; the mean of the squares of the deviations from the mean.

vector A DNA molecule, capable of replication, into which a gene or DNA segment is inserted by recombinant DNA techniques; a cloning vehicle.

viral oncogene A class of genes found in certain viruses that predispose to cancer; viral oncogenes are the viral counterparts of cellular oncogenes. *See also* **cellular oncogene, oncogene.**

virulent phage A phage or virus species capable only of a lytic cycle; contrasts with temperate phage.

virus An infectious intracellular parasite able to reproduce only inside living cells.

V region One of multiple DNA sequences coding for alternative amino acid sequences of part of the variable region of an antibody molecule. *See also* **variable region.**

V-type position effect A type of position effect in *Drosophila*, characterized by discontinuous expression of one or more genes during development and usually resulting from chromosome breakage and rejoining such that euchromatic genes are repositioned in or near centromeric heterochromatin.

wildtype The most-common phenotype or genotype in a natural population; also, a phenotype or genotype arbitrarily designated as a standard for comparison.

wobble The acceptable pairing of several possible bases in an anticodon with the base in the third position of a codon.

χ^2 *See* **chi-square.**

X chromosome A chromosome associated with sex determination that is present in two copies in the homogametic sex and in one copy in the heterogametic sex.

xeroderma pigmentosum An inherited defect in the repair of ultraviolet-light damage to DNA, associated with extreme sensitivity to sunlight and multiple skin cancers.

X-linked inheritance The pattern of hereditary transmission of genes located in the X chromosome; usually evident from the production of nonidentical classes of progeny from reciprocal crosses.

YAC *See* **yeast artificial chromosome.**

Y chromosome The sex chromosome that is present only in the heterogametic sex; in mammals, the male-determining sex chromosome.

yeast artificial chromosome A cloning vector in yeast that can accept very large fragments of DNA; a chromosome introduced into yeast derived from such a vector and containing DNA from another organism.

zinc finger A structural motif found in many DNA-binding proteins, in which a fingerlike projection entraps a zinc ion.

zygote The product of the fusion of a female and a male gamete in sexual reproduction; a fertilized egg.

zygotene The substage of meiotic prophase I during which homologous chromosomes synapse.

zygotic genes Genes that control early development through their expression in the zygote.

These references are for the student who either wants more information or needs an alternative explanation for the material presented in this book. For the former, I recommend the review articles and scientific reports; for the latter, the textbooks and the *Scientific American* articles. A few "classic" papers are listed for those who would like to know how scientific information is obtained; these papers are generally more advanced than textbooks.

Chapter 1

Carlson, E. A. 1987. *The Gene: A Critical History*, 2d ed. Saunders.

Dunn, L. C. 1965. A *Short History of Genetics*. McGraw-Hill.

Judson, H. F. 1979. *The Eighth Day of Creation: The Makers of the Revolution in Biology*. Simon and Schuster.

Mendel, G. 1981. Experiments in plant hybridization. (Translation.) In *Genetics: A Scientific American Reader*, ed. C. I. Davern. Freeman.

Olby, R. C. 1966. *Origins of Mendelism*. Constable.

Srb, A., R. Owen, and R. Edgar. 1965. *General Genetics*. Freeman.

Stern, C., and E., Sherwood. 1966. *The Origins of Genetics: A Mendel Source Book*. Freeman.

Chapter 2

Chandley, A. C. 1988. Meiosis in man. *Trends in Genetics* 4: 79.

Crow, J. F. 1983. *Genetics Notes*. 8th ed. Burgess.

Hyams, J. S., and B. R. Brinkley, eds. 1989. *Mitosis*. Academic.

McIntosh, J. R., and K. L. McDonald. 1989. The mitotic spindles. *Scientific American*, October.

McKusick, V. A. 1965. The royal hemophilia. *Scientific American*, August.

Sokal, R. R., and F. J. Rohlf. 1969. *Biometry*. Freeman.

Srb, A., R. Owen, and R. Edgar. 1965. *General Genetics*. Freeman.

Sturtevant, A. H. 1965. A *Short History of Genetics*. Harper and Row.

Voeller, B. R., ed. 1968. *The Chromosome Theory of Inheritance: Classical Papers in Development and Heredity*. Appleton-Century-Crofts.

Chapter 3

Creighton, H. S., and B. McClintock. 1931. A correlation of cytological and genetical crossing over in *Zea mays*. *Proceedings of the National Academy of Sciences, USA* 17: 492.

Fincham, J. R. S., P. R. Day, and A. Radford. 1979. *Fungal Genetics*. Blackwell.

Levine, L. 1971. *Papers on Genetics*. Mosby.

Sturtevant, A. H., and G. W. Beadle. 1962. *An Introduction to Genetics*. Dover.

Voeller, B. R. ed. 1968. The Chromosome Theory of Inheritance: Classical Papers in Development and Heredity. Appleton-Century-Crofts.

White, R., and J.-M. Lalouel. 1988. Chromosome mapping with DNA markers. *Scientific American*, February.

Chapter 4

Avery, O., C. MacLeod, and M. McCarty. 1944. Studies on the chemical nature of the substance inducing transformation of pneumococcal types. *Journal of Experimental Medicine* 79: 137.

Hershey, A. D., and M. Chase. 1952. Independent functions of viral protein and nucleic acid in growth of bacteriophage. *Journal of General Physiology* 36: 39.

Hotchkiss, R. D., and E. Weiss. 1956. Transformed bacteria. *Scientific American*, November.

Kelley, T., ed. 1988. *Eukaryotic DNA Replication*. Cold Spring Harbor Laboratory.

Kornberg, A. 1980. *DNA Replication*. Freeman.

Meselson, M., and F. Stahl. 1958. The replication of DNA in *Escherichia coli*. *Proceedings of the National Academy of Sciences, USA* 44: 671.

Mirsky, A. 1968. The discovery of DNA. *Scientific American*, June.

Mullis, K. B. 1990. The unusual origin of the polymerase chain reaction. *Scientific American*, April.

Radman, M., and R. Wagner. 1988. The high fidelity of DNA duplication. *Scientific American*, August.

Watson, J. D. 1968. *The Double Helix*. Athenaeum.

Watson, J. D., and F. H. C. Crick. 1953. Genetic implications of the structure of deoxyribonucleic acid. *Nature* 171: 964.

Watson, J. D., and F. H. C. Crick. 1953. Molecular structure of nucleic acid: a structure for deoxyribonucleic acid. *Nature* 171: 737.

Chapter 5

Bauer, W. R., F. H. C. Crick, and J. H. White. 1980. Supercoiled DNA. *Scientific American*, July.

Berg, D., E., and M. M. Howe. 1989. *Mobile DNA*. American Association for Microbiology.

Bickmore, W. A., and A. T. Sumner. 1989. Mammalian chromosome banding: an expression of genome organization. *Trends in Genetics* 5: 144.

Bukhari, A. I., J. Shapiro, and S. Adhya. 1977. *DNA Insertion Elements, Plasmids, and Episomes*. Cold Spring Harbor Laboratory.

Cairns, J. 1966. The bacterial chromosome. *Scientific American*, January.

Federoff, N. 1984. Transposable genetic elements in maize. *Scientific American*, June.

Green, M. M. 1980. Transposable elements in *Drosophila* and other diptera. *Annual Review of Genetics* 14: 109.

Isenberg, I. 1979. Histones. *Annual Review of Biochemistry* 48: 159.

Keller, E. 1981. McClintock's maize. *Science 81*, August.

Kornberg, R. D., and A. Klug. 1981. The nucleosome. *Scientific American*, February.

McClintock, B. 1965. Control of gene action in maize. *Brookhaven Symposium on Quantitative Biology* 18: 162.

McGhee, J., and G. Felsenfeld, 1980. Nucleosome structure. *Annual Review of Biochemistry* 40: 1115.

Zakian, V. 1989. Structure and function of telomeres. *Annual Review of Genetics* 23: 579.

Chapter 6

Borgaonkar, D. S. 1984. *Chromosomal Variation in Man: A Catalogue of Chromosomal Variants and Anomalies*, 4th ed. Liss.

Carson, H. L. 1970. Chromosome tracers of the origin of the species. *Science* 168: 1414.

Curtis, B. C., and D. R. Johnson. 1969. Hybrid wheat. *Scientific American*, May.

Epstein, C. J. 1988. Mechanisms of the effects of aneuploidy in mammals. *Annual Review of Genetics* 22: 51.

Garber, E. D. 1972. *Cytogenetics: An Introduction*. McGraw-Hill.

Guerrero, I. 1987. Proto-oncogenes in pattern formation. *Trends in Genetics* 3: 269.

Haseltire, W. A. and F. Wong-Stahl. 1988. The molecular biology of the AIDS virus. *Scientific American*, October.

Hsu, T. H. 1979. *Human and Mammalian Cytogenetics*. Springer-Verlag. 1979.

Manning, C. H., and H. O. Goodman. 1981. Parental origin of chromosomes in Down's syndrome. *Human Genetics* 59: 101.

Rowley, J. D. 1983. Human oncogene locations and chromosome aberrations. *Nature* 301: 290.

Sparks, R. S., D. E. Comings, and C. F. Fox. 1977. *Molecular Human Cytogenetics*. Academic.

Stebbins, G. L. 1971. *Chromosome Evolution in Higher Plants*. Addison-Wesley.

Stewart, G. D., T. J. Hassold, and D. M. Kurnit. 1988. Trisomy 21: molecular and cytogenetic studies of nondisjunction. *Advances in Human Genetics* 17: 99.

Swanson, C. P., and P. Webster. 1977. *The Cell*. Prentice-Hall.

Weinberg, R. A. 1988. Finding the anti-oncogene. *Scientific American*, September.

White, M. J. D. 1977. *Animal Cytology and Evolution*. Cambridge University Press.

Chapter 7

Gillham, N. 1978. *Organelle Heredity*. Raven.

Goodenough, U., and R. P. Levine. 1970. The genetic activity of mitochondria and chloroplasts. *Scientific American*, November.

Grivell, L. A. 1983. Mitochondrial DNA. *Scientific American*, March.

Jacobs, H. T., and D. M. Lonsdale. 1987. The selfish organelle. *Trends in Genetics* 3: 337.

Laughnan, J. R., and S. Gabay-Laughnan. 1983. Cytoplasmic male sterility in maize. *Annual Review of Genetics* 17: 27.

Preer, J. P., Jr. 1971. Extrachromosomal inheritance: hereditary symbionts, mitochondria, chloroplasts. *Annual Review of Genetics* 5: 361.

Sager, R. 1965. Genes outside the chromosome. *Scientific American*, January.

Sager, R. 1972. *Cytoplasmic Genes and Organelles*. Academic.

Sturtevant, A. H. 1923. Inheritance of the direction of coiling in Limnaea. *Science* 58: 269.

Wallace, D. C. 1989. Mitochondrial mutations and neuromuscular disease. *Trends in Genetics* 5: 9.

Wolf, K., and L. Del Gindice. 1988. The variable mitochondrial genome of *Ascomycetes*. *Advances in Human Genetics* 25: 186.

Chapter 8

Allison, A. C. 1956. Sickle cells and evolution. *Scientific American*, August.

Ayala, F. 1978. The mechanisms of evolution. *Scientific American*, September.

Bodmer, W. F., and L. L. Cavalli-Sforza. 1976. *Genetics, Evolution, and Man*. Freeman.

Cavalli-Sforza, L. L. 1969. Genetic drift in an Italian population. *Scientific American*, August.

Cavalli-Sforza, L. L. 1974. The genetics of human populations. *Scientific American*, September.

Crow, J. F., and M. Kimura. 1970. *An Introduction to Population Genetics Theory*. Harper and Row.

Dobzhansky, T. 1941. *Genetics and the Origin of the Species*. Columbia University Press.

Eckhardt, R. B. 1972. Population genetics and human origins. *Scientific American*, January.

Falconer, D. S. 1981. *Introduction to Quantitative Genetics*, 2d ed. Longman.

Fincham, J. R. S. 1983. *Genetics*. Jones and Bartlett.

Hardy, G. 1908. Mendelian proportions in a mixed population. *Science* 28: 49.

Harris, H., and D. A. Hopkinson. 1972. Average heterozygosity in man. *Journal of Human Genetics* 36: 9.

Hartl, D. L. 1988. *A Primer of Population Genetics*, 2d ed. Sinauer.

Hartl, D. L., and A. G. Clark. 1989. *Principles of Population Genetics*, 2d ed. Sinauer.

Kimura, M. 1983. *The Neutral Theory of Molecular Evolution*. Cambridge University Press.

Lewontin, R. C. 1974. *Genetic Basis of Evolutionary Change*. Columbia University Press.

Li, W.-H., and D. Graur. 1991. *Fundamentals of Molecular Evolution*. Sinauer.

Nei, M. 1987. *Molecular Evolutionary Genetics*. Columbia University Press.

Neufeld, D. J., and N. Colman. 1990. When science takes the witness stand. *Scientific American*, May.

Spiess, E. B. 1977. *Genes in Populations*. Wiley.

Stebbins, G. L. 1977. *Processes of Organic Evolution*, 3d ed. Prentice-Hall.

White, R., and J.-M. Lalouel. 1988. Chromosome mapping with DNA markers. *Scientific American*, February.

Wright, S. 1978. *Variability Within and Among Natural Populations*. Evolution and the Genetics of Populations, vol 4. University of Chicago.

Chapter 9

Bodmer, W. F., and L. L. Cavalli-Sforza. 1976. *Genetics, Evolution, and Man*. Freeman.

Bouchard, T. J., Jr., D. T. Lykken, M. McGue, N. L. Segal, and A. Tellegen. 1990. Sources of human psychological differences: the Minnesota study of twins reared apart. *Science* 250: 223.

Crow, J. 1957. Genetics of insect resistance to chemicals. *Annual Review of Entomology* 2: 227.

Devor, E. J,. and C. R. Cloninger. 1989. The genetics of alcoholism. *Annual Review of Genetics* 23: 19.

East, E. M. 1910. Mendelian interpretation of inheritance that is apparently continuous. *American Naturalist* 44: 65.

Falconer, D. S. 1981. *Introduction to Quantitative Genetics*, 2d ed. Longman.

Feldman, M. W., and R. C. Lewontin. 1975. The heritability hang-up. *Science* 190: 1163.

Foster, H. L. 1965. Mammalian pigment genetics. *Advances in Genetics* 13: 311.

Hartl, D. L., and A. G. Clark. 1989. *Principles of Population Genetics*, 2d ed. Sinauer.

Hedrick, P. W. 1983. *Genetics of Populations*. Jones and Bartlett.

Mather, K. 1943. Polygenic inheritance and natural selection. *Biological Reviews* 18: 32.

Pirchner, F. 1983. *Population Genetics in Animal Breeding*, 2d ed. Plenum.

Smith, J. M. 1978. The evolution of behavior. *Scientific American*, September.

Chapter 10

Adelberg, E. A., ed. 1966. *Papers on Bacterial Genetics*. Little Brown.

Birge, E. A. 1981. *Bacterial and Bacteriophage Genetics*. Springer-Verlag.

Campbell, A. 1976. How viruses insert their DNA into the DNA of the host cell. *Scientific American*, December.

Clowes, R. D. 1975. The molecules of infectious drug resistance. *Scientific American*, July.

Edgar, R. S., and R. H. Epstein. 1965. The genetics of a bacterial virus. *Scientific American*, February.

Hayes, W. 1968. *The Genetics of Bacteria and Their Viruses*. Wiley.

Hopwood, D. A., and K. E. Chater, eds. 1989. *Genetics of Bacterial Diversity*. Academic.

Kleckner, N. 1981. Transposable genetic elements. *Annual Review of Genetics* 15: 341.

Low, K. B., and R. Porter. 1978. Modes of genetic transfer and recombination in bacteria. *Annual Review of Genetics* 12: 249.

Neidhardt, F. C., J. L. Ingraham, K. B. Low, B. Magasanik, and H. E. Umbarger. 1987. *Escherichia coli and Salmonella typhimurium: Cellular and Molecular Biology* (2 volumes). American Society for Microbiology.

Novick, R. P. 1980. Plasmids. *Scientific American*, December.

Sambrook, J., E. F. Fritsch, and T. Maniatis. 1989. *Molecular Cloning: A Laboratory Manual*, 2d ed. Cold Spring Harbor Laboratory.

Scaife, J., D. Leach, and A. Galizzi. 1986. *Genetics of Bacteria*. Academic.

Stent, G. S., and R. Calendar. 1978. *Molecular Genetics: An Introductory Narrative*. Freeman.

Watanabe, T. 1967. Infectious drug resistance. *Scientific American*, December.

Zinder, N. 1958. Transduction in bacteria. *Scientific American*, November.

Chapter 11

Barrell, B. G., A. T. Bankier, and J. Drouin. 1979. A different genetic code in human mitochondria. *Nature* 282: 189.

Beadle, G. W. 1948. Genes of men and molds. *Scientific American*, September.

Bearn, A. G. 1956. The chemistry of hereditary disease. *Scientific American*, December.

Chambon, P. 1981. Split genes. *Scientific American*, May.

Cold Spring Harbor Laboratory. 1966. *The Genetic Code*. Cold Spring Harbor Symposium on Quantitative Biology, 31.

Crick, F. H. C. 1962. The genetic code. *Scientific American*, October.

Crick, F. H. C. 1966. The genetic code. *Scientific American*, October.

Crick, F. H. C. 1979. Split genes and RNA splicing. *Science*, 204: 264.

Crick, F. H. C., L. Barnett, S. Brenner, and R. J. Watts-Tobin. 1961. General nature of the genetic code for proteins. *Nature* 192: 1227.

Dickerson, R. E. 1972. The structure and history of an ancient protein. *Scientific American*, April.

Gorini, L. 1966. Antibiotics and the genetic code. *Scientific American*, April.

Lake, J. 1981. The ribosomes. *Scientific American*, August.

Miller, O. L., Jr. 1973. The visualization of genes in action. *Scientific American*, March.

Nathans, J. 1989. The genes for color vision. *Scientific American*, February.

Nirenberg, M. 1963. The genetic code. *Scientific American*, March.

Ross, J. 1989. The turnover of messenger RNA. *Scientific American*, April.

Taylor, J. H., ed. 1965. *Selected Papers on Molecular Genetics*. Academic.

Watson, J. D., N. H. Hopkins, J. W. Roberts, J. A. Steitz, and A. M. Weiner. 1987. *Molecular Biology of the Gene*, 4th ed. (2 volumes). Benjamin/Cummings.

Yanofsky, C. 1967. Gene structure and protein structure. *Scientific American*, May.

Chapter 12

Brown, D. 1981. Gene expression in eukaryotes. *Science* 211: 667.

Darnell, J. E. 1982. Variety in the level of gene control in eukaryotic cells. *Nature* 297: 365.

Felsenfeld, G., and J. McGhee. 1982. Methylation and gene control. *Nature* 296: 602.

Gilbert, W., and M. Ptashne. 1970. Genetic repressors. *Scientific American*, June.

Guarente, L. 1984. Yeast promoters: positive and negative elements. *Cell* 36: 799.

Holliday, R. 1989. A different kind of inheritance. *Scientific American*, June.

Khoury, G., and P. Gruss. 1983. Enhancer elements. *Cell* 33: 83.

Kolata, G. 1981. Gene regulation through chromosome structure. *Science* 214: 775.

Kolata, G. 1984. New clues to gene regulation. *Science* 224: 58.

Lewin, B. 1980. *Eukaryotic Chromosomes*. Gene Expression, vol. 2. Wiley.

Maniatis, T., and M. Ptashne. 1976. A DNA operator-repressor system. *Scientific American*, January.

Miller, J., and W. Reznikoff, eds. 1978. *The Operon*. Cold Spring Harbor Laboratory.

O'Malley, B. W., and W. T. Schroeder. 1976. The receptors of steroid hormones. *Scientific American*, February.

Ptashne, M. 1989. How gene activators work. *Scientific American*, January.

Smith, C. W. J., J. G. Patton, and B. Nadal-Ginerd. 1989. Alternative splicing in the control of gene expression. *Annual Review of Genetics* 23: 527.

Ullman, A., and A. Danchin. 1980. Role of cyclic AMP in regulatory mechanisms of bacteria. *Trends in Biochemical Sciences* 5: 95.

Weisbrod, S. 1982. Active chromatin. *Nature* 297: 289.

Yanofsky, C. 1981. Attenuation in the control of expression of bacterial operons. *Nature* 289: 751.

Chapter 13

Anderson, K. V. 1987. Dorsal-ventral embryonic pattern genes of *Drosophila*. *Trends in Genetics* 3: 91.

Artavanis-Tsakonas, S. 1988. The molecular biology of the *Notch* locus and the fine tuning of differentiation in *Drosophila melanogaster*. *Trends in Genetics* 4: 95.

Bellve, A. R., and H. J. Vogel, eds. 1991. *Molecular Mechanisms in Cellular Growth and Differentiation*. Academic.

Capecchi, M. R., ed. 1989. *The Molecular Genetics of Early Drosophila and Mouse Development*. Cold Spring Harbor Laboratory.

Davidson, E. H. 1986. *Gene Activity in Early Development*. 3d ed. Academic.

De Robertis, E. M., G. Oliver, and C. V. E. Wright. 1990. Homeobox genes and the vertebrate body plan. *Scientific American*, July.

Gaul, U., and H. Jackle. 1990. Role of gap genes in early *Drosophila* development. *Advances in Genetics* 27: 239.

Irish, V. 1987. Cracking the *Drosophila* egg. *Trends in Genetics* 3: 303.

Kaufman, T. C., M. A. Seeger, and G. Olsen. 1990. Molecular and genetic organization of the *Antennapedia* gene complex of *Drosophila melanogaster*. *Advances in Genetics* 27: 309.

Loomis, W. F., ed. 1987. *Genetic Regulation of Development*. Wiley.

Sternberg, D. W. 1990. Genetic control of cell type and pattern formation in *Caenorhabditis elegans*. *Advances in Genetics* 27: 63.

Wilkins, A. S. 1987. *Genetic Analysis of Animal Development*. Wiley.

Willmitzer, L. 1988. The use of transgenic plants to study plant gene expression. *Trends in Genetics* 4: 13.

Wood, W. B., ed. 1988. *The Nematode Caenorhabditis elegans*. Cold Spring Harbor Laboratory.

Wright, T. R. F., ed. 1990. *Genetic Regulatory Hierarchies in Development*. Academic.

Chapter 14

Ames, B. W. 1979. Identifying environmental chemicals causing mutations and cancer. *Science* 204: 587.

Celis, J. E., and J. D. Smith. 1979. *Nonsense Mutations and tRNA Suppressors*. Academic.

Crow, J. F., and C. Denniston. 1985. Mutation in human populations. *Advances in Human Genetics* 14: 59.

Denniston, C. 1982. Low-level radiation and genetic risk estimation in man. *Annual Review of Genetics* 16: 329.

Derring, R. A. 1962. Ultraviolet radiation and nucleic acid. *Scientific American*, December.

Devoret, R. 1979. Bacterial tests for potential carcinogens. *Scientific American*, August.

Drake, J. W. 1970. *The Molecular Basis of Mutation.* Holden-Day.

Friedberg, E. C. 1985. *DNA Repair.* Freeman.

Hanawalt, P. C., and R. H. Haynes. 1967. The repair of DNA. *Scientific American*, February.

Haseltine, W. A. 1983. Ultraviolet light repair and mutagenesis revisited. *Cell* 33: 13.

Lambert, M. E., J. F. McDonald, and I. B. Weinstein, eds. 1988. *Eukaryotic Transposable Elements as Mutagenic Agents.* Cold Spring Harbor Laboratory.

Little, J. W., and D. W. Mount. 1982. The SOS regulatory system of *E. coli. Cell* 29: 11.

Low, K. B., ed. 1989. *The Recombination of Genetic Material.* Academic.

Muller, H. J. 1955. Radiation and human mutation. *Scientific American*, November.

Sherman, F. 1982. Suppression in the yeast *Saccharomyces cerevisiae.* In *The Molecular Biology of the Yeast Saccharomyces*, eds. J. Strathern, E. Jones, and J. Broach. Cold Spring Harbors Laboratory.

Sigurbjornsson, B. 1971. Induced mutations in plants. *Scientific American*, January.

Singer, B., and J. T. Kusmierek. 1982. Chemical mutagenesis. *Annual Review of Biochemistry* 51: 655.

Sugimura, T., S. Kondo, and H. Takebe, eds. 1982. *Environmental Mutagens and Carcinogens.* Liss.

Wills, C. 1970. Genetic load. *Scientific American*, March.

Chapter 15

Bach, F. H., and J. J. Van Rood. 1976. The major histocompatibility complex: genetics and biology. *New England Journal of Medicine* 295: 806.

Ballantyne, J., G. Sensabaugh, and H. Witkowski, eds. 1989. *DNA Technology and Forensic Science.* Cold Spring Harbor Laboratory.

Baltimore, D. 1981. Somatic mutation gains its place among the generators of diversity. *Cell* 26: 295.

Bell, J. I., J. A. Todd, and H. O. McDevitt. 1989. The molecular basis of HLA-disease association. *Advances in Human Genetics* 18: 1.

Brown, D. D. 1973. The isolation of genes. *Scientific American*, August.

Caskey, C. T., and D. C. Robbins, eds. 1982. *Somatic Cell Genetics.* Plenum.

Davidson, R. L. 1973. *Somatic Cell Hybridization: Studies on Genetics and Development.* Addison-Wesley.

Ephrussi, B. 1972. *Hybridization of Somatic Cells.* Princeton University Press.

Ephrussi, B., and M. C. Weiss. 1969. Hybrid somatic cells. *Scientific American*, April.

Golde, D. W., and J. C. Gasson. 1988. Hormones that stimulate the growth of blood cells. *Scientific American*, July.

Grey, H. M., A. Sette, and S. Buus. 1989. How T cells see antigen. *Scientific American*, November.

Harris, H. 1970. *Cell Fusion.* Harvard University Press.

Honjo, T., F. W. Alt, and T. H. Rabbitts, eds. 1989. *Immunoglobulin Genes.* Academic.

Leder, P. 1982. The genetics of antibody diversity. *Scientific American*, May.

Marx, J. 1981. Antibodies: getting their genes together. *Science* 212: 1015.

McKusick, V. A. 1971. The mapping of human chromosomes. *Scientific American*, April.

Puck, T., and F. T. Kao. 1982. Somatic cell genetics and its application to medicine. *Annual Review of Genetics* 16: 225.

Ruddle, F. H. 1981. A new era in mammalian gene mapping: somatic cell genetics and recombinant DNA methodologies. *Nature* 294: 115.

Ruddle, F. H., and R. S. Kucherlapati. 1974. Hybrid cells and human cells. *Scientific American*, July.

Smith, K. A. 1990. Interleukin-2. *Scientific American*, March.

Tonegawa, S. 1983. Somatic generation of antibody diversity. *Nature* 302: 575.

Watkins, M. W. 1966. Blood group substances. *Science* 152: 172.

Young, J. D., and Z. A. Cohen. 1988. How killer cells kill. *Scientific American*, January.

Chapter 16

Abelson, J., and E. Butz. 1980. Recombinant DNA. *Science* 209: 1317.

Anderson, W. F., and E. G. Diacumakos. 1981. Genetic engineering in mammalian cells. *Scientific American*, July.

Baulcombe, D. 1989. Strategies for virus resistance in plants. *Trends in Genetics* 5: 56.

Broome, S., and W. Gilbert. 1978. Immunological screening method to detect specific translation products. *Proceedings of the National Academy of Sciences, USA* 5: 2746.

Brown, D. D. 1973. The isolation of genes. *Scientific American*, August.

Capecchi, M. R. 1989. The new mouse genetics: altering the genome by gene targeting. *Trends in Genetics* 5: 70.

Chilton, M.-D. 1983. A vector for introducing new genes into plants. *Scientific American*, June.

Cohen, S. N. 1975. The manipulation of genes. *Scientific American*, July.

Cooke, H. 1987. Cloning in yeast: an appropriate scale for mammalian genomes. *Trends in Genetics* 3: 173.

Curtiss, R. 1976. Genetic manipulation of microorganisms: potential benefits and hazards. *Annual Review of Microbiology* 30: 507.

Davies, K. E., ed. 1988. *Genome Analysis: A Practical Approach.* IRL.

Davies, K. E., and S. M. Tilghman, eds. 1990. *Genetic and Physical Mapping.* Cold Spring Harbor Laboratory.

Drlica, K. 1984. *Understanding Gene Cloning*. Wiley.

Emery, A. E. H. 1984. *An Introduction to Recombinant DNA*. Wiley.

Gilbert, W., and L. Villa-Kramanoff. 1980. Useful proteins from recombinant bacteria. *Scientific American*, April.

Hacket, P. B., J. A. Fuchs, and J. W. Messing. 1984. *An Introduction to Recombinant DNA Techniques*. Benjamin/Cummings.

Luria, S. E. 1970. The recognition of DNA in bacteria. *Scientific American*, January.

Mertz, J., and R. Davis. 1972. Cleavage of DNA: RI restriction enzyme generates cohesive ends. *Proceedings of the National Academy of Sciences, USA* 69: 3370.

Pestka, S. 1983. The purification and manufacture of human interferons. *Scientific American*, August.

Roberts, R. J. 1980. Restriction and modification enzymes and their recognition sequences. *Gene* 8: 329.

Sambrook, J., E. F. Fritsch, and T. Maniatis. 1989. *Molecular Cloning: A Laboratory Manual*, 2d ed. Cold Spring Harbor Laboratory.

Seeberg, P. H., et al. 1978. Synthesis of growth hormone by bacteria. *Nature* 276: 795.

Smith, D. H. 1979. Nucleotide sequence specificity of restriction enzymes. *Science* 205: 455.

Stephens, J. C., M. L. Cavanaugh, M. L. Gradie, M. L. Mador, and K. K. Kidd. 1990. Mapping the human genome: current status. *Science* 250: 237.

Verna, I. M. 1990. Gene therapy. *Scientific American*, November.

Wu, R., ed. 1979. Recombinant DNA. Methods in Enzymology, vol. 68.

Chapter 1

1-1. The strain or variety must be homozygous for all genes that affect the trait.

1-2. Genotype *Aa* gives *A* and *a* gametes, *Bb* gives *B* and *b*, and *AaBb* gives *AB*, *Ab*, *aB*, and *ab*.

1-3. Multiply the number of possible gametes formed for each gene (one for each homozygous gene and two for each heterozygous gene). For genotype *AA Bb Cc Dd Ee*, there are a total of $(1)(2)(2)(2)(2) = 16$ possible gametes. With *n* homozygous genes and *m* heterozygous genes, the number of possible gametes is $1^n 2^m$, which equals 2^m.

1-4. With dominance, there are two phenotypes, corresponding to the genotypes *RR* or *Rr* (dominant) and *rr* (recessive), in the ratio 3 : 1. With no dominance, there are three phenotypes, corresponding to the genotypes *RR*, *Rr*, and *rr*, in the ratio 1 : 2 : 1.

1-5. Incomplete dominance of the alleles of a single gene. One homozygous genotype is black, the other is splashed-white, and the heterozygote is slate blue.

1-6. 1/2; 1/2. The probability argument is that each birth is independent of all the preceding ones, and so the particular combination of girls and boys in a particular family has no influence on the sexes of future children.

1-7. Because the probability of each child's being a girl is 1/2, and the two children are independent, the probability of two girls in a row is $(1/2)(1/2) = 1/4$. The probability of a girl and a boy (not necessarily in that order) equals $2(1/4)(1/4) = 1/2$. The reason for the factor of two in this case is that the sexes may be in either of two possible orders: girl-boy or boy-girl. Each of these possibilities has a probability of 1/4, and so the total is $1/4 + 1/4 = 1/2$.

1-8. (a) The parent with the dominant phenotype must carry one copy of the recessive allele and hence must be heterozygous. (b) Because some of the progeny are homozygous recessive, both parents must be heterozygous. (c) The parent with the dominant phenotype could be either homozygous dominant or heterozygous; the occurrence of no homozygous recessive offspring encourages the suspicion that the parent may be homozygous dominant but, because there are only two offspring, heterozygosity cannot be ruled out.

1-9. An *A b* gamete can be formed only if the parent is *AA Bb*, which has probability 1/2 and, in this case, 1/2

of the gametes are *A b*; overall, the probability of an *A b* gamete is $1/2 \times 1/2 = 1/4$. *A B* gametes derive from *AA BB* parents with probability 1 and from *AA Bb* parents with probability 1/2; overall, the probability of an *A B* gamete is $1/2 \times 1 + 1/2 \times 1/2 = 3/4$. (Alternatively, the probability of an *A B* gamete may be calculated as $1 - 1/4 = 3/4$, because *A b* and *A B* are the only possibilities.)

1-10. (a) Two phenotypic classes are expected for the A-*a* pair of alleles (*A–* and *aa*), two for the *B-B* pair (*B–* and *bb*), and three for the *R-r* pair (*RR*, *Rr*, and *rr*), yielding a total number of phenotypic classes of $2 \times 2 \times 3 = 12$. (b) The probability of an *aa bb RR* individual is 1/4 (*aa*) $\times$ 1/4 (*bb*) $\times$ 1/4 (*RR*) = 1/64. (c) Homozygosity may occur for either allele of each of the three genes, giving such combinations as *AA BB rr*, *aa bb RR*, *AA bb rr*, and so forth. Because the probability of homozygosity for either allele is 1/2 for each gene, the proportion expected to be homozygous for all three genes is $(1/2)(1/2)(1/2) = 1/8$.

1-11. The probability of a heterozygous genotype for any one of the genes is 1/2, and so for all four together the probability is $(1/2)^4 = 1/16$.

1-12. Because both parents have solid coats but produce some spotted offspring, they must be *Ss*. With respect to the A-*a* pair of alleles, the female parent (tan) is *aa* and, because there are some tan offspring, the genotype of the black male parent must be *Aa*. Thus, the parental genotypes are *Ss aa* (solid tan female) and *Ss Aa* (solid black male).

1-13. Because one of the children is deaf (genotype *dd*, in which *d* is the recessive allele), both parents must be heterozygous, *Dd*. The progeny genotypes expected from the mating *Dd* $\times$ *Dd* are 1/4 *DD*, 2/4 *Dd*, and 1/4 *dd*. The son is not deaf and hence cannot be *dd*. Among the nondeaf offspring, the genotypes *DD* and *Dd* are in the proportions 1 : 2, and so their relative probabilities are 1/3 *DD* and 2/3 *Dd*. Therefore, the probability that the normal son is heterozygous is 2/3.

1-14. (a) Because the trait is rare, it is reasonable to assume that the affected father is heterozygous, *Hh*. Half of his gametes contain the *H* allele, and so the probability is 1/2 that the son received the allele and will later develop the disorder. (b) We do not know whether the son is heterozygous *Hh*, but the probability is 1/2 that he is; and, if he is heterozygous, half of his gametes will contain the *H* allele. Therefore, the overall probability that his child has the *H* allele is $(1/2)(1/2) = 1/4$.

1-15. Both parents must be heterozygous (*aa*), because each has an albino (*aa*) parent. Therefore, the probability of an albino child is 1/4, and the probability of two homozygous recessive children is (1/4)(1/4) = 1/16. The probability that at least one child is an albino is the probability that exactly one is albino plus the probability that both are albinos. The probability that exactly one is an albino is 2(3/4)(1/4) = 6/16, in which the factor of 2 comes from the fact that there may be two birth orders: normal-albino or albino-normal. Therefore, the overall probability of at least one albino child equals 1/16 + 6/16 = 7/16. [Alternatively, the probability of at least one albino child may be calculated as 1 minus the probability that both are nonalbino, or $1 - (3/4)^2 = 7/16$.]

1-16. Compatible transfusions are A donor with A or AB recipient, B donor with B or AB recipient, AB donor with AB recipient, and O donor with any recipient.

1-17. (a) The cross *RR BB* × *rr bb* produces F$_1$ progeny of genotype *Rr Bb*, which have red kernels. (b) Because there is independent segregation (independent assortment), the F$_2$ genotypes are *R– B–*, *R– bb*, *rr B–*, and *rr bb*. These genotypes are in the proportions 9 : 3 : 3 : 1, respectively, and they produce kernels that are red, brown, brown, and white, respectively. Therefore, the F$_2$ plants have red, brown, or white seeds in the proportions 9/16 : 6/16 : 1/16.

1-18. Because the genes segregate independently, they can be considered separately. For the *Cr-cr* pair of alleles, the genotypes of the zygotes are *Cp Cp*, *Cp cp*, and *cp cp* in the proportions 1/4, 1/2, and 1/4, respectively. However, the *Cp Cp* zygotes do not survive, and so among the survivors the genotypes are *Cp cp* (creeper) and *cp cp* (noncreeper), in the proportions 2/3 and 1/3, respectively. For the *W-w* pair of alleles, the progeny genotypes are *W–* (white) and *ww* (yellow), in the proportions 3/4 and 1/4, respectively. Altogether, the phenotypes and proportions of the surviving offspring can be obtained by multiplying (2/3 creeper + 1/3 noncreeper) × (3/4 white + 1/4 yellow), which yields 6/12 creeper, white + 2/12 creeper, yellow + 3/12 noncreeper, white + 1/12 noncreeper, yellow. Note that the proportions sum to 12/12 = 1, which serves as a check.

1-19. Colored offspring must have the genotype *C– ii*; otherwise, color could not be produced (as in *cc*) or would be suppressed (as in *I–*). Among the F$_2$ progeny, 3/4 of the offspring have the *C–* genotype and 1/4 of the offspring have the *ii* genotype. Because the genes undergo independent assortment, the overall expected proportion of colored offspring is 3/4 × 1/4 = 3/16.

1-20. The 9 : 3 : 3 : 1 ratio is that of the genotypes *A– B–*, *A– bb*, *aa B–*, and *aa bb*. The modified 9 : 7 ratio implies that all of the last three genotypes have the same phenotype. The F$_1$ testcross is between the genotypes *Aa Bb* × *aa bb*, and the progeny are expected in the proportions 1/4 *Aa Bb*, 1/4 *Aa bb*, 1/4

aa Bb, and 1/4 *aa bb*. Again, the last three genotypes have the same phenotype, and so the ratio of phenotypes among progeny of the testcross is 1 : 3.

1-21. For the child to be affected, both III-1 and III-2 must be heterozygous *Aa*. For III-1 to be *Aa*, individual II-2 must be *Aa* and must transmit the recessive allele to III-1. The probability that II-2 is *Aa* equals 2/3, and then the probability of transmitting the *a* allele to III-1 is 1/2. Altogether, the probability that III-1 is a carrier equals (2/3)(1/2) = 1/3. This is also the probability that III-2 is a carrier. Given that both III-1 and III-2 have genotype *Aa*, the probability of an *aa* offspring is 1/4. Overall, the probability that both III-1 and III-2 are carriers and that they have an *aa* child is (1/3)(1/3)(1/4) = 1/36. The numbers are multiplied because each of the events is independent. (The reason for the 2/3 is as follows: because II-3 is affected, the genotypes of I-1 and I-2 must be *Aa*. Among nonaffected individuals from this mating (individuals II-2 and II-4), the ratios of *AA* and *Aa* are 1/4 : 2/4, and so the probability of *Aa* is 2/3.)

1-22. For any individual offspring, the probabilities of black *B–* and white *bb* are 3/4 and 1/4, respectively. (a) The order white-black-white occurs with probability (1/4)(3/4)(1/4) = 3/64. The order black-white-black occurs with probability (3/4)(1/4)(3/4) = 9/64. Therefore, the probability of either the order white-black-white or the order black-white-black equals 3/64 + 9/64 = 3/16. The probabilities are summed because the events are mutually exclusive. (b) The probability of exactly two white and one black equals $3 \times (1/4)^2 (3/4)^1 = 9/64$. The factor of 3 is the number of birth orders of two white and one black (there are three because the black offspring must be either the first, second, or third), and the rest of the expression is the probability that any one of the birth orders occurs (in this case, 3/64).

1-23. The probability of all boys is $(1/2)^4 = 1/16$. The probability of all girls is also 1/16, and so the probability of either all boys or all girls equals 1/16 + 1/16 = 1/8. The probability of equal numbers of the two sexes is $6 \times (1/2)^4 = 3/8$. The factor of 6 is the number of possible birth orders of two boys and two girls (that is, BBGG, BGBG, BGGB, GGBB, GBGB, GBBG).

1-24. Black and splashed white are the homozygous genotypes and slate blue the heterozygote. The probabilities of each of these phenotypes from the mating between two heterozygotes are 1/4 black, 1/2 slate blue, and 1/4 splashed white. The probability of occurrence of the particular birth order black-slate blue-splashed white is (1/4)(1/2)(1/4) = 1/32, but altogether there are six different orders in which exactly one of each type could be produced. (The black offspring could be first, second, or third and, in each case, the remaining phenotypes could be in either of two orders.) Consequently, the overall probability of one of each phenotype, in any order, is $6 \times (1/32) = 3/16$.

1-25. The probability that one or more offspring are *aa* equals 1 minus the probability that all are A−. In the mating *Aa* × *Aa*, the probability that all of *n* offspring will be *Aa* equals $(3/4)^n$, and the question asks for the value of *n* such that $1 - (3/4)^n > 0.95$. Solving this inequality yields $n > \log(1 - 0.95)/\log(3/4) = 10.4$. Therefore, 11 is the smallest number of offspring for which there is a greater than 95 percent chance that at least one will be *aa*. As a check, note that $(3/4)^{10} = 0.056$ and $(3/4)^{11} = 0.042$, and so with 10 offspring the chance of one or more *aa* equals 0.944 and with 11 offspring the chance of one or more *aa* equals 0.958.

Chapter 2

2-1. DNA replication occurs during the S period within the interphase stage of mitosis and within the interphase-I stage of meiosis. (Note that chromosome replication does *not* occur during the interphase II of meiosis.) Chromosomes condense and first become visible in the light microscope during prophase of mitosis and prophase I of meiosis.

2-2. Each chromosome present in telophase of mitosis is identical with one of a pair of chromatids present in the preceding metaphase that are held together at the centromere. In this example, there are 23 *pairs* of chromosomes present after telophase, or $23 \times 2 = 46$ chromosomes altogether. Therefore, at the preceding metaphase there were $46 \times 2 = 92$ chromatids.

2-3. In leptotene, the chromosomes first become visible as *thin threads*. In zygotene, they become *paired threads* because of synapsis. The chromosomes continue to condense to become *thick threads* in pachytene. In diplotene, the sister chromatids become evident and each chromosome is seen to be a *paired thread* consisting of the sister chromatids. (Early signs of the bipartite nature of each chromosome are evident in pachytene.) In diakinesis, the homologous chromosomes repulse each other and *move apart*, and they would fall apart entirely were they not held together by the chiasmas.

2-4. The chromosome number. Each centromere defines a single chromosome; that is, two chromatids are considered part of a single chromosome as long as they share a single centromere. As soon as the centromere splits during anaphase, each chromatid is considered to be a chromosome in its own right. Meiosis I reduces the chromosome number from diploid to haploid because the daughter nuclei contain the haploid number of chromosomes (each chromosome consisting of a single centromere connecting two chromatids.) In meiosis II, the centromeres split, and so the number of chromosomes is kept equal.

2-5. (a) 20 chromosomes and 40 chromatids; (b) 20 chromosomes and 40 chromatids; (c) 10 chromosomes and 20 chromatids.

2-6. (a) For each centromere, there are two possibilities (for example, A or a); so altogether there are $2^7 = 128$ possible gametes. (b) The probability of a gamete's receiving a particular centromere designated by a capital letter is 1/2 for each centromere, and each centromere segregates independently of the others; hence, the probability for all seven simultaneously is $(1/2)^7 = 1/128$.

2-7. The wheat parent produces gametes with 14 chromosomes and the rye parent gametes with 7, and so the hybrid plants have 21 chromosomes.

2-8. The crisscross occurs because the X chromosome in a male is transmitted only to his daughters. The expression is misleading because any X chromosome in a female can be transmitted to a son or a daughter.

2-9. Because there is an affected son, the phenotypically normal mother must be heterozygous. This implies that the daughter has a probability of 1/2 of being heterozygous.

2-10. The *Bar* mutation is an X-linked dominant. Because the sexes are affected equally, some association with the sex chromosomes is suggested. Mating (a) provides an important clue. Because a male receives his X chromosome from his mother, the wildtype phenotype of the sons suggests that the *Bar* gene is in the X chromosome. The fact that all daughters are affected is also consistent with X-linkage, provided that the *Bar* mutation is dominant. Mating (b) confirms the hypothesis, because the females from mating (a) would have the genotype *Bar*/+ and so would produce the indicated progeny.

2-11. (a) Cross is *v*/Y × +/+, in which Y represents the Y chromosome; progeny are 1/2 +/Y males (wildtype) and 1/2 *v*/+ females (phenotypically wildtype, but heterozygous). (b) Mating is *v*/*v* × +/Y; progeny are 1/2 *v*/Y males (vermilion) and 1/2 *v*/+ females (phenotypically wildtype, but heterozygous). (c) Mating is *v*/+ × +/Y; progeny are 1/4 *v*/+ females (phenotypically wildtype), 1/4 +/+ females (wildtype), 1/4 *v*/Y males (vermilion), and 1/4 +/Y males (wildtype). (d) Mating is *v*/+ × *v*/Y; progeny are 1/4 *v*/*v* females (vermilion), 1/4 *v*/+ females (wildtype), 1/4 *v*/Y males (vermilion), and 1/4 +/Y males (wildtype).

2-12. All the females will be *v*/v^+; *bw*/*bw* and will have brown eyes; all the males will be *v*/Y; *bw*/*bw* and will have white eyes.

2-13. Because the sex-chromosome situation is the reverse of that in mammals, female chickens are XY and males XX, and females receive their X chromosome from their father. Therefore, the answer is to mate a gold male (*ss*) with a silver female (*S*). All the male progeny will be *Ss* (silver plumage) and the female progeny will be *s* (gold plumage), and these are easily distinguished.

2-14. (a) The mother is a heterozygous carrier of the mutation, and therefore the probability that a daughter is a carrier is 1/2. The probability that both daughters are carriers is $(1/2)^2 = 1/4$. (b) If the daughter is not heterozygous, the probability of an affected son is 0 and, if the daughter is heterozygous, the probability of an affected son is 1/2; the overall probability is therefore 1/2 (that is, the chance that the daughter is a carrier) $\times$ 1/2 (that is, the probability of an affected son if the daughter is a carrier) = 1/4.

2-15. Let A and a represent the normal and mutant X-linked alleles. Y represents the Y chromosome. The genotypes are as follows: I-1 is Aa (because there is an affected son), I-2 is AY, II-1 is AY, II-2 is aY, II-3 is Aa (because there is an affected son), II-4 AY, III-1 aY.

2-16. Twenty, because, in females, there are five homozygous and ten heterozygous genotypes and, in males, there are five hemizygous genotypes.

2-17. The Punnett square shows the outcome of the cross. The X and Y chromosomes from the male are denoted in red, the attached-X chromosomes (yoked Xs) and Y from the female in black. The red X chromosome also contains the w mutation. The zygotes in the upper left and lower right corners (shaded) fail to survive because they have three X chromosomes or none. The surviving progeny consist of white-eyed males $\hat{X}$XY and red-eyed females XXY in equal proportions. Attached-X inheritance differs from the typical situation in that the *sons*, rather than the daughters, receive their fathers' X chromosomes.

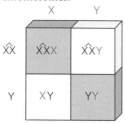

2-18. The genotype of the female is Cc, in which c denotes the allele for color blindness. In meiosis, the c-bearing chromatids undergo nondisjunction and produce an XX-bearing egg of genotype cc. Fertilization by a normal Y-bearing sperm results in an XXY zygote with genotype cc, which results in color blindness.

2-19. The female produces 1/2 X-bearing eggs and 1/2 O-bearing eggs, and the male produces 1/2 X-bearing sperm and 1/2 Y-bearing sperm. Random combinations result in 1/4 XX, 1/4 XO, 1/4 XY, and 1/4 YO, and the last class dies because no X chromosome is present. Among the survivors, the expected progeny are 1/3 XX females, 1/3 XO females, and 1/3 XY males.

2-20. (a) II-2 must be heterozygous, because she has an affected son, which means that half of her sons will be affected. The probability of a nonaffected child is therefore 3/4, and the probability of two nonaffected children is $(3/4)^2 = 9/16$. (b) Because the mother of II-4 is heterozygous, the probability that II-4 is heterozygous is 1/2. If she is heterozygous, half of her

sons will be affected. Therefore, the overall probability of an affected child is $1/2 \times 1/4 = 1/8$.

2-21. (a) This is like asking for the probability of the sex distribution GGGBBB (in that order), which equals ($1/2^6 = 1/64$). (b) This is like asking for the probability of the sex distribution either GGGBBB or BBBGGG, which equals $2(1/2)^6 = 1/32$.

2-22. $[4!/2!2!](1/2)^2(1/2)^2 = 3/8 = 0.375$; $[8!/4!4!](1/2)^4(1/2)^4 = 0.273$.

2-23. $(3/4)^4 = 81/256 = 0.316$; $[4!/3!1!](3/4)^3(1/4)^1 = 108/256 = 0.422$.

2-24. With 1 : 1 segregation, the expected numbers are 125 in each class, and so χ^2 equals $(140 - 125)^2/125 + (110 - 125)^2/125 = 3.6$, and there is one degree of freedom because there are two classes of data. In Figure 2-16, the associated P value is approximately 6 percent, which means that a fit at least as bad would be expected 6 percent of the time even if Mendelian segregation did occur. Therefore, on the basis of these data, there is no justification for rejecting the hypothesis of 1 : 1 segregation. On the other hand, the observed P value is close to 5 percent, which is the conventional level for rejecting the hypothesis, and so the result should not inspire great confidence.

2-25. The total number of progeny equals 800 and, on the assumption of a 9 : 3 : 3 : 1 ratio, the expected numbers equal 450, 150, 150, and 50, respectively. The χ^2 value equals $(462 - 450)^2/450 + (167 - 150)^2/150 + (127 - 150)^2/150 + (44 - 50)^2/50 = 6.49$, and there are three degrees of freedom because there are four classes of data. From Figure 2-16, the P value is approximately 0.09, which is well above the conventional rejection level of 0.05. Consequently, the observed numbers provide no basis for rejecting the hypothesis of a 9 : 3 : 3 : 1 ratio.

Chapter 3

3-1. Because the progeny are wildtype (they complement each other), both m_1 and m_2 must be heterozygous, which indicates that the genotype is $m_1 +/+ m_2$. This implies that m_1 and m_2 are mutations in different genes.

3-2. (a) $A b$, $a B$, $A B$, and $a b$. (b) $A b$ and $a B$.

3-3. The nonrecombinant gametes are $A B$ and $a b$, and the recombinant gametes are $A b$ and $a B$. The nonrecombinant gametes are always more frequent than the recombinant gametes.

3-4. The gametes from the $A B/a b$ parent are $A B$ and $a b$; those from the other parent are $A b$ and $a B$. The offspring genotypes are expected to be $A B/A b$, $A B/a B$, $a b/A b$, and $a b/a B$, in equal numbers. The phenotypes are $A-B-$, $A-bb$, and $aa B-$, which are expected in the ratio 2 : 1 : 1.

3-5. The *cis* configuration is *a b/A B*.

3-6. There is no crossing-over in male *Drosophila*, and so the gametes will be *A B* and *a b* in equal proportions. In the female, each of the nonrecombinant gametes *A B* and *a b* is expected with a frequency of $(1 - 0.05)/2 = 0.475$, and each of the recombinant gametes *A b* and *a B* is expected with a frequency of $0.05/2 = 0.025$.

3-7. (a) 17; (b) 1; (3) in diploids, there is one linkage group per pair of homologous chromosomes, and so the number of linkage groups is $42/2 = 21$.

3-8. Recombination at a rate of 6.2 percent implies that double crossovers (and other multiple crossovers) occur at a negligible frequency; in such a case, the distance in map units equals the frequency of recombination, and so the map distance between the genes is 6.2 map units.

3-9. Each meiotic cell produces four products—namely, *A B* and *a b*, which contain the chromatids not participating in the crossover, and *A b* and *a B*, which contain the chromatids that do take part in the crossover. Therefore, the frequency of recombinant gametes is 50 percent. Because one crossover occurs in the region between the genes in every cell, the occurrence of multiple crossing-over will not affect the recombination frequency as long as the chromatids participating in each crossover are chosen at random.

3-10. The *aa bb* genotype requires an *a b* gamete from each parent. From the *A b/a B* parent, the probability of an *a b* gamete is $0.16/2 = 0.08$ and, from the *A B/a b* parent, the probability of an *a b* gamete is $(1 - 0.16)/2 = 0.42$; so the probability of an *aa bb* progeny is $0.08 \times 0.42 = 0.034$, or 3.4 percent.

3-11. All the female gametes are *bz m*. The male gametes and the progeny that arise from them are *bz⁺M* (black males, frequency $[1 - 0.06]/2 = 47$ percent), *bz m* (bronze females, frequency 47 percent), *bz⁺m* (black females, frequency $0.06/2 = 3$ percent), and *bz M* (bronze males, frequency 3 percent).

3-12. The most-frequent gametes from the double heterozygous parent are the nonrecombinant gametes, in this case *Gl ra* and *gl Ra*. The genotype of the parent was therefore *Gl ra/gl Ra*. The recombinant gametes are *Gl Ra* and *gl ra*, and the frequency of recombination is $(6 + 3)/(88 + 103 + 6 + 3) = 4.5$ percent.

3-13. There is no recombination in the male, and so the male gametes are $+ +$ and *b cn*, each with a probability of 0.5. A map distance of 8 units means that the recombination frequency between the genes is 0.08. The female gametes are $+ +$, *b cn* (the nonrecombinants, each of which occurs at a frequency of $[1 - 0.08]/2 = 0.46$) and $+ cn$ and $b +$ (the recombinants, each of which occurs at a frequency of $0.08/2 = 0.04$). When the male and female gametes are combined at

random, their frequencies are multiplied, and the progeny and their frequencies are as follows:

DI $+ +/+ +$	0.23	$+ +/b\,cn$	0.23
$b\,cn/+ +$	0.23	$b\,cn/b\,cn$	0.23
$b +/+ +$	0.02	$b +/b\,cn$	0.02
$+ cn/b\,cn$	0.02	$+ cn/+ +$	0.02

Note that $b\,cn/+ +$ is the same as $+ +/b\,cn$, and so the overall frequency of this genotype is 0.46. With regard to phenotypes, the progeny are wildtype (73 percent), black cinnabar (23 percent), black (2 percent), and cinnabar (2 percent).

3-14. The most-frequent classes of gametes are the nonrecombinants, and so the parental genotype was *A B c/ a b C*. The least-frequent gametes are the double recombinants and the allele that distinguishes them from the nonrecombinants (in this case, *C* and *c*) is in the middle. Therefore, the gene order is A–C–B.

3-15. Comparison of the first three numbers implies that *r* lies between *c* and *p*, and the last two numbers imply that *s* is closer to *r* than to *c*. The map is therefore $c-10-r-3-p-5-s$. From this you would expect the recombination frequency between *c* and *p* to be 13 percent and that between *c* and *s* to be 18 percent. The observed values are a little smaller because of double crossovers.

3-16. (a) The nonrecombinant gametes (the most-frequent classes) are *y v⁺ sn* and *y⁺ v sn⁺*, and the double recombinants (the least-frequent classes) indicate that *sn* is in the middle. Therefore, the correct order of the genes is *y–sn–v*, and the parental genotype was *y sn v⁺/y⁺ sn⁺ v*. The total progeny is 1000. The recombination frequency in the *y–sn* interval is $(108 + 5 + 95 + 3)/1000 = 21.1$ percent, and the recombination frequency in the *sn–v* interval is $(53 + 5 + 63 + 3)/1000 = 12.4$ percent. The genetic map is $y-21.1-sn-12.4-v$. (b) On the assumption of independence, the expected frequency of double crossovers is $0.211 \times 0.124 = 0.0262$, whereas the observed is $(5 + 3)/1000 = 0.008$. The coincidence is the ratio $0.008/0.0262 = 0.31$, and so the interference is $1 - 0.31 = 0.69$, or 69 percent.

3-17. The nonrecombinant gametes are *c wx Sh* and *C Wx sh*, and the double recombinants indicate that *sh* is in the middle. The parental genotype was therefore *c Sh wx/C sh Wx*. The frequency of recombination in the *c–sh* region is $(84 + 20 + 99 + 15)/6708 = 3.25$ percent, and that in the *sh–wx* region is $(974 + 20 + 951 + 15)/6708 = 29.2$ percent. The genetic map is therefore $c-3.25-sh-29.2-wx$.

3-18. Let *x* be the observed frequency of double crossovers. Then, single crossovers in the *a–b* interval occur with frequency $0.15 - x$ and single crossovers in the *b–c* interval occur with frequency $0.20 - x$. The observed recombination frequency between *a* and *c* equals the sum of the single crossover frequencies (because double are undetected), which implies that $0.15 - x + 0.20 - x = 0.31$, or $x = 0.02$. The expected frequency of double crossovers equals $0.15 \times 0.20 = 0.03$, and the coincidence is therefore $0.02/0.03 = 0.67$. The interference equals $1 - 0.67 = 33$ percent.

3-19. (a) On the assumption of independence, $(0.30)(0.10)/2 = 0.015$. (Division by 2 is necessary because *sh wx gl* gametes constitute only one of the two classes of double recombinants.) (b) If the interference is 60 percent, the coincidence equals $1 - 0.60 = 0.40$. The observed frequency of doubles is therefore 40 percent as great as with independence, and so the expected frequency of *sh wx gl* gametes is $0.015 \times 0.40 = 0.006$.

3-20. *b* and *st* are unlinked, as are *hk* and *st*. The evidence is the $1 : 1 : 1 : 1$ segregation observed. For *b* and *st*, the comparisons are $243 + 10$ (black, scarlet) versus $241 + 15$ (black, nonscarlet) versus $226 + 12$ (nonblack, scarlet) versus $235 + 18$ (nonblack, nonscarlet). For *hk* and *st* the comparisons are $226 + 10$ (hook, scarlet) versus $235 + 15$ (hook, nonscarlet) versus $243 + 12$ (nonhook, scarlet) versus $241 + 18$ (nonhook, nonscarlet). On the other hand, *b* and *hk* are linked, and the frequency of recombination between them is $(15 + 10 + 12 + 18)/1000 = 5.5$ percent.

3-21. The frequency of *dpy-21 unc-34* recombinants is expected to be $0.24/2 = 0.12$ among both eggs and sperm, and so the expected frequency of *dpy-21 unc-34/dpy-21 unc-34* zygotes is $0.12^2 = 1.44$ percent.

3-22. The genes must be linked because, with independent assortment, the frequency of dark eyes $(R-P-)$ would be $9/16$ in both experiments. The first experiment is a mating of $R\,P/r\,p \times r\,p/r\,p$. The dark-eyed progeny represent half the nonrecombinants, and so the recombination frequency r can be estimated as $(1 - r)/2 = 628/(628 + 889)$, or $r = 0.172$. The second experiment is a mating of $R\,p/R\,p \times r\,p/r\,p$. In this case, the dark-eyed progeny represent half the recombinants, and so r can be estimated as $r/2 = 86/(86 + 771)$, or $r = 0.201$. It is not uncommon for repeated experiments or crosses carried out in different ways to yield somewhat different estimates of the recombinant frequency—in this problem, 0.17 and 0.20. The most-reliable value is obtained by averaging the experimental values—in this case, $(1/2)(0.171 + 0.201) = 0.186$.

3-23. Consider row 1: there are $-$ entries for 3 and 7, and so 1, 3, and 7 form one complementation group. Row 2: all $+$, and so 2 is the only representative of a second complementation group. Row 3: mutation already classified. Row 4: there is a $-$ for 6 only and no other $-$ in column 4, and so 4 and 6 constitute a third complementation group. Row 5: $-$ only for 9 and no other $-$ in column 5, and so 5 and 9 form a fourth complementation group. Rows 6 and 7: mutations already classified. Row 8: all $+$, and so 8 is the only member of a fifth complementation group. In summary, there are five complementation groups as

follows: group 1 (mutations 1, 3, and 7); group 2 (mutation 2); group 3 (mutations 4 and 6); group 4 (mutations 5 and 9); group 5 (mutation 8).

3-24. Consider each gene in relation to first- and second-division segregation. Gene *a* gives 1766 asci with first-division segregation and 234 with second-division segregation; the frequency of second-division segregation is $234/(1766 + 234) = 0.117$, which implies that the distance between *a* and the centromere is $0.117/2 = 5.85$ percent. Gene *b* gives 1780 and 220 first-division and second-division segregations, for a frequency of second-division segregation of $220/1780 + 220) = 0.110$. The distance between *b* and the centromere is therefore $0.110/2 = 5.50$ map units. If we consider *a* and *b* together, there are 1986 PD asci, 14 TT asci, and no NPD. Because NPD $<<$ PD, genes *a* and *b* are linked. By applying Equation (1), we find that the frequency of recombination between *a* and *b* is $(1/2)14/2000 = 0.35$ percent. A comparison of this distance with the gene–centromere distances calculated earlier results in the map a–0.35–b–5.50–centromere.

3-25. Considered individually, gene *c* yields 90 first-division segregation asci and 10 $(1 + 9)$ second-division segregation asci, and so the frequency of recombination between *c* and the centromere is $(10/100) \times 1/2 = 5$ percent. Gene *v* yields 79 first-division and 21 $(20 + 1)$ second-division asci, and so the frequency of recombination between *v* and the centromere is $(21/100) \times 1/2 = 10.5$ percent. If the genes are taken together, the asci include 34 PD, 36 NPD, and 30 $(20 + 1 + 9)$ TT. Because NPD = PD, the genes are unlinked. In summary, *c* and *v* are in different chromosomes with *c* 5 map units from the centromere and *v* 10.5 map units from the centromere.

Chapter 4

4-1. DNA: adenine, guanine, thymine, cytosine. Pairs: A with T, and G with C. The sugar is deoxyribose. A nucleoside is a base attached to a sugar. A nucleotide is a nucleoside with one or more phosphate groups attached.

4-2. One phosphate group per base in DNA. Each precursor nucleotide contains three phosphates.

4-3. One end terminates with a 3′-OH group (that is, a free hydroxyl on the 3′ carbon of the sugar), and the other terminates with a 5′-P group (that is, a free phosphate group on the 5′ carbon of the sugar).

4-4. A somatic cell normally contains two times as much DNA as a gamete, corresponding to the feature that meiosis reduces the chromosome number by half.

4-5. A template strand, a primer, DNA polymerase, and the four nucleoside triphosphates (dATP, dGTP, dCTP, dTTP). The 5′ phosphate (—P) of the incoming nucleotide reacts with the 3′ hydroxyl (—OH) of the elongating strand.

4-6. They are antiparallel in the sense that, at the end of a double helix, one strand terminates with a 3′-OH group and the other stand terminates with a 5′-P group.

4-7. Complementary strand is 5′-T-C-C-G-A-G-3′.

4-8. A nuclease is an enzyme capable of breaking a phosphodiester bond. An endonuclease can break any phosphodiester bond, whereas an exonuclease can only remove a terminal nucleotide.

4-9. Movement along the template strand is from the 3′ end to the 5′ end, because consecutive nucleotides are added to the 3′ end of the growing chain. In double-stranded DNA, one strand is replicated continuously and the other stand is replicated in short segments that are later joined.

4-10. DNA polymerase joins a 5′-triphosphate with a 3′-OH group, and the outermost two phosphates are released; the resulting phosphodiester bond contains one phosphate. DNA ligase joins a single 5′-P group with a 3′-OH group.

4-11. The enzyme has polymerizing activity, it is a 3′-5′ exonuclease (the proofreading function), and it is a 5′-3′ exonuclease.

4-12. Neither RNA polymerases nor primases require a primer to get synthesis started.

4-13. Smaller molecules move faster because they can penetrate the pores of the gel more easily.

4-14. (a) In rolling-circle replication, one parental strand remains in the circular part and the other is at the terminus of the branch. Therefore, only half of the parental radioactivity will appear in progeny. (b) If we assume no crossing-over between progeny DNA molecules, one progeny phage will be radioactive. If a single crossover occurs, two particles will be radioactive. If crossing-over is frequent and occurs at random, the radioactivity will be distributed among virtually all the progeny.

4-15. Rolling-circle replication must be initiated by a single-stranded break; θ replication does not need such a break.

4-16. (a) Because 18 percent is adenine, 18 percent is also thymine, for a total of 36 percent for [A] + [T]. (b) Because [A] + [T] = 36 percent, [G] + [C] equals 64 percent, or [G] = [C] = 32 percent each. Overall, the molecule is 64 percent GC. Thus, [C] is 32 percent.

4-17. (a) The distance between nucleotide pairs is 3.4 Å or 0.34 nm. The length of the molecule is 34 μm, or 34 × 10³ nm. Therefore, the number of nucleotide pairs equals $(34 \times 10^3)/0.34 = 10^5$. (b) There are ten nucleotide pairs per turn of the helix, and so the total number of turns equals $10^5/10 = 10^4$.

4-18. Because [A] is not equal to [T], and [G] is not equal to [C], the DNA cannot be double stranded and is probably single stranded.

4-19. (a) First, note that the convention for writing polynucleotide sequences is to put the 5′ end at the left. Thus, the complementary sequence to AGTC (which is really 5′-AGTC-3′) is 3′-TCAG-5′, written in conventional format as GACT (that is, 5′-GACT-3′). Hence, the frequency of CT is the same as that of its complement AG, which is given as 0.15. Similarly, AC = 0.03, TC = 0.08, and AA = 0.10. (b) If DNA had a parallel structure, the 5′ ends of the strands would be together, as would the 3′ ends, and so the sequence complementary to 5′-AGTC-3′ would be 5′-TCAG-3′. Thus, AG = 0.15 implies that TC = 0.15, and similarly the other observed nearest neighbors would imply that CA = 0.03, CT = 0.08, and AA = 0.10.

4-20. Remember that the group at the growing end of the leading strand is always a 3′-OH group because DNA polymerases can add nucleotides to only such a group. Because DNA is antiparallel, the opposite end of the leading strand is a 5′-P group. Therefore, (a) 3′-OH; (b) 3′-OH; (c) 5′-P.

4-21. Rolling-circle replication begins with a single-strand break that produces a 3′-OH group and a 5′-P group. The polymerase extends the 3′-OH terminus and displaces the 5′-P end. Because the DNA strands are antiparallel, the strand complementary to the displaced strand must be terminated by a 3′-OH group.

4-22. (a) The first base in the DNA that is copied is a T, and so the first base in the RNA is an A. RNA grows by addition to the 3′-OH group, and so the 5′-P end remains free. Therefore, the sequence of the eight-nucleotide primer is 5′-AGUCAUGC- 3′. (b) The C at the 3′ end has a free -OH; the A at the other end has a free *tri*phosphate. (c) Right to left, because this direction would require discontinous replication of the strand shown.

4-23. (a) The cloned SalI fragment is 20 kb in length and there are no SalI sites within it. (b) There is a single EcoRI site 7 kb from one end. (c) There are two HindIII sites within the fragment. (d) The HindIII sites must flank the EcoRI site, dividing the 7-kb EcoRI fragment into 3 and 4 kb and the 13-kb EcoRI fragment into 5 and 8 kb. The 3-kb and 8-kb fragments must be adjacent, because digestion with SalI and HindIII alone produces a 11-kg fragment. Except for the left-to-right orientation, the inferred restriction map must be as follows:

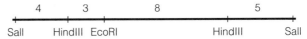

4-24. The bands in the gel result from incomplete chains whose synthesis was terminated by incorporation of the dideoxynucleotide indicated at the top. The smaller fragments are at the bottom of the gel, the larger ones at the top. Because DNA strands elongate by addition to their 3′ ends, the strand synthesized in the sequencing reactions has the sequence, from bottom to top, of 5′-ACTAGAGACCATGATCCTGTGATGAAT-AGC-3′. The template strand is complementary in sequence, and antiparallel, and so its sequence is 3′-TGATCTCTGGTACTAGGACACTACTTAT-CG-5′.

Chapter 5

5-1. Histones interact with deoxyribonucleic *acid*, which is indeed a weak acid. Each nucleosome contains two molecules of each of four distinct types of histone, forming the histone octamer, plus a single H1 histone in the linker. H1 is unpaired.

5-2. There is always a 1 : 1 : 1 : 1 ratio of histones H2A, H2B, H3, and H4 in all eukaryotic cells because of the universality of the histone octamer. The ratio of H1 to the others varies slightly.

5-3. No, many moderately and highly repetitive DNA sequences are not transposable elements; for example, the repetitive sequences found in telomeres and centromeres. The tipoff is a unique sequence occurring between two nearly identical sequences in direct or inverted orientation, which form the termini of the transposable element. Transposable elements are typically flanked by a short duplication of host sequence.

5-4. The 1.3-kb insertion is probably a transposable element. (In fact, it is — one designated *mariner*.)

5-5. Transposition in somatic cells could cause mutations that kill the host or decrease the ability of the host to reproduce, which reduces the chance of transmission of the transposable element to the next generation. Restriction of activity to the germ line lessens these potentially unfavorable effects.

5-6. Insertion is initiated by a staggered cut in the host DNA sequence, and the overhanging single-stranded ends are later filled in by a DNA polymerase. This process creates a direct repeat.

5-7. *E. coli*: 4,700,000 bp × 3.4 A°/bp × 10^{-7} mm/A° = 1.6 mm. Human beings: 3 × 10^9 bp × 3.4 A°/bp × 10^{-7} mm/A° = 1020 mm — that is, approximately one *meter*.

5-8. DNA structure is a long, thin thread, and its total length in most cells greatly exceeds the diameter of a nucleus, even allowing for a large number of fragments per cell. This forces you to conclude that each DNA molecule must be folded back on itself repeatedly.

5-9. Matching of chromomeres, which are locally folded regions of chromatin. Polytene chromosomes do not divide because the cells in which they exist do not divide.

5-10. Yes, but less than the number found in an equivalent amount of euchromatin.

5-11. The telomeres are at the tips of linear chromosomes. It is not possible to have a centromere exactly at the tip because each tip must terminate with a telomere to be genetically stable. A chromosome in the shape of a ring has no telomere.

5-12. No, because the *E. coli* chromosome is circular. There are no termini.

5-13. The repeated sequences are located at the ends of the element. They may be in direct or inverted orientation, depending on the particular transposable element.

5-14. (a) There are 10 base pairs for every turn of the double helix. With four turns of unwinding, 4 × 10 = 40 base pairs are broken. (b) Each twist compensates for the underwinding of one full turn of the helix; hence, there will be four twists.

5-15. The supercoiled form is in equilibrium with an underwound form having many unpaired bases. At the instant that a segment is single-stranded, S1 can attack it. Because a nicked circle lacks the strain of supercoiling, there is no tendency to have single-stranded regions, and so a nicked circle is resistant to S1 cleavage.

5-16. The rate of movement in the gel is determined by overall charge and by the ability of the molecule to penetrate the pores of the gel. Supercoiled and relaxed circles migrate at different rates because they have different conformations.

5-17. (a) Renaturation is concentration dependent because it requires complementary molecules to collide by chance before base pairs can form. (b) Here, the length of the GC tracts are important. Molecule 2 has a long GC segment, which will be the last region to separate. Hence molecule 1 has the lower temperature for strand separation.

5-18. The hybrid molecules that can be produced from single-stranded DNA fragments from species A and B are AA, AB, BA, and BB, and these are expected in equal amount. The molecules with a hybrid (^{14}N/^{15}N) density are AB and BA, and these account for 5 percent of the total. Together, AA and BB must make up another 5 percent, and so a total of 10 percent of the base sequences are common to the two species.

5-19. In a direct repeat, the sequence is repeated in the same 5′-to-3′ orientation. In an inverted repeat, the sequence is repeated in the reverse orientation and on the opposite strand, which is necessary to preserve the correct 5′-to-3′ orientation in view of the antiparallel nature of DNA strands. The figures below show a direct and inverted repeat of the example sequence. The dashed lines represent unspecified DNA sequences between the repeats.

Direct repeat

5'–C C A G G T G T A C A–3'·········5'–C C A G G T G T A C A–3'
3'–G G T C C A C A T G T–5'·········3'–G G T C C A C A T G T–5'

Inverted repeat

5'–C C A G G T G T A C A–3'·········5'–T G T A C A C C T G G–3'
3'–G G T C C A C A T G T–5'·········3'–A C A T G T G G A C C–5'

5-20. All such mutations are probably lethal. Because the amino acid sequence of histones is virtually the same in all organisms, this suggests that functional histones are extremely intolerant of amino acid changes.

5-21. The relatively large number of bands and the diverse locations in the genome among different flies suggests very strongly that the DNA sequence is a transposable element of some kind.

5-22. In the first experiment, the nucleosomes form at random positions in different molecules. In the second experiment, the protein X binds to a unique sequence in all the molecules, and the nucleosomes form by sequential addition of histone octamers on both sites of the X-binding site. Because nuclease attacks only in the linker regions, the cuts are made in small regions that are highly localized.

Chapter 6

6-1. A chromosome with two centromeres (dicentric); a chromosome with no centromere (acentric).

6-2. An inversion that includes the centromere, but with breakpoints that are not symmetrical around the centromere, moves the centromere from a central location in a metacentric chromosome to a noncentral location, forming a submetacentric chromosome. A metacentric chromosome can be formed from two acrocentric chromosomes that have long arms about equal in length by a Robertsonian translocation, which essentially fuses the acrocentrics at the centromere.

6-3. Gametes from a tetraploid are diploid. Possible gametes from $AAaa\ BBbb$ are $AABB$, $AABb$, $AAbb$, $AaBB$, $AaBb$, $Aabb$, $aaBB$, $aaBb$, and $aabb$.

6-4. An autopolyploid series is formed by the combination of identical sets of chromosomes. Therefore the chromosome numbers must be even multiples of the haploid number, or 10 (diploid), 20 (tetraploid), 30, 40, and 50.

6-5. Species S is an allotetraploid of A and B formed by hybridization of A and B, after which the chromosomes in the hybrid became duplicated. The unpaired chromosomes in the S × A cross are the 12 chromosomes from B, and the unpaired chromosomes in the S × B cross are the 14 chromosomes from A.

6-6. The chromosomes underwent replication with no cell division, resulting in an allotetraploid.

6-7. The only 45-chromosome karyotype found at appreciable frequencies in spontaneous abortions is 45,X, and so 45,X is the probable karyotype. Had the fetus survived, it would have been a 45,X female with Turner syndrome.

6-8. Because the X chromosome in the 45,X daughter contains the color-blindness allele, the 45,X daughter must have received the X chromosome from her father through a normal X-bearing sperm. The nondisjunction must therefore have occurred in the mother, resulting in an egg cell lacking any X chromosomes.

6-9. The mother has a Robertsonian translocation that includes chromosome 21. The child with Down syndrome has 46 chromosomes, including two copies of the normal chromosome 21 plus an additional copy attached to another chromosome (the Robertsonian translocation). This differs from the usual situation in which Down syndrome children have trisomy 21 (karyotype 47, +21).

6-10. The inversion has the sequence ABEDCFG, the deletion ABFG. The possible translocated chromosomes are (a) ABCDETUV and MNOPQRSFG or (b) ABCDESRQPONM and VUTFG. One of these possibilities includes two monocentric chromosomes and the other includes a dicentric and an acentric. Only the translocation with two monocentrics is genetically stable.

6-11. Because the deletion eliminates all wildtype genes in the deleted region, the phenotypic effects of recessive alleles in the homologous chromosome become manifest. This expression of recessive phenotypes is what is meant by "uncovering" a recessive allele.

6-12. The order is $a\ e\ d\ f\ c\ b$ or the other way around.

6-13. Genes b, a, c, e, d, and f are located in bands 1, 2, 3, 4, 5, and 6, respectively. The reasoning is as follows. Because deletion 1 uncovers any gene in band 1 but the other deletions do not, the pattern, $-++++$ observed for gene b puts b in band 1. Deletions 1–3 uncover band 2 but deletions 4–5 do not, and so the pattern $---++$ observed for gene a localizes gene a to band 2. Genes in band 3 are uncovered by deletions 1–4 but not 5, and so the pattern $----+$ implies that gene c is in band 3. Band 4 is uncovered by deletions 3–5 but not 1–2, which means that gene e with the pattern $++---$ is in band 4. Genes in band 5 are uncovered by deletions 4–5 but not 1–3, and so the pattern $+++--$ observed for gene d places it in band 5. Genes in band 6 are uncovered by deletion 5 but not 1–4, and so the pattern $++++-$ puts gene f in band 6.

6-14. The most likely explanation is an inversion. In the original strain, the inversion is homozygous, and so no problems arise in meiosis. The F_1 is heterozygous, and crossing-over within the inversion produces the dicentrics and acentrics. Because the crossover products are not recovered, the frequency of recombination is greatly reduced in the chromosome pair in which the inversion is heterozygous, and so the inversion is probably in chromosome 6.

6-15. Starting with chromosome (c), compare the chromosomes pair by pair to find those that differ by a single inversion. In this manner, you can deduce that inversion of the i–d–c region in (c) gave rise to (d), inversion of the h–g–c–d region in (d) gave rise to (a), and inversion of the c–d–e–f region in (a) gave rise to (b). Because you were told that (c) is the ancestral sequence, the evolutionary ancestry is (c) → (d) → (a) → (b).

6-16. The group of four synapsed chromosomes implies that a reciprocal translocation has occurred. The group of four includes both parts of the reciprocal translocation and their nontranslocated homologs.

6-17. Translocation heterozygotes are semisterile because segregation from the four synapsed chromosomes in meiosis frequently produces aneuploid gametes, which have large parts of chromosomes missing or present in excess. The aneuploid gametes fail to function in plants, and in animals they result in zygote lethality. The translocation homozygote is fully fertile because the translocated chromosomes undergo synapsis in pairs, and segregation is completely regular. In the cross translocation homozygote × normal homozygote, all progeny are expected to be semisterile.

6-18. In this male, there was a reciprocal translocation between the Curly-bearing chromosome and the Y chromosome. Because there is no crossing-over in the male, all Y-bearing sperm that give rise to viable progeny contain Cy and all X-bearing sperm that give rise to viable progeny contain Cy^+.

6-19. In this male, the tip of the X chromosome containing y^+ became attached to the Y chromosome, giving the genotype $y X/y^+ \cdot Y$, in which the raised dot designates which yellow allele is attached to the X or Y. The gametes are $y X$ and $y^+ \cdot Y$, and so all of the offspring are either yellow females or wildtype males.

6-20. The $A B c D E$ chromosome results from two-strand double crossing-over in the $B–C$ and $C–D$ regions. The double recombinant chromosome still has the inversion.

6-21. The unusual yeast strain is not completely haploid but is disomic for chromosome 2. Segregation of markers on this chromosome, like $his7$, is aberrant, but segregation of markers on other chromosomes is completely normal.

Chapter 7

7-1. Mitochondria carry out respiratory metabolism. Chloroplasts absorb the light required in photosynthesis and synthesize glucose from carbon dioxide and water.

Lower eukaryotic cells can usually survive without mitochondria by using a carbon source, such as glucose, that can be metabolized anaerobically. Cells of higher eukaryotes probably cannot survive without mitochondria. Plant cells that are nonphotosynthetic (for example, those in the roots) do not contain chloroplasts. However, green plants need chloroplasts in their photosynthetic cells to survive.

7-2. A variegated phenotype is a mosaic of patches of different colored tissue.

7-3. Maternal inheritance is through extranuclear factors transmitted through the cytoplasm. Maternal effects are determined by the nuclear genotype of the mother, typically through effects on the egg or (in mammals) the developing fetus.

7-4. Segregational petites imply that certain nuclear genes are needed for the maintenance of mitochondria.

7-5. Inheritance is strictly maternal, and so yellow is probably inherited cytoplasmically. A reasonable hypothesis is that the gene for yellow is contained in chloroplast DNA.

7-6. Pollen is not totally devoid of cytoplasm, and occasionally it contains a chloroplast that is transmitted to the zygote.

7-7. This trait is maternally inherited, and so the phenotype is that of the mother. The progeny are (a) green, (b) white, (c) variegated, and (d) green.

7-8. With X-linked inheritance, half of the F_2 males (1/4 of the total progeny) would be white. The observed result is that all of the F_2 progeny are green.

7-9. If the daughter chloroplasts segregate at random, both of them will go to the same pole half the time. For all four products to contain a descendant chloroplast, the segregation has to be perfect in three divisions, which occurs with probability $(1/2)^3 = 1/8$. Therefore, the probability that at least one cell lacks a descendant chloroplast equals $1 - 1/8 = 7/8$.

7-10. The m^- chloroplast is preferentially lost and the m alleles segregate $1 : 1$. The expected result is $1/2$ $a^+b^-m^+$ and $1/2$ $a^+b^- m^-$.

7-11. Petite, because neither parent contributes functional mitochondria.

7-12. Because every tetrad shows uniparental inheritance, it is likely that the trait is determined by factors transmitted through the cytoplasm, such as mitochondria. The $4 : 0$ tetrads result from diploids containing only mutant factors, and the $0 : 4$ tetrads result from diploids containing only wildtype factors.

7-13. In autogamy, one product of meiosis becomes a homozygous diploid nucleus, and so half of the Aa cells that undergo autogamy become AA and half become aa.

7-14. The mating pairs are 1/4 AA × AA, 1/2 AA × aa, and 1/4 aa × aa. They produce only AA, Aa and aa progeny, respectively, and so the progeny are 1/4 AA, 1/2 Aa, and 1/4 aa. In autogamy, the Aa genotypes become homozygous for either A or a, and so the expected result of autogamy is 1/2 AA and 1/2 aa.

7-15. The only functional pollen is a B. All the progeny of the AA bb plants are Aa Bb and all those of the aa BB plants are aa BB.

7-16. The F_1 females will be heterozygous for nuclear genes, and possibilities (1) and (2) imply that these females should have some viable sons. The all-female progeny supports the hypothesis of cytoplasmic inheritance.

7-17. Because the progeny coiling is determined by the genotype of the mother, the rightward coiling snail has genotype ss. To explain the rightward coiling, the mother of this parent snail would have contained the + allele and must in fact have been +/s.

Chapter 8

8-1. Mutation, migration, natural selection, and random genetic drift. In the absence of these processes, the allele frequencies do not change.

8-2. Fixation means that the allele frequency equals 1; loss means that it is 0.

8-3. Random genetic drift refers to changes in allele frequency that happen by chance in small populations. It occurs because the zygotes that are actually formed are but a small sample of a potentially infinite number, and most samples do not exactly match the population from which they come but differ to a greater or lesser extent because of chance. The ultimate fate of an allele when only random genetic drift is operating is either fixation or loss. Random changes in allele frequency are larger in small populations because statistical fluctuations are greater in smaller samples.

8-4. Because most of the alleles are in heterozygotes when the allele is rare.

8-5. In both cases, 0.

8-6. Adults 2/3 Aa and 1/3 aa. It is not necessary to assume random mating because the only fertile matings are Aa × Aa. These produce 1/4 AA, 1/2 Aa, and 1/4 aa zygotes, but the AA zygotes die, leaving 2/3 Aa and 1/3 aa adults.

8-7. The harmful recessive will have a smaller equilibrium frequency in the X-linked case because the allele is exposed to selection in hemizygous males.

8-8. The 10 AA individuals contain 20 A alleles, the 15 Aa contain 15 A and 15 a, and the 4 aa contain 8 a. Altogether there are 35 A and 23 a alleles, for a total of 58. The allele frequency of A is 35/58 = 0.603 and that of a is 23/58 = 0.397.

8-9. The number of alleles of each type are: A, 2 + 2 + 5 + 1 = 10; B, 5 + 12 + 12 + 10 = 39; C, 10 + 5 + 5 + 1 = 21. The total number is 70, and so the allele frequencies are: A, 10/70 = 0.14; B, 39/70 = 0.56; and C, 21/70 = 0.30.

8-10. Allele 5.7 frequency = (2 × 9 + 21 + 15 + 6)/200 = 0.30; allele 6.0 frequency = (2 × 12 + 21 + 18 + 7)/200 = 0.35; allele 6.2 frequency = (2 × 6 + 15 + 18 + 5)/200 = 0.25; allele 6.5 frequency = (2 × 1 + 6 + 7 + 5)/200 = 0.10.

8-11. If the genotype frequencies satisfy the Hardy-Weinberg principle, they should be in the proportions p^2, $2pq$, and q^2, respectively. In each case, p equals the frequency of AA plus one-half the frequency of Aa, and $q = 1 - p$. The allele frequencies and expected genotype frequencies with random mating are (a) p = 0.50, expected: 0.25, 0.50, 0.25; (b) p = 0.635, expected: 0.403, 0.464, 0.133; (c) p = 0.7, expected: 0.49, 0.42, 0.09; (d) p = 0.775, expected: 0.601, 0.349, 0.051; (e) p = 0.5 expected: 0.25, 0.50, 0.25. Therefore, only (a) and (c) fit the Hardy-Weinberg principle.

8-12. p^2 = 0.09, and so $p = (0.09)^{1/2}$ = 0.30. All other alleles have a combined frequency of 1 − 0.30 = 0.70.

8-13. The frequency of the recessive allele is q = (1/14,000)$^{1/2}$ = 0.0085. Hence, p = 1 − 0.0085 = 0.9915, and the frequency of heterozygotes is $2pq$ = 0.017 (about 1 in 60).

8-14. Because the frequency of homozygous recessives is 0.10, the frequency of the recessive allele q is $(0.10)^{1/2}$ = 0.316. This means that p = 1 − 0.316 = 0.684, and so the frequency of heterozygotes is 2(0.316)(0.684) = 0.43. Because the proportion 0.90 of the individuals are nondwarfs, the frequency of heterozygotes among nondwarfs is 0.43/0.90 = 0.48.

8-15. (a) Because 0.60 can germinate, the proportion 0.40 that cannot are the homozygous recessives. Thus, the allele frequency q of the recessive is $(0.4)^{1/2}$ = 0.63. This implies that p = 1 − 0.63 = 0.37 is the frequency of the resistance allele. (b) The frequency of heterozygotes is $2pq$ = 2(0.37)(0.63) = 0.46. The frequency of homozygous dominants is $(0.37)^2$ = 0.14, and the proportion of the surviving genotypes that are homozygous is 0.14/0.60 = 0.23.

8-16. The bald and nonbald alleles have frequencies of 0.3 and 0.7, respectively. Bald females are homozygous and occur in the frequency $(0.3)^2$ = 0.09, and nonbald females occur with frequency 0.91. Bald males are homozygous or heterozygous and occur with frequency $(0.3)^2$ + 2(0.3)(0.7) = 0.51, and nonbald males occur with frequency 0.49.

8-17. $I^A I^A$, $(0.16)^2$ = 0.026; $I^A I^O$, 2(0.16)(0.74) = 0.236; $I^B I^B$, $(0.10)^2$ = 0.01; $I^B I^O$, 2(0.10)(0.74) = 0.148; $I^O I^O$, $(0.74)^2$ = 0.548; $I^A I^B$, 2(0.16)(0.10) = 0.032. The phenotype frequencies are: O, 0.548; A, (0.026 +

0.236) = 0.262; B, (0.01 + 0.148) = 0.158; AB, 0.032.

8-18. (a) There are 17 A and 53 B alleles, and the allele frequencies are: A, 17/70 = 0.24; B, 53/70 = 0.76. (b) The expected genotype frequencies are: AA, $(0.24)^2$ = 0.058; AB, 2(0.24)(0.76) = 0.365; BB, $(0.76)^2$ = 0.578.

8-19. The frequency in males implies that q = 0.2. The expected frequency of the trait in females is q^2 = 0.0004. The expected frequency of heterozygous females is $2pq$ = 2(0.98)(0.02) = 0.0392.

8-20. The frequency of the yellow allele y is 0.2. In females, the expected frequencies of wildtype (+/+ and y/+) and yellow (y/y) are 0.8^2 + 2(0.8)(0.2) = 0.96 and 0.2^2 = 0.04, respectively. Among 1000 females, there would be 960 wildtype and 40 yellow. The phenotype frequencies are very different in males. In males, the expected frequencies of wildtype (+/ Y) and yellow (y/ Y) are 0.8 and 0.2, respectively, in which Y represents the Y chromosome. Among 1000 males, there would be 800 wildtype and 200 yellow.

8-21. The frequency of heterozygotes in an inbred population is smaller by the amount $2pqF$, in which F is the inbreeding coefficient.

8-22. The expected genotype frequencies with random mating are given in the answer to Problem 8-11. Inbreeding is suggested by a deficiency of heterozygotes. The suggestive populations are (d) with expected heterozygote frequency 0.349 and (e) with expected heterozygote frequency 0.5.

8-23. The allele frequency is the square root of the frequency of affected offspring among unrelated individuals, or $(8.5 \times 10^{-6})^{1/2} = 2.9 \times 10^{-3}$. With inbreeding the expected frequency of homozygous recessives is $q^2(1 - F) + qF$, in which F is the inbreeding coefficient. When $F = 1/16$, the expected frequency of homozygous recessives is $(2.9 \times 10^{-3})^2(1 - 1/16) + (2.9 \times 10^{-3})(1/16) = 1.9 \times 10^{-4}$; when $F = 1/64$, the value is 5.3×10^{-5}.

8-24. (a) Here, F= 0.66, p = 0.43, and q = 0.57. The expected genotype frequencies are: AA, $(0.43)^2(1 - 0.66)$ + (0.43)(0.66) = 0.347; AB, 2(0.43)(0.57)(1 − 0.66) = 0.167; BB, $(0.57)^2(1 - 0.66)$ + (0.57)(0.66) = 0.487. (b) With random mating, the expected genotype frequencies are: AA, $(0.43)^2$ = 0.185; AB, 2(0.43)(0.57) = 0.490; BB, $(0.57)^2$ = 0.325.

8-25. Use Equation 8-4 with n = 40, p_0/q_0 = 1 (equal amounts inoculated), and p_{40}/q_{40} = 0.35/0.65. The solution for w is w = 0.9846, which is the fitness of strain B relative to strain A, and the selection coefficient against B is 1 − 0.9846 = 0.0154.

8-26. (1) A fixed; (2) A lost; (3) stable equilibrium; (4) A lost.

8-27. An Aa female produces 1/2 A and 1/2 a gametes, and these yield heterozygous offspring if they are fertilized by a or A, respectively. The overall probability of a heterozygous offspring is $1/2q$ + $1/2p$ = $1/2(p + q)$ = 1/2.

8-28. The frequency of each allele is $1/n$ and there are $n(n − 1)/2$ possible heterozygotes, and so the total frequency of heterozygotes is $n(n − 1)/2 \times 2 \times 1/n \times 1/n = 1 - 1/n$.

Chapter 9

9-1. A trait in which the phenotypes differ in magnitude rather than falling into a small number of discrete classes. Quantitative traits are typically determined by multiple genetic and environmental factors acting together. Continuous traits are those, such as height, that are measured on a continuous scale. Meristic traits are countable, such as eggs per hen per year. Threshold traits are either present or absent, but their presence or absence is determined by an underlying liability, which is a continuous trait.

9-2. Genotypic variance is the variance in phenotype resulting from differences in genotype among individual organisms. Environmental variance is the variance in phenotype resulting from differences in environment among individual organisms.

9-3. Variance due to genotype-environment interaction results from nonadditive effects of genotypic and environmental factors. Variance due to genotype-environment association results from nonrandom occurrence of genotypes in the different environments. The latter is the more easily controlled.

9-4. Genotype-environment interaction.

9-5. The standard deviation is the square root of the variance. The variance of a distribution equals zero when all individuals have the same phenotype.

9-6. (a) The mean or average. (b) The one with the greater variance is broader. (c) 68 percent within one standard deviation of the mean, 95 percent within two.

9-7. Broad-sense heritability is the proportion of the phenotypic variance attributable to differences in genotype, or $H^2 = \sigma_g^2/\sigma_t^2$, which includes dominance effects and interactions between alleles. Narrow-sense heritability is the proportion of the phenotypic variance due only to additive effects, or σ_a^2/σ_t^2, which is used to predict the resemblance between parents and offspring in artificial selection. If all allele frequencies equal 1 or 0, both heritabilities equal zero.

9-8. The F_1 provides an estimate of the environmental variance. The variance of the F_2 is determined by the genotypic variance and the environmental variance.

9-9. Estimates as follows: mean, 104/ 10 = 10.4; variance, 24.4/(10 − 1) = 2.71; standard deviation $(2.71)^{1/2}$ = 1.65.

9-10. An IQ of 130 is two standard deviations above the mean, and so the proportion with a higher IQ equals $(1 - 0.95)/2 = 2.5$ percent; 85 is one standard deviation below the mean, and so the proportion with a lower IQ equals $(1 - 0.68)/2 = 16$ percent; the proportion above 85 equals $1 - 0.16 = 84$ percent, which can be calculated alternatively as $0.50 + 0.68/2 = 0.84$.

9-11. (a) To determine the mean, multiply each value for the pounds of milk produced (using the midpoint value) by the number of cows and divide by 304 (the total number of cows) to obtain a value of the mean of 4032.9. The estimated variance is 693,633.8, and the estimated standard deviation, which is the square root of the variance, is 832.8. (b) The range that includes 68 percent of the cows is the mean minus the standard deviation and the mean plus the standard deviation, or $4033 - 833 = 3200$ to $4033 + 833 = 4866$. (c) Rounding to the nearest 500 yields $3000-5000$. The observed number in this range is 239; the expected number is $(0.68)(304) = 207$. (d) For 95 percent of the animals, the range is within two standard deviations or $2367-5698$. Rounding off to the nearest 500 yields $2500-6000$. The observed number is 296, which compares well to the expected number of $(0.95)(304) = 289$.

9-12. (1) Because from 15 to 21 is the range of one standard deviation from the mean, the proportion is 68 percent. (2) Because from 12 to 24 is the range of two standard deviations from the mean, the proportion is 95 percent. (3) A phenotype of 15 is one standard deviation below the mean; because a normal distribution is symmetric about the mean, the proportion with phenotypes more than one standard deviation below the mean is $(1 - 0.68)/2 = 0.16$. (4) Those with 24 leaves or more are more than two standard deviations above the mean, and so the proportion is $(1 - 0.95)/2 = 0.025$. (5) Those with from 21 to 24 leaves are between one and two standard deviations above the mean, and the expected proportion is $0.16 - 0.025 = 0.135$ (because 0.16 have more than 21 leaves and 0.25 have more than 24 leaves).

9-13. (a) The genotypic variance is 0 because all F_1 mice have the same genotype. (b) With additive alleles, the mean of the F_1 mice will equal the average of the parental strains.

9-14. Among the F_1, the total variance equals the environmental variance, which is given as 1.46. Among the F_2, the total variance, given as 5.97, is the sum of the environmental variance (1.46) and the genotypic variance, and so the genotypic variance equals $5.97 - 1.46 = 4.51$. The broad-sense heritability is the genotypic variance (4.51) divided by the total variance (5.97), or $4.51/5.97 = 0.76$.

9-15. From the F_1, $\sigma_e^2 = 2^2 = 4$; from the F_2, $\sigma_g^2 + \sigma_e^2 = 3^2 = 9$; hence $\sigma_g^2 = 9 - 4 = 5$, and so $H^2 = 5/9 = 56$ percent.

9-16. Using Equation 9-4, the minimum number of genes affecting the trait is $D^2/8\sigma_g^2 = 44$.

9-17. Use Equation 9-6 with $M^* - M = 50$ grams in each generation of selection, with M initially being 700 grams. The expected gain in each generation of selection is $0.8 \times 50 = 40$ grams, and so after five generations the expected mean weight is $700 + 5 \times 40 = 900$ grams.

9-18. Use Equation 9-6 with $M = 10.8$, $M^* = 5.8$, and $M' = 8.8$. The narrow-sense heritability is estimated as $h^2 = (8.8 - 10.8)/(5.8 - 10.8) = 0.40$.

9-19. (a) In this case, $M = 10.8$, $M^* = 5.8$, and $M' = 9.9$. The estimate of h^2 is $(9.9 - 10.8)/(5.8 - 10.8) = 0.18$. (b) The result is consistent with that in Problem 9-18 because of the differences in the variance. In Problem 9-18, the additive variance was $0.4 \times 4 = 1.6$. In the present case, the phenotypic variance is 9. If the additive variance did not change, then the expected heritability would be $1.6/9 = 0.18$, which is observed. However, the problem has been idealized somewhat in that in real examples the additive variance would also be expected to change with a change in environment.

9-20. The offspring are genetically related as half-siblings, and Table 9-3 says that the theoretical correlation coefficient is $h^2/4$.

9-21. Table 9-3 gives the correlation coefficient between first cousins a $h^2/8$, and so $h^2 = 8 \times 0.09 = 72$ percent.

9-22. The mean weaning weight of the 20 individuals is 81, and the standard deviation is 13.8. The mean of the upper half of the distribution can be calculated by use of the relation given in the problem, which says that the mean of the upper half of the population is $81 + 0.8 \times 13.8 = 92$. Using Equation 9-6, we find the expected weaning weight of the offspring to be $81 + 0.2(92 - 81) = 83.2$.

9-23. One approach is to use Equation 9-4 with $D = 2n - 0 = 2n$, from which $n = (2n)^2/8\,\sigma_g^2$, or $\sigma_g^2 = n/2$. A more-direct approach uses the fact that, with unlinked, additive genes, the total genetic variance is the summation of the genetic variance resulting from each gene. With one gene, the genetic variance equals $1/2$, and so with n genes the total genetic variance equals $n/2$.

Chapter 10

10-1. The selected marker is originally present in the Hfr parent and is necessary for recombinant F^- progeny to survive. The counterselected marker is present in the F^- parent and prevents survival of the Hfr parent. Without selection and counterselection, the small number of recombinant cells could not be identified among the large number of Hfr and F^- parental cells.

10-2. Generalized transduction can give recombinants for any genetic marker regardless of its position in the chromosome. Specialized transduction gives recombinants for only a restricted set of genes near the prophage insertion site.

10-3. Minimal medium containing galactose and biotin: the galactose would require the presence of *gal*⁺ and the biotin would allow both *bio*⁺ and *bio*⁻ to grow.

10-4. The *E. coli* genome contains approximately 4500 kilobase pairs, which implies approximately 45 kb per minute. Because the λ genome is 50 kb, the genetic length of the prophage is approximately one minute (more precisely, 45/50 = 0.9 minutes).

10-5. Infect an *E. coli* λ lysogen with P1. The λ lysogen contains λ prophage, and so some of the resulting P1 phage will include the part of the chromosome containing the prophage.

10-6. After circularization and integration at the bacterial *att* site, the map is ...*att D E F A B C att*....

10-7. The specialized-transducing particles originate from aberrant prophage excision.

10-8. Plate on minimal medium lacking leucine (selects for Leu⁺) but containing streptomycin (selects for Str-r). *leu*⁺ is the selected marker and *str-r* the counterselected marker.

10-9. One plaque per phage. One plaque, because the plating was done before lysis, and so the progeny phage are confined to one tiny area.

10-10. The T2 plaques will be turbid because T2 fails to lyse the resistant bacteria. Plaques made by the T2*h* mutant will be clear because the mutant can lyse both the normal and the resistant cells.

10-11. The mean number of phages per cell equals 1. If you know the Poisson distribution, then it is apparent that the proportion of uninfected cells is $e^{-1} = 0.37$. If you do not know the Poisson distribution, then the answer can be calculated as follows. The probability that a bacterial cell escapes infection by a particular phage is $1 - 10^{-6}$, and the probability that it escapes infection by all phages is $P_0 = (1 - 10^{-6})^{10^6}$. Therefore, $\ln P_0 = 10^6 \times \ln(1 - 10^{-6}) = -1$, and so $P_0 = e^{-1} = 0.37$.

10-12. Apparently *h* and *tet* are closely linked, and so recombinants containing the *h*⁺ allele of the Hfr tend to also contain the *tet-s* allele of the Hfr, and these recombinants are eliminated by the counterselection for *tet-r*.

10-13. All receive the *a*⁺ allele, but whether or not a particular *b*⁺*str-r* recombinant contains *a*⁺ depends on the positions of the exchanges.

10-14. Depending on the size of the F′, it could be F′*g*, F′*g h*, F′*g h i*, and so forth, or F′*f*, F′*f e*, F′*f e d*, and so forth.

10-15. The first selection is for Met⁺ and so *lac*⁺ must have been transferred. The probability that a marker that is transferred is incorporated into the recombinants is about 50 percent, and so about 50 percent of the recombinants would be expected to be *lac*⁺. The second selection is for Lac⁺, and *met*⁺ is a late marker. The great majority of mating pairs will spontaneously break apart before transfer of *met*⁺, and so the frequency of *met*⁺ recombinants in this experiment will be close to zero.

10-16. The order 500 *his*⁺ 250 *leu*⁺, 50 *trp*⁺ implies that the genes are transferred in the order *his leu trp*. The *met* mutation is the counterselected marker that prevents growth of the Hfr parent. The medium containing histidine selects for *leu*⁺*trp*⁺, and the number is small because both genes must be incorporated by recombination.

10-17. The three possible orders are (1) *pur−pro−his*, (2) *pur−his−pro*, and (3) *pro−pur−his*. The predictions of the three orders are: (1) Virtually all *pur*⁺*his*⁻ transductants should be *pro*⁻, but this is not true. (2) Virtually all *pur*⁺*pro*⁻ transductants should be *his*⁻, but this is not true. (3) Some *pur*⁺*pro*⁻ transductants will be *his*⁻, and some *pur*⁺*his*⁻ transductants will be *pro*⁻ (depending on where the exchanges occur). Order (3) is the only one that is not contradicted by the data.

10-18. (a) Use $m = 2 - 2d^{1/3}$, in which m is map distance in minutes and d cotransduction frequency. With the values of d given, the map distances are 0.74, 0.41, and 0.18 minutes, respectively. (b) Use $d = [1 - (m/2)]^3$. With the values of m given, cotransduction frequencies are 0.42, 0.12, and 0, respectively. For markers greater than two minutes apart, the frequency of cotransduction equals zero. (c) Map distance *a−b* equals 0.66 minutes, *b−c* equals 1.07 minutes, and these are additive, giving 1.73 minutes for the distance *a−c*. Predicted cotransduction frequency for this map distance is 0.002.

10-19. It is probable that the *amp-r* gene in the natural isolate was present in a transposable element. Transposition into λ occurs in a small proportion of cases and, when these infect laboratory strain, the transposable element can transpose into the chromosome before the infecting λ is lost.

10-20. The *tet-r* gene in the phage was included in a transposable element. Transposition occurred in the lysogen to another location in the *E. coli* chromosome, and when the prophage was lost the antibiotic resistance remained.

10-21. Both the order of times of entry and the level of the plateaus imply that the order of gene transfer is *a b c d*. The times of entry are obtained by plotting the number of recombinants of each type versus time and extrapolating back to the time axis. The values are *a*, 10 minutes; *b*, 15 minutes; *c*, 20 minutes; *d*, 30 minutes. The low plateau value for the *d* gene probably results from close linkage with *str-s*, because in this case the *d* allele in the F′ strain would tend to be inherited along with the *str-r* allele.

Chapter 11

11-1. The template is the coding strand of the DNA. Because the 5′ end of the RNA transcript is synthesized first, the coding strand of DNA is transcribed in the 3′ → 5′ direction. The 5′ end of the mRNA is translated first, and the amino (–NH$_2$) end of the polypeptide is the first synthesized.

11-2. Start codon is AUG, which codes for methionine. Stop codons are UAA, UAG, and UGA. The probability of a start codon is 1/64, and the probability of stop codon is 3/64. The average distance between stop codons is 64/3 = 21.3 codons.

11-3. In prokaryotes, the start codon is the first AUG after a ribosome-binding sequence; in eukaryotes, the start codon is usually the nearest AUG to the 5′ cap in the RNA.

11-4. Answer (3), there are no terminating tRNA molecules.

11-5. The mRNA sequence is 5′- AAAAUGCCCUUAAU-CUCAGCGUCCUAC-3′, and translation begins with the first AUG, coding for methionine. The amino acid sequence coded by this region is Met-Pro-Leu-Ile-Ser-Ala-Ser-Tyr.

11-6. With G at the 5′ end the first codon is GUU, which codes for valine, and with a G at the 3′ end the last codon is UUG, which codes for leucine.

11-7. Codons are read from the 5′ end of the mRNA, and the first amino acid in a polypeptide chain is at the amino terminus. If the G were at the 5′ end, the first codon would be GAA, coding for glutamic acid at the amino terminus of the polypeptide. If the G were at the 3′ end, the last codon would be AAG, coding for arginine at the carboxyl terminus. Thus, the G was added at the 5′ end.

11-8. GUG codes for valine and UGU codes for cysteine. Therefore, an alternating polypeptide containing only valine and cysteine would be made.

11-9. The artificial mRNA can be read in any one of three reading frames, depending on where translation starts. The reading frames are GUC GUC GUC ..., UCG UCG UCG ..., and CGU CGU CGU .... The first codes for polyvaline, the second polyserine, and the third polyarginine.

11-10. The possible codons, their frequencies, and the amino acids coded are as follows:

AAA	$(3/4)^2$	=	0.421	Lys
AAC	$(3/4)^2(1/4)$	=	0.141	Asn
ACA	$(3/4)^2(1/4)$	=	0.141	Thr
CAA	$(3/4)^2(1/4)$	=	0.141	Gln
CCA	$(3/4)(1/4)^2$	=	0.047	Pro
CAC	$(3/4)(1/4)^2$	=	0.047	His
ACC	$(3/4)(1/4)^2$	=	0.047	Thr
CCC	$(1/4)^3$	=	0.015	Pro

The amino acids in the random polymer and their frequencies are therefore lysine (42.1 percent), asparagine (14.1 percent), threonine (18.8 percent), glutamine (14.1 percent), histidine (4.7 percent), and proline (6.2 percent).

11-11. Methionine has 1 codon, histidine 2, and threonine 4, and so the total number is $1 \times 2 \times 4 = 8$. The sequence AUGCAYACN encompasses them all. Because arginine has six codons, the possible number of sequences coding for Met-Arg-Thr is $1 \times 6 \times 4 = 24$. The sequences are AUGCGNACN (16 possibilities) and AUGAGRACN (8 possibilities).

11-12. In an overlapping code, a single base change affects more than one codon, and so changes in more than one amino acid should often occur (but not always, because single amino acid changes can result from redundancy in the code). For this reason, an overlapping code was eliminated from consideration.

11-13. Because the pairing of codon and anticodon is antiparallel, it is convenient to write the anticodon as 3′-UAI-5′. The first two codon positions are therefore 5′-AU-3′, and the third is either A, U or C (because each can pair with I). The possible codons are 5′-AUA-3′, 5′-AUU-3′, and 5′-AUC-3′, all of which code for isoleucine.

11-14. In theory, the codon 5′-UGG-3′ (tryptophan) could pair with either 3′-ACC-5′ or 3′-ACU-5′. However, 3′-ACU-5′ can also pair with 5′-UGA-3′, which is a chain-termination codon. If this anticodon were used for tryptophan, an amino acid would be inserted instead of terminating the chain. Therefore, the other anticodon (3′- ACC-5′) is the one used.

11-15. (a) The deletion must have fused the amino-coding terminus of the *B* gene with the carboxyl-coding terminus of the *A* gene. The noncoding strand must therefore be oriented 5′–*B*–*A*–3′. (b) The number of bases deleted must be a multiple of three; otherwise, the carboxyl terminus would not have the correct reading frame.

11-16. (a) The noncoding strand is given, and so the wildtype reading frame of the mRNA is AUG CAU CCG GGC UCA UUA GUC U..., which codes for Met-His-Pro-Gly-Ser-Leu-Val- .... (b) Mutant X mRNA reading frame is AUG GCA UCC GGG CUC AUU AGU CU..., which codes for Met-Ala-Ser-Gly-Leu-Ile-Ser-Leu-.... (c) Mutant Y mRNA reading frame is AUG CAU CCG GGC UCU UAG UCU ..., which codes for Met-His-Pro-Gly-Ser (the UAG is a termination codon). (d) The double-mutant mRNA reading frame is AUG GCA UCC GGG CUC UUA GUC U ..., which codes for Met-Ala-Ser-Gly-Leu-Leu-Val- .... Note that the double mutant differs from wildtype only in the region between the insertion and deletion mutations, in which the reading frame is shifted.

11-17. Mutation X can be explained by a single base change in the start codon for the first Met, in which case translation would begin with the second AUG codon. Mutation Y can be explained by a single base change that converts a codon for Tyr (UAU or UAC) into a chain-termination codon (UAA or UAG). The tripeptide sequence is Met-Leu-His.

11-18. The probability of a correct amino acid at a particular site is 0.999, and so the probability of all 300 amino acids being correctly translated is $(0.999)^{300} = 0.74$, or approximately 3/4.

11-19. Each loop represents an intron, and so the total number is seven introns.

11-20. If transcribed from left to right, the coding strand is the bottom one, and the mRNA sequence is AGA CUU CAG GCU CAA CGU GGU, which codes for Arg-Leu-Gln-Ala-Gln-Arg-Gly. To invert the molecule, the strands must be interchanged in order to preserve the correct polarity, and so the sequence of the inverted molecule is

5'-AGA ACGTTGAGCCTGAAG GGT-3'
3'-TCT TGCAACTCGGACTTC CCA-5'

This codes for an mRNA with sequence AGA ACG UUG AGC CUG AAG GGU, which codes for Arg-Thr-Leu-Ser-Leu-Lys-Gly.

Chapter 12

12-1. With negative regulation, transcription occurs unless prevented by a repressor; with positive regulation, transcription does not occur unless enabled by an activator.

12-2. A repressor combines with an inducer and in this state is unable to bind with the operator, and so transcription occurs; in the absence of inducer, the repressor binds with the operator and transcription is prevented. An example is the *lac* repressor. An aporepressor combines with an effector and in this state is able to bind with the operator and prevent transcription; in the absence of the effector, the aporepressor cannot bind with the operator, and so transcription occurs. An example is the *trp* aporepressor.

12-3. An autoregulated gene regulates its own expression. With positive autoregulation, there is positive feedback in which the gene product stimulates further induction; with negative autoregulation, there is negative feedback in which the gene product inhibits further induction. Positive autoregulation is used to amplify weak induction signals and negative autoregulation to prevent overproduction.

12-4. Constitutive.

12-5. The enzymes are expressed constitutively because glucose is metabolized in virtually all cells. However, the levels of enzyme are regulated to prevent runaway synthesis or inadequate synthesis.

12-6. The hemoglobin mRNA has a very long lifetime before being degraded.

12-7. (a) Transcription. (b) RNA processing. (c) Translation.

12-8. Need not be near because the repressor is diffusible. The *trp* repressor is one example in which the repressor gene is located quite far from the structural genes.

12-9. The mutant gene should bind more of the activator protein or bind it more efficiently. Therefore, the mutant gene should be induced with lower levels of the activator protein or expressed at higher levels than the wildtype gene or both.

12-10. (a) β-Galactosidase (*lacZ* product), lactose permease (*lacY* product), and the transacetylase (*lacA* product). (b) The promoter region becomes inaccessible to RNA polymerase. (c) Repressor synthesis is constitutive.

12-11. At 37°C, both phenotypes are Lac⁻; at 30°C, the *lacY⁻* is Lac⁺ and the *lacZ⁻* is Lac⁻.

12-12. Inducers combine with repressors and prevent them from binding with the operator regions of the respective operons.

12-13. The cAMP concentration is low in the presence of glucose. Both types of mutation are phenotypically Lac⁻. The cAMP-CRP binding does not interfere with repressor binding because the DNA binding sites are distinct.

12-14. No; there is often a polar effect in which the relative amount of each protein synthesized decreases toward the 3′ end of the mRNA.

12-15. With a repressor, the inducer is usually an early (often the first) substrate in the pathway, and the repressor is inactivated by combining with the inducer. With an aporepressor, the effector molecule is usually the product of the pathway, and the aporepressor is activated by the binding.

12-16. An attenuator is not a binding site for any protein, and RNA synthesis does not begin at an attenuator. An attenuator is strictly a potential termination site for transcription.

12-17. Steroid hormones pass through the cell to the nucleus where they combine with transcriptional activator proteins and stimulate transcription.

12-18. The α cell would be unable to switch to the **a** mating type.

12-19. The **a**1 protein functions only in combination with the α2 protein in diploids; hence, the *Mata'* haploid cell would appear normal. However, the diploid *Mata'/MATα* has the phenotype of an α haploid. The reason is that the genotype lacks an active **a**1−α2 complex to repress the haploid-specific genes and α1, and the α1 gene product activates the α-specific genes.

12-20. The mutant diploid *MATa/MATα* has the phenotype of an α haploid. As in Problem 12-19, the genotype lacks an active a1-α2 complex to repress the haploid-specific genes and α1, and the α1 gene product activates the α-specific genes.

12-21. In the absence of lactose or glucose, two proteins are bound—the *lac* repressor and CRP-cAMP. In the presence of glucose, only the repressor is bound.

12-22. (a) The strain is constitutive and has either the genotype *lacI⁻* or the genotype *lacO^c*. (b) The second mutant must be *lacZ⁻* because no β-galactosidase is made and, because permease synthesis is induced by lactose, the mutant must have a normal repressor-operator system. Therefore, the genotype is *lacI⁺ lacO⁺ lacZ⁻lacY⁺*. (c) Because the partial diploid is regulated, the operator in the operon with a functional *lacZ* gene must be wildtype. Hence, the first mutant must be *lacI⁻*.

12-23. The *lac* genes are transferred by the Hfr cell and enter the recipient. No repressor is present in the recipient initially, and so *lac* mRNA is made. However, the *lacI* gene is also transferred to the recipient, and soon afterward the repressor is made and *lac* transcription stops.

12-24. The mutant clearly lacks a general system that regulates sugar metabolism, and the most likely possibility is the cAMP-dependent regulatory system. Several mutations can prevent activity of this system. For example, the mutant could make either a defective CRP protein (which either fails to bind to the promoter or is unresponsive to cAMP) or an inactive adenyl cyclase gene and be unable to synthesize cAMP.

12-25. (a) No enzyme synthesis; *cis*-dominant because lack of RNA polymerase binding to the promoter prevents transcription of the adjacent genes. (b) Enzymes synthesized; *cis*-dominant because failure to bind the repressor cannot prevent constitutive transcription of the adjacent genes. (c) Enzymes synthesized; recessive because the defective repressor can be complemented by a wildtype repressor in the same cell. (d) Enzymes synthesized; recessive because the mutant repressor can be complemented by a wildtype repressor in the same cell. (e) No enzyme synthesis; dominant because the repressor is always activated and all *his* transcription is prevented.

12-26. The wildtype mRNA reads AAA CAC CAC CAU CAU CAC CAU CAU CCU GAC, which codes for Lys-His-His-His-His-His-His-His-Pro-Asp. The mutant mRNA reads AAA ACA CCA CCA UCA UCA CCA UCA UCC UGA C, which codes for Lys-Thr-Pro-Pro-Ser-Ser-Pro-Ser-Ser (the UGA is a normal stop codon). The expected phenotype of the mutant is His⁻ because of lack of attenuation by low levels of histidine. That is, translation of the attenuator occurs (preventing efficient transcription of the operon) even when the level of histidine within the cell is very low. However, low levels of proline or serine will prevent translation of the attenuator and allow transcription of the histidine operon to occur.

Chapter 13

13-1. Autonomous developmental fates are controlled by genetically programmed changes within cells. Fates determined by positional information are controlled by interactions with morphogens or with other cells. Cell transplantation and cell ablation (destruction) are two kinds of surgical manipulation that help distinguish between them.

13-2. Polarity of the oocyte refers to regional differentiation along the principal axes (anterior–posterior and dorsal–ventral).

13-3. Plants lack a germ line in the sense of specialized cells in a discrete part of the body set off early in development and destined to become reproductive cells. Instead, many cells in the mature plant have the capacity to differentiate into reproductive cells.

13-4. Totipotent means that a single cell is capable of regenerating an entire organism—that is, all the genetic information necessary for normal development is still present, and the developmental commitment is reversible. Some cells in mature plants are not totipotent—for example, the floral meristems.

13-5. Lineage mutations affect the developmental fates of one or more cells within a group of cells related by descent that normally undergo a defined pattern of development. Lineage mutations are detected by the occurrence of abnormal types or numbers of cells at characteristic positions within the embryo.

13-6. The *Drosophila* blastoderm is the flattened hollow ball of cells formed by cellularization of the syncytium formed by the early nuclear divisions. The fate map is a diagram of the blastoderm showing the locations of cells and their developmental fates. The existence of the fate map does not mean that development is autonomous, because many developmental decisions are made according to positional information in the blastoderm. In fact, most *Drosophila* developmental decisions are nonautonomous.

13-7. A female that is homozygous for a maternal-effect lethal produces defective eggs resulting in inviable embryos. However, the homozygous genotype is itself viable.

13-8. Cells normally destined to die adopt some other developmental fate, with the result that there are supernumerary copies of one or more cell types.

13-9. In a homeotic mutation, certain organs develop in a manner characteristic of that in neighboring segments, and therefore the lineage effect is one of transformation.

13-10. A sister-cell-segregation defect would cause both daughter cells to differentiate as B or both to differentiate as C. A parent-offspring-segregation defect would cause both daughter cells to differentiate as A.

13-11. Heterochronic mutations affect the timing of developmental events relative to each other; hence, a mutation that uniformly slows or accelerates development is not a heterochronic mutation.

13-12. Because the *Drosophila* oocyte contains all the transcripts needed to support the earliest stages of development; the mouse oocyte does not.

13-13. In a loss-of-function mutation, the genetic information in a gene is not expressed in some or all cells. In a gain-of-function mutation, the genetic information is expressed at inappropriate times or in inappropriate cells. Both mutations can occur in the same gene: some alleles may be loss of function and others gain of function. However, a particular allele must be one or the other, and so the mutations cannot occur together in the same allele.

13-14. Because the gene is necessary, a loss-of-function mutation will prevent the occurrence of the proper developmental fate. However, the allele will be recessive because a normal allele in the homologous chromosome will allow the developmental pathway to occur.

13-15. Because the gene is sufficient, a gain-of-function mutation will induce the developmental fate to occur. The allele will be dominant because a normal allele in the homologous chromosome will not prevent the developmental pathway from occurring.

13-16. Developmental abnormalities in gap mutations are across contiguous regions, defining a gap in the pattern of normal development. Developmental abnormalities in pair-rule mutations are in even- or odd-numbered segments or parasegments, and so the abnormalities are in horizontal stripes along the embryo.

13-17. The proteins required for cleavage and blastula formation are translated from mRNAs present in the mature oocyte, but transcription of zygotic genes is required for gastrulation.

13-18. The material required for rescue is either mRNA transcribed from the maternal o^+ gene or a protein product of the gene that is localized in the nucleus.

13-19. The transplanted nuclei respond to the cytoplasm of the oocyte and behave like the oocyte nucleus at each stage of development.

13-20. XDH activity is needed only at an early stage in development for the eyes to have wildtype pigmentation. The data are explained by a maternal effect in which the *mal$^+$/mal; ry/ry* females in the cross transmit enough *mal$^+$* gene product to allow brief activity of XDH in the *ry$^+$/ry* embryos. In theory, the *mal$^+$* product could be a transcription factor that allows *ry$^+$* transcription but, in fact, the *mal$^+$* gene product participates in the synthesis of a cofactor that combines with the *ry$^+$* gene product to make an active XDH enzyme.

Chapter 14

14-1. Such individuals are somatic mosaics and can be explained by somatic mutations in the pigmented cells of the iris of the eye or in their precursor cells.

14-2. Causes of a nonreverting mutation include (1) deletion of the gene, (2) multiple nucleotide substitutions in the gene, and (3) complex DNA rearrangement with more than one break in the gene. Deletions do not revert because the genetic information in the gene is missing, and the others do not revert because the nucleotide substitutions or DNA rearrangements are unlikely to be reversed precisely in one step.

14-3. Transitions are AT → GC, GC → AT, TA → CG, and CG → TA (total four). Transversions are AT → TA, AT → CG, GC → CG, GC → TA, TA → AT, TA → GC, CG → GC, and CG → AT (total eight).

14-4. All nucleotide substitutions in the second codon position result in either an amino acid replacement or a terminator codon.

14-5. Both UUR and CUR code for leucine, and both CGR and AGR code for arginine (R stands for any purine, either A or G). Hence, in each of these codons, one of the possible substitutions in the first codon position does not result in an amino acid replacement. Even when an amino acid replacement does occur, the mutant protein may still be functional because more than one amino acid at a given position in a protein molecule may be compatible with normal or nearly normal function.

14-6. Any single nucleotide substitution creates a new codon with a random nucleotide sequence, and so the probability that it is a chain-termination codon (UAA, UAG, or UGA) is 3/64, or approximately 5 percent.

14-7. The proofreading function of DNA polymerase, which consists of recognition of an incorrectly paired base and its removal by 3′ → 5′ exonuclease activity.

14-8. A frameshift mutagen will also revert frameshift mutations.

14-9. Exactly three. Otherwise, there would be a frameshift and all downstream amino acids would be different.

14-10. No, dominance and recessiveness are characteristics of the phenotype. They depend on the function of the gene and on which attributes of the phenotype are examined.

14-11. Photoreactivation and excision repair.

14-12. (a) Development arrests at A at both the high and the low temperatures. (b) Development arrests at B at both the high and the low temperatures. (c) Development arrests at A at the high temperature, and it arrests at B at the low temperature. (d) Development arrests at B at the high temperature and arrests at A at the low temperature.

14-13. The mutation frequency in the Lac$^-$ culture is at least $0.90 \times 10^{-8} = 9 \times 10^{-9}$. Because the mutation rate to a nonsense suppressor is independent of the presence of a suppressible mutation (in this case, the Lac$^-$ mutation), the mutation rate in the original Lac$^+$ culture is also at least 9×10^{-9}. The qualifier *at least* is important because not all nonsense suppressors may suppress the particular Lac$^-$ mutation (that is, some suppressors may insert an amino acid that still results in a nonfunctional enzyme).

14-14. (a) 5×10^{-5} is the probability of a mutation per generation, and so $1 - 5 \times 10^{-5}$ is the probability of no mutation per generation. The probability of no mutations in 10,000 generations equals $(1 - 5 \times 10^5)^{10,000} = 0.61$. (b) The average number of generations until a mutation occurs is $1/(5 \times 10^{-5}) = 20,000$.

14-15. At 100 rads, the total mutation rate is double the spontaneous mutation rate. Because a dose of 100 rads increases the rate by 100 percent, then a dose of 50 rads increases it by 50 percent and one of 10 rads by 10 percent.

14-16. Amino acid replacements at many positions in the A protein do not eliminate its ability to function; only amino acid replacements at critical positions that do eliminate tryptophan synthetase activity are detectable as Trp$^-$ mutants.

14-17. The possible lysine codons are AAA and AAG, and the possible glutamic acid codons are GAA and GAG. Therefore, the mutation results from an AT → GC transition in the first position of the codon.

14-18. Six amino acid replacements are possible because the codons UAY (Y stands for either pyrimidine, C or U) can mutate in a single step to codons for Phe (UUY), Ser (UCY), Cys (UGY), His (CAY), Asn (AAY), or Asp (GAY).

14-19. Single nucleotide substitutions that can account for the amino acid replacements are (1) Met (AUG) to LEU (UUG), (2) Met (AUG) to Lys (AAG), (3) Leu (CUN) to Pro (CCN), (4) Pro (CCN) to Thr (ACN), (5) Thr (ACR) to Arg (AGR), in which R stands for any purine (A or G) and N for any nucleotide. The substitutions in 1, 2, 4, and 5 are transversions and that in 3 is a transition. Because 5-bromouracil primarily induces

transitions, the Leu → Pro replacement is expected to be the most frequent.

14-20. The sectors are explained by a nucleotide substitution in only one strand of the DNA. Replication of this heteroduplex yields one daughter DNA molecule with the wildtype sequence and another with the mutant sequence. Cell division produces one Lac$^+$ and one Lac$^-$ cell, located side by side. Because the cells do not move on the agar surface, the Lac$^+$ cells form a purple half colony and the Lac$^-$ cells form a pink half colony.

14-21. Amino acids with codons one step away from the nonsense codon, because their tRNA genes are the most likely to mutate to nonsense suppressors. For UGA, these are UGC (Cys), UGU (Cys), UGG (Trp), UUA (Leu), UCA (Ser), AGA (Arg), CGA (Arg), and GGA (Gly).

14-22. (a) All progeny would be able to grow in the absence of arginine. (b) The genotype of the diploid is Su^+/Su^- arg^+/ arg^-, in which Su^+ indicates the ability to suppress and Su^- indicates inability to suppress. The resulting haploid spores are 1/4 $Su^+ arg^+$, 1/4 $Su^+ arg^-$, 1/4 $Su^- arg^+$, and 1/4 $Su^- arg^-$. The first three are Arg$^+$ and the last Arg$^-$, and so the ratio is 3 Arg$^+$: 1 Arg$^-$. (c) The diploid genotype is Su^+ $arg^-/Su^- arg^+$. The haploid spores are as listed in (b), but in this case they are in the proportions $0.05 : 0.45 : 0.45 : 0.05$. Therefore, 95 percent of the progeny are Arg$^+$ and 5 percent Arg$^-$.

14-23. The *a* and *b* gene products may be subunits of a multimeric protein. A few proteins with compensatory amino acid replacements can interact to give wildtype activity, but most mutant proteins cannot interact in this way with each other or with wildtype.

14-24. Some of the mutagenized P1 particles acquire mutations in the *a* gene and become a^-b^+. Therefore, the next step is to select b^+ transductants and screen them for those that are a^-. The procedure depends on close linkage.

14-25. The ACA codon must be changed into ATA. The first produces a UGU codon (Cys) and the second a UAU codon (Tyr). The oligonucleotide needed for this change is identical with that shown in Figure 14-20, except that the nucleotide in red must be an A.

Chapter 15

15-1. A somatic-cell hybrid is a cell formed by fusing cells from different species. They are produced at a low frequency in cells grown together in culture, but the frequency can be increased by various chemical treatments (for example, polyethylene glycol) or the use of Sendai virus.

15-2. HGPRT and TK can be selected in HAT medium to isolate cell fusions. One parental cell line lacks

HGPRT and the other lacks TK, and so only the cell hybrid produces both enzymes. Neither TK^- nor $HGPRT^-$ genotypes can grow in HAT medium.

15-3. An antigen is a substance able to elicit an immune response. An antibody is a circulating protein synthesized in response to an antigen and able to combine with it. Antigen presentation is a process in which a cell (often a macrophage), after ingesting an antigen, exhibits parts of it on the cell surface to stimulate other cells in the immune system that recognize the antigen.

15-4. Organisms are tolerant of antigens that are present in their own tissues or of antigens closely related to their own. Therefore, if a substance resembles an organism's own antigens, its immune system will not respond.

15-5. A germ-line cell is a cell that gives rise to gametes. A B cell secretes antibody. A T cell is a white blood cell that participates in the cellular immune response. The genetic constitutions are not identical because B cells have rearrangements of the antibody genes and T cells have rearrangements of the T-cell-receptor genes.

15-6. Helper T cells help stimulate B cells to produce antibody. Cytotoxic cells attack antigens directly.

15-7. Mismatched blood can result in a severe reaction triggered by clumping of transfused red blood cells. It is critical that the donor antigens also be present in the recipient, because this will prevent recipient antibodies from attacking donor cells. It is less important if the donor blood contains antibodies not present in the recipient, because the dilution of the antibodies in the recipient usually prevents any harm.

15-8. No, rejection of transplanted tissues or organs is primarily a T-cell function.

15-9. Rh^+ fetus and Rh^- mother. Firstborn children do not usually have the problem because the mother does not produce anti-Rh^+ antibody. However, if the mother had been exposed to Rh^+ cells somehow (for example, by transfusion), then a firstborn child can have the disease.

15-10. This depends on the genotype of the father, which could be $I^B I^B$ or $I^O I^B$. Because the genotype is not known, the possible blood type that the child could have is either type O or type B.

15-11. Not justified, because the husband could be $I^A I^O$, the wife $I^B I^O$, and the child $I^O I^O$.

15-12. There are four chains (two H chains and two L chains). All IgG molecules do not have the same amino acid sequence. The constant region has nearly the same amino acid sequence in all IgG molecules, whereas the amino acid sequence of the variable region differs in each type of IgG molecule.

15-13. The V and J segments join to form the variable region of the light chain, which is attached to the constant region. The variable portion of the heavy chain is formed by DNA joining of V, D, and J segments.

15-14. Because normal ABO antibodies do not cause hemolytic disease, the IgM antibodies must not cross the placenta.

15-15. (a) The $HGPRT$ gene is in the X chromosome. If the human cells were $HGPRT^+$, all clones would contain the human X chromosome, and they clearly do not. Consequently, the human cells must have been the TK^+ cells. (b) Because all selected clones must retain the TK^+ gene, and only 17q is present in all the clones, the TK gene must be present in 17q (the long arm of chromosome 17).

15-16. For each enzyme, match the horizontal patterns in the table of cell lines with vertical patterns for each chromosome in the table of chromosomes. For example, consider steroid sulfatase. Each + and − match in the X-chromosome column except for the entries p and q. In clone F, when only Xp is present, steroid sulfatase is also present. In clone E, when only Xq is present, the enzyme is absent. Thus, steroid sulfatase is in Xp. The locations of the other enzymes are as follows: phosphoglucomutase-3, 6q; esterase D, 13q; phosphofructokinase, 21; amylase, 1p; galactokinase, 17q.

15-17. Absence of the sequence in line B excludes 1, 6p, 9, 13, and 17; absence in line F excludes 2p, 12q, 17, 21, and Xp; absence in line G excludes 1q, 2, 9, 12, 13, 17. Overall, all of the chromosomes represented in the cell lines are excluded except 6q and Xq. The positive result in line D means that the hybridizing sequence *must* be present in Xq. It could also be present in 6q or in any of the other human chromosomes not present in this set of hybrids. It is possible for a DNA sequence to be present more than once in a genome, for example, the sequence could be a member of a multigene family or a transposable element.

15-18. (a) The mother is $se/se\ I^A/-$ and the child $Se/se\ I^B/-$. The dash must represent I^O because this is the only allele compatible with both the mother's and the child's blood groups. The father's genotype must be $Se/-\ I^B/-$, and in this case the dash could be I^A, I^B, or I^O. (b) The father's phenotype is either B, secretor or AB, secretor. (c) The child's genotype is $Se/se\ I^B/I^O$.

15-19. (a) The mother is dd. (b) Treatment will be ineffective because the mother already has anti-Rh^+ antibodies. (c) Only a dd fetus is not at risk, and the chance of this is 1/2.

15-20. Both males can be excluded. Male 1 is excluded because the baby must have received a C allele from its father. Male 2 is excluded because the baby must have received an M allele from its father.

15-21. An identical twin because identical twins must share the same genotype for all genes, including blood group and histocompatibility genes.

15-22. If the child carries only antigens inherited from its mother, then skin grafts from the child to the mother will be accepted. Otherwise, they will be rejected because of antigens that are inherited from the father and that the mother does not possess. If the child is a normal male, no tests are necessary to invalidate the claim. In this case, the child has a chromosome (the Y) not present in the mother, which must have come from the father.

15-23. The antibodies would be different. Combinatorial joining is a random process producing different antibodies having the same or very similar specificities. The IgM molecules may differ according to the particular V-J and V-D-J regions that became joined in the antibody-producing B cells, and also because of somatic mutation.

15-24. Diploid—2; triploid—3; trisomic—3 if the gene is in the trisomic chromosome, otherwise 2.

15-25. For each histocompatibility gene in which CH and C57 differ, the CH57 (F_1) animals will be heterozygous. All transplants from CH, C57, and CH57 will be accepted, because the donor tissue contains no antigens not also present in the CH57 recipient. However, transplants from wild mice or from a different inbred strain will be rejected, because the transplanted tissue will contain some antigens not present in CH57.

Chapter 16

16-1. The insertion of a DNA fragment from one organism into a vector molecule that allows propagation of the sequence in another organism.

16-2. The polymerase chain reaction (PCR) allows repeated replication of a specific DNA sequence by repeated rounds of denaturation, annealing with oligonucleotide primer sequences that flank the target sequence on either strand, and DNA synthesis. The procedure is used to amplify specific DNA sequences exponentially. To use PCR, enough of the target sequence must be known to be able to synthesize suitable primers.

16-3. Classical genetics studies phenotypes in order to identify mutant genotypes; reverse genetics makes mutant genotypes and studies the phenotypes. Reverse genetics refers to the use of directed mutagenesis of a cloned gene and its introduction back into a cell or living organism in order to study the phenotypic effects.

16-4. A YAC is a yeast artificial chromosome containing DNA from another organism. YAC clones are useful because the inserted segment of DNA can be very large.

16-5. The DNA content per average band equals about $10^8/5000 = 20$ kb, per average lettered subdivision equals approximately $10^8/600 = 167$ kb, and per average numbered section equals about $10^8/100 = 1$ Mb. A YAC with a 200-kb insert contains the DNA equivalent to about 10 salivary bands, 1.2 lettered subdivisions, and 0.2 numbered section.

16-6. A bacterial cloning vector must be able to be introduced into cells by some means, it must be able to replicate, and transformed cells must be selectable. It also helps if the vector contains convenient restriction sites for the insertion of DNA fragments.

16-7. The ends may be blunt or they may be cohesive with either a 3' overhang or a 5' overhang.

16-8. All restriction fragments have the same ends because all the restriction sites are the same and they are cleaved in the same places. Opposite ends of the same fragment must also be identical because restriction sites are palindromes.

16-9. No. The complement of 5'-GGCC is 3'-CCGG.

16-10. The insertion distrupts the gene, and sensitivity to the antibiotic shows that the vector has an inserted DNA sequence. The second antibiotic-resistance gene is needed to select transformants.

16-11. Reverse transcriptase can make DNA from an RNA template. It is used to clone DNA copies of RNA molecules, especially messenger RNA.

16-12. Comparison of the genomic and cDNA sequences tells you where the introns are in the genomic sequence. The genomic sequence contains the introns and the cDNA does not.

16-13. Bacterial cells do not normally recognize eukaryotic promoters and, if transcription does occur, the transcript will usually contain introns that the bacterial cell cannot remove. If these problems are overcome, the protein may still not be produced because the mRNA lacks a bacterial ribosome-binding site, or because the protein might require posttranslational processing, or because the protein might be unstable in bacterial cells and subject to degradation.

16-14. The finding is that the 3.6-kb and 5.3-kb fragments found in the free phage are joined to form an 8.9-kb fragment in the intracellular form. This suggests that the 3.6-kb and 5.3-kb fragments are terminal fragments of a linear molecule and that the intracellular form is circular.

16-15. A base substitution may destroy a restriction site and prevent two potential fragments from being separated or a deletion may eliminate a restriction site.

16-16. The gene probably contains an EcoRI site, and so is disrupted in the process of cloning, but not a HindIII site.

16-17. (a) The *tet-r* gene is not cleaved with BgIII, and so addition of tetracycline to the medium requires that the colonies be Tet-r and hence contain the plasmid. (b) Tet-r Kan-r and Tet-r Kan-s colonies will grow. (c) Tet-r Kan-s colonies contain inserts within the cleaved *kan* gene.

16-18. An insertion of two bases is generated within the codon, resulting in a frameshift. Therefore, all colonies should be Lac⁻.

16-19.

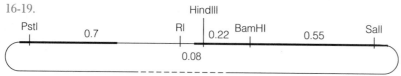

Rest of plasmid between PstI and SalI = 2.45 kb

homeotic genes, 356, 357–358
homeotic mutations, 343
homogametic, 38
homologous chromosomes, 31
    chromatid exchange in, 35
homothallism, 315
homozygous, 6
hormonal regulation, 320–321
hormones, 320
hot spots, 372
housekeeping genes, 318
human blood groups. *See* blood groups
humans
    chromosomes of, 150–153
    genome of, 123
    pedigree analysis in, 13
    sex determination in, 42–43
    trisomy in, 150–151
Huntington disease, 13, 438
hybrid cells, 401–403
hybrids, 2, 146
hybrid vigor, 225
hypoxanthine-guanine phosphoribosyl transferase, 402

identical twins, 221–222
imaginal disks, 351
immune response, 403–406
immunity, 255, 403
immunogenetics, 403–404
immunoglobulins, 409
inborn errors of metabolism, 14
inbreeding, 14, 193–194
    effects of, 40, 194
inbreeding coefficient, 193
inbreeding depression, 225
incomplete penetrance, 21
independent assortment, principle of, 6–9, 36
independent events, 11–13
individual selection, 223
induced mutations, 364, 374–379
inducers, 301
inducible transcription, 302–304
influenza, 87
influenza virus, 438
inheritance
    cytoplasmic, 167
    extranuclear, 167
    maternal, 168, 170
    Mendelian, 10–13
    quantitative, 208–213
    X-linked, 39–40
initiation, 102, 272, 280
    RNA primer and, 108
initiation factors, 280
inosine, 290
insecticides, 370
insertional inactivation, 428

insertion sequences, 256
in situ hybridization, 131, 132
insulin, 434
integrase, 253
integrated, 240, 252
intelligence, race and, 229–230
intercalation, 376
interference, 67
intergenic suppression, 381, 382–384
interphase, 29
interrupted-mating technique, 244
intervening sequences, 276
intragenic suppression, 381–382
introns, 276
inversion, 157–158
inverted repeats, 134
in vitro experiments, 104
ionizing radiation, mutagenicity of, 377–378
IQ scores
    genetic and cultural effects on, 228–229
    race and, 229–230
IS elements, 256
isoleucine, chemical structure of, 267

Jacob, François, 305
joining (J) regions, 411
Jukes study, 227

Kallikaks study, 227
kappa particles, 175
karyotype, 150
killer phenomenon, 175
kilobase pairs (kb), 123
Klinefelter syndrome, 153
Krüppel, 353

lac mutants, 302
lac operon, 305
    positive regulation of, 306–307
lactose metabolism, 302–308
lactose permease, 302
lagging strand, 107
lariat structure, 276
leader, 274
leader polypeptide, 310
leading strand, 107
Lederberg, Esther, 369
Lederberg, Joshua, 239, 369
leptotene, 34
Lesch-Nyhan syndrome, 437
leucine, chemical structure of, 267
leukemias, 159–160
light (L) chain, 409
Limnaea peregra, 178–180
lineage diagram, 341

somatic-cell hybrids, and gene mapping, 398–401
somatic cells, chromosomes of, 28
somatic mutations, 364
SOS repair, 380
Southern blotting, 114, 187
specialized transducing phage, 247
spermatocytes, 33
spina bifida, 209
splice acceptor, 276
splice donor, 276
spliceosomes, 277
splicing
   alternative, 326–327
   in the origin of T-cell receptors, 412
spontaneous abortion, chromosome abnormalities and, 153
spontaneous mutations, 364, 368–374
spores, 33, 68, 71, 358
Stahl, Franklin, 99–100
standard deviation, 211
start codons, 281
statistically significant, 46
sterility, cytoplasmic male, 174
Stern, Curt, 61
sticky ends, 421
stop codons, 282
Streptococcus pneumoniae, 88, 238
streptomycin, 237, 243, 371
sublineages, 343
submetacentric, 144
subunits, protein, 269
supercoiling, 123–124
suppressor mutations, 381–385
suppressor tRNA, 383
symbionts, cytoplasmic transmission in, 175–178
synapsis, 34, 158

tandem duplication, 156
TATA box, 272
T cells, 405, 412
T DNA, 433
telomerase, 138
telomeres, 137
   structure of, 136–138
telophase, 29, 31
telophase I, 34, 36
telophase II, 37
temperate phage, 249
temperature-sensitive mutations, 364
template, 104
template strand, 272
termination, 272, 280
termination factors, 280
testcross, 9
   three-point, 64
testis-determining factor (TDF), 151
tetrad analysis, 68–74
tetrads, 35, 69–71
   ordered, 71–73

unordered, 69–71
Tetrahymenae, 440
   regulation in, 312
   splicing in, 278
   telomere of, 138
tetraploidy, 145–146
tetratype, 69
theta replication, 101
30-nm fiber, 128
3'-OH, 93
three-point cross, 64
three-point testcross, gene mapping from, 64–67
threonine, chemical structure of, 267
threshold traits, 209
thymidine kinase, 402
thymine, 91
time-of-entry mapping, 243–246
TK, 402
tobacco mosaic virus, 87
tolerance, 413
topoisomerases, 102
total variance, 216
totipotent, 359
traits
   continuous, 209
   distribution of, 209–212
   meristic, 209
   multifactorial, 208
   quantitative, 208, 218–222
   threshold, 209
trans configuration, 56
   regulation of, 300–302, 318–326
transcriptional activator proteins, 319–320
transcriptional enhancers, 321–323
transducing phage, 247, 255–256
   generalized, 247
   specialized, 247
transduction, 236, 247–249
transfer RNA, 280, 288–289
   suppressor, 383
   uncharged, 282, 288
transformation, 89, 236, 423
   bacterial, 238–239
   germ-line, in animals, 432–433
transformation mutation, 343
transgenic animals, 433
trans-heterozygote, 75–76
transition mutation, 365
translation, 266, 279–283
translational control, 327–328
translation units, 292
translocation, 151, 158–159
transposable-element mutagenesis, 367–368
transposable elements, 134–136, 256–259
transposase, 135, 256
transposition, 134, 256–259
transposons, 256, 258–259
transposon tagging, 258
transversion mutation, 365